WITHDRAWN

Techniques in Somatic Cell Genetics

Techniques in Somatic Cell Genetics

Edited by

JERRY W. SHAY

The University of Texas Health Science Center at Dallas
Dallas, Texas

PLENUM PRESS • NEW YORK AND LONDON

Library of Congress Cataloging in Publication Data

Main entry under title:

Techniques in somatic cell genetics.

Includes bibliographical references and index.
1. Cytogenetics–Technique. I. Shay, Jerry W. II. Title: Somatic cell genetics. [DNLM:
1. Cytological techniques. 2. Cytogenetics. 3. Hybrid cells. QH 585 T255]
QH441.T4 1982   574.87'3223   82-9848
ISBN 0-306-41040-0   AACR2

©1982 Plenum Press, New York
A Division of Plenum Publishing Corporation
233 Spring Street, New York, N.Y. 10013

All rights reserved

No part of this book may be reproduced, stored in a retrieval system, or transmitted in any form or by any means, electronic, mechanical, photocopying, microfilming, recording, or otherwise, without written permission from the Publisher

Printed in the United States of America

Contributors

Abdullatif A. Al-Bader Department of Pathology, Faculty of Medicine, Kuwait University, Kuwait

Renato Baserga Department of Pathology and Fels Research Institute, Temple University School of Medicine, Philadelphia, Pennsylvania 19140

Horst Binding Botanisches Institut, Christian-Albrechts-Universität, Kiel, Federal Republic of Germany

Arthur Bollon Department of Medical Genetics, Wadley Institutes of Molecular Medicine, Dallas, Texas 75235

Clive L. Bunn Department of Biology, University of South Carolina, Columbia, South Carolina 29208

W. N. Choy Division of Biophysics, Johns Hopkins University School of Hygiene and Public Health, Baltimore, Maryland 21205

Mike A. Clark Department of Cell Biology, University of Texas Health Science Center, Dallas, Texas 75235. Present address: Wistar Institute of Anatomy and Biology, Philadelphia, Pennsylvania 19104

Andrew H. Crenshaw, Jr. Department of Anatomy, University of Tennessee Center for the Health Sciences, Memphis, Tennessee 38163

Carlo M. Croce Wistar Institute of Anatomy and Biology, Philadelphia, Pennsylvania 19104

Vaithilingam G. Dev Department of Medical Genetics, University of South Alabama College of Medicine, Mobile, Alabama 36617

Claus-Jens Doersen Department of Microbiology, California College of Medicine, University of California—Irvine, Irvine, California 92717. Present address: Division of Biology, California Institute of Technology, Pasadena, California 91125

Mary Eichelberger Section of Genetics, Laboratory of Viral Carcinogenesis, National Cancer Institute, Frederick, Maryland 21701

Jerome M. Eisenstadt Department of Human Genetics, Yale University School of Medicine, New Haven, Connecticut 06510

R. E. K. Fournier Department of Microbiology, and The Comprehensive Cancer Center, University of Southern California School of Medicine, Los Angeles, California 90033

Godfrey S. Getz Departments of Pathology and Biochemistry, University of Chicago, Chicago, Illinois 60637

T. V. Gopalakrishnan Department of Pediatrics, Johns Hopkins University School of Medicine, Baltimore, Maryland 21205

A. Graessmann Institut für Molekular Biologie und Biochemie, der Freien Universität Berlin, Berlin, West Germany

M. Graessmann Institut für Molekular Biologie und Biochmie, der Freien Universität Berlin, Berlin, West Germany

Margaret J. Hightower Department of Microbiology, State University of New York at Stony Brook, Stony Brook, New York 11794. Present address: Cold Spring Harbor Laboratory, Cold Spring Harbor, New York 11724

Nancy Hsiung Department of Biochemical Sciences, Princeton University, Princeton, New Jersey 08544

Robert T. Johnson Department of Zoology, University of Cambridge, Cambridge, CB2 3EJ, England

Johan F. Jongkind Department of Cell Biology and Genetics, Medical Faculty, Erasmus University, Rotterdam, The Netherlands

Paul M. Keller Biophysics Program, Pennsylvania State University, University Park, Pennsylvania 16802

Kathleen L. Kornafel Department of Pathology, University of Chicago, Chicago, Illinois 60637

Raju Kucherlapati Department of Biochemical Sciences, Princeton University, Princeton, New Jersey 08544

Mary C. Kuhns Department of Human Genetics, Yale University School of Medicine, New Haven, Connecticut 06510. Present address: Abbott Laboratories, North Chicago, Illinois 60064

Andrejs Liepins Memorial University, Faculty of Medicine, St. Johns, Newfoundland, Canada A1B 3V6

Alban Linnenbach Wistar Institute of Anatomy and Biology, Philadelphia, Pennsylvania 19104. Present address: Department of Internal Medicine, Yale University School of Medicine, New Haven, Connecticut 06510

J. W. Littlefield Department of Pediatrics, Johns Hopkins University School of Medicine, Baltimore, Maryland 21205

Joseph J. Lucas Department of Microbiology, State University of New York at Stony Brook, Stony Brook, New York 11794

Gerd G. Maul Wistar Institute of Anatomy and Biology, Philadelphia, Pennsylvania 19104

O. Wesley McBride Laboratory of Biochemistry, National Cancer Institute, National Institutes of Health, Bethesda, Maryland 20205

W. Edward Mercer Department of Pathology, and Fels Research Institute, Temple University School of Medicine, Philadelphia, Pennsylvania 19140

John Morrow Department of Biochemistry, Texas Tech University Health Sciences Center, Lubbock, Texas 79410

A. M. Mullinger Department of Zoology, University of Cambridge, Cambridge CB2 3EJ, England

Leonard R. Murrell Department of Anatomy, University of Tennessee Center for the Health Sciences, Memphis, Tennessee 38163

Reinhard Nehls Botanisches Institut, Christian-Albrechts-Universität, Kiel, Federal Republic of Germany

Thomas H. Norwood Department of Pathology, University of Washington, Seattle, Washington 98195

Stephen J. O'Brien Section of Genetics, Laboratory of Viral Carcinogenesis, National Cancer Institute, Frederick, Maryland 21701

Demetrios Papahadjopoulos Cancer Research Institute, and Department of Pharmacology, University of California—San Francisco, San Francisco, California 94143

Stanley Person Biophysics Program, Pennsylvania State University, University Park, Pennsylvania 16802

Potu N. Rao Department of Developmental Therapeutics, University of Texas System Cancer Center, M. D. Anderson Hospital and Tumor Institute, Houston, Texas 77030

Martin C. Rechsteiner Department of Biology, University of Utah, Salt Lake City, Utah 84112

Tim L. Reudelhuber Department of Biochemistry, University of Texas Health Science Center, Dallas, Texas 75235. Present address: Centre National de la Recherche Scientifique, Strasbourg, France

Frank H. Ruddle Department of Biology, Yale University, New Haven, Connecticut 06511

Jerry W. Shay Department of Cell Biology, University of Texas Health Science Center, Dallas, Texas 75325

Saul J. Silverstein Department of Microbiology, Columbia University, New York, New York 10032

Janice M. Simonson Section of Genetics, Laboratory of Viral Carcinogenesis, National Cancer Institute, Frederick, Maryland 21701

Doris L. Slate Department of Biology, Yale University, New Haven, Connecticut 06511. Present address: Cancer Metastasis Research Group. Pfizer Central Research, Groton, Connecticut 06340

Eric J. Stanbridge Department of Microbiology, California College of Medicine, University of California—Irvine, Irvine, California 92717

Robert M. Straubinger Cancer Research Institute, and Department of Pharmacology, University of California—San Francisco, San Francisco, California 94143

Elton Stubblefield University of Texas System Cancer Center, Department of Cell Biology, M.D. Anderson Hospital and Tumor Institute, Houston, Texas 77030

Prasad S. Sunkara Merrell Research Center, Cincinnati, Ohio 45215

Ramana Tantravahi Cytogenetics Laboratory, Sidney Farber Cancer Institute, and Department of Medicine, Harvard Medical School, Boston, Massachusetts 02115

George E. Veomett School of Life Sciences, University of Nebraska—Lincoln, Lincoln, Nebraska 68588

Anton Verkerk Department of Cell Biology and Genetics, Medical Faculty, Erasmus University, Rotterdam, The Netherlands

Douglas C. Wallace Department of Genetics, School of Medicine, Stanford University, Stanford, California 94305

Josef Weibel Wistar Institute of Anatomy and Biology, Philadelphia, Pennsylvania 19104

Lamont G. Weide Department of Cell Biology, University of Texas Health Science Center, Dallas, Texas 75235

Wayne Wray Department of Cell Biology, Baylor College of Medicine, Houston, Texas 77030

Woodring Erik Wright Department of Cell Biology and Internal Medicine, University of Texas Health Science Center, Dallas, Texas 75235

Carol J. Zeigler Department of Pathology, University of Washington, Seattle, Washington 98195

Michael L. Ziegler Department of Pathology, and College of Dentistry, University of Kentucky Medical Center, Lexington, Kentucky 40506

Preface

Somatic cell genetics is an exciting and rapidly expanding field of research. Since descriptions of the major experimental techniques in the field are scattered throughout various journals and other publications, there is a real need for a single reference source for both established investigators and students in the field. In addition, technical reports are frequently abridged such that many researchers are discouraged from attempting to adopt the appropriate methodology. This book, therefore, describes in detail the many recent technical advances in such areas of somatic cell genetics as transfer mediated by liposomes, erythrocyte ghosts, chromosomes, microcells, mitochondria, and isolated nuclear DNA. These techniques have increased our understanding of the organization and regulation of eukaryotic cells.

The production of antibiotic-resistant cell lines and their use in studying cytoplasmic inheritance are also included. Evidence for the cytoplasmic regulation of nuclear gene expression in eukaryotic cells is rapidly accumulating following the characterization of cytoplasmic mutations. The production of nuclear-coded mutations, their use in standard cell hybridization, and recent advances in techniques for fusing whole cells or cell components are also described.

I have not included the rapidly expanding hybridoma field in the scope of the book since this has been the subject of a recent comprehensive review by R. H. Kennett, T. J. McKearn, and K. B. Bechtol (*Monoclonal Antibodies*, Plenum Press, 1980). I would like to thank Kirk Jensen and his colleagues at Plenum Press for their support and Drs. Rosalie Ber and Marguerite Stauver for editorial assistance. Finally, I wish to acknowledge all the contributors without whom this book would not have been possible.

Jerry W. Shay

Contents

Chapter 1
Selection of Purine and Pyrimidine Nucleoside Analog Resistance in Mammalian Cells
John Morrow

1. Introduction	1
2. Historical Background	2
3. Methodology	3
3.1. Tissue Culture Flasks	3
3.2. Culture Reagents	3
3.3. Screening for Contaminants	4
3.4. Cell Lines	5
3.5. Metabolic Cooperation	5
3.6. Cloning of Variants	6
3.7. Sample Protocols	6
References	7

Chapter 2
Techniques for Using HAT Selection in Somatic Cell Genetics
W. N. Choy, T. V. Gopalakrishnan, and J. W. Littlefield

1. Introduction	11
1.1. Selective Systems in Somatic Cell Genetics	12
1.2. Principle of HAT Selection	12
2. Selection of Somatic Cell Hybrids with HAT Medium	14
3. Modifications of HAT Selection	14
3.1. AA Medium	15
3.2. HAM Medium	15
3.3. GAMA Medium	15
4. Modifications by Incorporation of Additional Selective Agents	15
4.1. Ouabain	15
4.2. Polyene Antibiotics	16
4.3. Diptheria Toxin	16
5. Applications of HAT Selection	16
5.1. Selection of Revertants of Drug-Resistant Mutants Deficient in HGPRT and TK	17

5.2. Gene Mapping 17
5.3. Gene Transfer 18
5.4. Selection of Hybridomas 18
References 19

Chapter 3
Techniques for Decreasing the Toxicity of Polyethylene Glycol
W. Edward Mercer and Renato Baserga

1. Introduction 23
2. Effects of Excluding Ca^{++} Ions and of the Source of PEG on
   Hybrid Colony Yield 24
3. Possible Role of Ca^{++} Ions in Cytotoxicity 25
4. Effects on Fusion Index of the Time Interval between
   Plating and PEG Exposure 27
5. Lectin Enhancement of Suspension Fusion 28
   5.1. Suspension Fusion Procedure 29
   5.2. Post-fusion Plating in Conditioned Medium 30
   5.3. Effects of PHA Concentration on the Fusion Index and
        Size of Polykaryons 31
6. Summary and Conclusions 32
   References 33

Chapter 4
The Use of Dimethyl Sulfoxide in Mammalian Cell Fusion
Thomas H. Norwood and Carol J. Zeigler

1. Introduction 35
2. Chemical and Biologic Properties of Dimethyl Sulfoxide 36
3. Methods of Procedure 37
   3.1. Parameters of Chemical Cell Fusion 37
   3.2. Suggested Protocol—Monolayer Fusion 40
   3.3. Suggested Protocol—Suspension Fusion 41
4. Evaluation of Cytotoxicity 41
5. Discussion 42
   References 44

Chapter 5
The Selection of Heterokaryons and Cell Hybrids Using the Biochemical Inhibitors Iodoacetamide and Diethylpyrocarbonate
Woodring Erik Wright

1. Introduction 47
2. Choice of Agents 48
   2.1. General Approach 48
   2.2. Screening of Selective Agents 48

	2.3. Proof That Binucleates Are Heterokaryons	51
3.	Treatment Conditions	52
	3.1. General Considerations	52
	3.2. Cell Concentration during Treatment	53
	3.3. Dosage Dependence of Heterokaryon Rescue	53
	3.4. Time Dependence of Cell Rescue	57
4.	Conditions for Cell Fusion	57
5.	Plating Conditions following Cell Fusion	58
	5.1. Direct Plating	58
	5.2. Initial Low-Density Plating	58
	5.3. Ficoll Enrichment for Viable Cells	60
	5.4. Cell Concentration versus Density	60
	5.5. Protective Effects of Serum	60
	5.6. Advantages of Bacteriologic Dishes	61
6.	Summary of Selection Protocol	61
7.	Isolation of Cell Hybrids	62
8.	Discussion	64
	References	64

Chapter 6

Techniques for Enucleation of Mammalian Cells
George E. Veomett

1.	General Introduction	67
2.	Monolayer Techniques for Cellular Enucleation	68
	2.1. Coverslip (cs) Technique	68
	2.2. Flask Techniques	70
	2.3. Different Support Systems for Monolayer Cultures	73
	2.4. Factors Affecting Enucleation	73
	2.5. Problems Encountered in Enucleation of Cells	74
	2.6. Modification of the Growth Substrate	75
	2.7. Techniques for Purifying the Cellular Components	75
3.	Gradient Techniques	76
	3.1. Preparation of Gradients	76
	3.2. Enucleation	77
	3.3. Recovery of Cytoplasts and Karyoplasts	77
	References	78

Chapter 7

Nonselective Isolation of Fibroblast Heterokaryons, Hybrids, and Cybrids by Flow Sorting
Johan F. Jongkind and Anton Verkerk

1.	Introduction	81
2.	Vital Fluorescent Cell Labeling	82
	2.1. Fluorescent Polystyrene Beads	82
	2.2. Fluorescent Stearylamine	84

2.3. Hoechst Bis-Benzimidazole Dyes 84
3. Isolation of Fibroblast Heterokaryons 86
　　3.1. Outline 86
　　3.2. Cell Fusion and Heterokaryon Collection 86
4. Isolation of Hybrids 87
　　4.1. Outline 87
　　4.2. Flow Sorting of Tetraploid Cells 88
　　4.3. Single-Cell Cloning of Fibroblast Hybrids 90
5. Isolation of Fibroblast Cybrids 90
　　5.1. Outline 90
　　5.2. Isolation of Fluorescent Cytoplasts 90
　　5.3. Cytoplast × Whole Cell Fusion and Flow Sorting
　　　　 of Cybrids 93
6. Biochemical Micromethods on Sorted Cells 93
　　6.1. Rationale 93
　　6.2. 5000–10,000 Sorted Cells 95
　　6.3. Single-Sorted Cells 97
7. Concluding Comments 98
　　References 98

Chapter 8

Techniques for Monitoring Cell Fusion: The Synthesis and Use of Fluorescent Vital Probes (R18, F18)

Paul M. Keller and Stanley Person

1. Introduction 101
2. Description of Membrane Probes 102
3. Syntheses of F18 and R18 103
4. Preparation of F18- or R18-Labeled Cells 104
5. Preparation of Labeled Virus 105
6. Measurement of Energy Transfer 105
7. Results of Fusion Experiments 106
8. Virus–Cell Interactions 106
9. Discussion 107
　　References 109

Chapter 9

Inheritance of Oligomycin Resistance in Tissue Culture Cells

Jerome M. Eisenstadt and Mary C. Kuhns

1. Introduction 111
2. Selection of Oligomycin-Resistant Mutants 112
3. Growth of Oligomycin-Resistant Mutants 113
4. Stability of Oligomycin-Resistant Mutants 113
5. Mitochondrial ATPase of Oligomycin-Resistant Cells 114
　　5.1. Preparation of Mitochondrial and Submitochondrial
　　　　 Particles 114

Contents

5.2. Assay of Mitochondrial ATPase Activity and
Sensitivity to Inhibitors 115
5.3. Use of [^{14}C]-DCCD for Analysis of the DCCD-Binding
Protein 116
6. Inheritance of Oligomycin Resistance 118
References 119

Chapter 10
Cytoplasmic Inheritance of Rutamycin Resistance in Mammalian Cells
Godfrey S. Getz and Kathleen L. Kornafel

1. Introduction 123
2. Mitochondrial Mutagenesis and Cytoplasmic Inheritance in
   Mammalian Cells 124
3. Oligomycin (Rutamycin)-Sensitive ATPase 126
4. Oligomycin-Resistant Mutants in Mammalian Cells 127
5. Rutamycin-Resistant Mouse Cells and Their ATPase 128
6. Pleiotropic Characteristics of Rutamycin-Resistant Mutants 130
   6.1. Glucose Dependence and Lactic Acid Production 130
   6.2. Respiratory Deficiency 132
7. Conclusions 134
References 135

Chapter 11
Erythromycin Resistance in Human Cells
Claus-Jens Doersen and Eric J. Stanbridge

1. Introduction 139
2. Selection of Erythromycin-Resistant Cell Lines 140
   2.1. Erythromycin Inhibition of Cell Proliferation
        Is pH Dependent 140
   2.2. Selection of ERY2301 141
   2.3. Selection of ERY2305 and ERY2309 142
   2.4. Isolation of D98-ERY^r 143
3. Characterization of Erythromycin-Resistant Cell Lines 143
   3.1. Cell Proliferation in the Presence of Erythromycin 143
   3.2. *In Vivo* Mitochondrial Protein Synthesis 144
   3.3. Cell-Free Mitochondrial Protein Synthesis 145
   3.4. Effects of Mycoplasma Contamination on the
        Erythromycin-Resistant Phenotype 149
4. Transfer of Erythromycin Resistance 149
   4.1. Cytoplasmic Transfer of Erythromycin Resistance 150
   4.2. Transfer of Erythromycin Resistance by Cell
        Hybridization 151
5. Conclusion 153
References 154

Chapter 12
Cytoplasmic Inheritance of Chloramphenicol Resistance in Mammaliam Cells
Douglas C. Wallace

1. CAP Inhibition of Bacterial and Mitochondrial Ribosomes 159
2. Isolation of Cytoplasmic CAP-Resistant Mutants 160
3. Identification of Cytoplasmic CAP^R Mutants 166
   3.1. Cytoplasmic Transfer of CAP Resistance 167
   3.2. Mitotic Segregation of CAP Resistance 170
   3.3. Elimination of CAP^R Determinants with R6G 170
4. Assignment of CAP Resistance to the Mitochondrial DNA 171
5. Molecular Basis of CAP Resistance 173
6. The Biochemistry of Cytoplasmic CAP Resistance 176
7. Cytoplasmic CAP Resistance in Interspecific Crosses 178
   References 182

Chapter 13
The Influence of Cytoplast-to-Cell Ratio on Cybrid Formation
Clive L. Bunn

1. Principles 189
   1.1. Uses of Cytoplast Fusions 189
   1.2. Properties of Cytoplasts 190
   1.3. Selection of Fusion Products 191
2. Procedures 192
   2.1. Polystyrene Bead Labeling 193
   2.2. Enucleation 194
   2.3. Cell–Cytoplast Fusion 194
   2.4. Efficiency of Cybrid Fusion and Proliferating Cybrid
        Formation with Varying Cytoplast:Cell Ratios 195
3. Comments: Cytoplasmic Dosage 197
   References 199

Chapter 14
Transformation of Mitochondrially Coded Genes into Mammalian Cells Using Intact Mitochondria and Mitochondrial DNA
Mike A. Clark, Tim L. Reudelhuber, and Jerry W. Shay

1. Introduction 203
2. Materials and Methods 203
   2.1. Cell Lines 203
   2.2. Mitochondria Isolation 204
   2.3. Procedures for Transformation with Intact Mitochondria 204
   2.4. Procedures for Transformation with Purified DNA 205

3. Results	205
4. Discussion	209
References	209

Chapter 15
Mitochondrial Influences in Hybrid Cells
Michael L. Ziegler

1. Introduction	211
2. Experimental Methods and Treatments with R6G	212
3. Induced Segregation of Mitochondrial Determinants by R6G	213
4. Elimination of Incompatibility in Interspecific Crosses	215
5. R6G and the Direction of Chromosome Loss	215
6. Summary	217
References	218

Chapter 16
Shedding of Tumor Cell Surface Membranes
Andrejs Liepins

1. Introduction	221
2. Materials and Methods	223
2.1. Cell Lines	223
2.2. Induction of Membrane Vesicle Shedding	223
2.3. Harvesting of Membrane Vesicles	223
2.4. Cell Size Distribution	224
2.5. Effect of Nucleotides on the Shedding of Cell Surface Membranes	224
2.6. Effects of Deuterium Oxide on the Shedding of Cell Surface Membranes	225
2.7. Electron Microscopy	225
3. Results and Discussion	225
3.1. Inductive Stimuli for MV Formation and Shedding	225
3.2. Morphologic Aspects of MV Shedding	226
3.3. Inhibition of MV Shedding	227
3.4. General Considerations of the MV Shedding Process	232
References	233

Chapter 17
Production of Microcytospheres
Gerd G. Maul and Josef Weibel

1. Introduction	237
2. Materials and Methods	237
3. Results	238
4. Discussion	241
References	242

Chapter 18
Isolation and Characterization of Mitoplasts
   Potu N. Rao, Prasad S. Sunkara, and Abdullatif A. Al-Bader

1. Introduction 245
2. Isolation of Mitoplasts 246
   2.1. Cells and Cell Synchrony 246
   2.2. Chemicals 247
   2.3. Extrusion of Chromosomes from Mitotic Cells 247
3. Characterization of the Mitoplasts 248
   3.1. Morphologic Features 248
   3.2. Metabolic Activity of the Mitoplasts 251
4. Induction of Premature Chromosome Condensation (PCC) with Mitoplasts 251
5. Conclusions 252
   References 254

Chapter 19
Nuclear Transplantation with Mammalian Cells
   Margaret J. Hightower and Joseph J. Lucas

1. Introduction 255
2. Preparation of Cytoplasts 256
   2.1. Method of Enucleation 256
   2.2. Separation of Cytoplasts and Whole Cells 257
   2.3. Some Characteristics of Cytoplasts 258
3. Preparation of Karyoplasts 258
   3.1. Method of Enucleation 258
   3.2. Purification of Karyoplasts 259
   3.3. Some Characteristics of Karyoplasts 260
4. Nuclear Transplantation 260
   4.1. Preparation of Sendai Virus 260
   4.2. Fusion of Cytoplasts and Karyoplasts 261
5. Identification of Hybrid Cells 262
   5.1. Drug and Toxin Sensitivities 262
   5.2. Immunofluorescent Staining of Fixed Cells 262
   5.3. Hoechst–Rhodamine Staining of Living Cells 263
6. Analysis of Hybrid Cells 265
   References 266

Chapter 20
Techniques for Purifying L-Cell Karyoplasts with Minimal Amounts of Cytoplasm
   Mike A. Clark and Jerry W. Shay

1. Introduction 269
2. Materials and Methods 270

    2.1. Cell Culture 270
    2.2. Tantalum Preparation 270
    2.3. Cell Enucleation Procedure 271
    2.4. Karyoplast Purification Using Ta 271
    2.5. Karyoplast Purification Using the Cell Sorter 272
    2.6. Karyoplast Viability 272
    2.7. Metabolic Activity Assays 272
    2.8. Cell Reconstruction 273
    2.9. Electron Microscopy 273
3. Results 273
    3.1. Results from the Ta Purification Procedure 274
    3.2. Results from the Cell Sorter-Purified Karyoplasts 276
    3.3. Results from Transmission Electron Microscopy 277
    3.4. Use of Purified Karyoplasts for Cell Reconstructions 277
4. Discussion 279
   References 280

Chapter 21
Techniques for Isolating Nuclear Hybrids
Lamont G. Weide, Mike A. Clark, and Jerry W. Shay

1. Introduction 281
2. Methods and Results 281
    2.1. Cell Lines and Media 281
    2.2. Cell Enucleation and Fusion 282
    2.3. Isolation of Intraspecific Nuclear Hybrids;
         Genetic Selection 282
    2.4. Isolation of Nuclear Hybrids; Physical–Genetic Selection .. 282
    2.5. Interspecific Fusion Product Survival Is Dependent on
         Cytoplasmic Effects 286
    2.6. Restriction Enzyme Analysis of Mitochondrial DNA
         from Nuclear Hybrids 286
3. Discussion 287
   References 288

Chapter 22
Monolayer Enucleation of Colcemid-Treated Human Cells and Polyethylene Glycol 400-Mediated Fusion of Microkaryoplasts (Microcells) with Whole Cells
Andrew H. Crenshaw, Jr. and Leonard R. Murrell

1. Introduction 291
2. Micronucleation of Human Somatic Cells with Colcemid 293
    2.1. Microkaryoplasts Containing Condensed Chromosomes 293
    2.2. Micronucleate Cells Containing Decondensed Chromatin 297

3. Isolation and Purification of Human Microkaryoplasts 299
   3.1. Microkaryoplasts Produced by Abnormal Cytokinesis 299
   3.2. Microkaryoplasts Produced by Cytochalasin B Enucleation
        of Micronucleatic Cells 299
4. Fusion of Microkaryoplasts to Whole Cells with
   Polyethylene Glycol 301
5. Conclusion 301
   Appendix 302
   A1. Titration of Colcemid Dose 302
   A2. Isolation of Microkaryoplasts Containing Condensed
       Chromosomes 302
   A3. Isolation of Microkaryoplasts Containing Decondensed
       Chromosomes 303
   A4. Purification of Microkaryoplasts 304
   A5. Fusion of Microkaryoplasts to Whole Cells with
       Polyethylene Glycol 400 MW 304
   References 305

Chapter 23

Microcell-Mediated Chromosome Transfer

R. E. K. Fournier

1. Introduction 309
2. Micronucleation of the Donor Cells 310
   2.1. Prolonged Arrest Micronucleation 311
   2.2. Sequential Treatment Micronucleation 313
   2.3. Micronucleation—General Observations 315
3. Enucleation of Micronucleate Populations 315
   3.1. Monolayer Enucleation Technqiues 316
   3.2. Suspension Enucleation 319
   3.3. Enucleation—General Observations 320
4. Purification of Isolated Microcell Preparations 320
   4.1. Purification of Nonadherent Particles 321
   4.2. Purification by Membrane Filtration 321
   4.3. Purification by Unit Gravity Sedimentation 323
   4.4. Microcell Purification—General Observations 322
5. Fusion of Microcells with Intact Recipients 323
   5.1. Suspension/Monolayer Fusion with Phytohemagglutinin-P
        and Polyethylene Glycol 323
   5.2. Suspension/Monolayer Fusion with Inactivated
        Sendai Virus 323
   5.3. Microcell Fusion—General Observations 324
6. Concluding Remarks 324
   References 326

Chapter 24
Techniques for Isolating Chromosome-Containing Minisegregant Cells
R. T. Johnson and A. M. Mullinger

1. Introduction 329
2. Standard Method for Production of Minisegregants from HeLa Cells 330
   - 2.1. Production of Mitotic Cells 330
   - 2.2. Induction of Extrusion Division in Mitotic HeLa Cells 330
   - 2.3. Separation of Minisegregant Cells According to Size 333
   - 2.4. Properties of Gradient Fractions 336
3. Modification of Basic Technique for Producing Minisegregants 339
4. The Use of Minisegregants in Somatic Cell Genetics 344
   - References 346

Chapter 25
Techniques to Isolate Specific Human Metaphase Chromosomes
Wayne Wray and Elton Stubblefield

1. Introduction 349
2. Methods 350
   - 2.1. Tissue Culture and Cell Lines 350
   - 2.2. Cell Synchronization 350
   - 2.3. Isolation Buffers 351
   - 2.4. Chromosome Isolation 351
   - 2.5. Chromosome Identification 352
   - 2.6. Chromosome Fractionation 358
   - 2.7. Flow Cytometry 359
   - 2.8. Chromosome Sorting of the Human Karyotype 360
   - 2.9. Isolation of Human Chromosome Group F 363
3. Applications for Isolated Chromosomes 365
   - 3.1. Chicken Gene Mapping 368
   - 3.2. Human Gene Mapping 368
   - References 372

Chapter 26
Techniques of Chromosome-Mediated Gene Transfer
O. Wesley McBride

1. Introduction 375
2. Methods for Metaphase Chromosome Isolation and Purification 376
   - 2.1. Chromosome Isolation at pH 3 376
   - 2.2. Chromosome Isolation at pH 7 378
3. Metaphase Chromosome Uptake 379
   - 3.1. Uptake in Suspension 379
   - 3.2. Calcium Phosphate-Precipitated Chromosome Uptake 380

4. Isolation and Analysis of Transformants 381
5. Applications and Discussion 382
   References 383

Chapter 27
Transfer of Macromolecules Using Erythrocyte Ghosts
Martin C. Rechsteiner

1. Introduction. Comparison of Injection Using Microneedles,
   Liposomes, and Red Blood Cell (RBC) Ghosts 385
2. Loading Macromolecules into RBCs 386
   2.1. Mechanism 386
   2.2. Quantitative Aspects 387
   2.3. Preparation of RBCs 388
   2.4. Preparation of Molecules for Loading 389
   2.5. Choice of Loading Procedure 389
   2.6. Preswell Loading Procedure 390
   2.7. Properties of Loaded RBCs 391
   2.8. Stability of Molecules within RBCs 391
3. Fusion of Loaded RBCs and Cultured Mammalian Cells 392
   3.1. Choice of Fusogen and Fusion Protocol 392
   3.2. Sendai-Mediated Fusion with Cells in Suspension 392
   3.3. PEG Fusion in Monolayer 393
   3.4. Removal of Unfused RBCs 393
   3.5. Expected Results 394
4. Identification and Enrichment for Microinjected Cells 394
5. Summary 397
   References 397

Chapter 28
Liposome-Mediated DNA Transfer
Robert M. Straubinger and Demetrios Papahadjopoulos

1. Introduction 399
2. Liposome Physical Properties and Preparation 400
   2.1. Lipids 401
   2.2. Liposome Preparation 401
   2.3. DNA Encapsulation 403
   2.4. Separation of Liposomes from Free Material 404
3. Liposome–Cell Interaction 405
   3.1. Influence of Vesicle Lipid Composition on
        Intracellular Delivery 406
   3.2. Incubation Conditions 407
   3.3. Improvement of Liposome Efficiency 407
4. Mechanism of Delivery 408
5. Protocol for Liposome–Cell Incubation 408

6. Future Prospects	409
References	410

Chapter 29
Techniques of DNA-Mediated Gene Transfer for Eukaryotic Cells
Arthur P. Bollon and Saul J. Silverstein

1. Introduction	415
2. Transformation of Yeast	416
2.1. Yeast DNA Transformation Protocol	416
2.2. Vectors for Cloning Genes in Yeast	418
3. Transformation of Mammalian Cells	420
3.1. The Components and Their Preparation	420
3.2. Isolation of Carrier DNA	420
3.3. Identification of Transformants	425
4. Potential Uses of Transformation of Eukaryotic Cells	425
References	426

Chapter 30
Viral-DNA Vectors in the Analysis of Mammalian Differentiation
Alban Linnenbach and Carlo M. Croce

1. Introduction	429
2. Isolation and Characterization of the Recombinant DNAs	430
3. Transformation of TK⁻ F9 Cells	430
3.1. Calcium Technique for DNA-Mediated Gene Transfer	430
3.2. HSV-1 tk Starch Gel Electrophoresis	431
4. Viral Antigens of Transformants before and after Differentiation	432
4.1. Retinoic Acid Induction	432
4.2. Indirect Immunofluorescence	433
4.3. Immunoprecipitation	434
5. Organization of the Plasmid Genome in Stem and Retinoic Acid-Treated Cells	435
5.1. Isolation of Cellular DNA	435
5.2. Nick Translation	435
5.3. Southern Blot Analysis	436
6. Transcription of the SV40 Genome in Stem and Differentiated Cells	440
6.1. Preparation of Cellular RNAs	440
6.2. RNA Transfer to Nitrocellulose	440
6.3. S1 Nuclease-Resistant Duplex Analysis of RNAs	442
7. Summary	444
References	445

Chapter 31
Detection of Specific DNA Sequences in Somatic Cell Hybrids and DNA Transfectants
Nancy Hsiung and Raju Kucherlapati

1. Introduction 449
2. Methods 450
   2.1. Isolation of DNA 450
   2.2. Analysis of DNA 453
   References 460

Chapter 32
Microinjection Turns a Tissue Culture Cell into a Test Tube
A. Graessmann and M. Graessmann

1. Introduction 463
2. Procedure 463
   2.1. General 463
   2.2. Preparation of Glass Microcapillaries 464
   2.3. Preparation of Glass Slides 465
   2.4. Cells 466
   2.5. Sample 466
   2.6. Microinjection 468
   References 470

Chapter 33
Techniques of Somatic Cell Hybridization by Fusion of Protoplasts
Horst Binding and Reinhard Nehls

1. Introduction 471
2. Plant Material 471
   2.1. Plant Species 471
   2.2. Cultivation of Plant Material before Protoplast Isolation 472
   2.3. Differentiation Stages of Cells Used for Protoplast Isolation .. 476
3. Protoplast Isolation Techniques 476
   3.1. Preparation of the Plant Material 477
   3.2. Preparations of Enzyme Solutions 477
   3.3. Enzyme Incubation 478
   3.4. Collection of Isolated Protoplasts 480
   3.5. Preparation of Protoplasts Lacking Particular Genetic Capacities 481
4. Protoplast Fusion 481
   4.1. Ca^{++}–High pH–Polyethylene Glycol (PEG) Techniques 481

4.2. Other Fusion Techniques 483
5. Techniques for the Regeneration of Fusion Products 483
  5.1. Culture Technique of Fusion Bodies 483
  5.2. Subculture of Regenerated Cell Clusters 485
  5.3. Formation of Shoots and Plants 486
6. Selection and Analysis of Fusion Products 486
  References 488

Chapter 34

Techniques for Chromosome Analysis

  Vaithilingam G. Dev and Ramana Tantravahi

1. Introduction 493
2. Harvesting and Chromosome Preparation 495
  2.1. Reagents 495
  2.2. Procedure 495
  2.3. Technical Notes 496
3. Quinacrine Banding (Q-Bands) 496
  3.1. Reagents 497
  3.2. Procedure 498
  3.3. Technical Notes 498
4. Giemsa Banding (G-Bands) 498
  4.1. Reagents 500
  4.2. Procedure 500
  4.3. Technical Notes 500
5. Reverse Banding with Chromomycin A_3/Methyl Green
  (R-Bands) 501
  5.1. Reagents 502
  5.2. Procedure 502
6. Constitutive Heterochromatin Banding (C-Bands) 503
  6.1. Reagents 505
  6.2. Procedure 505
  6.3. Technical Notes 506
7. Staining for Active Nucleolus Organizers 506
  7.1. Reagents 506
  7.2. Procedure 507
  7.3. Technical Notes 508
8. Giemsa-11 Staining 508
  8.1. Reagents 508
  8.2. Procedure 508
  8.3. Technical Notes 509
  References 510

Chapter 35
Genetic Analysis of Hybrid Cells Using Isozyme Markers as Monitors of Chromosome Segregation 513
Stephen J. O'Brien, Janice M. Simonson, and Mary Eichelberger

Chapter 36
Future Perspectives in Somatic Cell Genetics 525
Doris L. Slate and Frank H. Ruddle

Index 531

Chapter 1
Selection of Purine and Pyrimidine Nucleoside Analog Resistance in Mammalian Cells

JOHN MORROW

1. Introduction

Since its inception, somatic cell genetics has taken advantage of purine and pyrimidine analog resistance. In the initial stages of the development of this discipline variants resistant to these substances represented the most important and technically useful markers available in permanent cultured cell lines. Although in the past decade many new selective characters have been exploited, these substances are still widely used in genetic studies on somatic cells (Table I). Investigations in experimental mutagenesis (Clive et al., 1972), studies on DNA-mediated transformation (Scangos and Ruddle, 1981), gene mapping (McKusick, 1980), transport (Wohlhueter et al., 1978), somatic recombination (Rosenstraus and Chasin, 1978), and gene inactivation (Bradley, 1979) have taken advantage of these markers (Morrow, 1982).

The reason for the widespread utility of these variants results from several properties: (1) Azaguanine and thioguanine select for (in many cases) the loss of the enzyme hypoxanthine phosphoribosyl transferase (HPRT), which is controlled by an X-linked, and therefore hemizygous, gene (Shows and Brown, 1975; Westerveld et al., 1972). This facilitated the ease with which a recessive mutation could be isolated. (2) Purine and pyrimidine nucleosides are relatively stable substances; they occur in limited concentrations in serum, and they are not liberated by the breakdown of serum macromolecular components. Thus, interference or competition between analogs and their naturally occurring correlates is not a major obstacle (Peterson et al., 1976). (3) Although relatively high (when compared with microorganisms), the mutation rate is still low enough, in both the forward and reverse directions, to allow these variants to be used in a variety of genetic investigations (Morrow, 1970). (4) Mammalian cells are ordinarily prototrophic for purines

---

JOHN MORROW • Department of Biochemistry, Texas Tech University Health Sciences Center, Lubbock, Texas 79410.

Table I. Purine and Pyrimidine Nucleoside Analogs Used in the Selection of Drug-Resistant Variants[a]

Analog	Naturally occurring substance	Enzyme involved in resistance	Reference
6-Thioguanine	Guanine	HPRT	Sharp et al. (1973)
8-Azaguanine	Guanine	HPRT	Morrow (1970)
Trifluorothymidine	Thymidine	TK	Clive et al. (1979)
5-Bromodeoxyuridine	Thymidine	TK	Kit et al. (1963)
2,6-Diaminopurine	Adenine	APRT	Chasin (1974)
1-β-D-Arabino-furanosyl cytosine	Cytidine	dCK	De Saint Vincent and Buttin (1979)
6-Mercaptopurine	Guanine	HPRT	Wolpert et al. (1971)
2-Fluoroadenine	Adenine	APRT	Dickerman and Tischfield (1978)
8-Azaadenine	Adenine	APRT	Dickerman and Tischfield (1978)
Deoxyadenosine	Adenosine	dAK	Ullman et al. (1981)
6-Azauridine	Uridine	UK	Veselý and Čihák (1973)
5-Flourouridine	Uridine	UK	Medrano and Green (1974)
5-Iodo-2 deoxyuridine	Thymidine	TK	Veselý and Čihák (1973)
5-Azacytidine	Cytidine	dCK	Veselý and Čihák (1973)
5-Fluorouracil	Uracil	OP	Patterson (1980)

[a] Abbreviations: HPRT, hypoxanthine phosphoribosyltransferase; TK, thymidine kinase; APRT, adenine phosphoribosyl transferase; dCK, deoxycytidine kinase; dAK, deoxyadenosine kinase; UK, uridine kinase; OP, orotate phosphoribosyl transferase.

and pyrimidines and their addition to the medium is not required for cell survival. The use of a reverse selective technique in which auxotrophy is artificially induced in analog-resistant mutants through the use of aminopterin has proven to be extremely valuable for a host of investigations (Szybalski and Szybalska, 1962).

2. Historical Background

Interest in the purine and pyrimidine analogs grew out of their early role in cancer chemotherapy and the appearance of drug-resistant variants in tumor cell populations treated with these agents (Harris, 1964). A number of investigators in the late 1950s and early 1960s isolated resistant lines and demonstrated that they were in many cases stable, maintained their phenotype in the absence of the selecting agent, and owed their behavior to the diminution or absence of a phosphoribosyl transferase enzyme (in the case of the purine analogs) or of a nucleoside kinase (pyrimidine nucleoside analogs). Szybalski et al. (1962) carried out many of the early studies using these agents, and took advantage of a reverse selective system developed originally by Hakala (1957). Littlefield (1964b) exploited this system, the so-called HAT medium (hypoxanthine, aminopterin, thymidine) to isolate cell hybrids during the 1960s. His initial observations were followed by a host of studies from many laboratories throughout the world, in which every aspect of these characters was studied from the standpoint of genetics, biochemistry, quantitative mutagenesis, and, more recently, DNA transformation and gene transfer using whole chromosomes and other subcellular components. The recent cloning of the genes for adenine phosphoribosyl transferase (Lowy et al., 1980) and thymidine kinase (Pellicer et al., 1978) have made possible the investigation of a variety of significant questions related to the structure and regulation of eukaryotic genes (Scangos and Ruddle, 1981).

3. Methodology

3.1. Tissue Culture Flasks

Although glass prescription bottles are sometimes used for routine maintenance of cell lines, plastic tissue culture labware is employed for all experiments. Different brands have given identical results; currently we use Corning products, as the result of a state contract.

3.2. Culture Reagents

All routine cultivation and experimental procedures are carried out in Dulbecco's modification of Eagle's essential medium (DMEM). The medium is purchased from Kansas City Biologicals as a powder, dissolved in triple glass distilled water, and filtered through 0.22-μm Millipore filters using a stainless steel filtering apparatus which is sterilized with ethylene oxide. After filtration, the medium is checked for plating efficiency and sterility and stored in a refrigerator. Each batch is consumed in approximately 2 weeks. The complete medium is assembled through the addition of a 100× mixture of penicillin (final concentration is 100 units/ml) and streptomycin (100 μg/ml) which is kept frozen until used, and 10% serum. For routine culturing procedures we have adopted the use of a 30:70 mixture of fetal calf and calf serum, due to economic considerations. It was found that the great variability in the growth-supporting properties of calf serum precluded its use without addition of fetal calf serum. Serum fractions (especially fetal calf serum) contain unknown quantities of purines and pyrimidines which can influence the yield of mutants. For this reason, quantitative studies require the use of dialyzed serum. Commercial preparations have proven to be unsatisfactory and we prepare our own serum by dialysis for 48 hr against two changes of distilled water at 4°C.

Cell lines that grow attached to the culture vessels are transferred through the use of a trypsin–EDTA solution (0.5 g/liter; 0.2 g/liter) which has proven satisfactory even for lines that grow quite firmly attached to the culture vessel.

Analogs are prepared as 100× solutions. In order to facilitate their solubilization, it is necessary to use 1 M NaOH added drop by drop, after which the dissolved solution is brought up to proper volume in distilled water. HAT medium is prepared as a 100× stock solution, which is added to cultures containing standard growth medium. The final concentration is: aminopterin, 0.1 µg/ml; hypoxanthine, 5 µg/ml, thymidine, 5 µg/ml (Littlefield, 1964b). Stock solutions are not frozen and thawed, but are stored at 4°C.

3.3. Screening for Contaminants

Mycoplasma (PPLO) contamination is a chronic problem in the isolation of purine and pyrimidine nucleoside analog-resistant mutants (Clive et al., 1973). Contaminated cell lines can be drastically altered in their response to selective agents, in particular to HAT medium. It has been our experience and that of other workers that such contaminated cell lines will not grow in HAT medium, even when they possess the appropriate nucleoside phosphorylase or phosphoribosyl transferase enzyme. For this reason our cell lines are tested at biweekly intervals for mycoplasma by measuring the ratio of [^{14}C] uracil to [^{3}H] uridine uptake over a 60-min period (Schneider et al., 1974). Cell lines are harvested and replated in 60-mm plastic dishes at varying densities (5×10^5–2×10^6) of cells per dish. The cells are allowed to attach and grow in complete medium for 24 hr in a CO_2 incubator. At the end of this period the cells are washed once with 5 ml of serum-free DMEM. Next, 3 ml of serum-free DMEM containing 1 µCi/ml [G-^{3}H] uridine (3.66 mCi/mmole, New England Nuclear, Boston, Massachusetts) and 0.1 µCi/ml [2-^{14}C] uracil 2 (56.8 mCi/mmole; New England Nuclear) is added. Final concentrations of these isotopes in the medium are 2.7×10^{-1} and 1.8×10^{-3} mM for the uridine and uracil, respectively. Incubation is carried out for 1 hr at 37°C in a CO_2 incubator. At this time the radioactive medium is removed and the cells are rinsed with Ca^{++}- and Mg^{++}-free Hanks balanced salt solution. The cells are then dislodged with 3.0 ml of trypsin–EDTA. An equal volume of cold 20% trichloroacetic acid (TCA) is then added to the flasks and acid-insoluble material is collected on 24-mm Millipore (Bedford, Massachusetts) membrane filters (Type HA, 0.45-µm pore size) and washed thoroughly with 20 ml of cold 5% TCA prior to drying under an infrared heat lamp. Radioactivity is then assayed by placing the filter disks in scintillation shell vials containing 4.0 ml of a fluid consisting of toluene and 4.0 g/liter of Omnifluor (New England Nuclear). The samples are counted in a Beckman LS 230 liquid scintillation counter. Data are corrected for counting efficiency and expressed as disintegrations per minute (dpm). Incorporation of uracil into the TCA-insoluble fraction at levels above background is strongly indicative of contamination, and such cell lines are discarded. The results of such tests agree well with the assays based on other techniques, including the fluorescent staining method (Chen, 1977), microbiologic broth tests, and examination with the scanning electron microscope. Thus this procedure is rapid, accurate, and effective (Morrow et al., 1980).

A number of procedures for curing contaminated cell lines have been developed, and include: (1) Treatment with antibiotics such as tetracycline, which is effective against some strains of mycoplasma. (2) Injection of malignant contaminated cells into a histocompatible host (Van Diggelen et al., 1977). In those cases in which a tumor is produced, reisolation of uncontaminated cells is possible due to the ability of the immune system of the host to overcome the contaminating organisms. (3) Treatment with bromouracil, followed by exposure to visible light. The analog is incorporated into the mycoplasma DNA, and sensitizes the DNA to breakage by visible light. However, all of these methods are arduous, requiring long periods and extensive subcloning of the cells, and are not recommended except in cases involving irreplaceable cell lines.

3.4. Cell Lines

Tremendous variation in the ease with which drug-resistant mutants can be isolated can be observed in different cell lines. This may be due to differences in the number of copies of the gene present in the cell line under examination (Raskin and Gartler, 1978), or it may be the result of juxtaposition of genetic material next to structural genes which causes their suppression or inactivation (Morrow, 1977). The spontaneous frequency of resistant variants must be determined empirically, and may vary between four orders of magnitude. In some cases resort to mutagenic agents is required, whereas in others it is possible to select a highly resistant line through stepwise increments of the drug. Failure to obtain resistant variants in a line which is known to produce them is suggestive of mycoplasma contamination (see above). In certain cell lines (such as derivatives of the permanent human line, HeLa) we have never obtained thioguanine- or azaguanine-resistant variants without recourse to mutagenic agents.

When cells are selected with low levels of purine and pyrimidine nucleoside analogs, in many instances it is possible to select out resistant variants that occur with extremely high rates and are unstable and revert rapidly to wild type in the absence of selective pressure (Morrow, 1970; Fox and Radicic, 1978). The genetic and biochemical basis of such variants is not well understood, although in some instances it has been shown that they possess intermediate levels of the appropriate kinase or phosphorylase enzyme (Littlefield, 1964a). Because they are unstable and grow in HAT medium, they are not useful as markers in hybridization, transformation, or other experiments involving gene exchange.

3.5. Metabolic Cooperation

The drug resistance of a given phenotype can be obviated by the intimate presence of drug-sensitive cells. When enzyme-positive and enzyme-negative cells are in direct contact, the passage of analog nucleotides to the drug-resistant cells through intercellular gap junctions will result in their death. This is the result of transfer of fraudulent nucleotides and is especially pronounced in normal, diploid fibroblasts (DeMars and Held, 1972). The avoidance of this transfer through intercellular gap junctions requires that variants be selected at low cell densities (Goldfarb et al., 1974).

3.6. Cloning of Variants

In order to ensure genetic homogeneity, it is necessary that clonal isolates be obtained. Clones are isolated using stainless steel "penicylinders," which were developed originally for antibiotic testing. The cylinders (0.5 cm diameter) are rubbed in sterile Vaseline and placed over the clone. About 0.1 ml of trypsin is introduced into the cylinder with the aid of a Pasteur pipette. It is agitated vigorously and transferred to a fresh flask with 5 ml of growth medium already in the flask. After the cells have recovered and proliferated sufficiently, the procedure is repeated in order to assure cloning. In cases in which the clones are grown in flasks rather than petri dishes a small circle of plastic is removed with the aid of a hot cork borer, allowing access to the clone.

3.7. Sample Protocols

1. Isolation of a thioguanine-resistant V79 cell deficient in HPRT with a low reversion rate. V79 Chinese hamster cells are harvested with tryspin–EDTA solution and thoroughly dispersed with a finely drawn Pasteur pipette. The cells are centrifuged in a clinical centrifuge, washed in Hanks solution, diluted in several millimeters of Hanks, and counted on an electronic cell counter (Particle Data, Inc.). They are plated in plastic flasks in 30 μg/ml of thioguanine at 10^5 cells per 25-cm^2 flask or 3 $\times$ 10^5 cells per 75-cm^2 flask. The medium is replaced every 4 days. After 2 weeks, clones appear, and these are isolated, grown to large numbers, and tested for their reversion frequency. Cells are tested for reversion by plating 10^6 cells in a 75-cm^2 flask in HAT medium. The medium is changed every 2 days for the first week. If no revertants appear in the HAT medium within 2 weeks, the procedure is repeated. Nonreverting HPRT clones are reserved for further study.

2. Isolation of thymidine kinase-negative variants in the Friend leukemia cell line. Friend leukemia cells are grown in suspension using standard medium in screw-cap bottles. The cells are harvested and counted and diluted into fresh medium at a density of 10^4 cells/ml, approximately 50 ml total. Bromodeoxyuridine (BrdU) (1 μg/ml) is added, and the cells are grown at this concentration for approximately 1 week. Medium is changed by making 1:5 dilutions every 5 days. After 1 week the concentration of analog is doubled, and the procedure repeated. When the level of analog reaches 500 μg/ml, the

cells are cloned by diluting into microtiter plates, and clones are selected. They are tested for their reversion frequency in HAT medium by plating 10^5 per well in Linbro Multiwell dishes, which hold 1 ml volume in each of 25 wells. The HAT medium is added at the time the cultures are initiated. Among several isolates tested we were able to isolate one whose reversion frequency was less than 10^{-7}.

3. Induction of purine analog resistance following mutagen treatment. V79 Chinese hamster cells are harvested as above and several million cells are plated into a 75-cm^2 flask. After the cells have attached they are treated with 100–300 µg/ml of ethyl methane sulfonate (EMS). The EMS is prepared by dilution into Hanks solution and is used immediately (since it is rapidly hydrolyzed). Cells are exposed for 18 hr (approximately one cell division), the EMS is removed, and the flask is thoroughly rinsed with Hanks solution. The cells are harvested and a control is plated out at low densities to determine the cloning efficiency. The treated cells are grown for 4 days (96 hr) under nonselective conditions. It may be necessary during this time to split the cultures if they become too crowded. At the end of the lag period the cultures are harvested and plated into the selecting agent (in this instance 30 µg/ml thioguanine). The procedure described in protocol 1 is then followed.

4. Isolation of a double mutant cell line for hybridization to wild-type cells. A cell line that is resistant to ouabain (a Na$^+$/K$^+$ ATPase inhibitor) and also resistant to a pyrimidine nucleoside or purine analog is useful for hybridization to wild-type cells (Jha and Ozer, 1976). Since ouabain resistance is dominant, as is HAT resistance, hybrids can be selected in HAT medium plus ouabain. The double mutant parent is killed by the HAT medium, the wild type parent by the ouabain, and only the hybrids survive.

We accomplished selection of such a double mutant with the aid of the LMTK$^-$ cell line (an L cell isolated by stepwise increase of BrdU over the course of 1 year.) The LMTK$^-$ line has never been observed to revert, and may represent a deletion (Kit et al., 1963).

Ouabain resistance was attained using mutagenesis. LMTK$^-$ cells were plated into 25 cm^2 flasks at a density of 2×10^6 cells per flask. After 24 hr the medium was changed and 300 µg/ml of EMS was added. After 18 hr the medium was replaced. After 5 days of growth in mutagen-free medium, ouabain was added at concentrations of 1, 2, and 4 mM. Media were changed every 3–4 days. After 4 weeks, five colonies were detected in the flask with 4 mM ouabain. These colonies were isolated and then grown in ouabain-free medium, and after sufficient numbers of cells were grown, the plating efficiencies of these clones were obtained. The clones isolated in ouabain were found to be stably resistant, and were preserved for further investigations.

References

Bradley, W. E. C., 1979, Reversible inactivation of autosomal alleles in Chinese hamster cells, J. Cell. Physiol. **101**:325–340.

Chasin, L. A., 1974, Mutations affecting adenine phosphoribosyl transferase activity in Chinese hamster cells, *Cell* **2**:37–41.

Chen, T., 1977, In Situ detection of mycoplasma contamination in cell cultures by fluorescent Hoeschst 33258 Stain, *Exp. Cell Res.* **104**:255–262.

Clive, D., Flamm, W. G., Machesko, M. R., and Bernheim, N. J., 1972, A mutational assay system using the thymidine kinase locus in mouse lymphoma cells, *Mutat. Res.* **16**:77–87.

Clive, D., Flamm, W. G., and Patterson, J. B., 1973, Specific locus mutational assay systems for mouse lymphoma cells, Appendix II in: *Clinical Mutagens*, Volume 3 (A. Hollander, ed.), pp. 100–103, Plenum Press, New York.

Clive, D., Johnson, K., Spector, J., Batson, A., and Brown, M., 1979, Validation and characterization of the L5178Y/TK$^{+/-}$ mouse lymphoma mutagen assay system, *Mutat. Res.* **59**:61–108.

DeMars, R., and Held, K., 1972, The spontaneous azaguanine-resistant mutants of diploid human fibroblasts, *Humangenetik* **16**:87–110.

De Saint Vincent, B. R., and Buttin, G., 1979, Studies on 1-beta-D-arabinofuranosyl cytosine-resistant mutants of Chinese hamster fibroblasts: III. Joint resistance to arabinofurnanosyl cytosine and to excess thymidine: A semidominant manifestation of deoxycytidine triphosphate pool expansion, *Somat. Cell Genet.* **5**:67–82.

Dickerman, L. H., and Tischfield, J., 1978, Comparative effets of adenine analogs upon metabolic cooperation between Chinese hamster cells with different levels of adenine phosphoribosyltransferase activity, *Mutat. Res.* **49**:83–94.

Fox, M., and Radacic, M., 1978, Adaptational origin of some purine-analogue resistant phenotypes in cultured mammalian cells, *Mutat. Res.* **49**:275–296.

Goldfarb, P. S. G., Slack, C., Subak-Sharpe, J., and Wright, E., 1974, Metabolic cooperation between cells in tissue culture, in: *Symposia of the Society for Experimental Biology*, Vol. 28, pp. 463–484, Cambridge University Press, London.

Hakala, M. T., 1957, Prevention of toxicity of amethopterin for sarcoma 180 cells in tissue culture, *Science* **126**:255–256.

Harris, M., 1964, *Cell Culture and Somatic Variation*, Holt, Rinehart and Winston, New York.

Jha, K., and Ozer, H., 1976, Expression of transformation in cell hybrids. I. Isolation and application of density-inhibited Balb/3T3 cells deficient in hypoxanthine phosphoribosyltransferase and resistant to ouabain, *Somat. Cell Genet.* **2**:215–223.

Kit, S., Dubbs, D. R., Piekarski, L., and Hsu, T. C., 1963, Deletion of thymidine kinase activity from L cells resistant to bromodeoxyuridine *Exp. Cell. Res.* **31**:297–312.

Littlefield, J. W., 1964a, Three degrees of guanylic acid-inosinic acid pyrophosphorylase deficiency in mouse fibroblasts, *Nature* **203**:1142–1144.

Littlefield, J. W., 1964b, Selection of hybrids from matings of fibroblasts *in vitro* and their presumed recombinants, *Science* **145**:709–710.

Lowy, I., Pellicer, A., Jackson, J., Sim, G., Silverstein, S., and Axel, R., 1980, Isolation of transforming DNA: Cloning the hamster aprt gene *Cell* **22**:817–823.

McKusick, V. A., 1980, The anatomy of the human genome, *J. Heredity* **71**:370–391.

Medrano, L., and Green, H., 1974, A uridine kinase-deficient mutant of 3T3 and a selective method for cells containing the enzyme, *Cell* **1**:23–26.

Morrow, J., 1970, Genetic analysis of azaguanine resistance in an established mouse cell line, *Genetics* **65**:279–287.

Morrow, J., 1975, On the relationship between spontaneous mutation rates *in vivo* and *in vitro*, *Muta. Res.* **33**:367–372.

Morrow, J., 1977, Gene inactivation as a mechanism for the generation of variability in somatic cells cultivated *in vitro*, *Mutat. Res.* **44**:391–400.

Morrow, J., 1982, *Cell Genetics*, Academic Press, New York.

Morrow, J., Sammons, D., and Barron, E., 1980, Puromycin resistance in Chinese hamster cells: Genetic and biochemical studies of partially resistant, unstable clones, *Mutat. Res.* **69**:333–346.

Patterson, D., 1980, Isolation and characterization of 5-fluorouracil resistant mutants of Chinese hamster ovary cells deficient in the activities of orotate phosphoribosyltransferase and orotidine 5-monophosphate decarboxylase, *Somat. Cell Genet.* **6**:101–114.

Pellicer. A., Wigler, M., Axel, R., and Silverstein, S., 1978, The transfer and stable integration of the HSV thymidine kinase gene into mouse cells, *Cell* **14:**133–141.

Peterson, A. R., Krahn, D. F., Peterson, H., Heidelberger, C., Bhuyan, B. K., and Li, L. H., 1976, The influence of serum components on the growth and mutation of Chinese hamster cells in medium containing 8-azaguanine, *Mutat. Res.* **36:**345–356.

Raskind, W., and Gartler, S., 1978, The relationship between induced mutation frequency and chromosome dosage in established mouse fibroblasts lines, *Somat. Cell Genet.* **4:**491–506.

Reynolds, R., and Hetrick, F., 1969, Potential use of surface-active agents for controlling mycoplasma contamination in animal cell cultures, *Appl. Microbiol.* **17:**405–411.

Rosenstraus, M., and Chasin, L., 1978, Separation of linked markers in Chinese hamster cell hybrids: mitotic recombination is not involved, *Genetics* **90:**735–760.

Scangos, G., and Ruddle, F., 1981, Mechanisms and applications of DNA-mediated gene transfer in mammalian cells—A review, *Gene* **14:**1–10.

Schneider, E. L., Stanbridge, E., and Epstein, C., 1974, Incorporation of ^{3}H-Uridine and ^{3}H-Uracil into RNA, *Exp. Cell Res.* **84:**311–318.

Sharp, J. D., Capecchi, N., and Capecchi, M., 1973, Altered enzymes in drug resistant variants of mammalian tissue cultured cells, *Proc. Natl. Acad. Sci. USA* **70**(11):3145–3149.

Shows, T. D., and Brown, J., 1975, Human X-linked genes regionally mapped utilizing X-autosome translocations and somatic cell hybrids, *Proc. Natl. Acad. Sci. USA* **72:**(6): 2125–2129.

Szybalski, W., and Szybalska, E., 1962, Drug sensitivity as a genetic marker for human cell lines, *University of Michigan Medical Bulletin* **28:**277–293.

Szybalski, W., Szybalska, E., and Ragni, G., 1962, Genetic studies with human cell lines, *Natl. Cancer Inst. Monogr.* **7:**75–89.

Ullman, B., Levinson, B., Hershfield, M., and Martin, D., 1981, A biochemical genetic study of the role of specific nucleoside kinases in deoxy-adenosine phosphorylation by cultured human cells, *J. Biol. Chem.* **256**(2):848–852.

Van Diggelen, O., Shin, S., and Phillips, D., 1977, Reduction in cellular tumorigenicity after mycoplasma infection and elimination of mycoplasma from infected cultures by passage in nude mice, *Cancer Res.* **37:**2680–2687.

Veselý, J., and Čihák, A. 1973, Resistance of mammalian tumor cells toward pyrimidine analogues, a review *Oncology* **28:**204–226.

Westerveld, A., Visser, R., Freeke, M. A., and Bootsma, D., 1972, Evidence for linkage of PGK, HPRT, and GGPD in Chinese hamster cells studied by using a relationship between gene multiplicity and enzyme activity, *Biochem. Genet.* **7:**33–40.

Wohlhueter, R., Marz, R., Graff, J., and Plagemann, P., 1978, A rapid-mixing technique to measure transport in suspended animal cells: Application to nucleoside transport in Novikoff rat hepatoma cells, in: *Methods in Cell Biology*, Volume 20 (David Prescott, ed.), Academic Press, New York, pp. 211–236.

Wolpert, M., Damle, S., Brown, J., Sznycer, E., Agrawal, K., and Sartorelli, A., 1971, The role of phosphohydrolases in the mechanism of resistance of neoplastic cells to 6-thiopurines, *Cancer Res.* **31:**1620–1626.

Chapter 2

Techniques for Using HAT Selection in Somatic Cell Genetics

W. N. CHOY, T. V. GOPALAKRISHNAN, and J. W. LITTLEFIELD

1. Introduction

Somatic cell hydridization provides a unique system to analyze gene interactions between somatic cells through the formation of hybrids in intraspecific or interspecific crosses. With the development of conditional genetic markers, many phenotypic changes in cell hybrids, either permanent or transient, can be monitored in a quantitative fashion. Spontaneous fusions of cells are always infrequent and hybrids generated from such fusions are usually hard to identify. Improved fusion techniques and selection systems for cell hybrids in the past decade have circumvented many of the above difficulties.

To enhance the formation of cell hybrids, inactivated Sendai virus, a paramyxovirus, was first used as a fusing agent for the hybridization of Ehrlich's tumor cells (Okada, 1962), and of mouse and human cells (Harris and Watkins, 1965). Inactivation of the virus was usually required, although the viral DNA was found not to be involved in the fusion process (Okada and Tadokoro, 1962). This can be achieved either by a short UV pulse (Harris and Watkins, 1965) or by β-propiolactone (Neff and Enders, 1968). Until Pontecorvo introduced the use of polyethylene glycol (PEG) as a fusing agent for mammalian cells (Pontecorvo, 1975), Sendai virus was used customarily for hybridization studies. The advantages of using PEG instead of Sendai virus in regard to commercial availability and consistency of quality were recognized quickly. Extensive studies have been devoted to optimizing PEG fusion techniques on various cell types in recent years (Vaughan et al., 1976; Davidson and Gerald, 1976; Davidson et al. 1976; Norwood et al., 1976; O'Malley and Davidson, 1977; Schneiderman et al., 1979; Brake and Serra,

---

W. N. CHOY • Toxicology and Biological Evaluation Research, Western Regional Research Center, United States Department of Agriculture Berkeley, California 94710.   T. V. GOPALAKRISHNAN and J.W. LITTLEFIELD • Department of Pediatrics, Johns Hopkins University, School of Medicine, Baltimore, Maryland 21205.

1981; Rabinovitch and Norwood, 1981). PEG fusion is described in detail in Chapter 3, Mercer and Baserga, and Chapter 4, Norwood and Ziegler, this volume. Other membrane-active, PEG-related derivatives were also found recently to enhance cell fusion (Klebe and Mancuso, 1981).

1.1. Selective Systems in Somatic Cell Genetics

A number of selection techniques based on variations of cell morphology, growth characteristics, and genetic markers have been described which screen for "mutants," or "variants," in somatic cells (Clements, 1975; Chu and Powell, 1976; Siminovitch, 1976). Many such techniques, at least in principle, can be adapted for the selection of somatic cell hybrids. Primary cultures with limited life span or cells growing in suspension can be selected against by manipulation of tissue culture conditions. Such "half selections," that is, selective against only one parent, are usually incomplete and confined to a few specific cell types. New methods of hybrid selection, such as fluorescent probes (Keller and Person, Chapter 8), the flow cytometer (Jongkind and Verkerk, Chapter 7), or biochemical inhibitors (Wright, Chapter 5), will be discussed in this volume.

A powerful way to select hybrids is to eliminate parents with different genetic deficiencies by a combination of mutant selection techniques that allows hybrids with complementary functional genes to grow. This concept was introduced for eukaryotic cells by Littlefield (1964). It was first used to select for hybrids generated from two drug-resistant mouse strains, resistant to 8-azaguanine (8 AG) and resistant to 5-bromodeoxyuridine (BrdU), respectively, by using a selective medium containing hypoxanthine, aminopterin, and thymidine, commonly known as "HAT medium" (Littlefield, 1964, 1966). This method has since been widely used for somatic cell hybridization experiments. It has encouraged the development of a variety of selective media for other drug-resistant mutants (Kusano et al., 1971; Chan et al., 1975), auxotrophic mutants (Kao et al., 1969), and temperature-sensitive mutants (Zepp et al., 1971; Goldstein and Lin, 1972).

1.2. Principle of HAT Selection

The biosynthesis of purines and pyrimidines in mammalian cells is achieved by endogenous pathways supplemented by salvage pathways. Folic acid analogs, such as aminopterin or amethopterin (methotrexate), are stoichiometric inhibitors of folic acid reductase and can block the endogenous synthesis of purines (Fig. 1), one-carbon-transfer reactions, and the synthesis of certain amino acids. They also inhibit the conversion of deoxyuridylic acid to thymidylic acid in pyrimidine biosynthesis (Fig. 2). In the presence of aminopterin or amethopterin, normal cells can utilize preformed hypoxanthine and thymidine to synthesize the nucleic acid precursors through the purine and pyrimidine salvage pathways. Somatic cell variants lacking enzymes of the salvage pathways, however, cannot survive when the

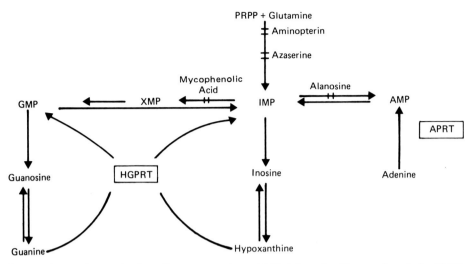

Figure 1. Purine biosynthesis and interconversions. AMP, Adenosine 5′-monophosphate; APRT, adenine phosphoribosyl transferase; GMP, guanine 5′-monophosphate; HGPRT, hypoxanthine-guanine-phosphoribosyl transferase; IMP, inosine 5′-monophosphate; PRPP, 5′-phosphoribosyl-1-pyrophosphate; XMP, xanthosine 5′-monophosphate.

endogeous synthesis is blocked. Based on earlier studies of Hakala (1957) and Werkheiser (1961), a selective medium containing hypoxanthine, aminopterin, and thymidine was designed to screen for rare revertants of hypoxanthine-guanine phosphoribosyltransferase-deficient (HGPRT⁻) mutants that have acquired the ability to convert hypoxanthine to inosinic acid (Szybalski et al., 1962). This HAT selective medium was further discovered to be extremely useful to select for somatic cell hybrids derived from parental

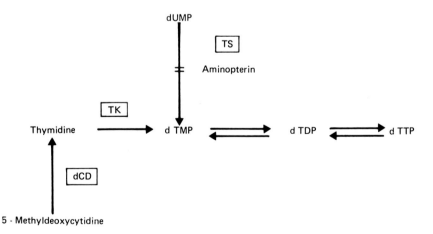

Figure 2. Pyrimidine biosynthesis and interconversions. dCD, Deoxycytidine deaminase; dUMP, deoxyuridine 5′-monophosphate; dTMP, deoxythymidine 5′-monophosphate; dTDP, deoxythymidine 5′-diphosphate; dTTP, deoxythymidine 5′-triphosphate; TK, thymidine kinase; TS, thymidylic synthetase.

variants defective in HGPRT and thymidine kinase (TK) activities. While the parental cells are killed by HAT, only hybrids that have inherited complementary unaltered HGPRT and TK genes from two parents are able to grow (Littlefield, 1964, 1966).

2. Selection of Somatic Cell Hybrids with HAT Medium

Since the first report on the use of HAT medium for hybrid selection (Littlefield, 1964), the recipe of HAT has been repeatedly modified for different cell types and applications. We wish to describe the system we have used frequently in this laboratory in recent years.

The drug-resistant parental cells used for hybridization experiments are always cleared of preexisting spontaneous revertants. Thus, mutants defective in HGPRT activity are cultured in 6-thioguanine (6TG) or 8AG-containing medium, and mutants defective in TK activity in BrdU-containing medium, for several generations before use. A few days before fusion, cells are withdrawn from drug-containing media and grown in regular medium to remove 6TG, 8AG, or BrdU. After cell hybridization, cells are either trypsinized and replated immediately in HAT in culture medium at a cell density of 1×10^5 cells in a 100-mm tissue culture dish, or they are fed with regular medium and selected with HAT medium on the following day. The latter method is recommended for experiments involving human fibroblasts. The HAT medium now used has the composition of 0.1 mM hypoxanthine, 0.01 mM amethopterin, and 0.02 mM thymidine. The HAT medium is changed on the third day after fusion and every 3 days thereafter, until hybrid clones appear.

If viable hybrids are to be recovered, the clones isolated should initially be grown in medium supplemented with hypoxanthine and thymidine at concentrations similar to HAT medium in order to overcome the residual toxic effect of the aminopterin block. After several generations of growth in hypoxanthine–thymidine-containing medium, hybrids can safely be cultured in regular drug-free medium.

3. Modifications of HAT Selection

Many factors, such as the toxicity of aminopterin, the serum purine concentration, and cellular variations in nucleic acid metabolism, can influence the efficiency of HAT selection. When dialyzed serum is used, glycine may be required for optimal cell growth, with a final concentration of 0.01–0.1 mM (Hakala, 1957). Such glycine-containing medium is referred to as THAG (Chu et al., 1969) or HATG medium. Also, azaserine, instead of folic acid analogs, has been used with hypoxanthine and thymidine to block endogenous purine synthesis in the selection of hybrids (Siniscalco et al., 1969; Liskay and Patterson, 1978). An extensive survey of the modifications of the original HAT recipes for different cell types has been reported (Fox and Boyle, 1976).

3.1. AA Medium

Other drug-resistant markers related to purine and pyrimidine metabolic pathways can be useful in selecting for cell hybrids. Kusano and co-workers (1971) developed a system to select for hybrids expressing the adenosine phosphoribosyl transferase (APRT) gene in human–mouse fusions. This is achieved by using the antibiotic alanosine to block the endogenous synthesis of adenylic acid (AMP) from inosinic acid (IMP). In the presence of adenine only cells or hybrids with functional APRT can grow (Fig. 1). The medium, consisting of adenine and alanosine each at a concentration of 50 μM, was designated AA medium (Kusano et al., 1971).

3.2. HAM Medium

Another modification of HAT medium was designed to be used with mutants defective in deoxycytidine deaminase (dCD). Cells deficient in dCD cannot convert 5-methyldeoxycytidine to thymidine and are killed by thymidine starvation when the endogenous pyrimidine synthesis pathway is blocked by aminopterin (Fig. 2). The medium used to select for hybrids with dCD activity consists of 0.1 mM hypoxanthine, 1 μM aminopterin, and 48 μM 5-methyldeoxycytidine; it is known as HAM medium (Chen et al., 1975). HAM selection resembles HAT selection, with dCD activity used for selection instead of TK activity.

3.3. GAMA Medium

A selective medium for somatic cell hybrids between HGPRT⁻ and APRT⁻ parents has been developed by using azaserine to block endogenous purine synthesis and mycophenolic acid to block the conversion of IMP to XMP in the AMP to GMP pathway (Fig. 1). In this medium, both adenine and guanine are required for cell survival, and parental cells deficient in either HGPRT or APRT activities are selectively eliminated. Only hybrids that have acquired complementary wild-type enzymes are able to grow. This medium, consisting of 0.2 mM guanine, 0.1 mM adenine, 6 μM mycophenolic acid, and 10 μM azaserine, was designated GAMA medium (Liskay and Patterson, 1978).

4. Modifications by Incorporation of Additional Selective Agents

4.1. Ouabain

Based on the observation that rodent cells are at least 10,000 times more resistant to ouabain toxicity than are human cells in culture, HAT medium containing ouabain was used to select for hybrids between normal human and

HGPRT⁻ rodent cells (Baker et al. 1974; Kucherlapati et al., 1975). This selective system can also be used to select for intraspecific hybrids between normal cells and ouabain-resistant mutants (Oua^r) (Jha and Ozer, 1976). Corsaro and Migeon (1978) suggested that double mutants carrying both HGRPT⁻ and Oua^r markers can theoretically serve as "universal hybridizers" when selected against HAT medium supplemented with ouabain (HOT medium). Such double mutants have been isolated in diploid human lymphoblast long-term cultures (Adelberg et al., 1975; Choy and Littlefield, 1980). The application of HOT selection was recently confirmed by studies using HGPRT⁻ Oua^r double mutants derived from euploid human fibroblast and HeLa cells (Weissman and Stanbridge, 1980).

4.2. Polyene Antibiotics

Mammalian cells differ in sensitivity to the structurally modified polyene antibiotics amphotericin B methyl ester (AME) and nystatin methyl ester (NME) (Fisher and Bryson, 1977). Resistance was found to be dominant in cell hybrids. Since Syrian hamster cells and HeLa cells are more sensitive to these polyene antibiotics than are mouse A9 cells, it was possible to select for somatic cell hybrids generated from fusions of normal Syrian hamster cells and HGPRT⁻ or TK⁻ mouse cells. HAT medium supplemented with AME (25–100 µg/ml) or NME (100–300 µg/ml), two media designated as HAT–AME and HAT–NME, respectively, were used (Fisher et al., 1978; Goldstein and Fisher, 1978). This HAT–polyene selection system, similar to the HOT selection system, requires that only one parent carry genetic markers.

4.3. Diphtheria Toxin

Mouse cells are more resistant to diphtheria toxin than are human cells over a 10,000-fold range (Dendy and Harris, 1973). This differential drug sensitivity relates to the presence of diptheria toxin receptors in human cells, which are lacking in the mouse. The genetic determinant of the toxin receptor has been mapped to human chromosome 5 (Creagen et al., 1979). In the presence of diphtheria toxin at a concentration of about 3 nM, human cells were selectively killed, while human–mouse hybrids lacking human chromosome number 5 survived (Ruddle, 1972). This selective mechanism can be coupled with HAT selection if HGPRT⁻ or TK⁻ mutants are involved in a hybridization experiment.

5. Applications of HAT Selection

Cell hybridization has been used extensively to study the genetics and differentiation of mammalian cells. The HAT system was the first selective system developed to isolate somatic cell hybrids, and has played an important role in the development of the field of somatic cell genetics. Significant contributions of the HAT system to contemporary biology will be discussed briefly.

5.1. Selection of Revertants of Drug-Resistant Mutants Deficient in HGPRT and TK

Mutants deficient in either HGPRT or TK activity are resistant to 6TG and 8AG and to BrdU, respectively, and are killed in HAT medium. Revertants sensitive to 6TG or BrdU can be selected by plating the mutants, with or without pretreatment with a mutagen, in HAT medium. In general, survivors can reexpress the enzyme activity and hence become sensitive again to 6TG or BrdU; often, however, the revertants can grow equally well in HAT and in the presence of these drugs. Survival in both has been attributed either to resistance to aminopterin (Fenwick, 1980) or modulation of enzyme activity by aminopterin. This was illustrated by the ability of hybrids formed between mouse A9 cells (HGPRT$^-$) and chick erythrocytes to express chick HGPRT exclusively only in the presence of HAT (Klinger and Shin, 1974). Similar modulation of TK activity was observed in the expression of herpes simplex virus TK and TK$^-$ mouse cells (Davidson et al., 1973). Reappearance of rodent HGPRT in hybrids between HGPRT$^-$ rodent cells and human cells has provided evidence for a regulatory locus for the HGPRT gene (Watson et al., 1972; Bakay et al., 1973; Bakay et al., 1978). Thus the HAT selective system has contributed significantly to our understanding of the HGPRT locus.

5.2. Gene Mapping

Another important contribution of the HAT selective system is the development of the concept of gene mapping using somatic cell hybrids (McKusick and Ruddle, 1977). Human–mouse hybrids selected and maintained in HAT medium segregate human chromosomes randomly and ultimately retain only the human chromosome coding for the salvage pathway enzyme which is required for cell survival and is deficient in the mouse cells. Techniques to differentiate between human and mouse chromosomes have made it possible to assign a gene to a specific human chromosome in somatic cell hybrids. The TK gene was the first gene mapped using this somatic cell hybridization technique, and was assigned ultimately to chromosome 17 (Boone et al., 1972). The HGPRT gene was mapped to the human X chromosome. Using translocations of human X chromosomes to various autosomes, linkage analysis and more precise mapping of genes have become possible. The HAT selective system has been an essential technique for the mapping of human genes.

5.3. Gene Transfer

The initial gene transfer experiments involving chromosomes utilized the HAT selective system to detect for the introduction of exogenous genes (McBride and Ozer, 1973; Graf et al., 1979; Lewis et al., 1980). The HGPRT gene was the first gene to be transferred by chromosomes into recipient mammalian cells. Later studies utilized the direct transfer of DNA segments (Wigler et al., 1977, 1978; Lester et al., 1980). This DNA-mediated system was extended subsequently to introduce selectable and nonselectable prokaryotic and eukaryotic genes into mouse L cells and other recipient cells by transfecting TK$^-$ cells with purified herpes simplex virus TK genes along with a nonselectable gene, followed by selection with HAT medium (Sabourin and Davidson, 1979; Wigler et al., 1979; Wold et al., 1979; Pellicer et al., 1980). In these "cotransfer" studies, a high proportion of the clones selected with HAT also incorporated the nonselectable gene. At least in two such systems, incorporation of mouse mammary tumor virus DNA and rate α_{2u} globulin genes into mouse L cells, the transferred genes functioned normally and were inducible with dexamethasone treatment (Hynes et al., 1981; Kurtz, 1981). This ability to cotransfer nonselectable genes with a selectable marker has provided a new approach to studies on gene regulation.

Gene transfer in cultured cells has also permitted the cloning of genes which can be selected for and which code for complex messages with only 5–10 copies per cell. For example, the chicken TK gene was cloned by transforming LMTK$^-$ cells with restriction enzyme-digested chicken DNA fragments ligated to a bacterial plasmid, PBR322. Then the TK gene was isolated from the transformed cells by a "plasmid rescue" procedure (Perucho et al., 1980). This approach has been extended to other systems (Lowy et al., 1980). Thus it is possible to clone a gene that codes for a selectable marker without necessitating the isolation of a cDNA probe.

5.4. Selection of Hybridomas

The HAT system has been the only selective system used so far to isolate "hybridomas" from fusions between cultured myeloma cells and spleen cells obtained from immunized hosts (Köhler and Milstein, 1975, 1976). The isolated hybrid clones secrete antibody of monoclonal nature. The hybridomas originally produced by Köhler and Milstein were derived from the spleen cells of mice immunized with either sheep red blood cells or trinitrophenyl lipopolysaccharide as antigens. Since then, a large variety of monoclonal antibodies specific for many antigenic components have been produced (Eisenbarth, 1981). Such specific antibodies provide new opportunities for the study of many problems in cell and developmental biology, and will undoubtedly also be powerful tools in diagnostic and clinical medicine.

References

Adelberg, E. A., Callahan, T., Slayman, C. W., and Hoffman, J. F., 1975, Ouabain-resistant mutants of mouse and human lymphocytes, *J. Gen. Physiol.* **66:**17a.

Bakay, B., Croce, C. M., Koprowski, H., and Nyhan, W. L., 1973, Restoration of hypoxanthine phosphoribosyltransferase activity in mouse IR cells after fusion with chick embryo fibroblasts, *Proc. Natl. Acad. Sci. USA* **70:**1998–2002.

Bakay, B., Graf, M., Carey, S., Nissinen, E., and Nyhan, W. L., 1978, Reexpression of HPRT activity following cell fusion with polyethylene glycol, *Biochem. Genet.* **16:**277–237.

Baker, R. M., Brunette, D. M., Mankovity, R., Thompson, L. H., Whitmore, G. F., Siminovitch, L., and Till, J. E., 1974, Ouabain resistant mutants of mouse and hamster cells in culture, *Cell* **1:**9–22.

Boone, C., Chen, T.-R., and Ruddle, F. H., 1972, Assignment of three human genes to chromosomes (LHD-A to 11, TK to 17, and IDH to 20) and evidence for translocation between human and mouse chromosomes in somatic cell hybrids, *Proc. Natl. Acad. Sci. USA* **69:**510–514.

Brake, C. and Serra, A., 1981, A simple method for fusing human lymphocytes with rodent cells in monolayer by polyethylene glycol, *Somat. Cell Genet.* **7:**109–116.

Chan, T. S., Long, C., and Green, H., 1975, A human–mouse somatic cell hybrid line selected for human deoxycytidine deaminase, *Somat. Cell Genet.* **1:**81–90.

Choy, W. N., and Littlefield, J. W., 1980, Isolation of diploid human lymphoblast mutants presumably homozygous for ouabain resistance, *Proc. Natl. Acad. Sci. USA* **77:**1101–1105.

Chu, E. H. Y., and Powell, S. S., 1976, Selective systems in somatic cell genetics, in: *Advances in Human Genetics*, Volume 7 (H. Harris and K. Kirschhorn, eds.), Plenum Press, New York, pp. 189–258.

Chu, E. H. Y., Brimer, P., Jacobson, K. B., and Merriam, E. V., 1969, Mammalian cell genetics. I. Selection and characterization of mutations auxotrophic for L-glutamine or resistant to 8-azaguanine in Chinese hamster cells in vitro, *Genetics* **62:**359–377.

Clements, G. B., 1975, Selection of biochemically variant cells in culture, in: *Advances in Cancer Research*, Volume 21 (G. Klein and S. Weinhouse, eds.), Academic Press, New York, pp. 273–390.

Corsaro, C. M., and Migeon, B. R., 1978, Gene expression in euploid human hybrid cells: Ouabain resistance is codominant, *Somat. Cell Genet.* **4:**531–540.

Creagan, R. P., Chen, S., and Ruddle, F. H., 1979, Genetic analysis of the cell surface: Association of human chromosome 5 with sensitivity to diphtheria toxin in mouse–human somatic cell hybrids, *Proc. Natl. Acad. Sci. USA* **72:**2237–2241.

Davidson, R. L., and Gerald, P. S., 1976, Improved techniques for the induction of mammalian cell hybridization by polyethylene glycol, *Somat. Cell Genet.* **2:**165–176.

Davidson, R. L., and Adelstein, S. J., and Oxman, M. N., 1973, Herpes Simplex Virus as a source of thymidine kinase for thymidine kinase-deficient mouse cells: Suppression and reactivation of the viral enzymes, *Proc. Natl. Acad. Sci. USA* **70:**1912–1916.

Davidson, R. L., O'Malley, K. A., and Wheeler, T. B., 1976, Polyethylene glycol-induced mammalian cell hybridization: Effect of polyethylene glycol molecular weight and concentration, *Somat. Cell Genet.* **2:**271–280.

Dendy, P. R., and Harris, H., 1973, Sensitivity to diphtheria toxin as a species-specific marker in hybrid cells, *J. Cell Sci.* **12:**831–837.

Eisenbarth, G. S., 1981, Application of monoclonal antibody techniques to biochemical research, *Anal. Biochem.* **111:**1–16.

Fenwick, R. G., 1980, Reversion of mutation affecting the molecular weight of HGPRT: Intragenic suppression and localization of X-linked genes, *Somat. Cell Genet.* **6:**477–494.

Fisher, P. B. and Bryson, V., 1977, Toxicity of nystatin and its methyl ester toward parental and hybrid mammalian cells, *In Vitro* **13:**548–556.

Fisher, P. B., Sisskin, E. E., and Goldstein, N. I., 1978, Selecting somatic cell hybrids with HAT media and nystation methyl ester *J. Cell Sci.* **32:**433–439.

Fox, M., and Boyle, J. M., 1976, Factors affecting the growth of Chinese hamster cells in HAT selection media, *Mutat. Res.* **35:**445–464.

Goldstein, N. I., and Fisher, P. B., 1978, Selection of mouse X hamster hybrids using HAT medium and a polyene antibiotic, *In Vitro* **14**:200–206.

Goldstein, S., and Lin, C. C., 1972, Survival and DNA repair of somatic cell hybrids after ultraviolet irradiation, *Nature New Biol.* **239**:142–145.

Graf, L. H., Urlaub, G., and Chasin, L. A., 1979, Transformation of the gene for hypoxanthine phosphoribosyltransferase, *Somat. Cell Genet.* **5**:1031–1044.

Hakala, M. T., 1957, Prevention of toxicity of amethopterin for sarcoma-180 cells in tissue culture, *Science* **126**:255.

Harris, H., and Watkins, J. F., 1965, Hybrid cells derived from mouse and man: Artificial heterokaryons of mammalian cells from different species, *Nature* **205**:640–646.

Hynes, N. E., Kennedy, N., Rahmsdorf, U., and Groner, B., 1981, Hormone-responsive expression of an endogenous proviral gene of mouse mammary tumor virus after molecular cloning and gene transfer into cultured cells, *Proc. Natl. Acad. Sci. USA* **78**:2038–2042.

Jha, K. K., and Ozer, H., 1976, Expression of transformation in cell hybrids. I. Isolation and application of density-inhibited BALB/3T3 cells deficient in hypoxanthine phosphoribosyltransferase and resistant to ouabain. *Somat. Cell Genet.* **2**:215–224.

Kao, F. T., Johnson, R. T., and Puch, T. T., 1969, Complementation analysis on virus-fused Chinese hamster cells with nutritional markers, *Science* **164**:312–314.

Klebe, R. J., and Mancuso, M. G., 1981, Chemicals which promote cell hybridization, *Somat. Cell Genet.* **7**:473–488.

Klinger, H. P., and Shin, S., 1974, Modulation of the activity of an avian gene transferred into a mammalian cell by cell fusion, *Proc. Natl. Acad. Sci. USA* **71**:1398–1402.

Köhler, G., and Milstein, C., 1975, Continuous cultures of fused cells secreting antibody of predefined specificity, *Nature* **256**:495–497.

Köhler, G., and Milstein, C., 1976, Derivation of specific antibody-producing tissue culture and tumor lines by cell fusion, *Eur. J. Immunol.* **6**:511–519.

Kucherlapati, R. S., Baker, R. M., and Ruddle, F. H., 1975, Ouabain as a selective agent in the isolation of somatic cell hybrids, *Cytogenet. Cell Genet.* **14**:362–363.

Kurtz, D., 1981, Hormonal inducibility of rat α_{2u} globulin genes in transfected mouse cells, *Nature* **291**:629–631.

Kusano, T., Long, C., and Green, H., 1971, A new reduced human–mouse somatic cell hybrid containing the human gene for adenine phosphoribosyltransferase, *Proc. Natl. Acad. Sci. USA* **68**:82–86.

Lester, S. C., Levain, S. K., Steglich, C., and DeMars, R., 1980, Expressions of human genes for adenine phosphoribosyltransferase and hypoxanthine-guanine phosphoribosyltransferase after genetic transformation of mouse cells with purified human DNA, *Somat. Cell Genet.* **6**:241–260.

Lewis, W. H., Srinwasan, P. R., Stokoe, N., and Siminovitch, L., 1980, Parameters governing the transfer of the genes for thymidine kinase and dihydrofolate reductase into mouse cells using metaphase chromosome or DNA, *Somat. Cell Genet.* **6**:333–347.

Liskay, R. M., and Patterson, D., 1978, Selection of somatic cell hybrids between HGPRT⁻ and APRT⁻ cells, in: *Methods in Cell Biology*, Volume XX (D. M. Prescott, ed.), Academic Press, New York, pp. 335–360.

Littlefield, J. W., 1964, Selection of hybrids from matings of fibroblasts in vitro and their presumed recombinants, *Science* **145**:709–710.

Littlefield, J. W. 1966, The use of drug-resistant markers to study the hybridization of mouse fibroblasts, *Exp. Cell Res.* **41**:190–196.

Lowy, I., Pellicer, A., Jackson, J. F., Sim, G. K., Silverstein, S., and Axel, R., 1980, Isolation of transforming DNA: Cloning the hamster aprt gene, *Cell* **22**:817–823.

McBride, O. W., and Ozer, H. L., 1973, Transfer of genetic information by purified metaphase chromosomes, *Proc. Natl. Acad. Sci. USA* **70**:1258–1262.

McKusick, V. A., and Ruddle, F. H., 1977, The status of the gene map of the human chromosomes, *Science* **196**:390–405.

Neff, J. M., and Enders, S. F., 1968, Polio virus replication and cytopathogenicity in monolayer hamster cultures fused with beta-propiolactone inactivated Sendai virus, *Proc. Soc. Exp. Biol. Med.* **127**:260–267.

Norwood, T. H., Zeigler, C. J., and Martin, G. M., 1976, Dimethyl sulfoxide enhances polyethylene glycol-mediated cell fusion, *Somat. Cell Genet.* **3**:263–270.

Okada, Y., 1962, Analysis of giant polynuclear cell formation caused by HVJ virus from Ehrlich's ascites tumor cells. I. Microscopic observation of giant polynuclear cell formation, *Exp. Cell Res.* **26**:98–107.

Okada, Y., and Tadokoro, J., 1962, Analysis of giant polynuclear cell formation caused by HVJ virus from Ehrlich's tumor cells. II. Quantitative analysis of giant polynuclear cell formation, *Exp. Cell Res.* **26**:108–118.

O'Malley, K. A., and Davidson, R. L., 1977, A new dimension in suspension fusion techniques with polyethylene glycol, *Somat. Cell Genet.* **4**:441–448.

Pellicer, A., Wagner, E. F., Kareh, A. E., Dewey, M. J., Renser, A. J., Silverstein, S., Axel, R., and Mintz, B., 1980, Introduction of a viral thymidine kinase gene and the human β-globin gene into developmentally multipotential mouse teratocarcinoma cells, *Proc. Natl. Acad. Sci. USA* **77**:2098–2102.

Perucho, M., Hanahan, D., Lipsich, L., and Wigler, M., 1980, Isolation of the chicken thymidine kinase gene by plasmid rescue, *Nature* **285**:207–210.

Pontecorvo, G., 1975, Production of mammalian somatic cell hybrids by means of polyethylene glycol treatment, *Somat. Cell Genet.* **1**:397–400.

Rabinovitch, P. S., and Norwood, T. H., 1981, Rapid kinetics of polyethylene glycol-mediated fusion, *Somat. Cell Genet.* **7**:281–288.

Ruddle, F. H., 1972, Linkage analysis using somatic cell hybrids, in: *Advances in Human Genetics*, Volume 3 (H. Harris and K. Hirschhorn, eds.) Plenum Press, New York, pp. 173–235.

Sabourin, D. J., and Davidson, R. L., 1979, Transfer of the herpes simplex thymidine kinase gene from human cells to mouse cells by means of metaphase chromosome, *Somat. Cell Genet.* **5**:159–174.

Schneiderman, S., Farber, J. L., and Baserga, R., 1979, A simple method for decreasing the toxicity of polyethylene glycol in mammalian cell hybridization, *Somat. Cell Genet.* **5**:263–270.

Siminovitch, L., 1976, On the nature of hereditable variation in cultured somatic cells, *Cell* **7**:1–11.

Siniscalco, M., Klinger, H. P., Eagle, H., Koprowski, H., Fujimoto, W. Y., and Seegmiller, J. E., 1969, Evidence for intergenic complementation in hybrid cells derived from two human diploid strains each carrying an X-linked mutation, *Proc. Natl. Acad. Sci. USA* **62**:793–799.

Szybalski, W., Szybalska, E. H., and Ragni, G., 1962, Genetic studies with human cell lines, *Natl. Cancer Inst. Monogr.* **7**:75–89.

Vaughan, V. L., Hansen, D., and Stradler, J., 1976, Parameters of polyethylene glycol-induced cell fusion and hybridization in lymphoid cell lines, *Somat. Cell Genet.* **2**:537–544.

Watson, B., Cromley, I. P., Gardiner, S. E., Evans, H. J., and Harris, H., 1972, Reappearance of murine hypoxanthine guanine phosphoribosyltransferase activity in mouse A9 cells after attempted hybridization with human cell lines, *Exp. Cell Res.* **75**:401–409.

Weissman, B., and Stanbridge, E. J., 1980, Characterization of ouabain resistant, hypoxanthine phosphoribosyltransferase deficient human cells and their usefulness as a general method for the production of human cell hybrids, *Cytogenet. Cell Genet.* **28**:277–239.

Werkheiser, W. C., 1961, Specific binding of 4-amino folic acid analogues by folic acid reductase, *J. Biol. Chem.* **236**:888–893.

Wigler, M., Silverstein, S., Lee, L. S., Pellicer, A., Cheng, Y. C., and Axel, R., 1977, Transfer of purified Herpes virus thymidine kinase gene to cultured mouse cells, *Cell* **11**:223–232.

Wigler, M., Pellicer, A., Silverstein, S., and Axel, R., 1978, Biochemical transfer of single-copy eukaryotic genes using total cellular DNA as donor, *Cell* **14**:725–731.

Wigler, M., Sweck, R., Sim, G. K., Wold, B., Pellicer, A., Lacy, E., Maniatis, T., Silverstein, S., and Axel R., 1979, Transformation of mammalian cells with genes from procaryotes and eucaryotes, *Cell* **16**:777–785.

Wold, B., Wigler, M., Lacy, E., Maniatis, T., Silverstein, S., and Axel, R., 1979, Introduction and expression of a rabbit β-globin gene in mouse fibroblasts, *Proc. Natl. Acad. Sci. USA* **76**:5684–5688.

Zepp, H. O., Conover, J. H., Hirschhorn, K., and Hodes, H. L., 1971, Human mosquito somatic cell hybrids induced by ultraviolet-inactivated Sendai virus, *Nature New Biol.* **229**:119–121.

Chapter 3
Techniques for Decreasing the Toxicity of Polyethylene Glycol

W. EDWARD MERCER AND RENATO BASERGA

1. Introduction

In recent years the use of the chemical fusogen polyethylene glycol (PEG) in place of inactivated Sendai virus (Harris and Watkins, 1965) has greatly facilitated somatic cell fusion experiments (Pontecorvo, 1975; Davidson and Gerald, 1976a). There are, however, some inherent problems associated with PEG-induced cell fusion. One problem is that PEG will effectively induce cell-to-cell fusion only within a very narrow range of concentrations (Davidson et al., 1976b), and as the optimum concentration of 50–55% is approached the cytotoxic effects from exposure to the fusogen become significant.

Another problem with PEG-induced fusion is that it is much less effective at promoting fusion when cells are in suspension than when the same cell types are attached to a substrate (Davidson and Gerald, 1976b; O'Malley and Davidson, 1977). Consequently, several techniques have appeared in the literature in recent years directed toward reducing the toxic effects of PEG solutions (Norwood et al., 1976; Mercer and Schlegel, 1979; Schneiderman et al., 1979), or for increasing the fusion efficiency of cells exposed to the fusogen in suspension (Vaughan et al., 1976; O'Malley and Davidson, 1977; Mercer and Schlegel, 1979; Sharon et al., 1980).

Schneiderman et al. (1979) introduced a Ca^{++}-free fusion protocol that greatly reduces the toxic effects associated with PEG-induced fusion. The use of the plant lectin phytohemagglutinin (PHA) in conjunction with low concentrations of PEG has been shown to enhance cell fusion in both monolayer and suspension cultures (Mercer and Schlegel, 1979; Schlegel and Mercer, 1980; McNeill and Brown, 1980; Poste and Nicolson, 1980). In this chapter we discuss the Ca^{++}-free fusion procedure and describe in detail several procedures that have been developed to reduce the cytotoxic effects of PEG-induced monolayer fusion and to increase the efficiency of suspension fusion.

W. EDWARD MERCER and RENATO BASERGA • Department of Pathology and Fels Research Institute, Temple University School of Medicine, Philadelphia, Pennsylvania 19140.

2. Effects of Excluding Ca^{++} Ions and of the Source of PEG on Hybrid Colony Yield

The yield of hybrid colonies after fusion of mammalian cells in monolayers with PEG 1000 is significantly increased if the cells are fused in Ca^{++}-free medium and kept in Ca^{++}-free medium for a short time after exposure to the fusogen (Schneiderman et al., 1979). The protective effect of Ca^{++} ion exclusion is much more apparent when Baker PEG from various sources is used instead of Koch-Light PEG, which is less toxic, even when used under standard monolayer fusion conditions (Mercer and Schlegel, unpublished results; Schneiderman et al., 1979) as described by Davidson and Gerald (1976a).

When Baker PEG diluted in Dulbecco's medium (1.8 mM Ca^{++}) was used to induce cell fusion between two temperature-sensitive hamster cell lines, tsAF8 (originally isolated by Meiss and Basilico, 1972) and K12 (isolated by Roscoe et al., 1973), the yield of hybrid colonies resulting from fusion was approximately 50 colonies/100 mm petri plate. If Baker PEG was used in the Ca^{++}-free fusion procedure, the yield of hybrid colonies was 100-fold greater. If Koch-Light PEG was used as the fusogen and Ca^{++} ions were present during and immediately following fusion, the yield of hybrid colonies was increased 80-fold over the yield obtained when Baker PEG was used.

The protective effect of Ca^{++} exclusion, so dramatically observed when Baker PEG was used as a fusogen, was not nearly as obvious when Koch-Light PEG was used in the Ca^{++}-free procedure. In this case the difference between Koch-Light PEG with or without Ca^{++} ions was only twofold. From these experiments the conclusion was drawn that Koch-Light PEG is, in general, less toxic to cells than Baker PEG even if Ca^{++} is excluded from the fusion protocol.

The effect of Ca^{++} exclusion on the yield of hybrid colonies was shown not to be directly attributed to an increase in the fusion efficiency, since the number of cells with two or more nuclei (polykaryons) was approximately the same whether or not fusion was carried out in the presence or absence of Ca^{++} ions. No attempt was made, however, to quantitate the effect of Ca^{++} ion exclusion on post-fusion viability.

Mercer and Schlegel (1980) have observed that exclusion of divalent cations, such as Mn^{++}, from the fusion solution and for a period immediately following exposure to the fusogen increased the viability of cells following fusion with human erythrocytes loaded with fluorescenated bovine serum albumin (Mercer et al., 1979). Schneiderman et al. (1979) have investigated the time required in the absence of Ca^{++} ions for protection of cells following exposure to Baker PEG 1000. When cells were fused and washed in Ca^{++}-free medium, then immediately exposed to regular Ca^{++}-containing medium (1.8 mM ca^{++}), the yield of hybrid colonies was increased only 2.7-fold over control fusions where Ca^{++} ions were present throughout, as in the standard monolayer fusion technique of Davidson and Gerald (1976a). However, when cells were allowed to remain in Ca^{++}-free medium for as little as 15 min after exposure to PEG in Ca^{++}-free medium the difference in hybrid colony yield increased over 60-fold. These results suggest that the protective effect of excluding Ca^{++} ions from the fusion protocol occurs during the post-fusion membrane stabilization period that follows removal of the fusogen (Fan et al., 1979).

3. Possible Role of Ca^{++} Ions in Cytotoxicity

In order to assess further the role that Ca^{++} ions play in post-fusion viability of cells following PEG exposure, we prepared solutions of Baker PEG 6000 as described by Davidson and Gerald (1976a) on a weight-to-weight basis (i.e., a 50% PEG solution contains 10 g of PEG to 10 ml of diluent) in Ca^{++}-free Dulbecco's medium or in Ca^{++}-free medium supplemented with either 1.0, 2.0, or 3.0 mM Ca^{++} ions, respectively. The pH of the resulting PEG solutions was adjusted to pH 7.5–8.0 using 1 N NaOH. Solutions to be used for fusion were prepared fresh on the day of the fusion and discarded afterward because we have routinely observed that during storage the pH of the solution can become extremely basic, especially when $NaHCO_3$-buffered medium is used as a diluent. Sharon et al. (1980) and Klebe and Mancuso (1981) have shown that extremes in pH of PEG solutions can effect cell viability and hybrid colony yield.

Swiss 3T3 mouse cells were cultured as previously described (Mercer and Schlegel, 1980). Monolayer fusions were performed by the standard method of Davidson and Gerald (1976a) or by the Ca^{++}-free method of Schneiderman et al. (1979). Cultures were exposed to PEG solutions for 1.0 or 2.0 min, then washed and incubated for 15–20 min in Ca^{++}-free Dulbecco's medium without serum before adding complete growth medium. At 16–18 hr post-fusion the number of polykaryons (i.e., the fusion index) and the viability of cultures were assessed as follows:

1. Quantitation of polykaryon formation was determined after fixing the cells in methanol and staining with Giemsa (Fisher) in Sorenson's buffer, pH 6.8 (Mercer and Schlegel, 1979), according to the formula described by Okada and Tadokoro (1962):

$$\text{Fusion index (FI)} = \left\{\frac{\text{No. of nuclei in polykaryons}}{\text{No. of nuclei in all cells}}\right\} \text{experimental cultures} - \left\{\frac{\text{No. of nuclei in polykaryons}}{\text{No. of nuclei in all cells}}\right\} \text{control cultures}$$

2. Viability of cultures was determined by taking advantage of the differential staining that viable cells exhibit when compared to dead cells (Mercer and Schlegel, 1979). Viable cells exhibit a normal pink nucleus with dark purple nucleoli and a blue cytoplasm, whereas dead cells exhibit little or no cytoplasmic staining and a distinct pyknotic nucleus which stains uniformly dark purple. Labeling with [^{3}H]thymidine for 24 hr following PEG

exposure has shown that pyknotic nuclei do not synthesize DNA (Mercer and Schlegel, unpublished results). Care was taken to exclude "hot spots" where toxicity was extremely high (as a result of inadequate PEG washing) from viability determinations. The viability index was determined according to the following formula:

$$\text{Viability index (VI)} = \left\{\frac{\text{No. of pyknotic nuclei}}{\text{No. of total nuclei}}\right\} \text{experimental cultures}$$
$$- \left\{\frac{\text{No. of pyknotic nuclei}}{\text{No. of total nuclei}}\right\} \text{control cultures}$$

The results of a typical experiment are given in Table I. As can be seen from the data contained in Table I, no significant increase in the fusion index was observed when Ca^{++} ions were included in the PEG solution, regardless of the time of PEG exposure. The fusion index did increase when cultures were exposed for longer periods of time; however, the fusion index was not significantly affected by having Ca^{++} ions present, i.e., under Ca^{++}-free conditions the fusion index increased eightfold between 1.0 and 2.0 min of PEG exposure; when 3.0 mM Ca^{++} ions were present during PEG exposure the increase in the fusion index between 1.0 and 2.0 min of exposure was fivefold. When Ca^{++} ions were present during PEG exposure the post-fusion viability was not significantly affected and no detectable loss of cells from the monolayer was apparent. However, even in the absence of Ca^{++} ions the viability decreased with longer times of PEG exposure, i.e., the difference in

Table I. Effects of Ca^{++} Ions on Fusion and Viability Indexes[a]

Experimental condition tested	Time of PEG exposure			
	1.0 min		2.0 min	
	FI	VI	FI	VI
Ca^{++}-free/Ca^{++}-free	0.036	0.97	0.28	0.72
1.0 mM Ca^{++}/Ca^{++}-free	0.033	0.91	0.25	0.70
2.0 mM Ca^{++}/Ca^{++}-free	0.026	0.87	0.29	0.68
3.0 mM Ca^{++}/Ca^{++}-free	0.041	0.82	0.21	0.67
1.0 mM Ca^{++}/1.0 mM Ca^{++}	ND	ND	0.31	0.34
2.0 mM Ca^{++}/2.0 mM Ca^{++}	ND	ND	0.27	0.21
3.0 mM Ca^{++}/3.0 mM Ca^{++}	ND	ND	0.30	0.18

[a] Swiss 3T3 mouse cells, grown in monolayer, were exposed to 50% Baker PEG-6000 for 1 or 2 min, utilizing either the complete exclusion of Ca^{++} ions, conditions where Ca^{++} ions were present in the medium used to dilute the PEG and then excluded from the medium following exposure to the fusogen, or where Ca^{++} ions were present throughout. The fusion index (FI) and the viability index (VI) were determined at 16–18 hr following exposure to the fusogen, as described in the preceding section, by counting at least 500 cells for each determination.

the viability index between cells exposed to PEG in Ca^{++}-free medium for 1.0 and 2.0 min was about 1.4-fold.

When the post-fusion Ca^{++} ion concentration was the same as that used to prepare the PEG solutions the toxic effects of Ca^{++} ions became apparent. The viability index for cells exposed for 2 min to PEG solutions containing 3.0 mM Ca^{++} or in Ca^{++}-free Dulbecco's medium was approximately the same (0.72 and 0.67, respectively) when the cells were placed in Ca^{2+}-free medium for a short time following exposure to PEG. However, there was a significant difference in viability when cells were exposed to PEG solutions containing 3.0 mM Ca^{++} ions and then placed in medium containing the same Ca^{++} ion concentration following PEG exposure. In this case the viability decreased by 3.7-fold. This result suggests that Ca^{++}-mediated cytotoxicity occurs during the post-fusion membrane stabilization period which follows initial exposure to the fusogen. This is consistent with the results of Schneiderman et al. (1979), who found that Ca^{++} ions added immediately following PEG exposure (when cells were fused under Ca^{++}-free conditions) resulted in a lower yield of hybrid colonies than when Ca^{++} ions were excluded for a short time after PEG exposure.

Phase-contrast microscopic observations of cells following exposure to PEG indicates that cells are killed very rapidly after removal of the fusogen. Opaque-lipid vesicles (Fan et al., 1980), which Norwood et al. (1976) have called "small microdroplets," are released during the post-fusion membrane stabilization period from the peripheral portion of the cell cytoplasm. If the release of these vesicles is extensive, nuclear pyknosis follows shortly thereafter. Qualitatively, we have observed that if Ca^{++} ions are excluded following PEG exposure, the release of lipid-rich vesicles into the surrounding medium is greatly retarded and post-fusion viability is improved. It is tempting to speculate that this phenomenom could be associated with the protective effects that Ca^{++} exclusion has in post-fusion viability following PEG exposure.

4. Effects on Fusion Index of the Time Interval between Plating and PEG Exposure

In the original monolayer fusion protocol, as described by Davidson and Gerald (1976a), cells to be fused in monolayers were grown for 24 hr in medium containing 10% serum prior to PEG exposure. Another difference between the Ca^{++}-free fusion protocol described by Schneiderman et al. (1979) and that described by Davidson and Gerald (1976a), aside from the exclusion of Ca^{++} ions, is the time interval after plating prior to PEG exposure. Schneiderman et al. (1979) exposed cells to PEG at 3 hr after plating. In this section we examine the effects of the time interval between plating and subsequent PEG exposure on polykaryon formation and post-fusion viability.

Swiss 3T3 mouse cells were plated into 35-mm petri plates containing a 22 mm^2 coverslip at 1.0×10^5 cells/plate in medium containing 10% serum. At 2, 4, 6, 8, 16, 20, and 24 hr after plating the monolayer cultures were exposed to

Table II. Effects of the Time Interval between Plating and Exposure to PEG on the Fusion and Viability Indexes[a]

Time after plating, hr	Fusion index	Viability index
2	0.41	0.92
4	0.32	0.87
6	0.27	0.91
8	0.20	0.93
16	0.05	0.89
20	0.08	0.80
24	0.07	0.95

[a] The effect of the time interval between initial plating and subsequent exposure to PEG on the fusion and viability indexes was determined after fusion in monolayer by the Ca^{++}-free protocol. Swiss 3T3 mouse cells were plated into 35-mm petri plates containing a 22 mm^2 coverslip at 1×10^5 cells/plate in medium containing 10% serum. At different times after plating the cells were exposed to 50% Koch-Light PEG 6000 for exactly 1 min. After 16–18 hr of growth the fusion and viability indexes were determined as described in the text.

50% Koch-Light PEG 6000 according to the Ca^{++}-free protocol of Schneiderman et al. (1979). At 16–18 hr after fusion the cultures were fixed and stained as described in the previous section and the fusion index and post-fusion viability were determined.

As can be seen in Table II, freshly plated cells are eightfold more susceptible to PEG-induced fusion than are cells that have had a 16–24-hr period of growth. A comparison of the fusion index of cultures shows that at early times after plating, i.e., 2–4 hr, more polykaryons are produced than at later times. At 6 hr the difference between the fusion index was only fourfold when compared with cultures fused at 20 or 24 hr. No significant difference in the fusion indexes was observed between cultures fused at 16, 20, or 24 hr, respectively. Whether the cultures were exposed to the fusogen at early times or at later times after plating, the post-fusion viability of cells was similar. The increased polykaryon formation in freshly plated cells can most likely be attributed to membrane perturbations that accompany trypsinization, since Hartmann et al. (1976) have shown that pretreatment of hen erythrocytes with proteolytic enzymes prior to PEG exposure greatly enhances fusion.

5. Lectin Enhancement of Suspension Fusion

The combined use of an agglutinin (PHA) and a fusogen (PEG) is especially useful when applied to nonadherent cells and subcellular fragments. This PHA/PEG fusion procedure has been used to replace Sendai virus as a fusogen in red cell-mediated microinjections, where macromolecules are transferred to recipient culture cells by fusion with nonnucleated red cells loaded by hypotonic lysis with the macromolecules to be transferred. Cytoplasts produced by cytochalasin-induced enucleation of mammalian cells (Carter, 1967) have been fused to whole cells in suspension using the PHA/PEG method (Mercer and Schlegel, 1982), and other subcellular components, such as microcells (Ege and Ringertz, 1974) have been fused to whole cells more efficiently using the PHA/PEG method (McNeill and Brown, 1980).

In experiments designed to study metastasis Poste and Nicolson (1980) employed this fusion method to fuse plasma membrane vesicles to mouse melanoma cells and demonstrated that enhanced fusion efficiency was a result of incorporating PHA into the system. Recently Szoka et al. (1981), using the strategy of lectin-enhanced, PEG-induced fusion have increased the efficiency of liposome-mediated transfer of macromolecules (Wilson et al., 1977) to recipient mammalian cells. The PHA/PEG fusion method has also been used to fuse primary liver cells and human amniocytes which are otherwise refractory to PEG fusion (Dr. D. Sammons, UpJohn Co., personal communication). In the next section we describe in detail some recent modifications that have been made in the PHA/PEG fusion method (Mercer and Schlegel, 1979) which, when used in conjunction with the Ca^{++}-free fusion procedure of Schneiderman et al. (1979), can be used to reduce cytotoxicity when cells are fused in suspension.

5.1. Suspension Fusion Procedure

Cells to be fused in suspension were removed from monolayer using a dilute (1:5) solution of 0.25% trypsin and 0.02% disodium EDTA Ca^{++}, Mg^{++}-free Hanks saline solution, and washed twice in growth medium containing 10% serum. The cell number was determined by hemocytometer count. A single-cell suspension of each respective cell type was prepared by passing the cells through an 18-gauge needle attached to a 10-ml syringe several times. Then, 2×10^6 cells of one type were mixed with 2×10^6 cells of another type (or 4×10^6 cells of the same type) in Ca^{++}-free Dulbecco's medium without serum (Schneiderman et al., 1979). Phytohemagglutinin (PHA-P, DIFCO), prepared as previously described (Mercer and Schlegel, 1979), was then added to give a final concentration of 5–50 $\mu g/ml$ (depending on the agglutinating ability of the cell type used), and the suspension was vortex-mixed for 30 sec at maximum agitation. The cell suspension was then transferred to a reciprocating water bath and incubated at 37°C for 10–15 min in order to allow agglutination of the cells to occur. The agglutinated cells were then centrifuged at 600 g in a Sorvall bench-top centrifuge at room temperature for 5 min. The PHA-containing medium was removed by aspiration and the "loose" cell pellet was resuspended in 10 ml of Ca^{++}-free medium using a Pasteur pipette, and centrifugation was repeated. Following centrifugation, the supernatant was removed by aspiration, taking care to remove all residual medium from the cell pellet.

The pellet was then resuspended by vortex-mixing and 1 ml of 44–46% Koch-Light PEG 6000 in Ca^{++}-free Dulbecco's medium, pH 7.5–8.0, was added with vortex-mixing. The cells were exposed to the PEG solution for 1–2 min, then 9 ml of Ca^{++}-free Dulbecco's medium was added and mixed to dilute the PEG. The mixture was immediately centrifuged at 600g for 3 min, and following centrifugation, the supernatant was aspirated from the cell pellet. The cell pellet was gently resuspended in 5 ml of Ca^{++}-free Dulbecco's medium and incubated for 15–20 min at room temperature to allow the cells to recover from fusion. The cells were then diluted in conditioned growth medium (i.e., medium which has been in contact for 3–4 days with the cell type being fused) to give 1×10^5 cells/ml. Then 4 ml of the cell suspension was transferred to each of 10–35-mm petri plates (in some cases containing 22 mm^2 glass coverslips). After the cells had attached (1–2 hr), the conditioned plating medium was removed, the monolayer was washed once with Hanks balanced saline solution, and fresh growth medium was added.

The percentage of viable suspension cells following exposure to PEG was determined just prior to plating by Trypan blue dye exclusion (Phillips, 1973). Staining was accomplished by resuspending the cells in 1 ml of Hanks balanced saline solution, and 0.1 ml of 0.4% Trypan blue stain solution (GIBCO) was added to the cell suspension. The mixture was dispersed with a Pasteur pipette and 10 μl of the suspension was placed on a hemocytometer and a viability count was made at 4 min after dispersing the cells. The viability index was determined according to the following formula:

$$\text{Viability index (VI)} = \left\{ \frac{\text{No. of nonstained cells}}{\text{No. of total cells}} \right\} \text{experimental cultures} - \left\{ \frac{\text{No. of nonstained cells}}{\text{No. of total cells}} \right\} \text{control cultures}$$

5.2. Post-fusion Plating in Conditioned Medium

When mammalian cells are fused in suspension, the proportion of cells with two or more nuclei can be classifed shortly after plating. Initially the cells appear as "grape like" clusters; however, within 15–20 min following PEG exposure the clusters coalesce into single spherical cells. Trypan blue dye exclusion shows that the majority of these cells are viable, yet when placed into fresh growth medium containing 10% serum, the plating efficiency is rarely above 20%. We have been able to significantly increase the plating efficiency of mouse 3T3 fibroblasts, chicken embryo fibroblasts, human diploid fibroblasts, and several other cell lines, by plating into conditioned medium following fusion in suspension (Mercer and Schlegel, unpublished results). We attribute this to the presence of surface glycoproteins contained within the conditioned medium, which have been shown to enhance the attachment of cells (Millis and Hoyle, 1978; Grinnell and Feld, 1979). Therefore, whenever we fuse cells in suspension, conditioned medium

(medium in contact with cells to be fused 3–4 days prior to fusion) is used to enhance initial attachment and spreading. When this practice is followed the plating efficiency of cells fused in suspension can be increased to above 50% (data not shown).

5.3. Effects of PHA Concentration on the Fusion Index and Size of Polykaryons

The enhancing effect on polykaryon formation by PHA can be seen from the results in Table III. As the concentration of PHA used to agglutinate suspension cells together was increased prior to exposure to PEG, the fusion index also was increased. When as little as 5 µg/ml of PHA was used to agglutinate cells together prior to PEG exposure, the fusion index was increased by three-fold. If 50 µg/ml of PHA was used, the fusion index was increased 11-fold over that of control cultures where no PHA was used. Enhanced fusion indexes were observed at all concentrations of PHA tested. As higher concentrations of PHA were used, viability of the cells decreased; the difference in viability between cells not exposed to PHA prior to fusion and those exposed to 50 µg/ml, was slightly more than 1.3-fold. We attribute this directly to the combined exposure to lectin and fusogen, since viability of cells exposed to PHA (50 µg/ml) in the absence of PEG exposure was approximately 95%.

Most likely the decrease in viability with increasing PHA concentration is a consequence of more fusion events, because increasing the PHA concentration increases the fusion index (Table III). These results also suggest that lectin enhancement of cell fusion could result solely from agglutinating cells together so as to bring the cytoplasmic membranes in close proximity in order that fusion can easily occur. However, membrane perturbation as a result of exposure to the lectin cannot be completely ruled out.

Table III. Effect of PHA Concentration on Fusion and Viability of 3T3 Cells.[a]

PHA concentration, µg/ml	Fusion index	Viability index
0	0.04	0.92
5	0.12	0.89
15	0.20	0.81
25	0.35	0.75
35	0.37	0.79
50	0.44	0.68

[a] Swiss 3T3 mouse cells were first agglutinated in suspension using different concentrations of PHA and then exposed to 45% Koch-Light PEG 6000 as described in the text. After exposure to the fusogen the viability index was determined by the Trypan blue dye exclusion test (Phillips, 1973), and the fusion index was determined at 16–18 hr after plating the fusion products in conditioned medium.

Table IV. Effect of PHA Concentration on Polykaryon Size[a]

PHA, µg/ml	Number of nuclei per cell											FI
	1	2	3	4	5	6	7	8	9	10	>10	
0	980	18	2									0.02
5	890	77	14	12	7							0.11
25	680	69	60	48	30	9	6	15	6	11	60	0.31
50	598	49	51	42	21	13	28	41	37	40	80	0.40

[a] The effect of PHA concentration on the number and size of polykaryons (Swiss 3T3 mouse cells) following exposure to 45% Koch-Light PEG 6000 as described in the text was determined at 16–18 hr after plating the fusion products in conditioned growth medium.

The determination of polykaryon formation according to the formula of Okada and Tadokoro (1962) given in the previous section can misrepresent the absolute number of cell-to-cell fusion events because it neglects to take into account the multiple fusions of trikaryons (cell with three nuclei) and higher order polykaryons (Ringertz and Savage, 1976). We therefore analyzed the frequencies of various classes of cells with two or more nuclei in order to determine whether a relationship exists between the size of the fusion products and PHA concentration. In these experiments Swiss 3T3 cells were fused in suspension as described in the previous section, and at 16–18 hr after plating, cells were classified according to the number of nuclei that they contained. The results are given in Table IV. As can be seen from the data in Table IV, the size of the fusion products was influenced by the concentration of PHA used to treat the cells prior to fusion. When the concentration of PHA exceeded 5 µg/ml, larger aggregates of agglutinated cells were produced and subsequent PEG exposure resulted in an increased number of fusion products with multiple nuclei. This result also suggests that the enhancement of polykaryon formation observed with increasing concentration of PHA (Mercer and Schlegel, 1979) results solely from the agglutinating properties of the lectin.

6. Summary and Conclusions

At present it is unclear why PEG solutions that are capable of producing somatic cell fusion are also toxic to cells. Our results suggest that Ca^{++} ions may be involved in cytotoxicity, perhaps during the post-fusion membrane stabilization period which follows PEG exposure, since the exclusion of Ca^{++} ions during this period retards the release of lipid-rich vesicles and increases post-fusion viability of the cells (Table I). However, an absolute determination of the effects of Ca^{++} exclusion on the release of lipid vesicles will have to await more precise quantitation.

It is also clear that several other factors can come into play that effect post-fusion viability. For example, the source and grade of PEG used to fuse cells can significantly affect cell viability and hybrid colony yield. Koch-Light

PEG purchased from Research Associates International, Elk Grove, Illinois, seems to be considerably less toxic than Baker PEG. In general we have observed that PEG 1000 is more toxic to cells, regardless of the source, than PEG 6000, probably because PEG 1000 is more fusogenic (Davidson and Gerald, 1976b). The pre-fusion culture conditions are also important, since freshly plated cells are more susceptible to PEG-induced fusion than are cells that have been grown for longer periods of time prior to fusogen exposure (Table II).

We have also demonstrated that PHA pretreatment of cells prior to fusion in suspension increases both the number and size of polykaryons (Tables III and IV; Mercer and Schlegel, 1979). These results suggest that the enhancing effect of PHA on PEG-induced fusion most likely occurs as a result of the agglutinating properties of the lectin. By employing the methods described in this chapter, the toxic effects of PEG-induced somatic cell fusion can be significantly reduced. When the Ca^{++}-free fusion protocol (Schneiderman et al., 1979) is used in conjunction with the PHA/PEG fusion method (Mercer and Schlegel, 1979), cells in suspension can be fused effectively and with a minimum of toxicity. To date, we have found that this suspension fusion procedure (as described in this chapter) is applicable to a wide variety of mammalian cell types from several different species.

References

Carter, S. B., 1967, Effects of cytochalasins on mammalian cells, *Nature* **213**:261–264.
Davidson, R. L., and Gerald, P. S., 1976a, Induction of mammalian somatic cell hybridization by polyethylene glycol, in: *Methods in Cell Biology*, Volume XV (D. M. Prescott, ed.), Academic Press, New York.
Davidson, R. L., and Gerald, P. S., 1976b, Improved techniques for the induction of mammalian cell hybridization by polyethylene glycol. *Somat. Cell Genet.* **2**:165–176.
Ege, T., and Ringertz, N. R., 1974, Preparation of microcells by enucleation of micronucleate cells, *Exp. Cell Res.* **87**:378–382.
Fan, V. S. C., McCammon, J. R., Ealy, G. T., and Burke, K. V., 1979, Process of membrane fusion, in: *XIth International Congress of Biochemistry Abstracts*, National Research Council of Canada, Ottawa, Ontario, p. 387.
Grinnell, F., and Feld, F. K., 1979, Initial adhesion of human fibroblasts in serum-free medium: Possible role of secreted fibronectin, *Cell* **17**:117–129.
Harris, H., and Watkins, J. F., 1965, Hybrid cells derived from mouse and man: Artificial heterokaryons of mammalian cells from different species, *Nature* **205**:640–646.
Hartmann, J. X., Galla, J. D., Emma, D. A., Kao, K. N., and Gamborg, O. L., 1976, The fusion of erythrocytes by treatment with proteolytic enzymes and polyethylene glycol, *Can. J. Genet. Cytol.* **18**:503–512.
Klebe, R. J., and Mancuso, M. G., 1981, Chemicals which promote cell hybridization, *Somat. Cell Genet.* **7**:473–488.
McNeill, C. A., and Brown, R. L., 1980, Genetic manipulation by means of microcell-mediated transfer of normal human chromosomes into recipient mouse cells, *Proc. Natl. Acad. Sci. USA* **77**:5394–5398.
Meiss, H. K., and Basilico, C., 1972, Temperature sensitive mutants of BHK 21 cells, *Nature New Biol.* **239**:66–68.
Mercer, W. E., and Schlegel, R. A., 1979, Phytohemagglutinin enhancement cell fusion reduces polyethylene glycol cytotoxicity, *Exp. Cell Res.* **120**:417–421.

Mercer, W. E., and Schlegel, R. A., 1980, Cell cycle re-entry of quiescent mammalian nuclei following heterokaryon formation, *Exp. Cell Res.* **128:**431–438.

Mercer, W. E., and Schlegel, R. A., 1982, Cytoplasts can transfer factor(s) which stimulate quiescent fibroblasts to enter S-phase, *J. Cell Physiol.* **110:**(in press).

Mercer, W. E., Terefinko, D. J., and Schlegel, R. A., 1979, Red cell-mediated microinjection of macromolecules into monolayer cultures of mammalian cells, *Cell Biol. Internatl. Rep.* **3:**265–270.

Millis, A. J. T., and Hoyle, M., 1978, Fibroblast-conditioned medium contains cell surface proteins required for cell attachment and spreading, *Nature* **271:**668–669.

Norwood, T. H.. Zeigler, C. J., and Martin, G. M., 1976, Dimethyl sulfoxide enhances polyethylene glycol-mediated somatic cell fusion, *Somat. Cell Genet.* **2:**263–270.

Okada, Y., and Tadokoro, J., 1962, Analysis of giant polynuclear cell formation caused by HVJ virus from Ehrlich's tumor cells. II. Quantitative analysis of giant polynuclear cell formation, *Exp. Cell Res.* **26:**108–118.

O'Malley, K. A., and Davidson, R. L., 1977, A new dimension in suspension fusion techniques with polyethylene glycol, *Somat. Cell Genet.* **3:**441–448.

Phillips, H., 1973, Dye exclusion tests for cell viability, in: *Tissue Culture: Methods and Application* (P. E. Kruse and M. K. Patterson, eds.), Academic Press, New York, pp. 406–408.

Pontecorvo, G., 1975, Production of mammalian somatic cell hybrids by means of polyethylene glycol treatment, *Somat. Cell Genet.* **1:**397–400.

Poste, G., and Nicolson, G. L., 1980, Arrest and metastsis of blood-borne tumor cells are modified by fusion of plasma membrane vesicles from highly metastatic cells, *Proc. Natl. Acad. Sci. USA* **77:**399–403.

Ringertz, N. R., and Savage, R. E., 1976, *Cell Hybrids*, Academic Press, New York, Chapter 4, pp. 41–43.

Roscoe, D. H., Read, M., and Robinson, H., 1973, Isolation of temperature sensitive mammalian cells by selective detachment, *J. Cell. Physiol.* **82:**325–331.

Schlegel, R. A., and Mercer, W. E., 1980, Red cell-mediated microinjection of quiescent fibroblasts, in: *Introduction of Macromolecules into Viable Mammalian Cells* (R. Baserga, C. Croce, and G. Rovera eds.), Alan R. Liss, New York, pp. 371–379.

Schneiderman, S., Farber, J. L., and Baserga, R., 1979, A simple method for decreasing the toxicity of polyethylene glycol in mammalian cell hybridization, *Somat. Cell Genet.* **5:**263–269.

Sharon, J., Morrison, S. L., and Kabat, E. A., 1980, Formation of hybridoma clones in soft agarose: Effect of pH and of medium, *Somat. Cell. Genet.* **6:**433–441.

Szoka, F., Magnusson, K., Wojcieszyn, J., Hou, Y., Derzko, Z., and Jacobson, K., 1981, Use of lectins and polyethylene glycol for fusion of glycolipid-containing liposomes with eukaryotic cells, *Proc. Natl. Acad. Sci. USA* **78:**1685–1689.

Vaughan, L. L., Hansen, D., and Stadler, J., 1976, Parameters of polyethylene glycol-induced cell fusion and hybridization in lymphoid cell lines, *Somat. Cell Genet.* **2:**537–544.

Wilson, F., Papahadjopoulos, D., and Taber, R. T., 1977, Biological properties of poliovirus encapsulated in lipid vesicles: Antibody resistance and infectivity in virus-resistant cells, *Proc. Natl. Acad. Sci. USA* **74:**3471–3475.

Chapter 4

The Use of Dimethyl Sulfoxide in Mammalian Cell Fusion

THOMAS H. NORWOOD and CAROL J. ZEIGLER

1. Introduction

Since Pontecorvo (1975) first demonstrated that polyethylene glycol (PEG) is an efficient fusogen of mammalian cells, its usefulness for this purpose has been demonstrated in a wide variety of cell types and organisms, including bacterial L forms (Yoshiyuki et al., 1979) and fungal protoplasts (Anné and Peberdy, 1976). There have been a number of publications describing refinements of the basic fusion protocol outlined in the earlier reports (Pontecorvo, 1975; Davidson and Gerald, 1976). Methods to fuse cells in suspension with PEG have been perfected (O'Malley and Davidson, 1977; Hales, 1977). Schneiderman et al. (1979) reported that decreasing the Ca^{++} concentration during and immediately following fusion diminished the toxicity of PEG, resulting in a higher recovery of proliferating hybrids. Another laboratory reported that the treatment of cultures with the lectin phytohemagglutinin prior to fusion enhanced the effectiveness of PEG at lower concentrations, thus decreasing the cytotoxicity of the procedure (Mercer and Schlegel, 1979). Norwood et al. (1976) reported that dimethyl sulfoxide (DMSO) enhanced fusion of human diploid cells over a range of concentrations of PEG. All of these refinements provide the investigator with greater maneuverability in developing chemically induced cell fusion systems suitable for his or her needs. In this chapter the role of DMSO in the optimization of mammalian cell fusion will be discussed. However, as will be discussed below, the mechanism of cell fusion remains to be elucidated. Therefore, it should be emphasized that the utility of DMSO or any other additive in facilitating mammalian cell fusion must be experimentally determined for any given experimental situation. We do not propose that DMSO is a panacea which will solve all problems encountered in cell fusion, but rather that it is one of a number of additives (Klebe and Mancuso, 1981) available to the investigator which enhance mammalian cell fusion.

THOMAS H. NORWOOD and CAROL J. ZEIGLER • Department of Pathology, University of Washington, Seattle, Washington 98195.

2. Chemical and Biologic Properties of Dimethyl Sulfoxide

Dimethyl sulfoxide (DMSO) is one of a number of polar, aprotic reagents which have been identified during the past three decades. It can act as either an oxidizing or a reducing agent. Because of these and other unique chemical properties, DMSO has been extensively studied. The chemistry of this interesting compound has been exhaustively reviewed by Szmant (1971).

DMSO has received great attention in both the scientific and lay communities because of its reputed wide range of biologic and medicinal properties. Many diverse pharmacologic actions of DMSO have been reported, including anti-inflammatory activity, anesthetic properties, inducement of diuresis, antagonism of platelet aggregation, and nonspecific enhancement of resistance to bacterial infection (Jacob, 1971). Because of the many and widely divergent claims regarding the pharmacologic and therapeutic properties of this compound, many of which are poorly substantiated, DMSO has come to be regarded as a panacea for a variety of conditions and ailments by certain segments of the public. Hence the role of DMSO as a therapeutic agent has been the subject of controversy in the medical community.

Some biologic properties of DMSO have been extensively studied. The capacity of this compound to penetrate tissue membranes, including the skin, is well documented (Kolb et al., 1967). Denko et al. (1967) has reported that ^{35}S-labeled DMSO is distributed to virtually every organ within 2 hr after application to the skin of rats. Most individuals who have worked with this compound are aware of this property because it causes a peculiar taste sensation within minutes after contact with the skin. Many studies have demonstrated that this compound also facilitates the cutaneous absorption of low-molecular-weight nonionized compounds (Sulzberger et al., 1967). Although DMSO is apparently of comparatively low toxicity (Mathew et al., 1980; Brown et al., 1963; Feinman et al., 1964), clearly, precautions should be taken to avoid contact with the skin, especially when handling toxic, carcinogenic, or mutagenic substances at the same time.

The cryoprotective properties of DMSO have been extensively studied and are well documented (Ashwood-Smith, 1971). This compound and glycerol are probably the two most common cryophylactic agents routinely used for the long-term cryopreservation of cells and tissues. The precise mechanism(s) of this protective action is not entirely clear. The fact that DMSO readily penetrates cells has led to the suggestion that this and similar cryoprotective agents act by decreasing the concentration of electrolytes in equilibrium with ice within the cell (Nash, 1966). The important observation for this discussion, which derived from the studies of the cryoprotective properties of DMSO, is that this chemical is of comparatively low cytotoxicity.

The capacity of DMSO at high concentrations to induce cell fusion was recognized by Ahkong et al. (1975) in studies on the mechanisms of membrane fusion in hen erythrocytes. It is one of a number of water-soluble compounds with fusogenic activity which have been used to probe the mechanism of the fusion of biological membranes (Maggio et al., 1976; Wilairat et al., 1978; Blow et al., 1979). However, DMSO came into widespread use as a tool in mammalian cell fusion following reports that it can enhance polyethylene glycol-mediated fusion (Norwood et al., 1976; Hales, 1977; Clark and Shay, 1978).

3. Methods of Procedure

3.1. Parameters of Chemical Cell Fusion

It must be noted that in any discussion of the techniques of cell fusion the determination of the optimum conditions for fusion in any given cell culture system can only be achieved by empirical manipulation of the physical and biologic parameters available to the investigator. Therefore, we emphasize that the fusion protocols outlined below will provide no more than a starting point for the investigator confronted with the problem of developing an efficient fusion system.

There are a variety of physical and biologic parameters which can be manipulated. The most important ones are:

Reagents. DMSO is available in a number of grades from a variety of manufacturers. We have routinely used the spectrophotometric-grade product provided by Schwartz/Mann (Orangeburg, New York). Most investigators have used PEG supplied by the J. T. Baker Company (Phillipsburg, New Jersey) and Koch-Light Ltd. (Colnbrook, U. K.). In our experience, the products from both companies are efficient fusogens. However, Schneiderman et al. (1979) have reported that the Koch-Light product, by virtue of its lower toxicity, produces a higher yield of proliferating hybrids.

PEG is commercially available in a wide range of molecular weights. With the exception of very high (20,000) and very low (200) molecular weight compounds, all of the commercial products have been demonstrated to induce fusion in mamalian cells (Klebe and Mancuso, 1981). It appears that the lower molecular weight compounds are more efficient fusogens. Davidson et al. (1976) have reported that PEG 1000 MW yielded the highest frequency of proliferating hybrids with the fusion protocol used in their laboratory. Other investigators have reported the greatest success with PEG 400 MW (Blow et al., 1978; Clark and Shay, 1978). We have observed that DMSO will enhance the efficiency of fusion in the 6000–1000 molecular weight range, but have not examined its effect with PEG 600 and 400 MW. It should be noted that the molecular weight at which most efficient fusion occurs may vary with different cell types. Indeed, Stadler et al. (1980) reported that PEG 1000 MW is toxic for a mouse lymphocyte line used in their laboratory. It has been almost uniformly observed that the optimum concentration of PEG is 50% (w/w) at all molecular weights. However, Blow et al. (1978) observed in fusion studies with human diploid cells that 55% (w/w) is the optimum concentration for PEG 400 and 600 MW.

Duration of exposure to the fusogen. The duration of exposure to PEG is potentially a useful variable for the investigator. The principal factor limiting the length of exposure to the concentrated PEG solution is cytotoxicity. For monolayer fusions most laboratories have adhered to the protocol suggested by Davidson and Gerald (1976): 50% (w/w) PEG for 1 min. In the case of suspension fusions longer periods have been reported to be necessary (Vaughan et al., 1976). Longer periods of treatment are tolerated at lower concentrations of the fusogen. Blow et al. (1978) reported that optimum fusion of human diploid fibroblasts with PEG 1500 MW was achieved by exposure for 5 min at a concentration 45% (w/w). With DMSO present in the solvent system the optimum treatment period for PEG solutions at lower concentrations is decreased.

pH. We have observed that optimum fusion occurs over a rather broad range of pH, from 7.4 to 8.0 (Table I). Other investigators have reported a similar observation regarding the effect of pH on the efficiency of fusion (Vaughan et al., 1976; Klebe and Mancuso, 1981). Again it should be cautioned that this parameter may be affected by the cell type and culture condition(s). Sharon et al. (1980) have reported that the optimum yield of hybridoma clone cultures in soft agarose is obtained when the fusion is carried out at pH 8.0–8.2. With the method by which we prepare the PEG solution, the pH with DMSO present ranges from 7.8 to 7.9. We have observed, however, that HeLa cells will fuse more efficiently at pH 8.0 (Table I).

Temperature. Temperature does not appear to be an important variable with PEG-mediated fusion. Successful fusion has been reported with the PEG solution at room temperature (Davidson et al., 1976) and at 37°C (Hales. 1977). We routinely allow the PEG to cool to about 45°C after autoclaving and before mixing with the solvent; fusions are carried out at 25–30°C.

Solvent systems. Since Pontecorvo first reported PEG to be an efficient fusogen for mammalian cells, bicarbonate-buffered medium without serum

Table I. Effect of pH and DMSO on Efficiency of Fusion Induced by PEG 1000 MW[a]

Percent PEG	pH	Percent DMSO	HeLA	3T3	Human diploid fibroblast
45	7.5	0	5.5 ± 0.3	17.2 ± 0.3	13.6 ± 1.0
45	7.5	10	21.3 + 2.9	25.2 ± 2.5	27.5 ± 3.4
45	8.0	0	12.2 ± 3.7	19.1 ± 1.0	10.5 ± 0.1
45	8.0	10	26.1 ± 1.8	29.0	29.7 ± 2.9
50	7.5	0	10.6 ± 2.4	15.1 ± 0.1	15.2 ± 1.2
50	7.5	10	30.4 ± 4.2	26.1 ± 3.3	24.3 ± 2.0
50	8.0	0	23.4 ± 1.1	17.1 ± 0.2	11.5 ± 1.4
50	8.0	10	37.1 ± 2.0	29.1 ± 2.0	29.4 ± 0.9

[a] The cells were cultured on 22-mm coverslips and the fusions were carried out according to the protocol described in the text. The values given are percent nuclei in bi- and polynucleate cells minus percent in control cultures not exposed to the fusogens. The results shown are the average (±high and low values) of determinations of the fusion index from two separate cultures.

Table II. Efficiency of Fusion of PEG 1000 MW
Dissolved in 0.15 M HEPES (pH 7.5)
with and without DMSO^a

	HeLa	3T3	Human diploid Fibroblast
0% DMSO	10.9	24.2 ± 0.6	13.1 ± 0.1
10% DMSO	24.2 ± 3.8	34.5 ± 0.1	26.7 ± 1.5

^{a}Culture and fusion technqiues as described in Table I. A concentration of 50% (W/W) was used in this study. Values given are percent nuclei in bi- and multinucleate cells as calculated in Table I.

has been universally used as the solvent for the reagent. Supplementation with DMSO (Norwood et al., 1976) and the use of Ca^{++}-free media (Schneiderman et al., 1979) are two modifications which have been reported to enhance the efficiencies of PEG-mediated cell fusion and the yield of hybrids, respectively. We reported that media buffered with HEPES (N-2-hydroxyethylpiperazine-N'-2-ethanesulfonic acid) are toxic to human diploid cells (Norwood et al., 1976). However, recently Klebe and Mancuso (1981) reported superior fusion when the PEG was dissolved in 0.15 M HEPES at pH 7.5. Preliminary studies carried out in our laboratory have confirmed that HEPES does enhance the efficiency of fusion in some cell types (Table II). Clearly, this solvent system should be given a trial in the development of any fusion system.

Culture conditions and cell type. There are a number of parameters related to culture conditions which may affect the efficiency of fusion. The most obvious is cell density. The most desirable density may not be that which will yield the highest frequency of fusion events. If one desires to carry out cytologic examination of the fused cells, the procedure should be carried out with subconfluent cultures so that the polynucleate cells can be more readily identified. The yield of polynucleate cells can be increased by carrying out the fusion in confluent cultures and then, following fusion, subcultivate at a lower density. However, we have observed that with some cell types (e.g., HeLa) the multinucleate cells are very susceptible to injury during trypsinization. Thus, this added step in the procedure may introduce unwanted variables, such as selective loss of specific subpopulations of cells.

The proliferative state of the culture and cell cycle stage have also been reported to influence the efficiency of fusion and the recovery of hybrids. Mercer and Schlegel (1979) have reported that supplementation with DMSO does not enhance fusion of quiescent (serum-deprived) 3T3 cells. In contrast, we have observed that DMSO does enhance the fusion efficiency in serum-deprived human diploid cell cultures (Norwood and Zeigler, unpublished observation). Hansen and Stadler (1977) have reported that mitotic cells in a mouse lymphoid line displayed greater fusion competence than interphase cells in the same culture. The same laboratory (Stadler et al., 1980) has more recently published data showing an inverse correlation between the doubling time of the lymphoid cells in culture and the fusion efficiency. Thus it may be that, with some cell types, the fusion efficiency may be compromised in extremely confluent cultures.

Fusion competence also varies with the cell type. We have observed that the addition of DMSO to the fusion solution will enhance, to varying degrees, the fusion efficiency of most cell types used in our laboratory. One exception is a mouse L-cell line, which displays a low fusion frequency following exposure to PEG. The addition of DMSO does not enhance the fusibility of these cells.

3.2. Suggested Protocol—Monolayer Fusion

The protocol we routinely use is a modification of that described by Davidson and Gerald (1976). The PEG is melted and sterilized via autoclaving for 10 min (liquid sterilization with slow exhaust). After allowing for cooling (to approximately 45°C) we mix the PEG with media containing 10% DMSO (w/v). We routinely use a 50% (w/w) PEG solution. If this proves toxic to the cells, the concentration of PEG is reduced to 45%. The molecular weight of the PEG used will in part be dictated by the cell type used. We have been using PEG 1000 MW. However, in view of the reports indicating the high efficacy of PEG 400 MW, it is recommended that this molecular weight be tried initially and if it proves to be unsatisfactory the next higher molecular weight be tested and so on. Certainly, the lower viscosity of the low molecular weight facilitates the rapid removal of the fusogen solution from the culture. The fusion procedure can be done either in petri dishes or, for cytologic preparation, on coverslips. As indicated above, the cell density can vary, depending upon the intended use of the fused cells. We have observed that in human diploid fibroblasts the most efficient fusion occurs in actively proliferating cultures which are just approaching confluency. The method of fusion varies slightly depending upon whether the cells are plated directly into petri dishes or onto coverslips previously placed in petri dishes. When the cells are plated in petri dishes the cultures are washed *thoroughly* three times by pipetting ample volumes of 1× Hanks basic salt solution into the culture vessel. The wash medium should be *completely* removed from the petri dish each time. This is particularly important in the last wash since residual medium will dilute the PEG solution. The cultures are then exposed to the fusogen solution for precisely 1 min. At the termination of the exposure period, the solution is rapidly removed from the cells, preferably using mechanical suction. The cultures are again washed three times in medium without serum but supplemented with 10% DMSO (w/v), and one time in medium supplemented with 10% serum. The cultures are then covered with medium containing 10% serum. In the case of cells cultivated on coverslips, the coverslip is removed from the culture vessel with sterile forceps and washed by dipping with agitation in Hanks basic salt solution previously placed in three 30-ml beakers. The residual wash medium is removed from the coverslip by touching it to the edge of a sterile 1 × 3 inch slide which has been placed in the last wash beaker. The coverslip is then immersed, with slight agitation, in the PEG solution for precisely 1 min. We also place one 1 × 3 inch slide in the beaker containing the PEG solution, anchored against one side at the base and leaning against the opposite wall at the top of the beaker. This maneuver facilitates recovery of the coverslip if it is inadvertently dropped in the solution. Following removal from the PEG solution, the coverslip is immediately washed three times by dipping successively with agitation in three 30-ml beakers containing medium with DMSO 10% (w/v) and once in a 30-ml beaker containing medium with 10% serum. The coverslip is then replaced in the petri dish, medium with serum is placed over the cells, and the cultures are reintroduced to the desired incubation temperature.

3.3. Suggested Protocol—Suspension Fusion

There have been a number of published reports outlining protocols for the fusion of cells in suspension (Vaughan et al., 1976; O'Malley and Davidson, 1977; Hales, 1977). We have developed a simplified procedure which has given satisfactory results with human diploid fibroblasts. The preparation of the reagents is identical to that described above. The cells to be fused are trypsinized from the monolayer and the trypsin is immediately inactivated with medium containing serum. The cells are then pelleted via centrifugation at 1000 rpm for 5 min in a desktop centrifuge, washed once by resuspension in medium without serum, and again pelleted via centrifugation. The wash medium is then *completely* aspirated. This last step is important, because any residual wash medim will dilute the PEG solution. Immediately following the wash procedure, 0.1–0.2 ml of the PEG solution is placed over the pellet, which is gently resuspended with a Pasteur pipette. The cell suspension is allowed to stand undisturbed for exactly 1 min, after which the PEG solution is rapidly diluted by the addition of 10 ml of medium with serum and the cells are *gently* distributed to culture vessels at the desired density. When distributing the cells, it is important to minimize shearing stress, which may disrupt polykaryons in the early stages of formation.

4. Evaluation of Cytotoxicity

PEG at the concentrations used for cell fusion can be cytotoxic. The degree of sensitivity to this fusogen varies with culture conditions and cell type. Probably the major cause of extensive cell injury results from excessively long exposure to the concentrated PEG solution. Minimization of cell toxicity during chemical fusion of mammalian cells is discussed in detail in Chapter 3 of this volume. We have observed that the presence of extensive cell injury can be documented via microscopic examination of the culture within 10–15 min after exposure to the fusogen. Cytotoxicity in human diploid cell cultures is manifested by the appearance of homogeneous blebs on the surface

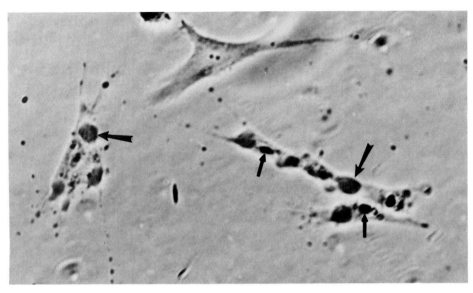

Figure 1. Phase contrast photomicrograph showing cell injury following exposure to PEG 1000 MW. Prominent homogeneous blebs are apparent on the surface of the injured cells (large arrows). Note the shrunken, pyknotic nuclei in the injured cells (small arrows). This morphologic appearance is in contrast to that of the apparently healthy cell at the top of the photomicrograph, which displays no blebbing and contains a pale nucleus with two prominent nucleoli (310×).

of the cells which gradually enlarge (Fig. 1). This is associated with the rapid onset of nuclear pyknosis. The nature of these blebs is unknown. Knutton (1979) observed similar blebbing in freeze-fracture electron microscopic studies of PEG-induced fusion in hen erythrocytes, and concluded, based on their morphologic appearance, that these structures are composed of lipids. The major point we wish to emphasize here is that cytotoxicity can be recognized soon after fusion, and microscopic examination of the culture approximately 30 min after completion of the procedure can save one many hours of frustration.

5. Discussion

The mechanism(s) of fusion of biologic membranes and how this fusion process is facilitated by certain chemical compounds remain unknown. A variety of oil-soluble and water-soluble chemicals have been demonstrated to have fusogenic activity (Blow et al., 1979). Klebe and Mancuso (1981) have reported that all these compounds must be present in high concentrations to achieve significant fusion of between 40 and 60%. They also reported that PEG of extremely high and low molecular weights were weakly fusogenic and concluded that other properties, in addition to the chemical structure of these agents, are involved in determining their fusogenicity. Also, these authors demonstrated that a variety of compounds can act synergistically with PEG in the facilitation of cell fusion.

There have been a number of morphologic studies of the events which occur during and following exposure to PEG. Freeze-fracture electron microscopic studies have revealed aggregation of intramembrane particles into discrete patches (Knutton, 1979; Robinson, et al., 1979). Scanning electron microscopic studies of mouse L cells treated with PEG and DMSO demonstrated a dramatic reduction in the number of microvilli on the cell surface, with a concomitant increase in the number of blebs on the cell surface (not to be confused with the "blebs" we have observed in injured human diploid fibroblast cultures). Based on these and other observations, these authors concluded that osmotic shock may be necessary for cell fusion. These studies, as well as transmission electron microscopic studies, all indicate that the fusion process is initiated very rapidly following exposure to PEG. Light microscopic and isotopic labeling studies carried out in our laboratory suggest that, in human diploid cells, fusion may be initiated only at the time of exposure to PEG and that subsequent cytoplasmic mixing is rapid (Rabinovitch and Norwood, 1981).

A number of hypotheses have been proposed to explain the fusogenic activity of certain chemicals. Most investigators have proposed that functional and structural alterations occur in the plasma membrane which permit aggregation of cells and thus close apposition of the membrane. Maggio et al. (1976) have emphasized that the decrease in the surface electrostatic potential of biologic membranes by fusogenic agents may be of critical importance. Blow et al. (1978, 1979) have suggested that changes in membrane permeability associated with calcium ion influx may be important. They demonstrated a good correlation with chemicals that promote cell fusion, including DMSO, and changes in calcium flux. However, the precise role of calcium in the fusion remains to be determined. The observation by Schneiderman et al. (1979) that cytotoxicity is reduced when Ca^{++}-free medium is used as a solvent suggests that, if calcium is important for cell fusion, it must be present at critical concentrations. Of even more interest with respect to DMSO is the suggestion by these authors (Blow et al., 1979) that alterations in the physical state of bulk water adjacent to the cell membrane may also be of importance. DMSO is known to have a strong affinity for water (Nash, 1966). Maroudas (1975) has suggested that DMSO (and glycerol) may act by altering the exclusion volume of membrane glycoproteins, which would permit the interaction of the lipid bilayers of adjacent cells. At this point in time we would speculate that the interaction between DMSO (and other chemical fusogens) and cell surface water may be of critical importance in the promotion of fusion of biologic membranes by these agents.

We have discussed in some detail the usefulness of DMSO as an additive which will enhance PEG-mediated cell fusion. However, we emphasize that the usefulness of this compound has limitations. It is less efficient with some cell types. Indeed, in some types of studies the use of DMSO may be contraindicated. For example, in studies of cell differentiation utilizing the technique of cell fusion, one should be very cautious regarding the use of DMSO, since it is known that this chemical will induce Friend leukemia cells to differentiate along the erythrocytic pathway (Friend et al., 1971). As we stressed in the introduction, DMSO is one of a number of agents now available to the investigator which can act synergistically with PEG in the enhancement of mammalian cell fusion.

Acknowledgment

Portions of the work described in this chapter were supported by U. S. Public Health Service grant AG-01751.

References

Anné, J., and Peberdy, J. F., 1976, Induced fusion of fungal protoplasts following treatment with polyethylene glycol, *J. Gen. Microbiol.* **92**:413–417.

Ahkong, Q. F., Fisher, D., Tampion, W., and Luiz, J. A., 1975, Mechanisms of cell fusion, *Nature* **253**:194–195.

Ashwood-Smith, M. J., 1971, Radioprotective and cryoprotective properties of DMSO, in: *Dimethyl Sulfoxide* (S. W. Jacob, E. R. Rosenbaum, and D. C. Wood, eds.), Marcel Dekker, New York, pp. 168–177.

Blow, M. J., Botham, G. M., Fisher, D., Grodall, A. H., Tilcock, C. P. S., and Luiz, J. A., 1978, Water and calcuim ions in cell fusion by poly (ethylene glycol), *FEBS Lett.* **94**:305–310.

Blow, M. J., Botham, G. M., and Lucy, J. A., 1979, Calcium ions and cell fusion. Effects of chemical fusogens on the permeability of erythrocytes to calcuim and other ions, *Biochem. J.* **182**:553–563.

Brown, U. K., Robinson, J., and Stevenson, D. E., 1963, A note on the toxicity and solvent properties of dimethyl sulfoxide, *J. Pharm. Pharmacol.* **15**:688–692.

Clark, M. A., and Shay, J. W., 1978, Scanning electron microscopic observations on the mechanism of somatic cell fusion using polyethylene glycol, *Scanning Electron Microscopy* **2**:327–332.

Davidson, R. L., and Gerald, P. S., 1976, Improved techniques for the induction of mammalian cell hybridization by polyethylene glycol, *Somat. Cell Genet.* **2**:165–176.

Davidson, R. L., O'Malley, K. A., and Wheeler, T. B., 1976, Polyethylene glycol-induced mammalian cell hybridization—effect of polyethylene glycol molecular weight and concentration, *Somat. Cell Genet.* **2**:271–280.

Denko, C. W., Goodman, R. M., Miller, R., and Donovan, T., 1967, Distribution of dimethyl sulfoxide ^{35}S in the rat, *Ann. N.Y. Acad. Sci.* **141**:77–84.

Feinman, H., Ben, M., and Levin, R., 1964, The toxicity of dimethylsulfoxide (DMSO) in primates *Pharmacologist* **6**:188.

Friend, C., Scher, W., Holland, J. G., and Sato, T., 1971, Hemoglobin synthesis in murine virus-induced leukemic cells *in vitro*: stimulation of erythroid differentiation by dimethyl sulfoxide, *Proc. Natl. Acad. Sci. USA* **68**:378–382.

Hales, A., 1977, A procedure for the fusion of cells in suspension by means of polyethylene glycol, *Somat. Cell. Genet.* **3**:227–230.

Hansen, D., and Stadler, J., 1977, Increased polyethylene glycol-mediated fusion competence in mitotic cells of a mouse lymphoid cell line, *Somat. Cell Genet.* **3**:471–482.

Jacob, S. W., 1971, Pharmacology of DMSO, in: *Dimethyl Sulfoxide* (S. W. Jacob, E. E. Rosenbaum, and D. C. Wood, eds), Marcel Dekker, New York, pp. 99–112.

Klebe, R. J., and Mancuso, M. G., 1981, Chemicals which promote cell hybridization, *Somat. Cell Genet.* **7**:473–488.

Knutton, S., 1979, Studies of membrane fusion III. Fusion of erythrocytes with polyethylene glycol, *J. Cell Sci.* **36**:61–72.

Kolb, K. H., Jaenicke, G., Kramer, M., and Schulz, P. E., 1967, Absorption, distribution, and elimination of labeled dimethyl sulfoxide in man and animals, *Ann. N.Y. Acad. Sci.* **141**:85–95.

Maggio, B., Ahkong, Q. F., and Lucy, J. A., 1976, Poly (ethylene glycol) surface potential and cell fusion, *Biochem. J.* **158**:647–650.

Maroudas, N. G., 1975, Polymer exclusion, cell adhesion and membrane fusion, *Nature* **254**:695–696.

Mathew, T., Karunanity, R., Yee, M. H., and Natarojan, P. N., 1980, Hepatotoxicity of dimethylformamide and dimethylsulfoxide at and above the levels used in some aflatoxin studies, *Lab. Invest.* **42**:257–262.

Maul, G. G., Steplewski, Z., Weibel, J., and Koprowski, H., 1976, Time sequence and morphological evaluations of cells fused by polyethylene glycol 6000, *In Vitro* **12**:787–796.

Mercer, W. E., and Schlegel, R. A., 1979, Phytohemagglutinin enhancement of cell fusion reduces polyethylene glycol cytotoxicity, *Exp. Cell Res.* **120**:417–421.

Nash, T., 1966, Chemical constitution and physical properties of compounds able to protect living cells against damage due to freezing and thawing, in: *Cryobiology* (H. T. Merryman, ed.), Academic Press, New York, pp. 172–211.

Norwood, T. H., Zeigler, C. J., and Martin, G. M., 1976, Dimethyl sulfoxide enhances polyethylene glycol-mediated cell fusion, *Somat. Cell Genet.* **2**:263.

O'Malley, K. A., and Davidson, R. L., 1977, A new dimension in suspension fusion techniques with polyethylene glycol, *Somat. Cell Genet.* **3**:441–448.

Pontecorvo, G., 1975, Production of mammalian somatic cell hybrids by means of polyethylene glycol treatment, *Somat. Cell Genet.* **1**:397–400.

Rabinovitch, P. S., and Norwood, T. H., 1981, Rapid kinetics of polyethylene glycol-mediated fusion, *Somat. Cell Genet.* **7**:281–287.

Robinson, J. M., Roos, D. S., Davidson, R. L., and Karnovsky, M. J., 1979, Membrane alterations and other morphological features associated with polyethylene glycol-induced cell fusion, *J. Cell Sci.* **40**:63–75.

Schneiderman, S., Farber, J. L., and Baserga, R., 1979, A simple method for decreasing the toxicity of polyethylene glycol in mammalian cell hybridization, *Somat. Cell Genet.* **5**:263–269.

Sharon, J., Morrison, S. L., and Kabat, E. A., 1980, Formation of hybridoma clones in soft agarose: Effect of pH and of medium, *Somat. Cell Genet.* **6**:435–441.

Stadler, J., Vaughan, V., and Hansen, J., 1980, Cell fusion impaired varients of a mouse lymphocyte cell line obtained by single-step selection for polyethylene glycol resistance, *Somat. Cell Genet.* **6**(1):127–137.

Sulzberger, M. B., Cortese, Jr., T. A., Fishman, L., Wiley, H. S., and Peyakovich, P. S., 1967, Some effects of DMSO on human skin *in vivo, Ann. N.Y. Acad. Sci.* **141**:437–450.

Szmant, H. H., 1971, Chemistry of DMSO, in: *Dimethyl Sulfoxide* (S. W. Jacob, E. E. Rosenbaum, and D. C. Woods, eds.), Marcel Dekker, New York, pp. 1–97.

Vaughan, L., Hansen, D., and Stadler, J., 1976, Parameters of polyethylene glycol-induced cell fusion and hybridization in lymphoid cell lines. *Somat. Cell Genet.* **2**:537–544.

Wilairat, P., Yuthavng, Y., and Khyngvanlert, R., 1978, Effect of membrane modification on cell fusion of hen erythrocytes induced by dimethyl sulfoxide, *Life Sciences* **22**:1993–1998.

Yoshiyuki, H., Kurono, M., and Kotai, S., 1979, Polyethylene glycol-induced fusion of L forms of *Staphylococcus aureaus. Biken J.* **22**:25–29.

Chapter 5

The Selection of Heterokaryons and Cell Hybrids Using the Biochemical Inhibitors Iodoacetamide and Diethylpyrocarbonate

WOODRING ERIK WRIGHT

1. Introduction

Most selective systems for isolating fusion products between different cell types rely on the preferential ability of the fusion products to grow in a special medium (reviewed in Ringertz and Savage, 1976). Because these selection systems require cell division, most experiments in somatic cell genetics have studied clones of cell hybrids rather than heterokaryons, which are the immediate product of the fusion of two different cells. However, there are many fundamental differences between hybrids and heterokaryons that make the latter an important subject of investigation. One of the most significant derives from the fact that most differentiated cells are slowly dividing or postmitotic. Restricting the analysis to only those fusion products capable of rapid cell division and clone formation thus automatically biases results against the expression of differentiated functions. This bias is particularly compelling if one examines the frequency with which heterokaryons give rise to growing hybrid clones. Although this varies widely with the cell combination employed and ranges from as much as one in three (Kao et al., 1969) to as few as one in 100,000 (Miller and Ruddle, 1976), an average value appears to be that about one in 100 heterokaryons gives rise to a hybrid clone. These clones thus represent a highly selected subset of all fusion products. Conclusions about the expression of differentiated functions in the clones must therefore be interpreted with caution.

Heterokaryons also provide a system in which different kinds of experimental information may be obtained than in the case of cell hybrids. For example, what are the regulatory changes occurring during the transition

---

WOODRING ERIK WRIGHT • Department of Cell Biology and Internal Medicine, The University of Texas Health Science Center, Dallas, Texas 75235.

between a precursor cell and its terminally differentiated progeny? Hybrid cell clones are an inappropriate tool for answering this question. Since the difference between the parental cells is one of only a few days or a few cell divisions, analyzing hybrid clones isolated after several weeks and many cell divisions have occurred is unlikely to provide meaningful information. However, heterokaryons are eminently suitable for this analysis. For example, we have demonstrated that factors present in heterokaryons formed by fusing postmitotic differentiated skeletal myoctyes to undifferentiated myoblasts induce the undifferentiated myoblasts to synthesize skeletal muscle proteins (Wright, 1981a, 1982). This chapter will describe a general technique that permits the isolation of heterokaryons between virtually all cell types, and thus makes heterokaryons a more accessible subject for somatic cell fusion investigations.

2. Choice of Agents

2.1. General Approach

The selective system using irreversible biochemical inhibitors for isolating highly enriched populations of heterokaryons from unfused parental cells and homokaryons is schematically illustrated in Fig. 1. In this model, each cell type contains eight hypothetical molecules necessary for its survival. Because of interlocking metabolic pathways and cellular reserves, the model postulates that a cell can survive long enough to resynthesize its missing components if any five of these molecules are present. Each cell type to be fused is treated with a different irreversible inhibitor which inactivates the indicated molecules. The unreacted inhibitor is then washed away and the cells are mixed and fused. The unfused parental cells and homokaryons only have three of the required eight molecules and thus die. However, even though molecules 1 and 2 were inactivated by both agents, the heterokaryons have six of the molecules. Since a cell can survive if any five molecules are present, the heterokaryons are able to recover from the treatment.

2.2. Screening of Selective Agents

The above model predicts that no cells should survive in simple mixtures of the two treated cell types, and that all of the cells surviving following fusion should have more than one nucleus. Using these criteria for success the following agents were initially screened for use in the selective system: iodoacetamide, N-ethylmaleimide, fluorodinitrochlorobenzene, and diethylpyrocarbonate. All possible combinations between the agents were tested. Iodoacetamide (IAM) reacting primarily with sulfhydryl groups, and diethylpyrocarbonate (DEPC), directed primarily against histidine residues, proved to be the most effective combination among this initial group of

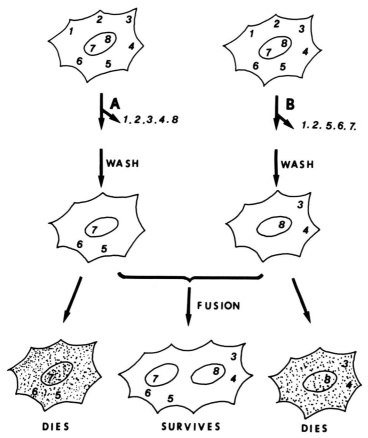

Figure 1. Schematic diagram of a selective system based on the use of irreversible biochemical inhibitors. The numbers represent enzymes or other molecules, and A and B represent agents that can irreversibly inactivate different subpopulations of these molecules. Assume that a cell can survive if any five of these eight molecules are intact. In an ideal case, the subpopulations affected by reagents A and B would be mutually exclusive, and perfect complementation would occur. In the case illustrated, some overlap occurs in that molecules 1 and 2 are inactivated by both agents. Nonetheless, the heterokaryon achieves a sufficient degree of complementation (six molecules) to permit it to survive long enough to resynthesize its full molecular repertoire. [Reproduced from Wright (1978), Academic Press, New York, with permission.]

inhibitors. After the conditions for the successful use of IAM and DEPC had been optimized, many additional inhibitors were tested. Table I lists agents that were screened by the ability of treated cells to rescue cells killed with either IAM or DEPC. Bifunctional inhibitors, pharmacologic agents, and toxins were used in an attempt to achieve greater specificity in the lethal lesion. Although most of these agents permitted some complementation and heterokaryon rescue, none proved superior to IAM or DEPC. The arginine reagent phenylglyoxal was reasonably effective and could replace IAM or DEPC if particular experimental conditions required that those inhibitors be avoided.

Table I. Agents Screened as Selective Agents

General inhibitors	General inhibitors	Toxins	Pharmacologic/herbicidal agents
N-Ethyl maleimide	Pepstatin A	Emetine	Captan
p-Mercuribenzoate	Phenylglyoxal	α-Amanitin	Dichlone
p-Chloromercuriphenyl sulfonic acid	2-Hydroxynitrobenzyl bromide	Actinomycin D	Ethacrynic acid
Bromoacetamide	2-Methoxynitrobenzyl bromide	Diptheria toxin	Astiban
Chloroacetamide	Ethidium bromide	Ochratoxin A	Pentamidine
Iodosobenzoate	Cordecepin	Patulin	Iproniazid
Phenylarsine oxide	Adenosine arabinoside	Penicillic acid	
Antimony potassium tartrate	Napthoquinone	Rubratoxin B	
			Bifunctional agents
Dimethyldithiocarbamic acid	Chloroquine	Verrucarin A	1,5-Difluoro-2,4-dinotrobenzene
Chloro-IPC	2,4-Dinitrofluorobenzene	Diacetoxyscirpenol	Fluoro-dinitrochlorobenzene
Diamide	Picryl sulfonic acid	Aflatoxin B_1, B_2	Bis-dimethylaminobenzhydryl
TPCK	Tetranitromethane	T-2 Toxin	Bis-(4-fluoro-3-nitrophenyl)-sulfone
TAME	Iodine	Citrinin	4-Bromphenacyl bromide
PMSF	N-acetylimidazole	Sterigmatocystin	Dimethylsuberimidate
		Luteoskyrin	M-Maleimidobenzyl-N-hydroxysuccinimide ester
			N,N'-O-phenylenedimaleimide
			Diisothiocyanostilbenedisulfonic acid (DIDS)

2.3. Proof That Binucleates Are Heterokaryons

Fusions between IAM/DEPC-treated cells resulted in populations in which 90–99% of the surviving cells were multinucleate. A variety of tests proved that these cells are in fact heterokaryons. 5-Bromodeoxyuridine (BrdU) suppresses the fluorescence of the nuclear dye Hoechst 33258 (Latt, 1973). A total of 98% of the rescued oligonucleated cells isolated after fusing

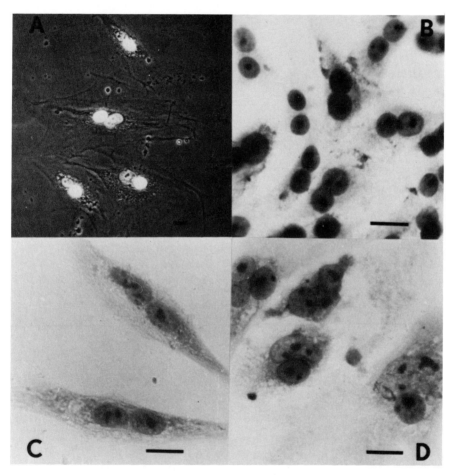

Figure 2. Selection of heterokaryons between cells treated with iodoacetamide and diethylpyrocarbonate. The dose of each agent used to selected against nonfused parental cells is shown in parentheses. Bars equal 20 μm. A. L_6 rat myoblasts (10 mM IAM) grown for four generations in 3 uM BrdU were fused to T984 teratoma-derived mouse myoblasts (0.008% DEPC). Cells were stained with 0.5 μg/ml 33258 Hoechst. A phase contrast image is superimposed upon the fluorescence exposure. The BrdU-treated L_6 nuclei fluoresce dimly, while the T984 nuclei fluoresce brightly. B. Homokaryons between $L5_{5219x}$ rat myoblasts (16 mM IAM) and $L5_{5219x}$ rat myoblasts (0.011% DEPC). C. Homokaryons between normal human skin fibroblasts (5 mM IAM) and normal human skin fibroblasts (0.007% DEPC). D. Heterokaryons between $L5_{5219x}$ rat myoblasts (15 mM IAM) and PCC3/A/1 mouse teratoma cells (0.008% DEPC). [Reproduced from Wright (1978), Academic Press, New York, with permission.]

rat myoblasts that had been grown in BrdU to thymidine-containing mouse myoblasts contained at least one dimly and one brightly fluorescing nucleus (Fig. 2a) (Wright, 1978). Figure 2 also illustrates a variety of heterokaryons made between different cell types. Rat nuclei stain much more intensely with Giemsa than do chick nuclei. Heterokaryons isolated by fusing rat and chick cells contain heterostaining nuclei (Wright, 1981a, 1982). Finally, if heterokaryons between rat myoblasts and clone 1 d mouse "fibroblasts" are cultivated until growing hybrid cells appear, both karyotype and isozyme analysis establish that the cells are in fact hybrids (Wright, 1978).

3. Treatment Conditions

3.1. General Considerations

Treatment at 0°C was chosen in order to minimize secondary degenerative effects that might occur between the time of treatment and the eventual fusion and rescue of the cells. Balanced salt solutions are used rather than complete medium in order to avoid the reaction of the inhibitors with components of the medium or serum. Although phosphate-buffered saline works well with most cell types, some cells are very sensitive to the absence of divalent cations during the treatment and washing steps. We thus use Hanks balanced salt solution (Hanks BSS). We chose 30 min as the shortest time of treatment consistent with reproducibility when multiple cell types or inhibitors were being manipulated.

Iodoacetamide stock solutions are dissolved in Hanks BSS and are considered self-sterilizing. Although we have found iodoacetamide to be stable in Hanks BSS at 4°C for at least 1 month, we nonetheless make fresh solutions for each experiment. Iodoacetamide is quite soluble in aqueous solutions and stocks are made to a convenient concentration. For example, if doses of 3.5, 4.6, and 5.9 mM IAM are to be used (a 1.3-fold interval between doses) in a volume of 50 ml, a 500 mM stock solution is made, so that 0.35, 0.46, and 0.59 ml added to the final volume gives the desired concentration.

DEPC is very unstable in aqueous solutions, hydrolyzing to ethanol and carbon dioxide with a half-life of approximately 15 min at 0°C. The release of CO_2 will cause a pH shift in bicarbonate-buffered solutions, but this is generally a small effect at the concentrations used. Stock solutions of DEPC 10% (v/v) in absolute ethanol can be divided among small gas vials and stored dessicated at −10°C for up to 2 months. Individual vials are used within 1 week to reduce the possibility of condensed moisture hydrolyzing the stock solution. Because of its short half-life in water, any intermediate dilutions are made in absolute ethanol, and portions of the final DEPC stock solution are added directly to the cell suspension at the start of the treatment. Care must be employed in measuring small volumes (e.g., 20–50 μl) of DEPC, since ethanol evaporating into the bore of the micropipettors used in most laboratories can reduce the volume of solution actually drawn into the pipette tip. The micropipettor should be equilibrated by filling and emptying the tip several times before the desired volume is withdrawn.

3.2. Cell Concentration during Treatment

The cell types to be tested are trypsinized, resuspended in serum-containing medium to neutralize the trypsin, counted, pelleted, and resuspended in cold Hanks BSS. The cells are then divided among the treatment tubes, which are held in an ice bath. At time 0, appropriate concentrations of stock reagents are added to each tube. The tubes are shaken each 10 min to prevent the cells from settling. After 30 min, the cells are rapidly pelleted in a refrigerated centrifuge (800g × 2 min), resuspended in Hanks BSS, pelleted, resuspended in complete medium, and plated. Surviving cells take approximately 24–36 hr to recover and attach. The cells are then fixed, stained, and counted. Figures 3A (iodoacetamide) and 3B (diethylpyrocarbonate) show the survival curves obtained at different cell concentrations during treatment. The point at which $10^{-4}\%$ (10^{-6}) survival was achieved is taken as the lethal dose, and this lethal dose is plotted as a function of cell concentration in Fig. 3C and 3D. It is clear that the lethal dose of IAM is highly dependent upon cell concentration, whereas the dose of DEPC is not. We believe this reflects the relative lipophilic nature of these two drugs (Wright, 1978). Thus it is important that a constant cell concentration be established for all treatments with iodoacetamide. We have arbitrarily chosen 3×10^5 cells/ml as a convenient value for our experiments. However, we treat cells with DEPC at any convenient concentration up to 10^6 cells/ml. In most fusion experiments, 15×10^6 cells are treated in a 50-ml conical centrifuge tube, pelleted, washed once with 25 ml of cold Hanks BSS, pelleted, then resuspended in a convenient volume (4–6 ml) of cold Hanks BSS. After control aliquots of cells have been plated to determine the lethal efficiency of each dose, the cells are mixed in appropriate combinations and pelleted for cell fusion with polyethylene glycol (see Section 6).

3.3. Dosage Dependence of Heterokaryon Rescue

Figure 4 shows the cell rescue (percent of original nuclei recovered as viable heterokaryons) after cells treated with DEPC were fused to cells treated with increasing concentrations of IAM. Each line shows the result obtained with a different dose of DEPC. The numbers beside each point indicate the purity (percent of nuclei in cells containing more than one nucleus) that was obtained at each point. It is clear that the higher the dose of IAM, the lower the efficiency of heterokaryon cell rescue. We believe this is partially due to overlap in the spectrum of molecules affected by each inhibitor. At higher doses, more overlap and more damage occur and the efficiency of rescue declines. The optimal dose of IAM would thus be the

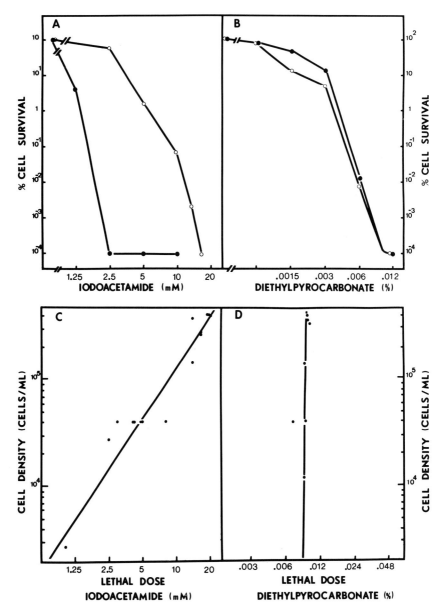

Figure 3. Cell density dependence of iodoacetamide killing. A. $L5_{5219x}$ rat myoblasts were treated at 0.27×10^5 (●) or 2.7×10^5 (○) cells/ml with various doses of iodoacetamide and the percent survival was determined for each dose. B. $L5_{5219x}$ rat myoblasts were treated at 0.12×10^5 (●) or 1.2×10^5 (○) cells/ml with various doses of diethylpyrocarbonate and the percent of survival was determined for each dose. C, D. Density dependence of the lethal dose. Dosage curves similar to those in A and B were established for various cell densities. The dose at which only $10^{-4}\%$ of the cells survived was taken as the "lethal dose." Each point represents the lethal dose from a different dosage curve; most points represent independent experiments (different days). Iodoacetamide killing is higly dependent on cell density during treatment, whereas diethylpyrocarbonate killing is not. [Reproduced from Wright (1978), Academic Press, New York, with permission.]

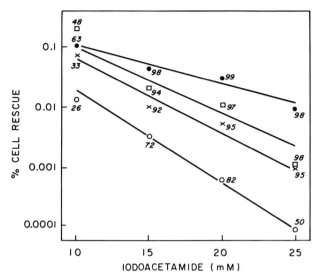

Figure 4. Dosage dependence of heterokaryon rescue. L5$_{5219x}$ rat myoblasts were treated with four different doses of iodoacetamide or four different doses of diethylpyrocarbonate. All 16 possible crosses were performed. The number of heterokaryons surviving in each cross was determined 3 days after fusion. The number beside each point is the purity obtained (the percent of surviving nuclei present in cells with more than one nucleus). (●) 0.01%; (□) 0.015%; (×) 0.02%; and (○) 0.025% diethylpyrocarbonate. A 100% killing in control aliquots was obtained at 20 mM iodoacetamide and 0.01% diethylpyrocarbonate. Cell rescue is highly dose dependent. [Reproduced from Wright (1978), Academic Press, New York, with permission.]

lowest dose that kills a sufficiently high proportion of the parental cells to yield a relatively pure population of heterokaryons. Because this optimal dose varies slightly from day to day according to the health of the cells, we routinely treat cells with two or three doses of IAM and two or three doses of DEPC and do all the relevant combinations. The experiment in Fig. 4 also indicates the importance of using *both* inhibitors at the lowest effective doses. The bottom line shows that at 0.25% DEPC the efficiency of rescue was so low that highly enriched populations were never obtained. The four fusions at 15 mM IAM show that in some situations increasing the dose of one agent can reduce the purity of the rescued heterokaryons. In these fusions, a relatively constant number of surviving parental cells becomes an increasing proportion of the population as the rescue declines, thus reducing the purity from 98% to 72%. Because of this factor, it is particularly important during the initial attempts to use biochemical selection to employ a wide range of doses in order to make it clear if an impure population is due to too high or too low a dose of a particular inhibitor. Most cells require between 3.5 and 7 mM IAM and between 0.0035 and 0.006% DEPC. Table II shows the approximate optimal doses for various cell types we have used. Although wider intervals can be used in initial experiments, we generally use a 1.25- to 1.3-fold interval between treatment doses.

Although we have not rigorously investigated the effects of prolonged

Table II. Approximate Optimal Lethal Dose for Different Cell Types

Cell type	Species	Primary culture, secondary culture, or cell line designation	Approximate optimal lethal dose	
			Iodoacetamide, mM	Diethylpyrocarbonate, %
Fibroblast	Human	Secondary (IMR-90)	6	0.006
	Chick	Secondary	4	0.0045
	Quail	Secondary	3.5	0.004
	Rat	Secondary	2	0.003
	Hamster	CHO	—	0.0025
	Mouse	Clone 1d	—	0.01
Skeletal myoblast	Rat	L6	6	0.0045
	Rat	L5$_{219x}$	15	0.01
	Rat	Primary	3.5	0.003
	Mouse	T984	5	0.008
	Chick	Primary	4	0.003
	Quail	Secondary	6	0.005
	Rabbit	Primary	—	0.0025
Melanoma	Mouse	B16	2.0	0.0035
Teratoma	Mouse	PCC4	—	0.0016

cell confluency, it has been our impression that the optimal doses are more reproducible if cells have been subcultivated no more than 3 days prior to use. When postmitotic cells are used, we feed them 2–3 days earlier (rather than subcultivating them) in order to achieve a similar effect.

3.4. Time Dependence of Cell Rescue

Secondary degenerative effects probably represent common mechanisms of cell death and are not likely to be complementable. All washes prior to the fusion of the two cell types are thus performed using cold Hanks BSS and the cells are kept in an ice bath in order to reduce cellular metabolism. Figure 5 shows that even 0°C does not prevent the onset of some degeneration, since the efficiency of rescue does decline progressively the longer the treated and washed cells are held at 0°C before initiating cell fusion.

4. Conditions for Cell Fusion

Although good rescue is obtained with inactivated Sendai virus, we have found polyethylene glycol (PEG) to be much more convenient. However, it is important to optimize the use of PEG, since most published procedures are relatively toxic to the cells. Although this can be tolerated in experiments using untreated cells, cells compromised by lethal biochemical inhibitors can be so damaged by poor PEG fusion conditions that heterokaryon rescue is prevented. The toxicity of PEG is reduced by using lower molecular weight polymers (Davidson et al., 1976) (e.g., PEG 1000 rather than PEG 6000), and by reducing the concentration of PEG necessary to produce adequate fusion by including dimethyl sulfoxide in the medium (Norwood et al., 1976). The procedure we use for optimizing the fusion conditions is as follows. Untreated

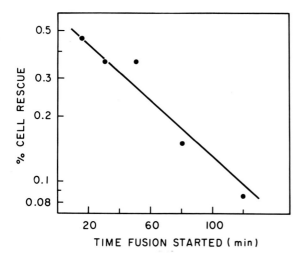

Figure 5. Time dependence of cell rescue. $L5_{5219x}$ rat myoblast cells were treated with either 15 mM iodoacetamide or 0.01% diethylpyrocarbonate, then mixed together and divided into aliquots. Fusion was begun at the indicated times following completion of the 30-min treatment. Cells were maintained at 0°C until fused. The efficiency of heterokaryon formation is approximately 2.5%. Thus, in this experiment, one in five heterokaryons were rescued at the earliest times. Cell rescue decreases with time as degenerative effects begin. [Reproduced from Wright (1978), Academic Press, New York, with permission.]

cells are fused at different PEG concentrations and plated overnight. The next morning, aliquots are fixed, stained, and counted to determine the efficiency of fusion. Other aliquots are trypsinized and counted to determine the number of cells able to reattach after PEG treatment. The product of the percent fusion and the percent viability gives the actual yield of fused cells. The least toxic PEG concentration producing an adequate heterokaryon yield is chosen for use.

For most cell combinations we use 35% PEG. First PEG 1000 (BDH, available in the U.S. from Gallard-Schlesinger Chem. Mfg. Corp., Carle Place, New York) is melted in a boiling water bath, then 2.7 ml is added to 5 ml of sterile HEPES-buffered, serum-free medium containing 15% DMSO (HEPES: N-2-hydroxyethylpiperazine-N'-2-ethanesulfonic acid). The medium is empirically buffered so that the final pH is near neutrality. For most purposes this solution can be considered sterile without further treatment. Although a precipitate does develop over time, the solution is active for at least 1 month when stored at 4°C.

Our fusion protocol is to pellet the mixed cells in a 13 × 100 mm sterile glass test tube fitted with a sleeve cap, pour off the supernatant, and then briefly recentrifuge the tube so that any liquid adhering to the walls pools at the bottom. The pellet is then aspirated to dryness and kept at 0°C until used. The actual fusion is performed by first allowing the cells to warm to room temperature for 1–2 min, then resuspending the cells in approximately 0.2 ml of 35% PEG per 10^7 cells. The cells are mixed by pipetting for 15 sec, then by rotating the test tube for an additional 30 sec. The reaction is stopped by dilution with 1 ml of complete medium, and the cells are then gently distributed into petri dishes. The PEG need not be removed if it is diluted more than 20-fold.

5. Plating Conditions following Cell Fusion

5.1. Direct Plating

Because only one in ten to one in 50 of the heterokaryons actually formed are rescued, the surviving cells attach at relatively low culture densities. Figure 6 shows the result of attempting to increase this density by plating increasing numbers of cells per cm² following cell fusion. Plating densities above 4×10^5 cells/cm² resulted in a decreased rescue, presumably as a result of the toxicity of products released from dying cells. Additional attempts to concentrate the rescued heterokaryons by trypsinizing them at 0°C were unsuccessful in that more than three-quarters of these newly recovered heterokaryons did not survive the trypsinization.

5.2. Initial Low-Density Plating

Heterokaryons take 24–36 hr to recover from the biochemical treatments and attach. Most of the toxic products from dying parental cells are released

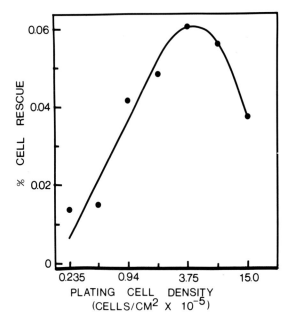

Figure 6. Plating density dependence of cell rescue. $L5_{5219x}$ rat myoblasts treated with 0.008% DEPC were fused to $L5_{5219x}$ rat myoblasts treated with 12 mM IAM. Following fusion the cells were plated at increasing cell densities. All aliquots were washed and fixed after 48 hr. The % cell rescue equals the number of nuclei surviving in polynucleated cells divided by the total number of original nuclei before treatment. The efficiency of cell rescue increases initially, but begins to decrease beyond 4×10^5 cells/cm². Nonetheless, the total number of surviving heterokaryons/cm² is greater even when plated at 1.5×10^6 cells/cm². [Adapted from Wright (1978), Academic Press, New York, with permission.]

during the first day (Wright, 1981b). It is possible to initially plate the cells at low density for 24 hr while the toxic products are being released into the medium, then to concentrate the still unattached heterokaryons and plate them at high density. The heterokaryons are initially plated on bacterial-grade petri dishes in order to prevent possible cell attachment. Figure 7 compares the number of heterokaryons that survive when cells are plated onto tissue-culture-grade 0.3-cm² microwells at increasing cell densities either

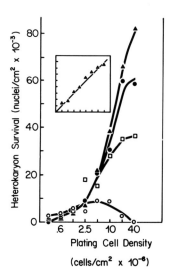

Figure 7. Surviving heterokaryon cell density as a function of plating density following various treatments. Embryonic chick myoblasts treated with 5.5 mM IAM were fused to L6 rat myoblasts treated with 0.005% DEPC. A total of 97% of the nuclei of cells surviving selection were in heterokaryons. (○) Cells plated at their final density immediately after being treated with biochemical inhibitors and fused. (□) Cells plated at 0.3×10^6 cells/cm² on bacterial-grade dishes following fusion, then pelleted and replated the next day. (●) Cells plated initially as above, but pelleted onto Ficoll–sodium diatrizoate (density 1.08, Lymphocyte Separation Medium, Litton Bionetics, Kensington, Maryland) after 24 hr. Only cells floating at the interface were replated. (▲) Cells plated initially as above, but left for 2 days before separating viable from nonviable cells. In contrast to Fig. 6, these data are plotted as the *number* of surviving heterokaryons/cm² rather than the *rate* of heterokaryon rescue. The insert shows the data for the cells separated on day 2 replotted on a log–log scale to show that plating at high cell densities did not significantly reduce the *rate* of heterokaryon survival under these conditions. [Adapted from Wright (1981b), Plenum Press, New York, with permission.]

immediately after heterokaryon formation (open circles) or after 1 day at low density on bacterial-grade dishes (open squares). This manipulation increases the number of surviving heterokaryons from about $8 \times 10^3/cm^2$ to almost $40 \times 10^3/cm^2$.

5.3. Ficoll Enrichment for Viable Cells

Whereas viable cells will band at the interface when centrifuged onto a Ficoll–sodium diatrizoate cushion, dead cells fail to exclude Ficoll and pellet (De Vries et al., 1973). The toxicity of dying cells can thus be minimized by removing many of them by centrifuging the recovered heterokaryons onto a Ficoll cushion. Figure 7 shows that an additional 2- to 4-fold gain in maximum plating density can be obtained by centrifuging the cells onto a Ficoll–sodium diatrizoate cushion either 1 day (closed circles) or 2 days (closed triangles) after heterokaryon formation. Performing this separation after 2 days is the most efficient in terms of the entire experimental protocol (see Section 6).

5.4. Cell Concentration versus Density

Figure 8 shows that it is the concentration of cells/ml rather than the density of cells/cm^2 that determines the toxicity of dying cells during the first day following heterokaryon formation. This suggests that diffusion is not limiting, and it is the volume of dilution of toxic products rather than the proximity of dying cells on the surface of the dish that regulates this effect.

5.5. Protective Effects of Serum

High serum concentrations provide some beneficial effects on cell rescue during the immediate period following heterokaryon formation. Figure 9

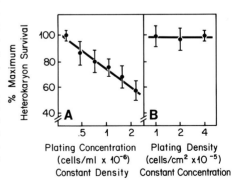

Figure 8. Initial cell concentration and initial cell density dependence of heterokaryon rescue. Chick myoblasts treated with 4.5 mM IAM were fused to L6 rat myoblasts treated with 0.0055% DEPC. A. Cells were initially plated at different concentrations in bacteriologic dishes at a constant density of 2×10^5 cells/cm^2. B. Cells were initially plated at different densities at a constant concentration of 8×10^5 cells/ml. After 24 hr, all aliquots were pelleted and replated at a constant density of 5×10^6 cells/cm^2, then fixed and counted the following day. Cell concentration is the critical factor during the initial recovery following biochemical selection. [Reproduced from Wright (1981b), Plenum Press, New York, with permission.]

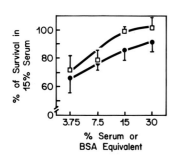

Figure 9. Serum dependence of heterokaryon rescue. L6 rat myoblasts treated with 6 mM IAM were fused to L6 rat myoblasts treated with 0.0055% DEPC. The heterokaryons were then plated at 1.5×10^6 cells/ml in bacterologic dishes containing different concentrations of fetal bovine serum (□). Alternatively, the cells were plated in medium supplemented with increasing concentrations of a serum substitute composed of 80 mg/ml bovine serum albumin (●). All of the BSA-supplemented aliquots also contained 3.75% fetal bovine serum in order to provide at least limited quantities of other serum factors. After 1 day, aliquots were transferred to tissue culture dishes, then fixed, stained, and counted the following day. High protein concentrations increase heterokaryon recovery following biochemical selection.

demonstrates that bovine serum albumin can partially substitute for serum. This implies that the effect may be a result of increased protein simply providing an alternate substrate for released lysozomal enzymes, thus protecting recovering heterokaryons from their toxic effects.

5.6. Advantages of Bacteriologic Dishes

Initially plating the heterokaryons on bacteriologic-grade dishes provides several important advantages. In addition to permitting a high final plating density as described above, it allows a much greater flexibility and precision in determining final plating conditions. In a typical experiment in which two doses of IAM and two doses of DEPC have been used, four fusion combinations are produced. If the fused cells are initially plated on bacterial dishes on day 0, aliquots can be transferred to tissue culture dishes on day 1, and the dead cells can be washed away and the attached cells fixed and stained on day 2. The purity and rescue for each fusion combination can then be determined. Since the remaining cells are still in suspension, it is then possible to mix any of the fusion combinations that are sufficiently pure into one uniform population rather than having to manipulate each one separately. Furthermore, since the number of rescued heterokaryons for that particular experiment can be calculated from the stained aliquots, the surface area needed to provide a desired heterokaryon cell density can also be calculated. The Ficoll enrichment for viable heterokaryons can then be performed if needed, or the cells can be simply pelleted and plated under appropriate conditions.

6. Summary of Selection Protocol

The following describes a typical experiment. First, 45×10^6 cells of each type to be fused are trypsinized, pelleted, and distributed at 300,000 cells/ml among three 50-ml conical centrifuge tubes containing 50 ml of cold Hanks BSS (thus 15×10^6 cells/tube). Sufficient amounts of inhibitors are added to make the final concentration 3.5, 4.6, or 5.9 mM IAM for cell type 1 and 0.0035%, 0.0046%, or 0.0059% DEPC for cell type 2. The tubes are mixed thoroughly each 10 min for 30 min, then pelleted at 2000 rpm at 4°C for 2 min. Working as rapidly as possible, we discard the supernatants and add 25 ml of cold Hanks BSS to each tube; the cells are then mixed and recentrifuged. The supernatant is again discarded and the cells resuspended in 6 ml of cold Hanks BSS. Two drops from each treatment are added to 2-cm^2 wells containing complete medium. These treatment controls will provide an indication of the efficiency of killing at that concentration of inhibitor (but without the superimposed toxicity of PEG treatment). The remaining cells from each treatment are divided among three small sterile test tubes, so that each of the nine fusion combinations in this 3×3 matrix receives appropriate IAM- and DEPC-treated cells. These tubes are then pelleted, the supernatant discarded, repelleted, aspirated to dryness, and kept on ice. Each tube containing 10^7 cells is then fused with 0.2 ml of 35% PEG 1000. While one tube is being fused, the next tube is transferred to room temperature to warm up. Each 45-sec fusion is stopped by the addition of 1 ml of medium, and the cells are then gently transferred to a 100-mm bacteriologic-grade petri dish containing 12 ml of medium containing 20% serum. The following day the cells are aspirated, and approximately 0.25 ml transferred to a 0.3-cm^3 well of a 96-well microtiter dish. The next day these aliquots are washed, fixed, and stained. The number of nuclei in mononucleates versus bi-or trinucleates is determined by counting a convenient number of fields, and the rescue calculated from appropriate conversion factors. Assume that the top two doses of each agent were adequate, so that the four fusion combinations when combined had a net purity of 95% and a rescue of 0.2%, and thus 80,000 rescued heterokaryon nuclei. If these cells were plated at 80,000 viable heterokaryon nuclei/cm^2, the plating density would be equivalent to 40×10^6 of the original cells/cm^2. This is sufficiently high (see Fig. 7) that a Ficoll enrichment is justified. The four fusions would be combined in one 50-ml tube, underlaid with 4 ml of Ficoll–sodium diatrizoate, and centrifuged at 600g for 5 min. The supernatant is carefully aspirated from the upper surface until 15 ml is left in the tube. The remaining medium will serve to dilute the Ficoll for the subsequent centrifugation. The cells floating at the interface and all of the medium, including the upper 2 ml of Ficoll, are then transferred to a 15-ml conical tube and centrifuged for 10 min at 400g. The supernatant is discarded, the pellet is washed with 5 ml of medium and repelleted, and the cells are then plated on approximately 1 cm^2 (for example, three microwells at 0.3 cm^2/well). The following morning most of the remaining dead cells are gently washed away and the recovered heterokaryons are fed fresh medium.

7. Isolation of Cell Hybrids

Although our primary interest has been to use this technique to investigate the interactions occurring in heterokaryons, we have also used it to make cell hybrids (Wright, 1978; Wright and Gros, 1981). However, it is important to recognize the limitations of biochemical selection in isolating hybrid clones. In a good experiment, 99% of the cells in a population might be heterokaryons, of which 70% might be binucleates. If the frequency of transition from heterokaryon to growing hybrid clone is 10^{-3}, then ten times as many contaminating parental clones may grow as hybrid clones. Consequently, its successful use in isolating cell hybrids is highly dependent upon as yet unknown factors determining the extent of heterokaryon cell division.

The genetic factors influencing heterokaryon cell division are apt to be most favorable in intraspecific cell crosses. Several other laboratories have successfully used biochemical selection to isolate intraspecific hybrids between human cells (Zakrzewski and Sperling, 1980), animal cells (Coleman et al., 1980), or plant protoplasts (Nehls, 1978). Two of these studies used treatment with irreversible inhibitors to select against only one parent. The second parent was untreated, and selected against by using genetic markers (Coleman et al., 1980) or by exploiting its contact-inhibited phenotype (Zakrzewski and Sperling, 1980).

We recommend the following approach to using biochemical selection for isolating cell hybrids. First ensure that the technique is being successfully employed, by fixing and staining aliquots to monitor the purity of the heterokaryons 2 days after heterokaryon formation. Sufficiently enriched populations can then be cloned directly or cultivated at high density before cloning. Diploid cells tend to clone poorly following traumatic manipulations. We suspect that heterokaryons between diploid cells might thus exhibit a very low cloning efficiency immediately following selection with irreversible biochemical inhibitors. Allowing them to recover at high cell densities for 1 week before cloning might thus dramatically improve hybrid clone formation.

If biochemical selection is used in a half-selective technique, one parental population survives at least temporarily. Thus the "purity" of the population can no longer be monitored by counting the proportion of nuclei in cells with more than one nucleus. Controls in which cells are mixed but not fused are therefore essential in order to determine if adequate doses of the irreversible inhibitor have been used. Since parental cell survival is determined by the combined effects of the inhibitor and PEG toxicity, the best mixture control is to fuse each cell type to itself and then mix them together. The lack of clones in this control and their presence when cells were mixed, and fused is then a fairly reliable indicator of hybrid formation. All putative hybrid clones must then be proven to be true hybrids by karyotypic or isozyme analysis.

The use of large intervals between inhibitor doses can be desirable in that it permits a wide range of concentrations to be employed. Although we most frequently use a 1.25-fold interval between doses for the isolation of heterokaryons, a 1.5-fold increase would probably be sufficient in a half-selective hybrid formation experiment. This is because the dosage dependence of cell rescue (Fig. 4) is apt to be relatively slight in a half-selective system in which a lethally treated cell is fused to a completely healthy untreated one. Large dosage intervals can thus probably be used without the risk of adjacent doses going from an inadequate treatment to one in which the cell has been so damaged that rescue is no longer possible.

8. Discussion

The use of irreversible biochemical inhibitors to select for highly enriched populations of heterokaryons provides several advantages over traditional hybrid selection systems. Some of these can be briefly summarized as follows:

1. Analyzing a population composed of thousands of independent fusion events avoids the bias introduced by studying a few clones selected for their ability to divide rapidly.

2. Cells at different developmental stages separated by only a few cell divisions can be fused and analyzed before their differences are erased by the multiple cell divisions that occur during clone formation.

3. Postmitotic differentiated cells can be used in somatic cell fusion studies since cell division is not a prerequisite for isolating fusion products.

4. Phenotypic changes requiring cell division can be determined by monitoring heterokaryon behavior over time as DNA synthesis is initiated.

5. Since heterokaryons are obtained as highly enriched populations, biochemical as well as cytochemical techniques can be used to analyze their behavior.

6. Many differentiated cell types gradually lose the capacity to express differentiated functions after prolonged time in culture (phenotypic drift). The use of heterokaryons avoids the possibility that such phenotypic drift has occurred during the 10–20 cell generations that occur before hybrid clones are analyzed.

7. Cells lacking selectable genetic markers can be used.

8. Normal diploid cells can thus be used. This avoids the ambiguity introduced by the tumorigenic properties and abnormal chromosomal constitution of established cell lines.

9. Ambiguities in interpreting the hybrid cell's phenotype produced by the loss of even a few chromosomes are avoided by studying heterokaryons before cell division has begun.

10. Finally and perhaps most importantly, since diploid cells can be used, the entire repertoire of human and animal mutations becomes amenable to analysis using cell fusion techniques.

Clearly, heterokaryons will not and should not replace cell hybrids as a subject for investigation. Rather, each should prove complementary in the kinds of information they can provide. Since irreversible biochemical inhibitors provide a method for isolating both heterokaryons and cell hybrids, they should be useful for the investigation of cellular behavior by somatic cell hybridization.

Note Added in Proof. The differential sensitivity of various species to oubain has long been exploited for isolating hybrid cells. We have recently found that ouabain can be successfully used in conjunction with irreversible agents to select for heterokaryons between chick and rat cells. Ouabain itself did not kill the chick cells rapidly enough to be useful. Plating the cells in 3×10^{-5} M ouabain after biochemical treatment, however, reduced the lethal iodoacetamide dose for the chick cells from 3.8 mM to 1 mM without affecting the lethal diethylpyrocarbonate dose for the rat cells. The use of the lower iodoacetamide dose resulted in a two- to fivefold increase in the rate of rescue of the chick–rat heterokaryons.

References

Coleman, J. R., Konieczny, S. F., Lawrence, J. B., Shaffer, M., and Coleman, A. W., 1980, Differentiation properties of homotypic hybrid cells derived from myogenic cell lines, *Eur. J. Cell Biol.* **22**:402.

Davidson, R. L., O'Malley, J. A., and Wheeler, T. B., 1976, Polyethylene glycol-induced mammalian cell hybridization: Effect of polyethelene glycol molecular weight and concentration, *Somat. Cell Genet.* **2**:271–280.

De Vries, J. E., van Benthem, M., and Rumke, P., 1973, Separation of viable from nonviable tumor cells by flotation on a Ficoll–Triosil mixture, *Transplantation* **15**:409–410.

Kao, F.-T., Johnson, R. T., and Puck, T. T. 1969, Complementation analysis on virus fused chinese hamster cells with nutritional markers, *Science* **164**:312–314.

Latt, S. A., 1973, Microfluorometric detection of deoxyribonucleic acid replication in human metaphase chromosomes, *Proc. Natl. Acad. Sci. USA* **70**:3395–3399.

Miller, R. A., and Ruddle, F. H., 1976, Pluripotent teratocarcinoma–thymus somatic cell hybrids, *Cell* **9**:45–55.

Nehls, R., 1978, The use of metabolic inhibitors for the selection of fusion products of higher plant protoplasts, *Mol. Gen. Genet.* **166**:117–118.

Norwood, T. H., Ziegler, C. J., and Martin, G. M., 1976, Dimethyl sulfoxide enhances polyethylene glycol-mediated somatic cell fusion, *Somat. Cell Genet.* **2**:263–270.

Ringertz, N., and Savage, R. E., 1976, *Cell Hybrids*, Academic Press, New York, pp. 147–158.

Wright, W. E., 1978, The isolation of heterokaryons and hybrids by a selective system using irreversible biochemical inhibitors, *Exp. Cell Res.* **112**:395–407.

Wright, W. E., 1981a, The synthesis of rat myosin light chains in heterokaryons formed between undifferentiated rat myoblasts and chick skeletal myocytes, *J. Cell Biol.* **91**:11–16.

Wright, W. E., 1982, The regulation of myosin light chain synthesis in heterokaryons between differentiated and undifferentiated myogenic cells, in: *Muscle and Cell Motility*, Volume II (R. M. Dowbin and J. W. Shay, eds.), pp. 177–184, Plenum Press, New York.

Wright, W. E., 1981b, The recovery of heterokaryons at high cell density following isolation using irreversible biochemical inhibitors, *Somat. Cell Genet* **7**:769–775.

Wright, W. E., and Gros, F., 1981, Coexpression of myogenic functions in L6 rat x T984 mouse myoblast hybrids, *Develop. Biol.* **86**:236–240.

Zakrzewski, S., and Sperling, K., 1980, Genetic heterogeneity of Fanconi's anemia demonstrated by somatic cell hybrids, *Hum. Genet.* **56**:81–84.

Chapter 6
Techniques for Enucleation of Mammalian Cells

GEORGE E. VEOMETT

1. General Introduction

The purpose of this chapter is to summarize succinctly a few techniques for enucleating vertebrate cells. It is hoped that the information given will enable investigators to choose and apply a technique for enucleation that is suitable to their cell lines and the problems they wish to investigate. Therefore, I have avoided an exhaustive, critical review of the literature, especially that literature concerning the properties of the subcellular fractions obtained and the results of investigations employing those subcellular fractions.

Carter (1967) initially described the effects of a number of fungal metabolites on mammalian cells in culture. A class of such metabolites inhibited cytokinesis and were named the cytochalasins. Cytochalasin B (CB) was one of the more prevalent and least expensive compounds. In addition, it "induced" the spontaneous enucleation of a low percentage of cells in some cultures. Techniques employing the combined treatment of cells with CB and exposure of the cells to a centrifugal field (Prescott et al., 1972; Wright and Hayflick, 1973) increased the percentage of cells enucleated and the number of types of cells which could be enucleated.

All of the techniques described in this chapter employ the treatment of cells with CB. This compound can be obtained from a number of different commercial sources (for example, Sigma Chemical Co., St. Louis, Missouri, and Aldrich Chemical Co., Milwaukee, Wisconsin). The lyophilized CB obtained from the commercial suppliers can be stored, desiccated, in the freezer ($-20°C$ or lower) for 6 months or longer without losing a detectable level of biologic activity. It is hydrophobic and not easily dissolved in aqueous media. However, the lyophilized powder is readily soluble either in dimethyl sulfoxide (DMSO) or absolute alcohol. Concentrated stock solutions (2–10 mg/ml) are prepared in either of these solvents and stored in the refrigerator or freezer. Aqueous solutions containing CB are prepared by dilution of the concentrated stock solutions.

GEORGE E. VEOMETT • School of Life Sciences, University of Nebraska—Lincoln, Lincoln, Nebraska 68588.

All of the techniques discussed here also involve subjecting the cells to a centrifugal force during their exposure to CB. The whole cells are restrained by some barrier [either the substratum for cell growth (monolayer techniques) or an isopycnic, liquid barrier (gradient techniques)] and the denser nuclei are forced away from the less dense cytoplasmic portion of the cells. The nuclei, encased in a small amount of cytoplasm, are initially attached to the bulk of the cytoplasm by "stalks" which sever under the stretching forces generated during centrifugation and spontaneously reseal. The nucleated fragments (karyoplasts) which are generated can be removed from either the bottom of the centrifuge tube or from an isopycnic layer of different density from that containing the cytoplasmic fragments (cytoplasts); the cytoplasts either remain attached to the solid substratum or in the isopycnic fluid layer.

The actual principles involved in both basic types of techniques of enucleation are therefore identical.

2. Monolayer Techniques for Cellular Enucleation

Cells growing as monolayer cultures can be enucleated while attached to the solid substrate. The substrate itself can be plastic, glass, nitrocellulose, and so on. Numerous techniques have been developed for such enucleations of monolayer cultures. The major requirement of these techniques is that the cytoplasm remain attached to the substratum during the enucleation process. Therefore, these techniques are best suited for cells that attach strongly to the substrate.

Cells that normally attach rather poorly can often be enucleated using these techniques if the substratum is modified to increase their adhesion. A few modifications of the substratum are discussed in Section 2.6.

For enucleation, the monolayer cell cultures are immersed in medium containing CB and subjected to centrifugation. The actual medium employed appears relatively unimportant and varies from normal growth medium (which we use in my laboratory) to balanced salt solutions containing CB. The omission of serum from the medium often increases the adhesion of cells and therefore may be beneficial for cell types that adhere poorly. A CB-containing medium which has been used for cellular enucleations can be collected, filter-sterilized, and reused. There is a slight decrement in the efficiency of enucleation and we routinely discard enucleation media that have been used five times.

2.1. Coverslip (cs) Technique

Plastic or heavy glass coverslips are a popular support for enucleation of cells in monolayer. Plastic cs are easily prepared, sterilized, and handled during the enucleation process. The number of cells processed is, of course, dependent on the size of the cs. The technique I describe here is essentially that developed by Prescott and Kirkpatrick (1973) and is one I use routinely.

One may modify this technique, for example, by altering the size of the cs. The size of the supporting dishes and centrifuge tubes, the volumes of CB media employed, and so on must then be adjusted accordingly.

2.1.1. Preparation of Plastic Coverslips

Plastic coverslips are prepared from the bottoms of plastic tissue cultureware. The inner surfaces of such cultureware have been treated to allow cellular attachment and the inner surfaces are used as the direct substrate for cell growth. The bottoms of the cultureware are marked to allow easy identification of inner and outer surfaces. Etching a nonsymmetric letter on the bottom of the cultureware with a diamond pencil or a glass marker is the most common practice.

The cs are punched from the bottom of the cultureware using a hot metal punch. The design of such a punch is shown in Fig. 1. (If the technique is not to be used repeatedly, a home-made punch can be devised from a metal cork-borer.)

The metal punch is heated in a flame and placed with pressure over a marked region of the cultureware. The heated metal melts the plastic and a disk is "punched out," and the disk is removed using forceps and/or a metal rod which can be inserted into the punch from the opposite end.

The cs are sterilized by exposure to UV light or immersion in 70% ethanol and subsequent flaming and/or drying under UV light. The cs is then used like any other cs for cell growth. The etched marking is placed down and cells allowed to attach to and grow on the "inner" surface of the disk.

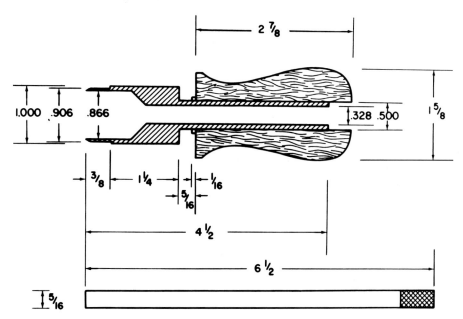

Figure 1. Design of metal punch for preparation of plastic coverslips. [From Prescott and Kirkpatrick (1973), with permission of authors and publisher.]

2.1.2. Enucleation

For enucleation, the cs is removed from the normal culture medium and inverted (i.e., cells facing bottom of tube) into a sterile centrifuge tube containing sterile CB medium. For 22–25-mm-diameter cs, the disks are placed in 40-ml, round-bottomed plastic centrifuge tubes for the Sorval SS 34 rotor containing 10 ml of prewarmed medium with CB at the appropriate concentration (see Section 2.4).

The centrifuge tubes are placed in a prewarmed rotor and the cells subjected to centrifugation. The centrifuge is prewarmed by spinning the SS 34 rotor at 15,000 rpm for 30 min with the appropriate temperature settings (see Section 2.4). The precise conditions for enculeation must be determined for each cell line. For L929 cells, we use 10 μg/ml CB, and 20 min centrifugation at 12,000 rpm (17,600g) at 35–37°C.

2.1.3. Recovery of Cytoplasts and Karyoplasts

After centrifugation, the cs are removed from the centrifuge tubes using sterile forceps and are inverted (i.e., cell side facing up) in fresh growth medium. The adherent cytoplasts are initially greatly distorted. Within 10–30 min, the cytoplasts resume a more normal cellular morphology. This recovery appears to be accelerated if the medium is changed after 5–10 min.

The percentage of enucleation can be determined by examining a cs by microscopy. Cellular material is fixed by addition of a suitable fixative such as 3:1 (three parts absolute alcohol to one part glacial acetic acid) or simply addition of absolute alcohol. If using 3:1 as a fixative, add it slowly to the medium in the dish, since the addition of the fixative directly to the cells is often too harsh, causing the cellular material to lyse and detach. After the addition of fixative, the nucleated and enucleated cells can be discerned readily under phase contrast microscopy. The fixed cs can also be dried, stained with Giemsa or 0.1% crystal violet in water, and examined by conventional light microscopy. We generally count 400 cells and cytoplasts to determine the percentage of enucleation.

The karyoplasts can be recovered from the CB medium and the "pellet" of cellular material at the bottom of the centrifuge tube. The pellet may be supended in the CB medium, the CB medium transferred to the conical centrifuge tube, and the cellular material collected by a second centrifugation. The initial karyoplast preparation is a mixture of karyoplasts, detached whole cells, and cytoplasmic fragments. The latter are recognizable by their smaller, more heterogeneous size. This initial karyoplast preparation can be partially purified (see Section 2.7). The purity of the karyoplast preparation is determined by plating a known number of cell fragments in normal growth medium and counting the number of whole cells 24 hr later.

2.2. Flask Techniques

A second monolayer technique which is quite useful is to use 25-cm^2 tissue culture flasks as the centrifuge tubes. No special treatment of the flasks is required, although the growth surfaces can also be modified (see Section 2.6). The tissue culture flasks are used in the ordinary manner for the growth of cells to be enucleated. The number of cells seeded, the duration of growth prior to enucleation, and the confluency of the culture are variable. Overly dense cultures are not recommended for enucleation because (a) whole cells may detach readily from the substrate and (b) the monolayer may "peel" from the substrate during the enucleation process.

The precise type of tissue culture flask seems to be unimportant. There are two general types of flask available: straight-necked flasks (e.g., Falcon 3012) and bent-neck designs. The straight-necked designs work well for both types of procedures (Sections 2.2.1 and 2.2.2), whereas the bent-neck flasks work best for procedures using acrylic inserts (Section 2.2.2).

For enucleation of cells growing normally in tissue culture flasks, the medium is removed, prewarmed CB-containing medium is added to completely fill the flask, and the flasks are sealed, either with the acrylic caps supplied by the manufacturer or with sterile rubber stoppers (size 000, for straight-neck flasks). The flasks are used as the centrifuge tube itself.

2.2.1. Simple Rotor Technique

The completely filled flasks can be used as centrifuge tubes in the GSA rotor for the Sorval RC-2B or RC-5 type centrifuges. The rotor is prewarmed by prespinning the empty rotor approximately 20 min at 10,000 rpm with the appropriate temperature settings. After prewarming of the centrifuge rotor, approximately 125–130 ml of water (water tempered to desired temperature) is added to each rotor well. One filled tissue culture flask is added to each well, oriented with the growth surface of the flask toward the center of rotation, and the centrifuge is started with the top of the rotor off. The rotor is spun at about 500–1000 rpm and prewarmed water is added to completely "fill" the rotor. The rotor is then stopped, the top replaced, the water in the rotor chamber removed with paper towels or a sponge, and the flasks centrifuged at 7500–9000 rpm for approximately 20 min.

The water added to rotor wells acts as a cushion for the culture flasks and inhibits the breakage of the flasks. The most sensitive area for breakage appears to be the neck of the flask. The use of rubber stoppers in lieu of the normal screw caps appears to cushion the flasks further and inhibit breakage. The rubber stoppers apparently can bend and reorient themselves while maintaining a tight seal and appear to relieve some of the pressures exerted on the neck of the flask.

2.2.2. Use of Acrylic Inserts

The tissue culture flasks can also be centrifuged in special acrylic inserts for the GSA rotor. The specifications and general design of such inserts is given in Fig. 2. Tempered water is added to the inserts and a filled, sealed culture flask added. The inserts themselves are sealed with the screw cap and placed into the wells of the rotor with the growth surface of the flask oriented

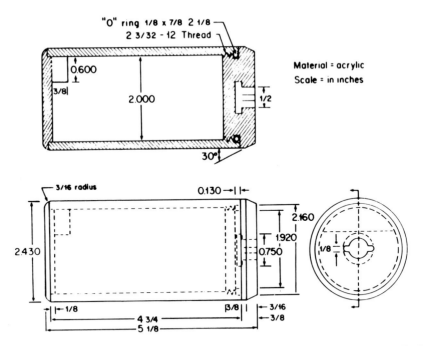

Figure 2. Design of acrylic inserts for use in enucleation techniques using tissue culture flasks. [From Shay and Clark (1979), with permission of authors and publisher.]

toward the center of rotation. As few as two samples may be run—all samples being matched by weight.

The inserts offer a number of advantages: (a) virtually any style 25-cm² culture flask works well, (b) as few as two samples can be run in the rotor, (c) the sealed units eliminate much of the water in the centrifuge, (d) breakage of flasks is generally quite low. The major disadvantage of the technique is that the construction of the inserts is costly and requires time.

2.2.3. Recovery of Cytoplasts and Karyoplasts

After centrifugation, the flasks are removed and the exteriors dried. The karyoplasts (nucleated fragments) can often be seen as a small pellet in the corner of the flask. Karyoplasts are also adherent to the inner surface of the front (top) side of the flask. The flasks are unsealed and the CB-containing medium collected (by simple pouring or by suction). The pellet of karyoplasts is allowed to sit in a small amount of medium and is then gently dispersed by pipetting. The top side of the flask is gently washed and the karyoplasts removed by pipetting. This karyoplast preparation is handled as described in Section 2.1.3.

The flasks contain the cytoplasts, which are treated in the same way as the cytoplasts obtained from the cs technique (Section 2.1.3).

2.3. Different Support Systems for Monolayer Cultures

The techniques described above illustrate the general principles involved in the enucleation of monolayer cultures. Numerous other types of cellular supports have been described and successfully used. Some of these support systems will be briefly described (a) to illustrate the range of possibilities and (b) to suggest alternative techniques if those described above are not satisfactory for one's purposes. The appropriate references, cited in each description, provide more details for each individual system.

1. Follett (1974) has described a technique in which 30-mm-diameter tissue culture dishes are used as the substrate for cell growth. These dishes can be inverted and fit into tubes for a rotor for the MSE ultracentrifuge. This system resembles the cs technique, except that the "cs" is the bottom of the tissue culture dish.

2. Rectangular glass slides (Poste, 1973), rectangular Lucite inserts to which glass cs have been attached with epoxy cement (Wright, 1973), and rectangular plastic sheets cut from the bottoms of tissue culture flasks (Lucas et al., 1976) can be used as substrates for cell growth. The Lucite inserts and plastic sheets can be prepared with one end molded to fit the rounded bottoms of centrifuge tubes. Cells are grown on these plates and sheets, using them as "cs." For enucleation, the plates are inserted on end into centrifuge tubes, which are then filled with CB medium; centrifugation is normally performed in swinging-bucket rotors. With glass slides, additional inserts for supporting the slide in the tube can be made (Poste, 1973).

3. Centrifuge tubes (Croce and Koprowski, 1973) or glass cylinders which fit snugly into centrifuge tubes (Wright, 1973) can be used for cell support. Nitrocellulose tubes for ultracentrifugation are treated briefly with concentrated sulfuric acid; this treatment alters the plastic and allows cells to attach and grow on the sides of the centrifuge tube. Cells are grown in the centrifuge tube or in the glass cylinders, using these supports as roller bottles. The centrifugation step is performed in swinging-bucket rotors with the tubes completely filled with CB medium.

2.4. Factors Affecting Enucleation

Several factors affect the enucleation of cells in monolayer. The required concentration of CB in the enucleation medium should be determined for each type of cell to be enucleated. There appears to be a critical minimal concentration required to obtain enucleation. This concentration tends to be in the range of 4–10 μg/ml CB for a wide variety of cell types. Below this critical concentration, little enucleation occurs.

A second factor is the centrifugal force that the cells experience. The enucleation is indeed "induced" and the level of enucleation is dependent on that force. In general, g forces in the range of 10,000–30,000g are required. The precise force requirement is dependent on the rotor, the geometry of the cell support (cells closer to the center of rotation experience lower g forces), and the type of cell being enucleated.

The length of time the cells experience the centrifugal force is also important. Generally times ranging from about 15 min to 1 hr are used. Too short a time and there is not much enucleation, too long a time and there may be much cellular detachment from the substrate (see Section 2.5).

The induced enucleation of vertebrate cells is also dependent on the temperature. Treatment of cells with CB and centrifugation at temperatures below 20°C do not yield large numbers of enucleates. Therefore, temperature is an additional variable and I suggest that initial enucleation conditions be performed at temperatures of 30–37°C.

The factors which I have noted here also interact. Thus, one can vary the degree of enucleation by (a) altering the CB concentration, (b) changing the effective g force by varying the rotor used and/or the speed of the centrifugation, and (c) altering the temperature. These factors should be considered in attempting to overcome any problems encountered in the enucleation of cells.

2.5. Problems Encountered in Enucleation of Cells

There are basically four types of problems encountered in enucleating cells in monolayer. The first is a physical breakdown of the substratum to which the cells are attached. For example, with the simple flask technique, the neck of the culture flasks appears to be the weakest portion of the flask and may crack or actually snap off during centrifugation. The use of neoprene stoppers in lieu of the plastic screw caps often corrects this problem. The use of the acrylic inserts perhaps is the most effective corrective procedure. With plastic cs and glass apparatus, the substrate may physically break. The use of thicker materials or special holders may correct these breakage problems.

Two problems often go hand in hand—viz. the failure of the cells to enucleate and the loss of cells from the solid substrate. This combined problem is often perplexing; if one raises the CB concentration or increases the g force, cells are lost from the substratum. If they are lowered to prevent loss of cells, the attached cells fail to enucleate. There appears to be a competition between enucleation and cell loss.

I can suggest three possible alterations of the enucleation conditions to correct these latter, combined problems. First, raise the concentration of CB *and* lower the temperature at which the centrifugation is performed. Under these conditions a satisfactory level of enucleation and cell loss may be obtained. The second possible solution is to alter the substrate (see Section 2.6) and a third is to use a suspension technique for enucleation (see Section 3).

A fourth problem is the contamination of cellular fragments with the unwanted cellular components, i.e., karyoplasts with cytoplasmic fragments and cytoplasts with nucleated components. Enrichment procedures for the various components have been devised (see Section 2.7).

2.6. Modification of the Growth Substrate

Many types of cells which attach to a growth substratum do not attach firmly enough for enucleation purposes. The growth substratum can be modified to increase the adhesion of cells. In many cases these modifications are sufficient to permit enucleation of monolayer cultures.

1. Collagen coating of plastic surfaces. The growth surfaces are covered for 5 min with a solution of collagen solubilized in dilute acetic acid (Bornstein, 1958). The surfaces are drained, dried, washed in balanced salt solution, and sterilized by exposure to UV light from a germicidal lamp. The collagen coating is not readily removed and assists in the adherence of cells.

2. Protein and/or lectin coating. Solutions of basic proteins (McKeehan and Ham, 1976) such as poly L-lysine or protamine sulfate or lectins such as concanavalin A (Gopalakrishnan and Thompson, 1975) are prepared at concentrations of 2–20 mg/ml in 0.15 M NaCl and filter-sterilized. The growth surfaces are covered with the sterile solutions for 10–30 min, the protein solutions removed, and growth surfaces washed three times with sterile balanced salt solutions and used for the growth of cells. Although the proteins and/or lectins are not covalently attached to the plastic, they do assist in cell adhesion and are not lost from the growth surfaces under normal conditions.

3. Covalent attachment of proteins and/or lectins. More protein and/or lectin can be attached to the plastic using a technique originally described by Edelman et al. (1971). Protein–lectin solutions are prepared as described above; after the protein–lectin solution has been added to the substrate, an equal volume of a water-soluble carbodiimide [1-cyclohexyl-3-(2-morpholinoethyl) metho-p-toluene sulfonate; (Aldrich Chem. Co.] in 0.15 M NaCl is also added and the mixture allowed to react at room temperature for 30 min. The concentration of the carbodiimide should be five times that of the protein. The growth surfaces are then processed and used as described above.

2.7. Techniques for Purifying the Cellular Components

Cytoplasts are generally contaminated with whole cells. For many purposes the degree of whole cell contamination may be sufficiently small to be overlooked. Cytoplasts, however, can also be partly purified. Cytoplasts are more resistant to the action of tryspin and a brief trypsinization may remove more of the whole cell contaminants. The contaminating whole cells can also be inactivated using chemical methods, such as treatment with mitomycin C, a DNA cross-linking agent (Wright and Hayflick, 1975), or physically separated on Ficoll gradients (Poste, 1972).

The karyoplast preparations are more heterogeneous; they contain karyoplasts, whole cells that have detached from the substrate, and cytoplasmic fragments. It appears that the higher the centrifugal force used to enucleate the cells, the more contamination. Whole cells can be removed by "plating" the karyoplast preparation. Whole cells attach to the substrate more rapidly. After a 90-min incubation at the normal growth temperature, the karyoplasts and small cytoplasmic fragments can be removed in the supernatant fluids. The karyoplasts can be separated from the cytoplasmic fragments by layering the preparation over a 1–6% Ficoll gradient (Zorn et al., 1979) or a 1–3% serum gradient and allowing the components to settle out at unit gravity. After 90 min, bands are apparent; the lower band is the enriched karyoplast preparation. It has also been noted that viable karyoplasts can be separated from the nonviable ones using Ficoll-isopaque (Zorn et al., 1979). Additional techniques for isolating karyoplasts with minimal cytoplasm and which do not regenerate into whole cells are described in Chapters 20 and 21.

3. Gradient Techniques

Vertebrate cells may also be enucleated in density gradients. In this case the cytoplasts are restrained from sedimenting by an isopycnic liquid layer; the karyoplasts are more dense and continue to sediment until they reach their isopycnic point. The continued sedimentation of the karyoplasts generates the stretching forces active on the cell, which eventually result in the separation of the karyoplasts from the cytoplasts. The physical distance separating the two liquid density layers is important in the enucleation process.

Two gradient techniques have been employed. One method utilizes colloidal silica as the density agent and the "gradient" is generated during the centrifugation process (Bossart et al., 1975). This method has not, apparently, been very popular. The interested reader should consult the reference cited.

The second method, originally used by Wigler and Weinstein (1975), has been used successfully by a number of investigators and is the method described here. In this method, the density agent is Ficoll.

3.1. Preparation of Gradients

Ficoll, a high-molecular-weight synthetic polysaccharide, is obtained from Pharmacia Fine Chemicals (Piscataway, New Jersey), Sigma Chemical Co. (St. Louis, Missouri), or other commercial sources. Concentrated 50% (w/w) stock solutions are prepared by dissolving Ficoll in an equal weight distilled water. The Ficoll often must be stirred overnight in the cold to effect solubilization. The Ficoll may also be dissolved at a higher concentration and dialyzed overnight to remove low-molecular-weight materials and adjusted to a 50% (w/w) concentration after dialysis.

Discontinuous gradients are prepared with the following Ficoll concentrations: 25, 17, 16, 15, and 12.5%. All solutions are prepared in MEM (minimal essential medium) modified for suspension culture (e.g., medium 410-1400 from Gibco) and all solutions contain 10 μg/ml CB. The working solutions can be prepared easily from the stock 50% w/w Ficoll and four-times concentrated spinner MEM as shown in Table I.

Table. I. Preparation of Stock Solutions for Gradient Enucleation

Ficoll solution, %	Stock 50% Ficoll, ml	4 × MEM, ml	H$_2$O, ml	CB (2 µg/ml stock), ml
25	25	12.5	12.25	0.25
17	17	12.5	20.25	0.25
16	16	12.5	21.25	0.25
15	15	12.5	22.25	0.25
12.5	12.5	12.5	24.75	0.25

The step gradient is prepared in sterile cellulose nitrate centrifuge tubes. The following volumes are suggested for tubes to fit the SW 41 rotor for the Beckman ultracentrifuge:

	Volume, ml
25%	2
17%	2
16%	0.5
15%	0.5
12.5%	2

(For other types of tubes, adjust the volumes accordingly.) The gradients are equilibrated at 37°C for 6–8 hr after preparation. I suggest gradients be prepared the afternoon of the day before use.

3.2. Enucleation

Approximately $(1.5–6) \times 10^7$ cells in 3 ml of 10–12.5% Ficoll are layered on the preformed step gradient; 3 ml of MEM for suspension culture is layered on top of the cells. The cells must be free of clumping, and vigorous pipetting of the cell suspension before addition to the gradient is recommended.

Cells in monolayer are removed from the substrate with 0.5 mM EDTA or trypsin, suspended by pipetting, collected by centrifugation, and washed once with MEM for suspension culture. The cells are then suspended at an appropriate density, pipetted to eliminate cellular clumping, and layered onto the gradient. Cells from suspension culture are also collected by centrifugation, and treated similarly.

Centrifugation is performed at 25,000 rpm in the SW 41 rotor for 60 min at 31°C in a Beckman ultracentrifuge. The centrifuge and rotor are prewarmed by prerunning at 25,000 rpm for 4 hr prior to use.

3.3. Recovery of Cytoplasts and Karyoplasts

After centrifugation, distinct layers of cellular material are seen in the centrifuge tubes. The cytoplasts tend to be located in the interface of 15–17%

Ficoll, the karyoplasts at the interface of 17–25%. The precise location of the components, however, must be determined for each cell type and appears to vary, depending on cell types (Bunn and Eisenstadt, 1979; McBurney and Strutt, 1979; Wigler and Weinstein, 1975). The bands of cellular material are removed with a Pasteur pipette or a syringe and cannula, diluted 1 to 20 in MEM for suspension culture, and collected by centrifugation. They can then be handled similarly to the cytoplast and karyoplast preparations described for monolayer techniques.

Generally, 30–70% of the starting number of cells can be removed as cytoplasts. Cytoplasts recover from CB well when the starting number of cells is that suggested above; with lower numbers of cells cytoplasts recover poorly, and with higher numbers, cell clumping inhibits enucleation.

Among the advantages of this system are: (a) a wide variety of cell types can be enucleated and (b) a large number of cytoplasts are generated. Among the disadvantages is the more complicated nature of the technique.

References

Bornstein, M. B., 1958, Reconstituted rat-tail collagen used as a substrate for tissue cultures on coverslips in Maximow slides and roller tubes, *Lab. Invest.* **7**:134–137.

Bossart, W., Loeffler, H., and Bienz, K., 1975, Encleation of cells by density gradient centrifugation, *Exp. Cell Res.* **96**:360–366.

Bunn, C. L., and Eisenstadt, J. M., 1977, Hybrid formation in mouse L cells: The influence of cytoplast-to-cell ratio, *Somat. Cell Genet.* **3**:335–341.

Carter, S. B., 1967, Effects of cytochalasin on mammalian cells, *Nature* **213**:261–264.

Croce, C. M. and Koprowski, H., 1973, Enucleation of cells made simple and rescue of SV 40 by enucleated cells made even simpler, *Virology* **51**:227–229.

Edelman, G. M., Rutishauser, U., and Millette, C. F., 1971, Cell fractionation and arrangement on fibers, beads and surfaces, *Proc. Natl. Acad. Sci. USA* **68**:2153.

Follett, E. A. C., 1974, A convenient method for enucleating cells in quantity, *Exp. Cell Res.* **84**:72–78.

Gopalarkrishnan, T. V., and Thompson, E. B., 1975. A method for enucleating cultured mammalian cells, *Exp. Cell Res.* **96**:435–439.

Lucas, J. J., Szekely, E., and Kates, J. R., 1976, The regeneration and division of mouse L-cell karyoplasts, *Cell* **7**:115–122.

McBurney, M. W., and Strutt, B., 1979, Fusion of embryonal carcinoma cells to fibroblast cells, cytoplasts and karyoplasts. Developmental properties of viable fusion products, *Exp. Cell Res.* **124**:171–180.

McKeehan, W. L., and Ham, R. G., 1976, Stimulation of clonal growth of normal fibroblasts with substrate coated with basic polymers, *J. Cell Biol.* **71**:727–734.

Poste, G., 1972, Enucleation of mammalian cells by cytochalasin B. I. Characterization of anucleate cells, *Exp. Cell Res.* **73**:273–286.

Poste, G., 1973, Anucleate mammalian cells: Applications in cell biology and virology, in *Methods in Cell Biology*, Volume VII (D. M. Prescott, ed.), Academic Press, New York, pp. 211–249.

Prescott, D. M., and Kirkpatrick, J. B., 1973, Mass enucleation of cultured animal cells, in: *Methods in Cell Biology*, Volume VII (D. M. Prescott, ed.), Academic Press, New York, pp. 189–202.

Prescott, D. M., Myerson, D., and Wallace, J., 1972, Enucleation of mammalian cells with cytocholasin B, *Exp. Cell Res.* **71**:480–485.

Shay, J. W., and Clark, M. A., 1979, Nuclear control of tumorgenicity in cells reconstructed by PEG-induced fusion of cell fragments, *J. Suparmol. Struct.* **11**:33–49.

Veomett, G., Shay, J. Hough, P. V. C., and Prescott, D. M., 1976, Large scale enucleation of mammalian cells, in: *Methods in Cell Biology*, Volume XIII (D. M. Prescott, ed.), Academic Press, New York, pp. 1–6.

Wigler, M. H., and Weinstein, I. B., 1975, A preparative method for obtaining enucleated mammalian cells, *Biochem. Biophys. Res. Commun.* **63**:669–674.

Wright, W. E., 1973, Production of mass populations of anucleated cytoplasms, in: *Methods in Cell Biology*, Volume VII (D. M. Prescott, ed.), Academic Press, New York, pp. 203–210.

Wright, W. E., and Hayflick, L., 1973, Enucleation of cultured human cells, *Proc. Soc. Exp. Biol. Med.* **144**:487–592.

Wright, W. E., and Hayflick, L., 1975, Nuclear control of cellular aging demonstrated by hybridization of anucleate and whole cultured normal human fibroblasts, *Exp. Cell Res.* **96**:113–121.

Zorn, G. A., Lucas, J. J., and Kates, J. C., 1979, Purification and characterization of regenerating mouse L929 karyoplasts, *Cell* **18**:659–672.

Chapter 7
Nonselective Isolation of Fibroblast Heterokaryons, Hybrids, and Cybrids by Flow Sorting

JOHAN F. JONGKIND and ANTON VERKERK

1. Introduction

The use of new cell combinations to study genetic interaction between somatic cells (e.g., fibroblast heterokaryons, hybrids, and cybrids) is hampered by the fact that no selection procedures are available for the short-term isolation of the specific fusion products. Moreover, the techniques for the isolation of cytoplasts (anucleate cells) and karyoplasts (minicells) [for a review see Ringertz and Savage (1976)] yield fractions which are contaminated considerably with nonenucleated cells. The use of these impure fractions as a fusing partner may give rise to difficulties in the interpretation of the results.

The development of flow sorters which are able to isolate cells on the basis of two different fluorescent parameters (Hulett et al., 1973) and the recent two-laser excitation sorters (Dean and Pinkel, 1978) have made it possible to circumvent the above problems. Fluorescent cell markers in combination with a fluorescence-activated cell sorter (FACS) have been used to isolate cytoplasts greater than 99% pure (Schaap et al., 1979, 1981). Moreover when two parent lines are labeled respectively with red and green fluorescent cell markers, the red–green fusion product can be isolated by flow sorting (Jongkind et al., 1979, 1980, 1981).

This chapter gives a detailed description of the techniques used in our laboratory to isolate specific fusion products between cells (heterokaryons, hybrids) and between cytoplasts and cells (cybrids). A generalized scheme for the variety of fusion products which can be isolated using these techniques is presented in Fig. 1.

Since the amount of fusion products which can be obtained with a cell sorter is dependent on the fusion index and the sorting velocity, the use of

---

JOHAN F. JONGKIND and ANTON VERKERK • Department of Cell Biology and Genetics, Medical Faculty, Erasmus University, Rotterdam, The Netherlands.

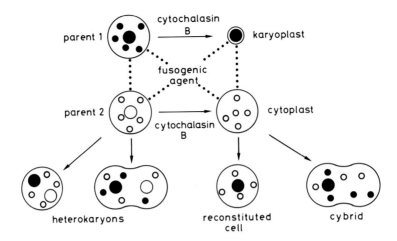

Figure 1. Generalized scheme for the production and flow sorting of somatic cell fusion products. The nuclei and cytoplasms are labeled with differential fluorescent probes. Fusogenic agents include inactivated Sendai virus and polyethylene glycol (PEG). [Slightly modified from Jovin and Arndt-Jovin (1980)].

biochemical microtechniques may be necessary. Therefore a section is added on the enzyme and protein assay of a small number of sorted cells.

2. Vital Fluorescent Cell Labeling

2.1. Fluorescent Polystyrene Beads

Nonfluorescent polystyrene beads of different sizes which are phagocytized by cells have been used to mark different cell types in order to identify the specific fusion products (Ege and Ringertz, 1975; Veomett et al., 1974).

For fluorescent cell labeling, fluorescent polystyrene beads of different colors can likewise be used to mark the cell populations. The type of fluorescent label and the intensity per bead are dictated by the characteristics of the flow sorter, its spectral laser lines, and its filters (see Section 3.1).

For our experiments we use commercially available fluorescent beads (Polysciences; red: 1.6 µm, no. 7769; green: 1.83 µm, no. 9847). The excitation and emission spectra of suspensions of these beads are presented in Fig. 2. Other fluorescent beads can be obtained from Duke Scientific Corp. (Palo Alto, California) and Serva (Heidelberg, Germany).

The assortment of fluorescent polystyrene beads which can be used as cell labels in flow cytometry is rather limited to the fluorescent dyes which are in use. Polystyrene microbeads with functional groups make it possible to attach different kinds of fluorescent molecules to the beads. For example, dansylchloride and primary amino-containing microspheres are conjugated by adding 200 µl of a 2.5% aminobead suspension (Polysciences; 1.42 µm, no. 7764) to 5 ml of a stain solution containing 1 mg/ml dansyl (1-dimethyl-

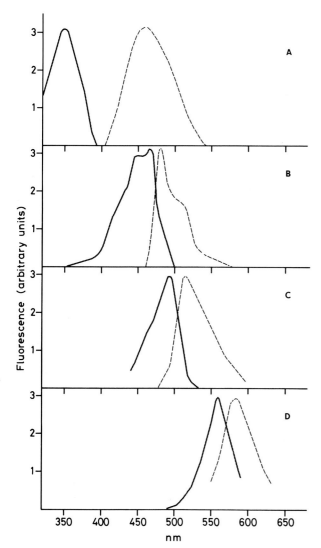

Figure 2. Uncorrected excitation (—) and emission (- - -) spectra of cells labeled with different fluorescent vital labels. The spectra were obtained using the method of Ohkuma and Poole (1978). A. Cells with Hoechst 33342. B. Cells with green fluorescent beads (1.83 μm, no. 9847). C. Cells with green fluorescent beads (0.52 μm, no. 7763 FITC–NH$_2$ bead). D. Cells with red fluorescent beads (1.6 μm, no. 7769).

aminonaphtalene-5 sulfonyl) chloride in 95% ethanol saturated with sodium bicarbonate (Rossilet and Ruch, 1968). The suspension is agitated overnight, and the beads are washed three times with water. After washing, the fluorescent beads are suspended in water.

For the preparation of weak green fluorescent microbeads(Section 3.1), small amino beads (200 μl of 2.5% solution; Polysciences, 0.52 μm, no. 7763) are suspended in 5 ml of 0.5 M bicarbonate buffer pH 9.5, containing 0.01% Tween 80 and 1.6 mg fluoresceine-isothiocyanate (FITC). Staining is carried out at room temperature for 24 hr in a rotating tube. To obtain a sterile bead suspension, the staining solution is sterilized by filtration (0.22 μm Millex, Millipore). The stock solution of amino beads is autoclaved prior to conjugation. Free FITC is removed by dialyzing the beads against water for 5 days at 4°C. The excitation and emission spectra of the fluorescent FITC–NH$_2$ beads are presented in Fig. 2.

Fluorescent labeling of human fibroblasts was carried out by culturing a proliferating cell culture for 2 days in Ham's F10 medium supplemented with 10% fetal calf serum (FCS) and antibiotics (100 μg of streptomycin and 100 U of penicillin per ml) containing ~2 × 10^7 beads/ml. In this incubation period the beads are taken up by phagocytosis. Cells were harvested with trypsin and washed with saline to remove the beads that were attached to the outer cell membrane. This procedure resulted in a suspension of cells with intracellular microbeads.

The influence of latex bead labeling on the proliferation rate of fibroblasts was assayed by protein determination on nonconfluent cultures. The results indicate that the labeling procedure with fluorescent beads impaired cell growth as compared with nonlabeled cells (Table I). Additional flow sorting had no effect (Jongkind et al., 1981).

2.2. Fluorescent Stearylamine

Fluorescent labeling of cells can also be accomplished by incubating the cells with fluorescent lipid molecules (Keller et al., 1977). The fluorescent lipids can be synthesized by conjugating stearylamine with FITC or tetramethylrhodamine-isothiocyanate (TRITC). Purification of the fluorescent conjugate from unbound dye can be done by thin-layer chromatagraphy. The procedure we used for the manufacture and purification of the fluorescent probes was adopted from Keller et al. (1977) (see also Chapter 8 in this book). The purified fluorescent stearylamine was dissolved in dimethyl sulfoxide (DMSO) and stored at 4°C.

Since not all cultured cells are able to phagocytose fluorescent microbeads, we tried fluorescent stearylamine as an alternative fluorescent probe for flow sorting (Jongkind et al., 1981). Fluorescent labeling of human fibroblasts was carried out by culturing the cells overnight in F10 medium to which was added 0.25 vol % of diluted fluorescent probe. The dilutions from the DMSO stock solutions which were used were made in water (FITC stearylamine $A_{470} = 0.05$; TRITC stearylamine $A_{550} = 0.17$).

Stearylamine labeling had a similar inhibitory effect on growth rate as polystyrene bead labeling (Table I).

2.3. Hoechst Bis-Benzimidazole Dyes

A number of bis-benzimidazole dyes have been developed by the Hoechst firm as chemotherapeutic agents (Loewe and Urbanietz, 1974). These dyes complex nonintercalatingly with DNA, shifting excitation and emission peaks and increasing quantum efficiency (Latt and Wohlleb, 1975; Latt and Stetten, 1976). In vital staining of cells there are differences between the

Table I. Growth of Fibroblasts after Labeling and Flow Sorting[a,b]

	Unlabeled	Latex		Stearylamine		Hoechst 33342[c]	
		Green	Red	Green	Red	Medium+	Medium−
Not sorted	1.00[d]	0.54	0.51	0.44	0.68	ND	ND
Sorted	0.80	0.54	0.57	0.51	0.54	1.00	0.12

[a] Modified from Jongkind et al. (1981) and Schaap et al. (1981).
[b] The growth was determined by protein determination after a 7-day culture period starting from a nonconfluent culture (50,000 cells/35 mm dish).
[c] In some of the dishes the medium was changed daily (medium +), while others were not changed (medium −). ND, not determined.
[d] Protein content of a 7-day treated culture as compared with a 7-day untreated unsorted control culture.

uptake of the different dyes by the cells. When Hoechst 33342 is used as a vital stain, it binds stoichiometrically with DNA (Arndt-Jovin and Jovin, 1977). Therefore the DNA content of individual cells can be measured after a staining period of 30–60 min by fluorescence measurements, and the cells in the G_1, S, and G_2 + M phases of the cell cycle can be isolated by flow sorting using the UV lines of an argon-ion laser. The viability of the stained and sorted cells is retained (Arndt-Jovin and Jovin, 1977). While Hoechst 33258 is unsuitable as a quantitative vital DNA stain, due to the slow uptake and absence of stoichiometry, both the ethoxy (H 33342) and methoxy (H 33662) derivatives can be used for quantitative DNA measurements in different types of cells (J. Vijg, unpublished observations). In our experiments we used Hoechst 33342 for the purification of DNA-free cytoplasts (see Section 5.2) and for the isolation of tetraploid hybrids (see Section 4.2).

Hoechst 33342 was prepared as a 1 mM stock solution in water (2 mg/ml H_2O). Cells in suspension were incubated for 1 hr in a medium containing 1 vol % of the stock solution (10 μM final concentration) at 37°C. After incubation, the cells were stored in ice before cell sorting. After sorting, fresh medium was added each day to prevent growth inhibition (see Table I). The spectra of the stained cells are presented in Fig. 2.

3. Isolation of Fibroblast Heterokaryons

3.1. Outline

Heterokaryons in a fusion mixture can be detected using nonfluorescent polystyrene beads of different sizes which are phagocytized by cells as a cell marker (Ege and Ringertz, 1975; Veomett et al., 1974). For the isolation of heterokaryons by fluorescence-activated cell sorting, differently stained fluorescent polystyrene beads can be used to mark the parent cell populations and to sort the heterofluorescent fusion product.

The kind of fluorescent label and the intensity per bead are dictated by the characteristics of the sorting machine, its spectral laser lines, and its fluorescence filters. When there is an overlap of the emission spectra, as with, e.g., FITC and TRITC, and when 488 nm is used as the excitation wavelength, the red fluorescent beads are not excited at their optimum (see Fig. 2a). Therefore a strong red fluorescent marker must be used in combination with a weak green (fluoresceine) label in order to get an optimum two-color separation of heterofluorescent (red–green) cells (Jongkind et al., 1981). In our experiments we made use of commercially available fluorescent beads (Polysciences; red: 1.6 μm, no. 7769; green: 0.52 μm) and NH_2 beads (Polysciences 7763) conjugated with FITC.

3.2. Cell Fusion and Heterokaryon Collection

Fibroblasts labeled with red or green fluorescent beads or stearylamine (see Sections 2.1 and 2.2) were fused using inactivated Sendai virus (Harris and Watkins, 1965) or polyethylene glycol (PEG) 1000 (Hales, 1977) as fusing agents. Fusions of polystyrene-labeled cells were cultured overnight in medium with 15% FCS to recover before sorting. When stearylamine was used as a fluorescent label, only a period of 9 hr was used, in order to avoid significant loss and exchange of label between the differently labeled fibroblasts. After this recovery period the cells were harvested by trypsinization and suspended in medium.

Separation was carried out with a FACS II cell sorter (Becton and Dickinson) equipped with an argon-ion laser (Spectraphysics 164-05) at the 488-nm line with a constant output of 100 mW (Jongkind et al., 1979). As a filter system we used a combination of K 510 and K 515 filters to block scattered laser light. The fluorescent light was divided by a dichroic mirror (TK 560) which passes wavelengths longer than 560 nm and reflects shorter wavelengths. In front of the "green" photomultiplier tube (PMT) an S 525 bandpass filter was inserted, while a K 580 filter was placed before the "red" PMT (all filters were from Schott, except TK 560, which was from Balzers). The tubing of the FACS was sterilized by flushing the sample system with 70% ethanol in water. The sheath fluid was kept sterile by inserting a bacterial filter (Millipore 022 μm) in the sheath line after the pressure vessel. When cell clumps were present in excessive amounts the fusion mixture was first dispersed through a 70-μm nozzle of the cell sorter.

For checking the purity of the sorted fractions, cells were analyzed with a velocity of 1000 cell/sec and the desired population of heterofluorescent cells was sorted on a coverslip (diameter 14 mm) in a four-well microculture plate (NUNC: well diameter 16 mm), and cultured overnight. The attached fibroblasts were fixed, stained, and mounted. As a mounting medium we used an aqueous solution of glycerine (50%)–gelatin (70%) to avoid solubilization of the polystyrene beads; L 15 mounting medium (Carl Zeiss) can be used as an alternative. After mounting, the cells can be identified by phase contrast microscopy due to the different size of the fluorescing polystyrene beads (Fig. 3). The results using different labels and fusing agents are summarized in Table II.

When heterokaryons are needed for enzyme analysis, the cells are sorted in medium-containing petri dishes which have been coated with 1% agar to prevent attachment of the cells during the long sorting period. They can be analyzed subsequently by microchemical methodology (see Section 6).

4. Isolation of Hybrids

4.1. Outline

Proliferating hybrids originate from fusion products of different cell types. Two kinds of methods are available for the isolation of proliferating hybrids of euploid cultures of fibroblasts. The first one is based on biochemical selection using mutant fibroblasts and special media that do not permit the growth of parental cells (Siniscalco et al., 1969; Nadler et al., 1970; Migeon et al., 1974; Corsaro and Migeon 1978a,b). The other is based on the nonselective isolation of hybrid clones either by stainless steel cylinder

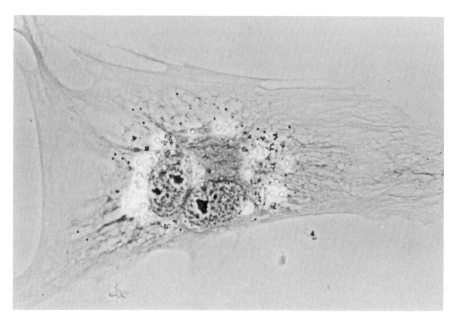

Figure 3. Heterokaryon after PEG-induced fusion, sorting, and culturing for 24 hr. The binucleated cell contains both large (red fluorescent) and small (green fluorescent) microspheres. (Jongkind et al., 1979.)

isolation (Hoehn et al., 1975, 1978; Bryant et al., 1978) or by unit gravity sedimentation (Chang et al., 1979).

In our experiments we used flow sorting as the "nonselective" method for isolation and cloning of hybrids from human fibroblasts. First the proliferating hybrids in a fusion mixture can be enriched by isolating the heterofluorescent fusion products by flow sorting (see Section 3). After a few weeks of culturing, these cells are again sorted using the DNA stain Hoechst 33342 (see Section 2.3) for the isolation of viable tetraploid cells. Individual cells of the double-sorted fusion population are subsequently cloned using the single-cell sorting modification of the cell sorter.

We used two human mutant DNA-repair-deficient cell strains (xeroderma pigmentosum) which become repair proficient after heterolog cell fusion. In this system the complementation can be studied on the single-cell level (De Weerd-Kastelein et al., 1972).

4.2. Flow Sorting of Tetraploid Cells

After PEG fusion of xeroderma pigmentosum complementation groups A and C and isolation of the heterofluorescent fusion product by flow sorting (see Section 3.2), the heterokaryon fraction (250,000 cells/35 mm dish) was cultured during 4–5 weeks until confluency. The cells were trypsinized and treated with 10 μm Hoechst 33342 (see Section 2.3).

Table II. Distribution of Viable Nucleated Cells after Fusion and Flow Sorting[a]

Fluorescent labeling	Fusion procedure	Percent gated[b]	n[c]	Percent monokaryons	Percent bikaryons	Percent trikaryons	Percent tetrakaryons
Latex	PEG	6	131	13.7	79.4	6.9	—
Latex	PEG	3.5	495	14.5	76.8	8.1	0.6
Stearylamine	Sendai	7	581	6.3	84.3	6.7	2.6

[a] Jongkind et al. (1981).
[b] Percentage of cells within the deflection window of the FACS.
[c] Number of counted attached cells after sorting and culturing for 24 hr.

The tetraploid fraction was sorted with a cell sorter (FACS II) equipped with a 5-W argon-ion laser (Spectraphysics 164-05) (Schaap et al., 1979) using the 351–364-nm lines at a constant output of 35 mW. Both KV and K445 (Schott) filters were used to block scattered laser light. A dot plot of the cellular DNA content of the cultured heterokaryon fraction is given in Fig. 4.

By determination of unscheduled DNA synthesis on the single-cell level (Bootsma et al., 1970), the percentage of hybrid (repair-proficient) mononuclear cells in the double-sorted population could be estimated to be 10–30% (Table III).

4.3. Single-Cell Cloning of Fibroblast Hybrids

The cell sorter was modified in order to sort individual cells (Parks et al., 1979) into the wells of a 96-well microtiter plate (Fig. 5). The wells contained a feeder layer of irradiated (4 krad Rö) DNA-repair-deficient cells (200 xeroderma type A cells/well).

Single individual tetraploid cells (see Section 4.2) were sorted into the wells of microtiter plates. From the 170 single-sorted cells 45 fibroblast clones appeared in the wells of the microtiter plates (cloning efficiency 27%): three clones (7%) were pure hybrid (repair-proficient) clones. These results were in accordance with those of the original double-sorted population (Table III, fusion IV).

5. Isolation of Fibroblast Cybrids

5.1. Outline

To investigate the separate role of nucleus and cytoplasm in the mechanism of complementation, use has been made of cybrids: fusion products of whole cells with enucleated cells of another mutant (de Wit-Verbeek et al., 1978; Jongkind et al., 1980).

The cybrids were isolated by two-color flow sorting of the heterofluorescent fusion products between green fluorescent cytoplasts and red fluorescent whole cells. The green fluorescent cytoplasts were obtained by enucleation of bead-containing cells, followed by flow sorting of the enucleated cells (cytoplasts). The purity was 99.7% (Schaap et al., 1981).

After fusion, the heterofluorescent fusion products can be isolated by flow sorting, and assayed for enzyme activity with microchemical methods (see Section 6).

The purity of the sorted cybrid fraction was tested by fusing green fluorescent control cytoplasts (HPRT$^+$) with red fluorescent control fibroblasts (HPRT$^-$) and determining the percentage of HPRT$^+$ cells (cybrids) by autoradiography.

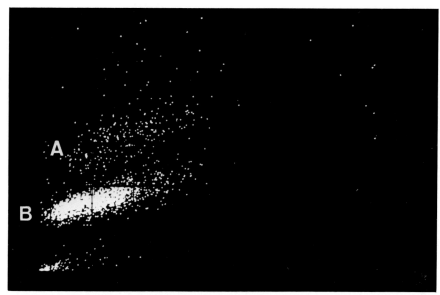

Figure 4. Flow analysis of the heterokaryon fraction after a culturing period of 5 weeks. The cells are stained with Hoechst 33342. Abscissa: scatter intensity per cell; ordinate: Hoechst fluorescence intensity per cell. A. Tetraploid cells. B. Diploid cells.

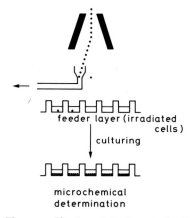

Figure 5. Cloning of single-sorted cells.

5.2. Isolation of Fluorescent Cytoplasts

Human skin fibroblasts were labeled with green fluorescent beads (Polysciences, 1.83 µm, no. 9847) in a bead-containing medium (see Section 2.1). Cells were harvested with trypsin, washed once with F10, and resuspended in 1 ml enucleation medium. The fibroblasts were preincubated in this medium for 30 min at 37°C. The enucleation medium consisted of a freshly prepared solution of 40% (v/v) Percoll (Pharmacia) in F10 containing 5 µg/ml cytochalasin B (Serva) (stock solution 1 mg/ml in DMSO). Then SW41

Table III. Mononuclear Hybrid Cells in a Double-Sorted Fusion Population of Xeroderma Variants[a]

	Mononuclear hybrids (Repair[+])[b]	Mononuclear cells (Repair[-])	Binuclear cells
Fusion[c] I	31%	69%	1%
Fusion II	12%	86%	1.7%
Fusion III	11%	83%	4.8%
Fusion[d] IV	12%	88%	0%

[a] XP25RO and XP21RO fibroblasts.
[b] Unscheduled DNA synthesis.
[c] Polyethylene glycol.
[d] This fusion is used for cloning experiments.

polyallomer centrifuge tubes were filled with 8.5 ml of enucleation medium (37°C) and the cell suspension was layered on top. Centrifugation was performed using a prewarmed SW 41 rotor at 37°C at 22,500 rpm (64,000g_{av}) for 20 min in a Beckman L5-65 ultracentrifuge. After centrifugation, the band at the bottom of the gradient was diluted five times with F10, centrifuged, and washed once with F10.

The washed enucleated cell mixture was stained with Hoechst 33342 for 30 min (see Section 2.3) and the fluorescent cytoplasts were isolated using a

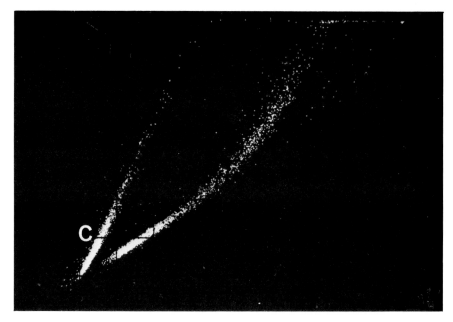

Figure 6. Flow analysis of the cell population after enucleation by density gradient centrifugation on a Percoll gradient in the presence of cytochalasin B. Cells were labeled with Hoechst 33342. Abscissa: Hoechst fluorescence intensity per cell; ordinate: green fluorescence intensity per cell. C. Cytoplasts. (Schaap et al., 1981.)

FACS II cell sorter (Fig. 6). A KV 399 filter was used to block scattered laser light. The fluorescent light was divided by a dichroic mirror (RKP 506) which passes wavelengths longer than 506 nm and reflects shorter wavelengths. In front of the "blue" photomultiplier tube (PMT), KP 490 and K 445 filters were inserted, while K 510 and K 515 filters were placed before the "green" PMT (all filters were from Schott, except RKP 506, which was from Balzers).

The intactness of the cytoplast membrane was tested by incubating the cytoplasts for 5 min with fluoresceine diacetate (Serva: 0.1 μg/ml F10) according to Rotman and Papermaster (1966). Since intact cytoplasts become fluorescent with this procedure, the percentage of intact fluorescent cytoplasts can be determined with fluorescence microscopy. In this test all sorted cytoplasts had intact plasma membranes (Schaap et al., 1981). Residual protein synthesis was tested by incubating sorted attached cytoplasts (1½ hr after sorting) with [^{3}H]leucine (10 μCi/ml, specific activity 54 Ci/mmole). After 1 hr incubation the cytoplasts were fixed with Bouin's fixative and autoradiography was performed using Ilford K2 emulsion. In our preparations [^{3}H]leucine incorporation could be observed in 90% of the sorted cytoplasts (Schaap et al., 1981).

5.3. Cytoplast × Whole Cell Fusion and Flow Sorting of Cybrids

The green fluorescent cytoplasts were fused with red fluorescent cells in a ratio 3:1 using PEG or Sendai virus (see Section 3.2). The fusion mixture was cultured overnight, harvested by trypsinization, and suspended in medium.

The isolation of the red–green heterofluorescent fusion products was carried out with a FACS II cell sorter (see Section 3.2). The yield of the sorted cybrid fraction was tested by fusing control cytoplasts (HPRT$^+$) with Lesch-Nyhan fibroblasts (HPRT$^-$). Twenty hours after fusion, heterofluorescent cells (cybrids) were sorted onto coverslips, incubated with [^{3}H]hypoxanthine (10 μCi/ml; specific activity 1 Ci/mmole) for 20 hr, fixed, and subjected to autoradiography (Ilford K2 emulsion). The hypoxanthine incorporation in the sorted heterofluorescent fraction indicated (Table IV) that by using Sendai virus as a fusing agent, at least 90% of the sorted cells were true cybrids (HPRT$^+$).

6. Biochemical Micromethods on Sorted Cells

6.1. Rationale

The amount of fusion product obtained with the flow sorting methods described is dependent on the fusion index of the cells, the sorting velocity, and the duration of sorting. In general these factors restrict the amount of fusion product to the range of 5000–10,000 cells. Since a fibroblast has a protein content of ~0.3 ng, biochemical determinations have to be done in micrograms of cell protein. Special micromethods have been designed to assay enzymes and substrates in these small amounts of cell material (Glick,

Table IV. HPRT Activity of Cybrids after Fusion and Flow Sorting[a,b]

Fusion procedure	Percent gated[c]	n[d]	Percent monokaryons		Percent bikaryons		Percent multikaryons	
			HPRT[+]	HPRT[−]	HPRT[+]	HPRT[−]	HPRT[+]	HPRT[−]
Sendai 250 HAU	4	206	85	6.8	5.3	1.0	1.0	—
PEG 1000	1.5	100	71	10	10	6	3	—

[a] Jongkind et al. (1980).
[b] HPRT, Hypoxanthine phosphoribosyl transferase. The presence of HPRT activity was determined by autoradiography. A cell was counted as HPRT[+] when it contained more than 200 grains. An HPRT[−] cell (Lesch-Nyhan) contained less than 20 grains. Fusion was between control (HPRT[+]) cytoplasts and Lesch-Nyhan (HPRT[−]) fibroblasts.
[c] Percentage of cells within the deflection window of the FACS.
[d] Number of counted cells after sorting and culturing for 24 hr.

1961; Lowry and Passonneau, 1972). The assays are based on fluorimetric methods for analysis of frozen-dried cells in the microgram range. These micromethods can be adjusted to *in vitro* cultured amniotic fluid cells to detect genetic metabolic diseases (Galjaard et al., 1974a,b).

The methods we used for the determination of specific enzyme activities utilized in Triton X100 extracts of small amounts (5000 cells) of attached *in vitro* cultured cells (van Diggelen et al., 1980; Jongkind et al., 1982). Protein was determined with a micromodification of the fluorescamine method (Udenfriend et al., 1972), while the enzymes were assayed in microvolumes with a microfluorimeter (Jongkind et al., 1974; de Josselin de Jong et al., 1980). Moreover, we developed ultramicromethods to assay the enzyme activity in single-sorted cells.

6.2. 5000–10,000 Sorted Cells

6.2.1. Cells in Suspension

6.2.1a. Homogenization. Sorted cells were washed twice with saline by centrifugation and the cell pellets were stored overnight at $-70°C$. The pellets were taken up in 20 μl Triton X-100 (0.25% in water; ~100 μg cell protein/ml). After triple freezing and thawing, the cell debris were precipitated in an Eppendorf 5412 centrifuge (3 min at maximum speed). The homogenization procedure was carried out at $4°C$. Protein and enzyme assays were carried out in the supernatant.

6.2.1b. Protein Assay. Protein was determined in the Triton extract by a micromodification of the fluorescamine method (Udenfriend et al., 1972). Three microliters of the Triton extract was mixed with 25 μl borate buffer (250 mM; pH 9) in polythene microtest tubes (Brand). Then 25 μl of a fresh florescamine solution (0.2 mg/ml dry acetone) was added under vigorous stirring on a Vortex mixer. The mixture was diluted with 200 μl water. The fluorescence of the sample was read in 200-μl glass cuvettes in a spectrofluorimeter (Perkin-Elmer fluorescence spectrophotometer MPF 2A: excitation 395 nm, emission 470 nm).

6.2.1c. Microenzyme Assay. For the measurement of enzyme activity in small numbers of sorted cells, we used fluorimetric methods of enzyme analysis (Lowry and Passonneau, 1972). Since the intensity of the fluorescence to be measured depends upon, among other factors, the concentration of the fluorescent product formed, low enzyme activities can be measured by reducing the incubation volume and by performing measurements in nanoliter to microliter volumes in microdroplets (Jongkind et al., 1974), capillaries (Galjaard et al., 1974a), or in wells of a Terasaki microtest plate (de Josselin de Jong et al., 1980). These fluorescent measurements can be done with a fluorescence microscope with epi-illuminiation equipped with a photomultiplier to detect and quantitate the emitted fluorescence light (Jongkind et al., 1974).

A typical example is given for one of the lysosomal enzymes, β-

galactosidase. This enzyme splits the nonfluorescent substrate 4-methyl-umbelliferyl-β-D-galactopyranoside into galactose and the fluorescent product 4-methyl-umbelliferone at acid pH. For the assay of β-galactosidase, 1 μl of the Triton extract ($\sim$100 μg cell protein/ml) is mixed with 2 μl of substrate solution and the incubation carried out in a well of a Terasaki microtest plate (Greiner). The samples were covered with oil (n-hexadecane-paraffin 4:6) to prevent evaporation during incubation for 1 hr on a dry bath (Fischer). At the end of the incubation period the samples were mixed with 10 μl of 0.5 M carbonate, pH 10.7, to stop the enzymatic reaction. The fluorescence of the samples was determined in a final volume of 13 μl in the original Terasaki tray, using a Leitz microfluorimeter. In this way we were able to study enzyme complementation in sorted hybrids (Jongkind et al., 1980). Many enzymes can be assayed simultaneously on the Triton extract, using other fluorimetric substrates for enzyme analysis (Jongkind et al., 1982).

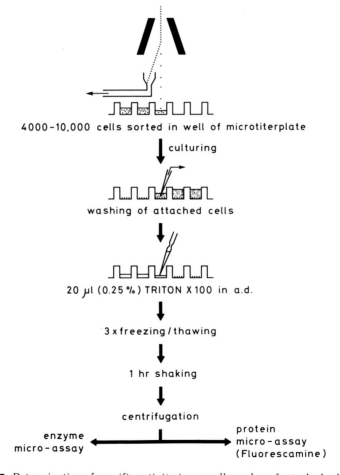

Figure 7. Determination of specific activity in a small number of attached cells.

6.2.2. Attached Cells

For the determination of enzyme activity in attached cells, which are sorted in the wells of 96-well microtiter plates (Fig. 7), 20 µl of Triton X-100 (0.25%) are added per well. The cells are detergent-extracted by shaking for 1 hr on a plate shaker at 4°C.

After freezing and thawing, the Triton homogenate (~20 µl) is transferred to a micropolythene tube (Brand) and the insoluble cell debris is centrifuged down in an Eppendorf centrifuge at maximum speed. The clear supernatant solution is used for protein and enzyme activity determinations (see Section 6.2.1).

6.3. Single-Sorted Cells

To determine the enzyme activity in single-sorted cells, cells are sorted singly (see Section 4.3) under droplets of oil (n-hexadecane-paraffin 10:1) which are deposited on a stretched Teflon foil (Jongkind et al., 1974). To a sorted cell with an accompanying droplet of 3 nl of sheath fluid, 8 nl of an enzyme substrate (see Section 6.2.1) was added with the help of ultramicro-construction pipettes. After incubation at 37°C in a dry bath (Fischer) the enzymatic reaction was stopped by pipetting 24 nl of stop-buffer into the 11-nl droplet. The fluorescence of the droplets (35 nl) was read in a microfluorimeter (Jongkind et al., 1974) (Fig. 8). Results on the β-galactosidase activity in single-sorted cells are presented in Table V.

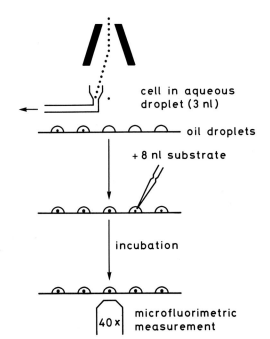

Figure 8. Ultramicroenzyme assay in single-sorted cells.

Table V. β-Galactosidase Activity in Human Fibroblasts after Single-Cell Sorting[a]

0.073	0.208	X = 0.215
0.094	0.216	SEM = 0.034
0.119	0.271	
0.134	0.316	
0.148	0.324	
0.166	0.528	
0.191		

[a] Enzyme activity is expressed as pmoles. 4-Methylumbelliferone (4MU) generated per cell per hour from 4-MU-β-galactoside. Measuring volume is 35 nl.

7. Concluding Comments

With flow sorting it is possible to isolate new cell combinations, such as heterokaryons, hybrids, and cybrids, without the use of mutant cells and selecting media.

The drawback of small numbers of sorted cells can be overcome by the use of biochemical micromethods.

Autofluorescence of cells interferes with the fluorescence of the intracellular label. Since in late passage fibroblast cultures a high percentage of autofluorescent cells is present (Jongkind et al., 1982), the use of early passage fibroblasts is recommended to avoid deviant sorting results.

Acknowledgments

We thank Henk Schaap (cytoplasts, cybrids), Jan Vijg (Hoechst dyes), and Jan de Wit (hybrid detection) for their cooperation. We acknowledge Piet Hartwijk and Pim Visser for the drawings and Tar van Os for his photographic work. Finally, we are very grateful for the excellent secretarial help of Rita Boucke.

References

Arndt-Jovin, D. J., and Jovin, T. M., 1977, Analysis and sorting of living cells according to deoxyribonucleic acid content, *J. Histochem. Cytochem.* **25:**585–589.

Bootsma, D., Mulder, M. P., Pot, F., and Cohen, J. A., 1970, Different inheritent levels of DNA repair replication in xeroderma pigmentosum cell strains after exposure to ultraviolet irradiation, *Mutat. Res.* **9:**507–516.

Bryant, E. M., Crouch, E., Bornstein, P., Martin, G. M., Johnston, P., and Hoehn, H., 1978, Regulaton of growth and gene activity in euploid hybrids between human neonatal-fibroblasts and epitheloid amniotic fluid cells, *Am. J. Hum. Genet.* **30:**392–405.

Chang, P. L., Joubert, G. I., and Davidson, R. G., 1979, Non-selective isolation of human somatic cell hybrids by unit-gravity sedimentation, *Nature* **278:**168–170.

Corsaro, C. M., and Migeon, B. R., 1978a, Gene expression in euploid human hybrid cells: Ouabain resistance is codominant, *Somat. Cell Genet.* **4:**531–540.

Corsaro, C. M., and Migeon, B. R., 1978b, Effect of intercellular communication on the selection of intraspecific human hybrids in HAT and ouabain, *Somat. Cell Genet.* **4:**541–551.

Dean, P. N., and Pinkel, D., 1978, High resolution dual laser flow cytometry, *J. Histochem. Cytochem.* **26:**622–627.

DeJosselin de Jong, J. E., Jongkind, J. F., and Ywema, H. R., 1980, A scanning inverted microfluorimeter with electronic shutter control for automatic measurements in micro-test plates, *Anal. Biochem.* **102:**120–125.

De Weerd-Kastelein, E. A., Keizer, W., and Bootsma, D., 1972, Genetic heterogeneity of xeroderma pigmentosum demonstrated by somatic cell hybridization, *Nature New Biol.* **238:**80–83.

De Wit-Verbeek, H. A., Hoogeveen, A., and Galjaard, H., 1978, Complementation studies with enucleated fibroblasts from different variants of β-galactosidase deficiency, *Exp. Cell Res.* **113:**215–218.

Ege, T., and Ringertz, N. R., 1975, Viability of cells reconstituted by virus-induced fusion of minicells with anucleate cells, *Exp. Cell Res.* **94:**469–473.

Galjaard, H., van Hoogstraten, J. J., De Josselin de Jong, J. E., and Mulder, M. P., 1974a, Methodology of the quantitative cytochemical analysis of single or small numbers of cultured cells, *Histochem. J.* **6:**409–429.

Galjaard, H., Hoogeveen, A., Keijzer, W., De Wit-Verbeek, E., and Vlek-Noot, C., 1974b, The use of quantitative cytochemical analysis in rapid prenatal detection and somatic cell genetic studies of metabolic diseases, *Histochem. J.* **6:**491–509.

Glick, D., 1961, *Quantitative Chemical Techniques of Histo and Cytochemistry*, John Wiley and Sons, New York.

Hales, A., 1977, A procedure for the fusion of cells in suspension by means of polyethylene glycol, *Somat. Cell Genet.* **3:**227–230.

Harris, H., and Watkins, J. F., 1965, Hybrid cells derived from mouse and man, *Nature* **205:**640–646.

Hoehn, H., Bryant, E. M., Johnston, P., Norwood, T. H., and Martin, G. M., 1975, Non-selective isolation, stability and longevity of hybrids between normal human somatic cells, *Nature* **258:**608–610.

Hoehn, H., Bryant, E. M., and Martin, G. M., 1978, The replicative life spans of euploid hybrids derived from short-lived and long-lived human skin fibroblast cultures, *Cytogenet. Cell Genet.* **21:**282–295.

Hulett, H. R., Bonner, W. A., Sweet, R. G., and Herzenberg, L. A., 1973, Development and application of a rapid cell sorter, *Clin. Chem.* **19:**813–816.

Jongkind, J. F., Ploem, J. S., Reuser, A. J. J., and Galjaard, H., 1974, Enzyme assays at the single cell level using a new type of microfluorimeter, *Histochemistry* **40:**221–229.

Jongkind, J. F., Verkerk, A., and Tanke, H., 1979, Isolation of human fibroblast heterokaryons with two-colour flow sorting (FACS II), *Exp. Cell Res.* **120:**444–448.

Jongkind, J. F., Verkerk, A., Schaap, G. H., and Galjaard, H., 1980, Non-selective isolation of fibroblast-cybrids by flow-sorting, *Exp. Cell Res.* **130:**481–484.

Jongkind, J. F., Verkerk, A., Schaap, G. H., and Galjaard, H., 1981, Flow-sorting in studies on metabolic and genetic interaction between human fibroblasts, *Acta Path. Microbiol. Scand. (A) Suppl.* **274:**164–169.

Jongkind, J. F., Verkerk, A., Visser, W. J., and Van Dongen, J. M., 1982, Isolation of autofluorescent "aged" human fibroblasts by flow sorting. Morphology, enzyme activity and proliferative capacity, *Exp. Cell Res.*, (in press).

Jovin, T. M., and Arndt-Jovin, D. J., 1980, Cell separation, *Trends Biochem. Sci.* **5:**214–219.

Keller, P. M., Person, S., and Snipes, W., 1977, A fluorescent enhancement assay of cell fusion, *J. Cell Sci.* **28:**167–177.

Latt, S. A., and Stetten, G., 1976, Spectral studies on 33258 Hoechst and related bisbenzimidazole dyes useful for fluorescent detection of desoxyribonuceic acid synthesis, *J. Histochem. Cytochem.* **24:**24–33.

Latt, S. A., and Wohlleb, J. C., 1975, Optical studies on the interaction of 33258 Hoechst with DNA, chromatin and metaphase chromosomes, *Chromosoma* **52:**297–316.

Loewe, H., and Urbanietz, J., 1974, Basisch substituierte 2,6-Bis-benzimidanolderivate, eine neue chemotherapeutisch aktive Körperklasse, Arzneim. Forsch. **24:**1927–1933.

Lowry, O. H., and Passonneau, J. V., 1972, *A Flexible System of Enzymatic Analysis*, Academic Press, New York.

Migeon, B. R., Norum, R. A., and Corsaro, C. M., 1974, Isolation and analysis of somatic hybrids derived from two human diploid cells, Proc. Natl. Acad. Sci. USA **71:**937–941.

Nadler, H. L., Chacko, C. M., and Rachmeler, M., 1970, Interallelic complementation in hybrid cells derived from human diploid strains deficient in galactose-1-phosphate uridyl transferase activity, Proc. Natl. Acad. Sci. USA **67:**976–982.

Ohkuma, S., and Poole, B., 1978, Fluorescence probe measurement of the intra lysosomal pH in living cells and the perturbation of pH by various agents, Proc. Natl. Acad. Sci. USA **75:**3327–3331.

Parks, D. R., Bryan, V. M., Oi, V. T., and Herzenberg, L. A., 1979, Antigen-specific identification and cloning of hybridomas with a fluorescence-activated cell sorter, Proc. Natl. Acad. Sci. USA **76:**1962–1966.

Ringertz, N. R., and Savage, R. E, 1976, *Cell Hybrids*, Academic Press, New York.

Rossilet, A., and Ruch, F., 1968, Cytofluorimetric determination of lysine with dansylchloride, J. Histochem. Cytochem. **16:**459–466.

Rotman, B., and Papermaster, B. W., 1966, Membrane properties of living mammalian cells as studied by enzymatic hydrolysis of fluorogenic esters, Proc. Natl. Acad. Sci. USA **55:**134–141.

Schaap, G. H., Van der Kamp, A. W. M, Ory, F. G., and Jongkind, J. F., 1979, Isolation of anucleate cells using a fluorescence activated cell sorter (FACS II), Exp. Cell Res. **122:**422–426.

Schaap, G. H., Verkerk, A., Van der Kamp, A. W. M., Ory, F. G., and Jongkind, J. F., 1981, Separation of cytoplasts from nucleated cells by flow sorting using Hoechst 33342, Acta Patho. Microbiol. Scand. (A) Suppl. **274:**159–163.

Siniscalco, M., Klinger, H. P., Eagle, H., Koprowski, H., Fujimoto, W. Y., and Seegmiller, J. E., 1969, Evidence for intergenic complementation in hybrid cells derived from two human diploid strains each carrying an X-linked mutation, Proc. Natl. Acad. Sci. USA **62:**793–799.

Udenfriend, S., Stein, S., Böhlen, P., Dairman, W., Leimgruber, W., and Weigele, M., 1972, Fluorescamine: A reagent for assay of aminoacids, peptides, proteins, and primary amines in the picomole range, Science **178:**871–872.

Van Diggelen, O. P., Galjaard, H., Sinnott, M. L., and Smith, P. J., 1980, Specific inactivation of lysosomal glycosidases in living fibroblasts by the corresponding glycosylmethyl-p-nitrophenyltriazenes, Biochem. J. **188:**337–343.

Veomett, G., Prescott, D. M., Shay, J., and Porter, K. R., 1974, Reconstruction of mammalian cells from nuclear and cytoplasmic components separated by treatment with cytochalasin B, Proc. Natl. Acad. Sci. USA **71:**1999–2002.

Chapter 8
Techniques for Monitoring Cell Fusion: The Synthesis and Use of Fluorescent Vital Probes (R18, F18)

PAUL M. KELLER and STANLEY PERSON

1. Introduction

The ability to directly measure membrane fusion is useful both in the study of the fusion process itself and in experiments involving membrane fusion, such as the production of heterokaryons, liposome–cell interactions, and viral envelope–cell interactions.

The currently accepted model of biologic membranes envisions the membrane as a two-dimensional fluid layer of lipids and proteins (Singer and Nicolson, 1972). Measurements of the lateral diffusion of lipid molecules have yielded diffusion coefficients on the order of 10^{-8} cm^2/sec (Hubbell and McConnell, 1969; Kornberg and McConnell, 1971; Devaux and McConnell, 1972; Schlessinger et al., 1977). Thus a direct way to monitor membrane fusion is to follow the rapid intermixing of the lipid phases of membranes following fusion, as has been done for fluorescent-labeled proteins by Frye and Edidin (1970).

This chapter describes such an assay based on resonance energy transfer between two fluorescent lipid probes. In addition it describes the synthesis and use of these two fluorescent probes, which partition into biologic membranes and may be used to specifically label the membranes of cells or liposomes.

Resonance energy transfer has been documented by the use of compounds with donor and acceptor groups separated by known distances (Latt et al., 1965; Stryer and Haugland, 1967; Gabor, 1968; Haugland et al., 1969; Becker et al., 1975). In order for resonance energy transfer to occur, two conditions must be met; the emission spectrum of the donor molecule must overlap the absorption spectrum of the acceptor molecule, and the molecules must be in close proximity, on the order of 50 Å.

PAUL M. KELLER and STANLEY PERSON • Biophysics Program, Pennsylvania State University, University Park, Pennsylvania. 16802.

In practice, membrane fusion was monitored by measuring the energy transfer between the two fluorescent probes each of which was partitioned into a specific membrane population. The two membrane populations were caused to fuse, the donor probe (F18) was specifically excited, and the fluorescence of the acceptor probe (R18) was observed. The increase in fluorescence as the probes mixed, due to energy transfer, was used as a measure of fusion.

In the experiments described here virus-induced cell fusion from within of human embryonic lung (HEL) cells using a plaque morphology mutant (syn-20) of herpes simplex virus type 1 (HSV-1) was studied. Thus HSV-1 mutant causes a normal infection of HEL cells and also causes extensive cell fusion. A description of the cell and virus growth procedures can be found in Person et al. (1976). It should be noted that this same assay has been used to monitor liposome–liposome and liposome–cell fusion (Deamer and Uster, 1980).

2. Description of Membrane Probes

In order to meet the criteria necessary for energy transfer and membrane partitioning the probes were synthesized by the covalent linkage of fluorescein or rhodamine to a saturated hydrocarbon chain 18 carbon atoms long (Fig. 1). The hydrocarbon chain causes the molecule to partition into the lipid region of biologic membranes. Once partitioned into a membrane, the probes do not readily diffuse out. The probes quantitatively remain in the membranes for periods of the order of 24 hr (Keller et al., 1977).

The emission spectrum of the donor probe (F18) overlaps the absorption spectrum of the acceptor probe (R18), satisfying the spectral conditions necessary for resonance energy transfer. In addition, their absorption and emission spectra are such that it is possible to preferentially excite F18 at 460 nm and selectivity observe the fluorescence of R18 at wavelengths greater than 600 nm (Fig. 2), thus allowing one to detect resonance energy transfer.

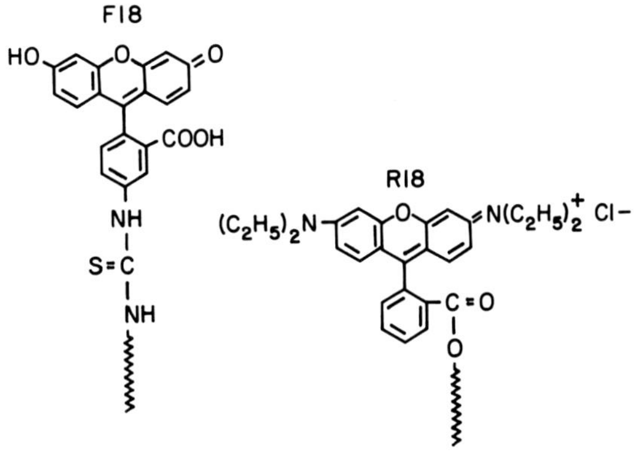

Figure 1. The structure of F18 and R18.

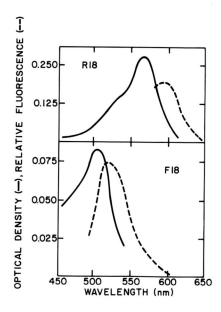

Figure 2. Absorption and fluorescence spectra of 1 × R18 and 1 × F18 in TBS. Absorption is given in units of optical density, the fluorescence in relative units. The fluorescence spectra were corrected for photomultiplier tube sensitivity. F18 can be excited at 460 nm without exciting R18; the fluorescence spectrum of F18 overlaps the absorption spectrum of R18; and the emission of R18 can be observed above 600 nm without observing F18 omission.

3. Syntheses of F18 and R18

The F18 was synthesized by reacting 0.5 g of fluorescein isothiocyanate (Sigma Chemical Co., St. Louis, Missouri) with 0.5 g of octadecylamine (Aldrich Chemical Co., Milwaukee, Wisconsin) in 10 ml dry pyridine for 4 days. The reaction mix (F18) was streaked across the bottom of a number of 2-mm-thick, 20 × 20 cm Silica Gel G plates (Analabs, Inc., North Haven, Connecticut). The pyridine was allowed to evaporate and the plates developed in chloroform/acetone/methanol/acetic acid/water (50/20/10/10/5). This separates unreacted octadecylamine from the unreacted fluorescein isothiocyanate and the reaction product F18, which had larger R_f values. This was verified by ninhydrin staining. After the plates had dried they were rotated 180° and developed in water/pyridine (4/1), causing the unreacted fluorescein isothiocyanate to move, leaving the F18 behind. As controls, unreacted octadecylamine and unreacted fluorescein isothiocyanate were run on separate plates. After the second development the F18 streak was scraped from the plate and dissolved in ethanol, and the chromatography steps were repeated. The F18 streak was then scraped from the plate, dissolved in ethanol, and centrifuged at 7500g three times to remove the Silica Gel G. The ethanol was then allowed to evaporate and the product stored as a dry powder at +20°C. When needed, F18 was added to ethanol to give a final concentration of 1 mg of F18/ml ethanol.

The R18 was synthesized by taking a suspension of 1.7 g of Rhodamine B (Aldrich Chemical) in 10 ml dry benzene and treating it with 0.3 ml of dry pyridine. To this mixture 0.27 ml of thionyl chloride was added dropwise with stirring and cooling. After the addition of thionyl chloride the reaction vessel was sealed and stirred at room temperature for 12 hr. One gram of octadecanol was added and allowed to react for 12 hr at room temperature with stirring. The benzene was allowed to evaporate, and the reaction product was resuspended in ethanol. The solution was then streaked onto 2-mm-thick, 20 × 20 cm Silica Gel G plates and developed using the two solvent systems described for the purification of F18. The water/pyridine development was followed by development in ether to move the octadecanol away from the product, R18. The reddish streak (R18) at the bottom of the plates which did not move in water/pyridine or ether was scraped from the plates. The scrapings were resuspended in ethanol and the chromatographic separation repeated twice. The final R18 ethanol solution was centrifuged at 7500g three times to remove the silica gel, the ethanol allowed to evaporate, and the final product stored at +20°C. When needed, the R18 was resuspended in ethanol at a concentration of about 5 mg/ml.

4. Preparation of F18- or R18-Labeled Cells

The culture media and reagents and cell and virus growth procedures have been described previously (Person et al., 1976). We routinely labeled HEL cells by adding the probes suspended in ethanol to enough growth medium to produce absorbances of 0.085 at 506 nm (F18) or 0.260 at 565 nm (R18), and a final ethanol concentration of less than 0.2%. Cells in late logarithmic phase were trypsinized and seeded into 16-ounce prescription bottles at 7.5×10^6 cells per bottle. Six hours later the medium was decanted and 10 ml of either F18 or R18 labeling medium was added. The probes were allowed to partition into the cells for 12–14 hr at 37°C. The labeling medium was decanted and the cells were washed twice with 10 ml of a tricine-buffered saline (TBS). In our experiments the cells were infected with virus at this point. Cells were then harvested by trypsinization and the two labeled populations were mixed in equal proportions and seeded into 30-mm-diameter glass culture dishes (stock #1934-1230, Bellco Glass, Vineland, New Jersey). Usually 1×10^6 F18 and 1×10^6 R18 labeled cells in 2 ml of growth medium were used, forming a monolayer upon attachment. The petri dishes were sealed with a 5% solution of agarose and incubated at 38°C. It was necessary to carefully standardize the labeling procedure to obtain reproducible data, because resonance energy transfer varies as $1/R^6$, where R is the distance of separation between molecules.

The probe molecules at the concentrations used did not effect cell growth, cell attachment, or HSV-1 growth in cells (Keller et al., 1977). When Coulter counter fusion curves were determined for unlabeled cells or cells containing either F18 or R18 no difference in either the extent or rate of fusion was found (Keller et al., 1977). These results suggest that at the concentrations used the probes did not effect the experimental system. If higher concentrations of probe molecules are used, pertubations can occur (Keller, 1976).

In order for the fluorescence assay to measure membrane fusion it was necessary that the probes be exclusively located in the membrane region of the cell and only interact as a result of fusion. By following the fate of the probes of mixed F18 and R18 labeled cells using a fluorescence microscope it could be seen that the probes remained in their respective cell populations and only mixed as a result of fusion. When observed using a fluorescence microscope the probes appear spread throughout the cellular membranes. Their distribution was not limited to the plasma membrane (Keller et al., 1977). Thin-layer chromatography of Folch extracts (Folch et al., 1957) of cells labeled for 24 hr with F18 or R18 demonstrated that the cells had not degraded the probes.

5. Preparation of Labeled Virus

We labeled HSV-1 virions with F18 by incubating purified virions (Keller et al., 1977) in 8× the standard F18 labeling solution for 12 hr at room temperature. The virions were then pelleted to separate them from unadsorbed probe molecules (Keller et al., 1977). Other investigators (Leary and Notter, 1982) have labeled avian tumor viruses with R18 for virus adsorption studies. These investigators suspended R18 in dimethyl sulfoxide (DMSO), labeled the virions in 0.001 M R18 for 18 hr at 4°C, and separated the unadsorbed probe from the virus by ultracentrifugation on a sucrose gradient. They found no loss of infectivity due to labeling. The labeled virions were stored at −80°C.

6. Measurement of Energy Transfer

Fluorescence measurements were performed using a fluorometer which was designed and constructed to observe energy transfer between the two fluorescent probes when they were partitioned into a monolayer of cells. For a detailed description see Keller et al. (1977). The fluorometer operated by shining 460-nm light on the bottom of a 30-mm glass petri dish (thus exciting F18) and measuring light above 600 nm emitted from the dish (R18 fluorescence) using a photomultiplier tube for amplification. For a given petri dish there was a particular background fluorescence due to (a) light leakage through the fluorometer filters; (b) scattered light; (c) background fluorescence due to a small amount of excitation of R18 by 460-nm light and a small amount of fluorescence of F18 above 600 nm. This background was corrected for by taking the first three readings on a given petri dish before fusion and subtracting the average value from all subsequent readings. Since fluorescence is sensitive to temperature, pH, ionic strength, and viscosity, care was taken to keep all of these constant during an experiment. Although we constructed a fluorometer for these experiments, a commercially available fluorometer should be adaptable for such experiments.

7. Results of Fusion Experiments

The results of a typical cell fusion experiment are shown in Fig. 3. HEL cells were labeled with F18 or R18, infected with virus, harvested, mixed in equal proportions, and seeded into petri dishes. The fluorescence of R18 was then measured while exciting F18. When the cells were infected with a fusion-causing virus (syn-20), an increase in fluoresence was observed about 4 hr postinfection. The increase plateaued at about 8 hr postinfection, a time at which most of the cells had fused as measured by other fusion assays. Mock-infected cells demonstrated no change in fluorescence, while wild-type-infected cells demonstrated a small increase in fluorescence. Wild-type-infected cells also demonstrate a small amount of fusion when measured by a Coulter counter or microscopic assay (Person et al., 1976).

To demonstrate that the fluorescence of R18 only occurred as a result of mixing of the two fluorescent probes following fusion, the ratio of F18 to R18 syn-20-infected cells was varied. The total fluorescence increase was then measured 8 hr after infection, a time at which the cells have extensively fused. In can be seen (Fig. 4) that fluorescence enhancement only occurred when a mixture of R18- and F18-labeled cells were present, with no change in fluorescence if the infected cells contained only F18 or R18. To determine whether the fluorescence increase followed a $1/R^6$ dependence, as would be the case for resonance energy transfer, the number of probe molecules per cell was varied by varying the initial probe concentration from 0 to 1×, 1× being the standard probe concentration. Equal numbers of R18- and F18-labeled, syn-20-infected cells were mixed for each concentration and the increase in fluorescence measured 8 hr after infection. The relationship between concentration and quanta transferred (Fig. 5) gave a reasonable fit for resonance energy transfer with the probes distributed throughout a volume.

8. Virus–Cell Interactions

Since HSV-1 is an enveloped virus, it is possible to label the viral envelope with one fluorescent probe and to observe its interaction with cells

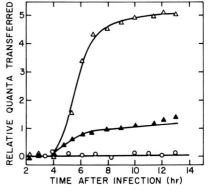

Figure 3. R18 fluorescence of mock infected, wild-type infected, and syn-20 infected cells. R18 fluorescence was measured using culture dishes containing 10^6 F18- and 10^6 R18-labeled cells for (○) mock infected, (▲) wild-type infected, and (△) syn-20 infected cells. Wild-type infected cells produced about 20% of the fluorescence increase relative to syn-20 infected cells. The increase in fluorescence for syn-20 infected cells began at a time when fusion was observed to occur in parallel experiments using a Coulter counter assay to moniter fusion. Each point represents the average relative quanta transferred using three culture dishes.

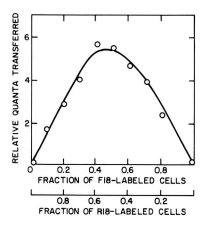

Figure 4. The maximum fluorescence enhancement as a function of the fraction of F18- or R18-labeled cells. Labeled cells were infected with syn-20 and the cells prepared for fluorescence measurements by the usual procedures. However, in these experiments, the fraction of F18-labeled cells to R18-labeled cells was varied. The fraction of F18-labeled cells increases from left to right; the fraction of R18-labeled cells increases from right to left. Fluorescence was measured 8 hr after infection, at a time when the fluorescence increase had normally plateaued.

containing the other fluorescent probe. F18-labeled virus was added to R18-labeled cells and the R18 fluorescence measured (Fig. 6). A rapid increase in fluorescence was observed which could be blocked by neutralizing antibodies.

9. Discussion

Two fluorescent membrane probes which can be used to label and to follow the mixing of the lipid regions of membranes have been developed.

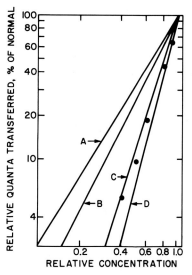

Figure 5. R18 fluorescence as a function of probe concentration. Labeled HEL cells were infected with syn-20 and the cells prepared for fluorescence measurements by the usual procedures. However, in these experiments, the concentrations of F18 and R18, although kept in the same proportions, were varied from 0 to 1.0, 1.0 representing the usual F18 and R18 probe concentrations. Fluorescence was measured 8 hr after infection, at a time when the fluorescence increase had normally plateaued; 100% fluorescence quanta transferred was taken as the fluorescence at the 1.0 probe concentration. The data points are indicated as ● and the lines were calculated as follows. Let N represent the number of probe molecules incorporated into cellular membranes and assume it is proportional to the number added to the growth medium. If the cellular probes are distributed in three dimensions (3D), the average separation between probes r is given by $r \propto (1/N)^{1/3}$, or $N \propto 1/r^3$. In two dimensions (2D), $N \propto 1/r^2$. Photon emission and reabsorption (PER) has a $1/r^2$ distance dependence, and resonance energy transfer (RET) has a $1/r^6$ distance dependence. Fluorescence enhancement (FE) is proportional to donor absorption (proportional to N) times the energy transfer from donor to acceptor (proportional to $1/r^2$ or $1/r^6$). For the 2D case $(FE)_{PER} \propto N^2$ and $(FE)_{RET} \propto N^4$. For the 3D case $(FE)_{PER} \propto N^{5/3}$ and $(FE)_{RET} \propto N^3$. The slopes of the lines in the figure are equal to the exponents of N. A, PER, 3D: B, PER, 2D; C, RET, 3D; D, RET. 2D.

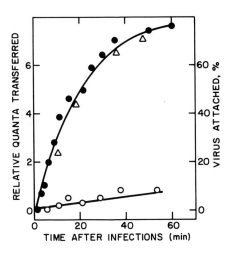

Figure 6. R18 fluorescence in a virus-cell system and the attachment of virus to host cells. F18-labeled syn-20 virus was added to 3×10^6 R18-labeled cells containing twice the usual probe concentration and the fluorescence increase measured as a function of time. Sufficient virus was added in 0.2 ml of growth medium to give about 10 PFU/cell. Data are shown for syn-20 virus labeled with F18 (●) and for the same preparations of virus incubated with HSV-1 antisera for 30 min prior to mixing the virus with labeled cells (○). Data for the attachment of syn-20 to HEL cells are also given (△).

Fluorescence enhancement by energy transfer has been shown to be a valid assay of membrane mixing, since:

1. The probes are hydrophobic and partition into nonpolar environments.
2. The probes at the concentrations used did not perturb the experimental system.
3. Fluorescence enhancement only occurs when both probes are present and the cells have fused. It does not occur when there is no fusion, or when there is fusion in the presence of only one probe.
4. The initial starting time, rate, and total amount of enhancement compare favorably with other assays of cell fusion.

The concentration dependence of fluorescence enhancement suggests that the increase is due to resonance energy transfer. The data best fit the kinetics for a three-dimensional array. There are several possible explanations for this. The probes presumably are on both sides of the membranes and transfer may be occurring across the bilayer. There could be transfer occurring between probes located in intracellular membrane structures where many layers are present. It is also possible that the fit is fortuitous, with some of the transfer being from photon emission and reabsorption and some from resonance energy transfer.

It should be noted that these same probes have been used to follow liposome–liposome and liposome–cell fusion (Deamer and Uster, 1980). These same authors have also developed other probe systems based on the same principles.

We found that it was also possible to label virus and to observe its interaction with labeled cells. It is not clear from the data whether the increase in fluorescence in this interaction was due to adsorption, phagocytosis, or fusion, since all three could bring the probe molecules in close proximity.

In addition to being used for energy transfer studies, probes such as F18 and R18 can be used as fluorescent "tags" to follow the fate of membranes. For example, other investigators have used R18-labeled enveloped virions to follow the kinetics of virus adsorption to single cells using a flow cytometer (Leary and Notter, 1982).

References

Becker, J. S., Oliver, J. M., and Berlin, R. D., 1975, Fluorescent techniques for following interactions of microtubule subunits and membranes. *Nature* **254**:152–154.

Deamer, D. W., and Uster, P. S., 1980, Liposome preparation methods and monitoring liposome fusion, in: *Introduction of Macromolecules Into Viable Mammalian Cells* (R. Baserga, C. Croce, and G. Rovera, eds.), Alan R. Liss, New York, pp. 205–220.

Devaux, P., and McConnell, H. M., 1972, Lateral diffusion in spin labeled phosphatidylcholine multilayers, *J. Am. Chem. Soc.* **94**:4475–4481.

Folch, J., Lees, M., and Sloane-Stanley, G., 1957, A simple method for the isolation and purification of total lipids from animal tissues, *J. Biol. Chem.* **226**:497–509.

Frye, L. D., and Edidin, M., 1970, The rapid intermixing of cell surface antigens after formation of mouse–human heterokaryons, *J. Cell Sci.* **7**:319–335.

Gabor, G., 1968, Radiationless energy transfer through a polypeptide chain, *Biopolymers* **6**:809–816.

Haugland, R. P., Yguerabide, J., and Stryer, L., 1969, Dependence of the kinetics of singlet–singlet energy transfer on spectral overlap, *Proc. Natl. Acad. Sci. USA* **63**:23–30.

Hubbell, W. L., and McConnell, H. M., 1969, Motion of steroid spin labels in membranes, *Proc. Natl. Acad. Sci. USA* **63**:16–22.

Keller, P. M., 1976, A fluorescence enhancement assay of membrane interactions, Ph.D. dissertation. The Pennsylvania State University.

Keller, P. M., Person, S., and Snipes, W., 1977, A fluorescence enhancement assay of cell fusion, *J. Cell Sci.* **28**:167–177.

Kornberg, R. D., and McConnell, H. M., 1971, Inside–outside transitions of phospholipids in vesicle membranes, *Biochemistry* **10**:1111–1120.

Latt, S. A., Cheung, H. T., and Blout, E. R., 1965, Energy transfer. A system with relatively fixed donor acceptor separation, *J. Am. Chem. Soc.* **87**:995–1003.

Leary, J. F. and Notter, M. F. D., 1982. Kinetics of virus adsorption to single cells using fluorescent membrane probes and multiparameter flow cytometry, *Cell Biophys.* **4**:63–76.

Person, S., Knowles, R. W., Read, G. S., Warner, S. C., and Bond, V. C., 1976, The kinetics of cell fusion induced by a syncytia-producing mutant of herpes simplex virus type I, *J. Virol.* **17**:183–190.

Schlessinger, J., Alxerod, D., Koppel, D. E., Webb, W. W., and Elson, E. L., 1977, Lateral transport of a lipid probe and labeled proteins on a cell membrane, *Science* **195**:307–309.

Singer, S. J., and Nicolson, G. L., 1972, The fluid mosaic model of the structure of cell membranes, *Science* **175**:720–731.

Stryer, L., and Haugland, R. P., 1967, Energy transfer: a spectroscopic ruler, *Proc. Natl. Acad. Sci. USA* **58**:719–726.

Chapter 9
Inheritance of Oligomycin Resistance in Tissue Culture Cells

JEROME M. EISENSTADT and MARY C. KUHNS

1. Introduction

The mitochondrial ATPase complex has been resolved into the soluble F_1, having the catalytic site for ATP synthesis, and the hydrophobic membrane-integrated F_0, involved in proton translocation (Senior, 1973; Koslov and Skulacher, 1977). The F_0 has been studied with the use of oligomycin and dicyclohexylcarbodiimide (DCCD), which inhibit enzyme activity of the complex by acting on a component of F_0, (Bulos and Racker, 1968; Roberton et al., 1968; Sebald et al., 1976, 1979; Kiehl and Hatefi, 1980).

Mutants having decreased sensitivity to oligomycin have been isolated and characterized in *Saccharomyces cerevisiae*, *Neurospora crassa*, *Aspergillus nidulans*, mouse fibroblasts, and Chinese hamster fibroblasts (Avner et al., 1973; Tzagoloff et al., 1975; Rowlands and Turner, 1973; Turner et al., 1979; Howell and Sager, 1979; Lichtor and Getz, 1978; Lagarde and Siminovitch, 1979; Kuhns and Eisenstadt, 1979; Breen and Scheffler, 1980). Analyses of the mode of inheritance of these oligomycin-resistant mutations have established the genetic location of two important membrane subunits. Subunit 6 of the mitochondrial ATPase is a mitochondrial gene product and has been mapped on the mitochondrial genome in yeast and *Aspergillus* (Hensgens et al., 1979; Sebald et al., 1979; Macino et al., 1980). Subunit 9, identified as the DCCD-binding protein, is also a mitochondrial gene product in yeast (Sebald et al., 1979; Criddle et al., 1979; Cattel et al., 1970; Sierra and Tzagoloff, 1973; Edwards and Urger, 1978; Jackl and Sebald, 1975). However, the DCCD-binding protein is a nuclear gene product in *Aspergillus* and *Neurospora* (Sebald et al., 1976; Turner et al., 1979; Edwards and Urger, 1978; Jackl and Sebald, 1975). Thus, oligomycin resistance has been associated with mutations in at least two subunits of the membrane component.

JEROME M. EISENSTADT • Department of Human Genetics, Yale University School of Medicine, New Haven, Connecticut 06510. MARY C. KUHNS • Department of Human Genetics, Yale University School of Medicine, New Haven, Connecticut 06510. Present address: Abbott Laboratories, North Chicago, Illinois 60064.

In addition, single amino acid substitutions in the DCCD-binding protein have been established for oligomycin-resistant mutants from *Saccharomyces cerevisiae* and *Neurospora crassa* (Sebald and Wachter, 1978).

In mammalian cells nuclear and cytoplasmic modes of inheritance of oligomycin resistance have been described (Howell and Sager, 1979; Lichtor and Getz, 1978; Lagarde and Siminovitch, 1979; Breen and Scheffler, 1980; Kuhns and Eisenstadt, 1981a), including mutants having mitochondrial ATPase with decreased sensitivity to oligomycin (Lichtor and Getz, 1978; Kuhns and Eisenstadt, 1979; Breen and Scheffler, 1980). Recently it has been shown that the DNA sequence for the DCCD-binding protein is absent in the mitochondrial genome of HeLa cells (Anderson et al., 1981) and that a mouse fibroblast mutant with nuclear inheritance of oligomycin resistance has an altered DCCD-binding protein (Kuhns and Eisenstadt, 1981b).

The study of the inheritance of oligomycin resistance has been of great value in determining the genetic location of mitochondrial ATPase subunits and in defining the genetic composition of the mitochondrial genome itself. Oligomycin-resistant mutations have also contributed to our understanding of the synthesis, assembly, and function of enzyme subunits.

The requirement for gene products encoded in both the nuclear and mitochondrial genomes offers an opportunity to study two interacting genetic systems normally operating in the eukaryotic cell. Finally, studies of mutations in the protein component on which oligomycin and DCCD act have furthered knowledge of the amino acid residues involved in proton translocation and may ultimately be of great importance in understanding the process of oxidative phosphorylation.

2. Selection of Oligomycin-Resistant Mutants

Difficulties in obtaining mutations in the mitochondrial genomes of mammalian cells have been attributed to the large number of mitochondria and mitochondrial genomes in the cell, where nonlethal mutations are rare and intermitochondrial competition may be great. Therefore, a physical means of reducing the number of mitochondrial genomes prior to mutagenesis was devised. The method was to isolate minicells (cells that contain the nucleus and 5–10% of the normal amount of cytoplasm and mitochondria), mutagenize them, and finally select in oligomycin. A total of 4×10^7 LM(TK$^-$) cells were enucleated by cytochalasin B treatment and were centrifuged in a discontinuous Ficoll gradient (Wigler and Weinstein, 1975) (Veomett, this volume, Chapter 6). The gradient, prepared in a Beckman SW-41 centrifuge tube, consisted of 2 ml of 25%, 2 ml of 17%, 0.5 ml of 16%, 0.5 ml of 15%, and 2 ml of 12.5% Ficoll. The Ficoll, dissolved in MEMS (minimal essential medium, suspension) growth medium, contained 10 µg/ml cytochalasin B. The cells were suspended in 3 ml of 10% Ficoll and layered on the top of the gradient with an overlay of MEMS to fill the centrifuge tube. The gradients were placed on a prewarmed SW-41 rotor (37°C) and centrifuged for 1 hr at 25,000 rpm at 35–37°C. Cell fractions were removed and examined under the microscope. The gradient fraction containing the minicells (the interface between the 17 and 25% layers) was diluted with α MEM (minimal essential medium, α modification) and whole cells contaminating the minicell fraction were removed by selective attachment to tissue culture flasks. The diluted minicell preparation was placed in a tissue culture growth flask for 2 hr to allow the whole cells to attach. Under these conditions minicells do not attach. After three 2-hr passages in the flasks, the unattached minicell fraction was plated in 25-cm^2 flasks and mutagenized with 300–700 µg/ml ethylmethane sulfonate for 18 hr. After 1 week in growth medium, the attached cells were harvested and replated at 4.5×10^4 or 7×10^5 cells in 25- or 75-mm^2 flasks in αMEM containing oligomycin (3–5 µg/ml). After 2–3 weeks the resulting colonies were picked, grown, and recloned.

3. Growth of Oligomycin-Resistant Mutants

The growth characteristics of oligomycin-resistant clones can be studied (Table 1). Figure 1 illustrates the results of a typical growth experiment in which one mutant (OLI 14) was compared to wild-type cells in the presence and absence of the antibiotic. In the absence of the antibiotic, growth of the mutant was identical to that of the wild type. In the presence of 5 µg/ml of oligomycin, growth of the wild type is completely inhibited. The mutant doubled every 1.5 days, compared to 1 day in the absence of oligomycin. Examination of a number of independently isolated mutants shows that, in the absence of drug, their growth rate is identical to that of the wild type. The mutant cells grow with doubling times of 1.5–1.6 days in 3 µg/ml oligomycin and 1.2–2.5 days in 5 µg/ml oligomycin. Such concentrations of oligomycin are inhibitory to the wild-type cells. Similar results can be obtained by measuring the cloning efficiency (the number of colonies present at 10–14 days divided by the number of cells plated and expressed as a percentage) of the wild-type and mutant cells in the presence and absence of oligomycin. The wild type does not produce clones at 3 or 5 µg/ml of oligomycin, while one mutant has a relative cloning efficiency (compared to the absence of drug) of 92% (3 µg/ml) and 79% (5 µg/ml).

4. Stability of Oligomycin-Resistant Mutants

To determine the stability of oligomycin resistance, the growth characteristics of mutants were studied after cultures had been grown in the absence

Table I. Criteria for the Selection of Oligomycin-Resistant Mutants

I. The mutant should grow and clone in concentrations of oligomycin that are inhibitory to the wild type.
II. The mutation should be stable over many generations in the absence of selection
III. (a) Oligomycin resistance should be demonstrable at the level of the mitochondrial ATPase from the mutant and in oligomycin-resistant progeny derived from inheritance studies.
(b) If an alteration in a mitochondrial ATPase subunit is detected, this alteration should be inherited along with oligomycin resistance.

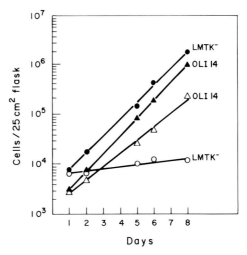

Figure 1. Growth of OLI 14 and wild type in the presence and absence of oligomycin. Wild type; no oligomycin (●); oligomycin 5 µg/ml (○); OLI 14 no oligomycin (▲), oligomycin 5 µg/ml (△). [From Kuhns and Eisenstadt (1979).]

of drug for 50–65 generations. During the first week in the absence of oligomycin, the population doubling times of the mutants remained at 1.5–2.0 days and eventually stabilized at 0.5–1 in the remaining time without selection (Kuhns and Eisenstadt, 1979).

All of the mutants continued to grow at the same rate as the wild type in the absence of oligomycin. Two classes of mutants emerged when challenged with 3 or 5 µg/ml oligomycin/ml after growth for 50 generations without selective pressure. Approximately one-half of the mutants examined comprised a stable class. The population doubling times of these mutants, whether originally isolated in 3 or 5 µg/ml oligomycin, remained at 1–2 days in oligomycin, and the cloning efficiencies remained high (greater than 45% of the value observed in the absence of drug). The other half of the mutants consisted of unstable clones with cloning efficiencies of 1% or less at the oligomycin concentration of the original selection. Although the unstable mutants showed a loss of resistance to the concentration of oligomycin used in the original selection procedure, 5 µg/ml, the mutants retained the ability to clone and grow at lower concentrations of oligomycin (3 µg/ml) with efficiencies 42–58% of those observed in the absence of oligomycin.

5. Mitochondrial ATPase of Oligomycin-Resistant Cells

5.1. Preparation of Mitochondrial and Submitochondrial Particles

Tissue culture cells were grown either in monolayer cultures using plastic roller bottles or in spinner cultures. For preparation of mitochondria, 10^8 cells or more were harvested and washed twice in buffer containing 0.13 M NaCl, 5 mM KCl, 1 mM $MgCl_2$, 2.5 mM Tris-HCl, pH 7.5. The following operations were carried out at 4°C. Cells were allowed to swell in buffer (1.5 mM $CaCl_2$, 10 mM NaCl, 10 mM Tris-HCl, pH 7.5) for 30 min. The cell suspension was homogenized with a Teflon pestle and motor-driven homogenizer and then adjusted to 0.25 M sucrose. The homogenate was centrifuged twice at 1500g for 5 min to sediment nuclei and unbroken cells. The supernatant was centrifuged at 9000g for 20 min and the crude mitochondrial pellet was suspended in TSC buffer (10 mM Tris-HCl, pH 7.5, 3 mM $MgCl_2$, 0.25 M sucrose) and stored at -20°C until used. Preparations can be stored in this manner for up to 1 year with no apparent loss of ATPase activity and no decrease in sensitivity to oligomycin of DCCD.

For preparation of submitochondrial particles, freshly prepared mitochondria were suspended in cold TSC buffer at 8 mg protein/ml and sonicated at 4°C with three to six 10-sec bursts of a Bronwill Biosonic IV sonicator (VWR). The volume was adjusted to 4 mg protein/ml with TSC and the suspension was centrifuged at 9500g for 10 min to sediment unbroken mitochondria. After a second centrifugation at 12,000g for 15 min to remove whole mitochondria and large debris, the supernatant was sedimented at 100,000g for 45 min. The pellet of submitochondrial particles was suspended in TSC to 5 mg protein/ml. The concentrated suspension was stored at -20°C and used within 1 week of preparation.

5.2. Assay to Mitochondrial ATPase Activity and Sensitivity to Inhibitors

ATPase activity was measured following the modified procedure of Pullman et al. (1960). Activity was measured at 37°C, at pH 8.5, in a coupled enzyme assay system containing 5 mM $MgCl_2$, 0.2 mM NADH, 5 mM phosphoenolpyruvate, seven units pyruvate kinase, 15 units lactic dehydrogenase, 5 μl absolute methanol with or without oligomycin, 5 mM Na_2 ATP, and 0.04 M Tris HCO_3 in a final volume of 1.0 ml. The reaction mixture was equilibrated for 5 min at 37°C. The reaction was then started by the addition of 25-100 μg mitochondrial protein. In assays for sensitivity of enzyme to DCCD, mitochondria were preincubated for 20 min at 37°C with DCCD in the above reaction mixture from which Na_2 ATP had been omitted. The reaction was begun by the addition of Na_2 ATP to a final concentration of 5 mM. Reaction rates were recorded on a recording spectrophotometer at 340 nm. Activity was calculated from reaction slopes and expressed as μmole ATP hydrolyzed/min/mg protein. Protein was determined using BioRad reagent with bovine gamma globulin as the standard. Figure 2 shows a typical sensitivity curve for mitochondria from wild-type mouse L cells and for mitochondria from an oligomycin-resistant mutant. Note the effect of culture conditions on the oligomycin sensitivity of enzyme from this mutant (Kuhns and Eisenstadt, 1979).

From the inhibitor sensitivity curve for each cell line and specific culture conditions, the concentration of inhibitor giving half-maximal activity was determined (I_{50}). Typical I_{50} values for a wild-type mouse L cell line LM(TK⁻)

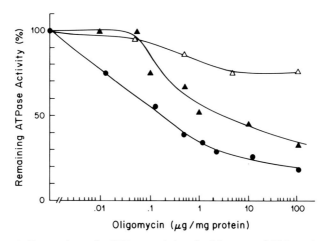

Figure 2. Effect of oligomycin on the ATPase activity of wild-type and OLI 14 mitochondria. Wild type (●); OLI 14 grown in the absence of oligomycin (△); OLI 14 grown in oligomycin 5 μg/ml (▲). Specific activities (μmole ATP hydrolyzed/min/mg protein): wild type, 0.331; OLI 14 grown without oligomycin 0.193; OLI 14 grown in oligomycin 5 μg/ml, 0.099.

and for an oligomycin-resistant mutant with a modified DCCD-binding protein (Kuhns and Eisenstadt, 1981c) are shown in Table II. I_{50} is expressed as μg or nmole of inhibitor/mg protein in the assay.

5.3. Use of [^{14}C]DCDD for Analysis of DCCD-Binding Protein

DCCD inhibits mitochondrial ATPase activity by acting on the same membrane component as oligomycin and is particularly useful in determining the nature of oligomycin resistance. DCCD can readily be obtained as a ^{14}C-labeled compound. It acts by binding covalently and irreversibly to a proteolipid subunit of the mitochondrial ATPase. Thus it is possible to study a specifically labeled protein subunit of the mitochondrial ATPase complex both in mutants and in the progeny of inheritance experiments. The following is a method of labeling the DCCD-binding protein of tissue culture cells.

Mitochondria at 3 mg protein/ml were incubated with [^{14}C]dicyclohexyl-carbodiimide (2–8 nmole/mg protein) at 37°C for 30 min. The labeled

Table II. Mitochondrial ATPase Parameters for Wild-Type and an Oligomycin-Resistant Strain

Strain	Culture conditions: oligomycin, μg/ml	ATPase activity, μmole/min/mg protein	I_{50}	
			Oligomycin[a]	DCCD[b]
LM(TK)⁻	0	0.331	0.27	0.8
OLI 14	0	0.193	1.8	4.0
	5	0.10	100	—

[a] μg/mg protein.
[b] nmole/mg protein.

mitochondria were centrifuged at 14,000g for 20 min to remove unbound label and then solubilized in the SDS buffer for further analysis by polyacrylamide gel electrophoresis or column chromatography.

The DCCD-binding protein has a very low molecular weight (8000 Daltons based on amino acid analysis) and readily aggregates into dimeric and tetrameric forms. To resolve this low-molecular-weight protein and to study the aggregates, column chromatography in SDS and urea provides an excellent system. After labeling with [^{14}C]DCCD and centrifugation as described above, mitochondria were solubilized at 2 mg protein/ml in 1% SDS 1% mercaptoethanol, 6 M urea, 0.01 M Na phosphate, pH 7.2, without heating. The samples (0.1–1.0 mg protein) were mixed with 1–2 mg unlabeled marker proteins such as myoglobin (mol. wt. 17,200) and cytochrome C (mol. wt. 11,700) and applied to a Bio Gel P-100 column (BioRad). A column with dimensions of 1.65 × 85 cm gave very high resolution. The sample was eluted with 0.1% SDS, 6 M urea, 0.1 M Na phosphate, pH 7.2, at a flow rate of about 4 ml/hr. Addition of 0.02% sodium azide to the elution buffer prevented microbial contamination of the bed. Marker positions were determined by absorbance of fractions at 280 nm. Fractions were then counted for radioactivity.

Figure 3 shows a typical elution pattern for the DCCD-binding protein of LM(TK$^-$). This method gives linear resolution of proteins in the molecular

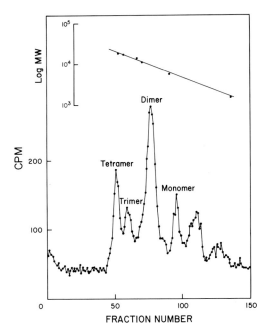

Figure 3. Chromatography of [^{14}C]DCCD-labeled mitochondria in SDS and urea. Labeled mitochondria were solubilized in 1% dodecylsulfate, 1% mercaptoethanol, 6 M urea, 0.01 M Na phosphate, pH 7.2. The sample (100 μg protein), mixed with added marker proteins, was applied to a P-100 column (1.6 × 85 cm) and eluted with 0.1% SDS, 6 M Urea, 0.1 M Na phosphate, pH 7.2, at a flow rate of 4 ml/hr. Inset shows the positions of markers: lactoglobulin (18,400), myoglobin (17,200), lysozyme (14,300), cytochrome C (11,700), insulin (6000), and bacitracin (1450).

weight range 18,300 to 1450. The average molecular weights of the monomer, dimer, and tetramer are 5300, 10,600, and 21,500, respectively, in this system. A common feature of the DCCD-binding proteins from many sources is an apparent molecular weight (determined by gel electrophoresis or chromatography in SDS) that is lower than the molecular weight calculated from amino acid analysis. The labeled material eluting with a molecular weight of 1500–2500 is nonspecifically associated [^{14}C]DCCD.

In an oligomycin-resistant mutant OLI 14 we found that all three forms of the DCCD-binding protein have a decreased molecular weight compared to the wild type (Kuhns and Eisenstadt, 1982). This difference between the mutant and wild-type protein is useful in confirming the results of inheritance experiments involving oligomycin resistance and in studying the expression of this gene in inter- and intraspecific hybrids.

6. Inheritance of Oligomycin Resistance

In mammalian cells the cytoplasmic location of a genetic determinant can be demonstrated by fusion of cytoplasts of the cell carrying the determinant with cells lacking the determinant (cybrid formation). The nuclear location of a genetic determinant can be demonstrated by fusion of karyoplasts of the cell carrying the determinant with cells lacking the determinant or by fusion of the cells (hybrid formation).

OLI 14 was enucleated as described previously and the cytoplasts were fused with LAN-2 cells. LAN-2 carries the nuclear marker for α-amanitin resistance and is sensitive to oligomycin and HAT (hypoxanthine–aminopterin–thymidine). Selection was done in α-amanitin and oligomycin as shown in Table III. No increase in the low frequency of background colonies in oligomycin occurred when OLI 14 cytoplasts were fused with LAN-2. None of the background colonies could be grown or cloned in oligomycin.

This experiment was repeated many times with identical outcomes. The results strongly suggest that oligomycin resistance in OLI 14 is not cytoplasmically inherited, but that it is nuclear in origin.

To test the validity of this conclusion, experiments to test directly for nuclear inheritance of oligomycin resistance were undertaken. If the gene for oligomycin resistance is located in the nucleus, resistance should be transferred by the karyoplast or minicell fraction from the mutant. The experiment was done by fusing the OLI 14 karyoplasts with LAN-2 and selecting for hybrids in HAT or HAT plus oligomycin. Neither parent can clone in HAT medium. Clones resistant to both HAT and oligomycin were found at high frequency (Table III). The cells of these clones were very large, had a chromosome composition characteristic of hybrid cells, and could be recloned in HAT medium containing 5 μg/ml oligomycin.

Direct hybridization experiments were carried out between OLI 142 (a subclone of OLI 14) and LAN-2. The frequency of clones arising in HAT plus oligomycin was 65% of that in HAT alone. Fusion of two oligomycin-sensitive strains resulted in hybrid clones in HAT but none in HAT and oligomycin.

Table III. Inheritance of Oligomycin Resistance in Cytoplast–Cell, Karyoplast–Cell, and Cell–Cell Fusions[a]

Fusion	Selection	Frequency, colonies/10^5 sensitive cells
Cytoplast × cell		
en OLI 14 × LAN-2	AM + Oligo 1	15.8
(Oligo^R, AM^S) (Oligo^S, AM^R)	AM + Oligo 2	8.5
LAN-2 (Oligo^S, AM^R)	AM + Oligo 1	20.4
	AM + Oligo 2	7.0
Karyoplast × cell		
kp OLI 14 × LAN-2	HAT + Oligo 1	12.6
(Oligo^R, HAT^S) (Oligo^S, HAT^S)	HAT + Oligo 2	13.7
Cell × cell		
OLI 142 × LAN-2	HAT	8.0
(Oligo^R, HAT^S) × (Oligo^S × HAT^S)	HAT + Oligo 3	5.2
LM(TK⁻) × LAN-2	HAT	1.7
Oligo^S, HAT^S) (Oligo^S × HAT^S)	HAT + Oligo 2	0.0
LAN-2	HAT	0.0
	HAT + Oligo 1	0.0

[a] Fusion was carried out with PEG. en denotes the cytoplast fraction and kp denotes the karyoplast fraction of an enucleated culture. Abbreviations: AM, α-amanitin; HAT, hypoxanthine–aminopterin–thymidine. Concentration of AM in selection media, 3.5 µg/ml. Concentrations of oligomycin shown in µg/ml.

The presence of a sensitive hybrid nucleus did not generate resistance to even low levels of oligomycin.

The positive transfer of oligomycin resistance with OLI 14 karyoplasts was demonstrable at the level of the mitochondrial ATPase. Ten clones resistant to HAT and oligomycin from the karyoplast–cell and cell–cell fusion experiments were examined. All of the clones assayed exhibited oligomycin- and DCCD-resistant ATPase activity. The source of this mutant enzyme must be OLI 14. Since colonies resistant to oligomycin and which contained mutant enzyme characteristics arose in the karyoplast–cell fusions and not in the cytoplast–cell fusions, we conclude that oligomycin resistance is under nuclear control in OLI 14. The resistance to oligomycin is, in all likelihood, the result of a mutation in the nuclear gene coding for the DCCD-binding protein of the mammalian mitochondrial ATPase.

These mutant cell lines can prove to be useful in the study of the genetic and biochemical interaction of the nuclear and mitochondrial genomes as well as in complementation analysis leading to gene assignment and mapping. The existence of a detectable variant protein will also permit detailed analysis of the interaction of the subunits of the complex multisubunit ATPase complex.

References

Anderson, S., Bankier, A. T., Barrell, B. G., de Bruijn, M. H. L., Coulson, A. R., Drouin, J., Eperon, I. C., Nierlich, D. P., Roe, B. A., Sanger, F., Schreier, P. H., Smith, A. J. H., Staden, R., and Young,

I. G., 1981, Sequence and organization of the human mitochondrial genome, *Nature* **290**:457–465.

Avner, P. R., Coen, D., Dujon, B., and Slonimski, P. P., 1973, Mitochondrial genetics. IV. Allelism and mapping studies of oligomycin resistant mutants in S. cerevisiae, *Mol. Gen. Genet.* **125**:9–52.

Breen, G. A. M., and Scheffler, I. E., 1980, Cytoplasmic inheritance of oligomycin resistance in Chinese hamster ovary cells, *J. Cell. Biol.* **86**:723–729.

Bulos, B., and Racker, E., 1968, Partial resolution of the enzymes catalyzing oxidative phosphorylation, *J. Biol. Chem.* **243**:3891–3900.

Bunn, C. L., Wallace, D. C., and Eisenstadt, J. M., 1977, Cytoplasmic inheritance of chloramphenicol resistance in mouse tissue culture cells, *Proc. Natl. Acad. Sci. USA* **71**:1681–1685.

Cattel, K. J., Knight, I. G., Lindop, G. R., and Beechey, R. B., 1970, The isolation of dicyclohexylcarbodiimide binding proteins from mitochondrial membranes, *Biochem. J.* **117**:1011–1013.

Criddle, R. S., Arulanadan, C., Edwards, T., Johnson, R., Scharf, S., and Enns, R., 1979, Investigation of the oligomycin binding protein in yeast mitochondrial ATPase, in: *Genetics and Biogenesis of Chloroplasts and Mitochondria* (T. Bucher, ed.), Elsevier North-Holland, Amsterdam, pp. 151–157.

Edwards, D. L., and Urger, B. W., 1978, Nuclear mutations conferring oligomycin resistance in *Neurospora crassa*, *J. Biol. Chem.* **253**:4254–4258.

Hensgens, L. A. M., Grivell, L. A., Borst, P., and Bos, J. L., 1979, Nucleotide sequence of the mitochondrial structural gene for subunit 9 of yeast ATPase complex, *Proc. Natl. Acad. Sci. USA* **765**:1663–1667.

Howell, N., and Sager, R., 1979, Cytoplasmic genetics of mammalian cells: Conditional sensitivity to mitochondrial inhibitors and isolation of new mutant phenotypes, *Somat. Cell Genet.* **5**:833–845.

Jackl, G., and Sebald, W., 1975, Identification of two products of mitochondrial protein synthesis associated with mitochondrial adenosine triphosphatase from *Neurospora Crassa*, *Eur. J. Biochem.* **54**:97–106.

Kiehl, R., and Hatefi, Y., 1980, Interaction of [^{14}C] dicyclohexylcarbodiimide with complex V (mitochondrial adenosine triphosphate synthetase complex), *Biochemistry* **19**:541–548.

Koslov, I. A., and Skulacher, V. P., 1977, H $\pm$ Adenosine triphosphate and membrane energy coupling, *Biochim. Biophys. Acta* **463**:29–89.

Kuhns, M. C., and Eisenstadt, J. M., 1979, Oligomycin-resistant mitochondrial ATPase from mouse fibroblasts, *Somat. Cell Genet.* **5**:821–832.

Kuhns, M. C., and Eisenstadt, J. M., 1981a, Nuclear inheritance of oligomycin resistance in mouse L cells, *Somat. Cell Genet.*, **7**:737–750.

Kuhns, M. C., and Eisenstadt, J. M., 1981b, A modified DCCD-binding protein in mouse fibroblast mutants with decreased sensitivity to oligomycin and dicychohexylcarbodiimide, *Fed. Proc.* **40**:1277a.

Kuhns, M. C., and Eisenstadt, J. M., 1982, The dicyclohexylcarbodiimide-binding protein of the mitochondrial ATPase from mouse L cells, submitted.

Lagarde, A. E., and Siminovitch, L., 1979, Studies on Chinese hamster ovary mutants showing multiple cross-reference to oxidative phosphorylation inhibitors, *Somat. Cell Genet.* **5**:847–871.

Lichtor, T., and Getz, G. S., 1978, Cytoplasmic inheritance of rutamycin resistance in mouse fibroblasts, *Proc. Natl. Acad. Sci. USA* **75**:323–328.

Macino, G., Scazzocchio, C., Waring, R. B., McPhail Berks, M., and Davies, R. W., 1980, Conservation and rearrangement of mitochondrial structural gene sequences, *Nature* **288**:404–406.

Pullman, M. E., Penefsky, H. S., Datta, A., and Racker, E., 1960, Partial resolution of the enzymes catalyzing oxidative phosphorylation, *J. Biol. Chem.* **235**:3322–3329.

Roberton, A. M., Holloway, C. T., Knight, I. G., and Beechey, R. B., 1968, *Biochem. J.* **108**:444–456.

Rowlands, R. T., and Turner, G., 1973, Nuclear and extranuclear inheritance of oligomycin resistance in *Aspergillus nidulans*, *Mol. Gen. Genet.* **126**:201–216.

Sebald, W., and Wachter, E., 1978, Amino acid sequence of the putative protophore of the energy-transducing ATPase complex, in: *Energy Conservation in Biological Membranes* (G. Schäfer, and M. Klingenberg, eds), Springer-Verlag, New York, pp. 228–236.

Sebald, W., Graft, T., and Wild, G., 1976, Cytoplasmic synthesis of the dicyclohexylcarbodiimide-binding protein in *Meurospora Crassa*, in: *Genetics and Biogenesis of Chloroplasts and Mitochondria* (T. Bucher, ed.), Elsevier North-Holland, Amsterdam, pp. 167–174.

Sebald, W., Wachter, E., and Tzagoloff, A., 1979, Identification of amino acid substitutions in the dicyclohexylcarbodiimide-binding subunit of the mitochondrial ATPase complex from oligomycin-resistant mutants of Saccharomyces cerevisiae, *Eur. J. Biochem.* **100**:559–607.

Senior, A. E., 1973, The structure of mitochondrial ATPase, *Biochim. Biophys. Acta* **301**:249–277.

Sierra, M. F., and Tzagoloff, A., 1973, Assembly of the mitochondrial membrane system. Purification of a mitochondrial product of the ATPase, *Proc. Natl. Acad. Sci. USA* **70**:3155–3159.

Turner, G., Imam, G., and Küntzel, H., 1979, Mitochondrial ATPase complex of *Aspergillus nidulans* and the dicyclohexylcarbodiimide-binding protein, *Eur. J. Biochem.* **97**:565–571.

Tzagoloff, A., Akai, A., and Needleman, R. B., 1975, Characterization of nuclear mutants of Saccharomyces cerevisiae with defects in mitochondrial ATPase and respiratory enzymes, *J. Biol. Chem.* **250**:8228–8235.

Wigler, M. H., and Weinstein, I. B., 1975, A preparative method for obtaining enucleated mammalian cells, *Biochem. Biophys. Res. Commun.* **63**:669–674.

Chapter 10
Cytoplasmic Inheritance of Rutamycin Resistance in Mammalian Cells

GODFREY S. GETZ and KATHLEEN L. KORNAFEL

1. Introduction

The yeast *Saccharomyces cerevisiae* has proven the most informative model organism for the study of mitochondrial biogenesis and genetics. This is attributed to its ability to grow anaerobically and therefore without mitochondrial respiration. It was its ability to grow in the absence of mitochondrial protein synthesis that initially led Slonimski and colleagues (Coen et al., 1970) to seek mutants resistant to inhibitors of this process. In the last 10–15 years a rich library of information on mitochondrial genetics has been accumulated, so that the mitochondrial genome has probably become the best understood of eukaryotic genomes. The yeast mitochondrial genome is 70–76 kb in size and codes for the two ribosomal RNAs, 24 tRNAs, cytochrome b, three cytochrome oxidase peptides, and at least two ATPase peptides, as well as at least one ribosomal (Var) protein (Schatz and Mason, 1974; Borst and Grivell, 1978; Locker and Rabinowitz, 1979; Tzagoloff et al., 1979). The vast majority of the mitochondrial proteins are specified by the nuclear genome, are synthesized in the cytoplasm, in some cases as larger precursors, and are specifically imported into the mitochondria (Schatz, 1979). Several but not all of the yeast mitochondrial genomes contain intervening sequences (Bos et al., 1978; Haid et al., 1979; Bonitz et al., 1980). Indeed the complete nucleotide sequence of most of these genes has now been determined. A substantial portion of the yeast mitochondrial genome appears to consist of spacer DNA, whose role, if any, is unclear. Recombination between mitochondrial DNA molecules has been unequivocally demonstrated (Lewin et al., 1979).

A variety of mitochondrial mutants have been described in yeast. Cytoplasmic petites, usually involving a partial to complete deletion of the

---

GODFREY S. GETZ • Departments of Pathology and Biochemistry, University of Chicago, Chicago, Illinois 60637.   KATHLEEN L. KORNAFEL • Department of Pathology, University of Chicago, Chicago, Illinois 60637.

mitochondrial genome, survive only in facultative anaerobic organisms (Locker et al., 1979). The incomplete deletions have been most valuable in mapping and sequencing segments of the mitochondrial genome in yeast (Locker et al., 1979). Other useful mutants are those resistant to antibiotics affecting mitochondrial protein synthesis (chloramphenicol, etc.), cytochrome b function (antimycin), or mitochondrial ATPase (oligomycin). Finally, Tzagoloff pioneered the study of localized genetic lesions (mit⁻) in the enzymatically active inner membrane proteins, namely three cytochrome oxidase peptides, cytochrome b, and at least two ATPase peptides (Tzagoloff et al., 1975, 1979). Similar classes of mutants affecting mitochondrial protein synthesis are the syn⁻ mutants. In contrast to the cytoplasmic petite or syn⁻ mutants, which interrupt mitochondrial protein synthesis and therefore affect all mitochondrial functions depending upon mitochondrial genetic input, the mit⁻ and antibiotic-resistant mutants generally express selective mitochondrial dysfunctions related to one of the mitochondrial gene products.

Despite the large difference in the sizes of the mitochondrial genomes of yeast (70–76 kb of DNA) (Borst and Grivell, 1978; Locker and Rabinowitz, 1979) and mammalian organisms (human mitochondrial DNA has 16,569 base pairs) (Anderson et al., 1981), until recently the genetic information was thought to be very similar. It remains to be established whether mammalian and yeast mitochondrial DNA do have similar information content. The human mitochondrial genome shows enormous economy of its sequences; the two ribosomal RNAs, 22 tRNAs, and 13 proteins are packed tightly with few or no uncoding bases between them. Of the 13 presumptive proteins encoded in human mitochondrial DNA, the identities of five are assigned—three cytochrome oxidase peptides, one ATPase peptide, and cytochrome b. The remaining eight reading frames are unassigned. The representation of similar coding units in the mitochondrial genomes of nonmammalian eukaryotes remains to be established.

2. Mitochondrial Mutagenesis and Cytoplasmic Inheritance In Mammalian Cells

Since mammalian cells are obligate aerobes and presumably require some mitochondrial function for survival, phenotypic markers of the mitochondrial genome are not as readily obtained as in the facultative anaerobe *Saccharomyces cerevisiae*. However, the dominant inheritance of antibiotic resistance markers such as chloramphenicol, erythromycin, antimycin, and oligomycin (or rutamycin) resistances has also been useful in mammalian cells. Indeed, recently (Blanc et al., 1981), it has been reported that chloramphenicol resistance in mouse cytoplasmic mutants is attributable to a single nucleotide base change in the 3' end of the large ribosomal RNA subunit gene within a ten nucleotide segment that is homologous in human and yeast mitochondrial large ribosomal RNAs as well as in the 23 S rRNA of *Escherichia coli*. Chloramphenicol resistance in two yeast mutants has also been positioned in this same conserved nucleotide sequence, resulting from single base changes (Dujon, 1980).

A very much larger redundancy of the mitochondrial genome in mammalian cells (Bogenhagen and Clayton, 1974; Williamson et al., 1977) contrasted with yeast has probably accounted for the greater difficulty of isolating mutants of mitochondrial DNA in the former cells. Consequently, a number of strategies have been employed to generate mutants of the mammalian mitochondrial genome. In none of the systems employed has the efficacy of the mutagenesis been unequivocally demonstrated. Two devices have been used to reduce the number of mitochondrial DNA targets for mutagenesis. The expression of the resistance phenotype requires that mutant mitochondrial DNA molecules should multiply, usually under selective pressure, to become the predominant proportion of mitochondrial genomes within the mutant cell. Wiseman and Attardi (1978) have found that at low concentrations ethidium bromide, frequently used as a mutagen in mammalian cells, will reduce the number of mitochondrial DNA molecules per cell by approximately tenfold. They have used human cells pretreated in this way (Wiseman and Attardi, 1979) to improve the yield of mitochondrially inherited mutants. In this case mutagenesis was with 2-amino purine and cycloheximide or with manganese, thought to be strong mitochondrial mutagens in yeast (Putrament et al., 1973; Shannon et al., 1973). Nevertheless, it has not been established that these mutagenesis protocols work in mammalian cells as they apparently work in yeast. Several apparently cytoplasmically inherited mutants were isolated which were deficient in mitochondrial protein synthesis, were resistant to chloramphenicol, and had reduced activities of cytochrome oxidase and mitochondrial ATPase.

An alternative approach has been used by Kuhns and Eisenstadt (1979) (see Chapter 9, this volume) who mutagenized minicells produced by cytochalasin B treatment of mouse L cells. This treatment resulted in karyplasts retaining only about 10% of cytoplasm and mitochondria. By this procedure plus mutagenesis with ethyl methane sulfonate, 14 oligomycin resistant mutants were obtained. In this procedure the gene(s) responsible for this variation has been localized to the nucleus.

We have used an alternative approach to isolate mitochondrial mutants (Lichtor and Getz, 1978). We have taken advantage of the selective retention of a mitochondria-specific thymidine kinase in a bromodeoxyuridine-resistant mouse cell line lacking the function of the soluble cytoplasmic thymidine kinase (Clayton and Teplitz, 1972). In such cells 5-bromodeoxyuridine is selectively incorporated into mitochondrial DNA (Berk and Clayton, 1974), following which long-wavelength ultraviolet irradiation may be expected to enhance the yield of mitochondrial mutants. This protocol has yielded rutamycin-resistant mutants and has since also been used to obtain antimycin-resistant mutants. Though the efficacy of this mutagenesis is not established, the fact that mutants were obtained with a higher frequency than 10^{-6} suggests that it might indeed be effective. No mutants were obtained without the bromodeoxyuridine substitution or without the ultraviolet irradiation. The rutamycin resistant colonies took over a month to grow up—a delay that might be expected while the one or few altered mitochondria multiply to repopulate the cell cytoplasm with mitochondria able to sustain metabolism in the presence of rutamycin.

That cytoplasmic mutants have been developed requires the demonstration that the functional defect can be expressed at the level of the isolated mitochondria *and* that this mutant phenotype can be transmitted in a cybridization between a recipient cell and the mutant cytoplasm. This has been shown to be the case for resistance to several antibiotics affecting mitochondrial protein synthesis and for resistance to rutamycin or oligomycin. It was first shown for chloramphenicol resistance by Bunn et al. (1974). Very recently alterations of mitochondrial DNA structure have been correlated with the acquisition of chloramphenicol resistance (Blanc et al., 1981; Giles et al., 1980).

Isolated mitochondria from rutamycin-resistant cells have an ATPase which is relatively resistant to rutamycin inhibition. Cytoplasts from rutamycin-resistant TK⁻ clones were fused with A9—a mouse cell resistant to 8-azaguanine—and the products were selected in rutamycin plus 8-azaguanine. Rutamycin-resistant cybrid clones are sensitive to 5-bromodeoxyuridine and HAT and have the single chromosomes set of A9. The ATPase of the isolated mitochondria of the cybrid is also resistant to rutamycin. A much lower frequency of rutamycin-resistant and 8-azaguanine-resistant hybrids was obtained by fusion of the nucleated rutamycin-resistant mutant with A9 cells. This is best explained by the loss of the X chromosome of the rutamycin-resistant partner, thus removing the source of the hypoxanthine guanine phosphoribosyl transferase from the hybrid. No rutamycin- and 8-azaguanine-resistant cell was obtained when A9 was fused either with enucleated or nucleated rutamycin-sensitive clone 1D, the parent of the rutamycin-resistant mutant. The only alternative to the likely conclusion that the cybrids result from the cytoplasmic transfer of genetic information conferring rutamycin resistance is a very high frequency of spontaneous mutation by A9 recipient cells to rutamycin resistance (at the level of mitochondrial ATPase)—a circumstance not observed in several control fusions. This unlikely alternative can therefore be essentially discounted.

3. Oligomycin (Rutamycin)-Sensitive ATPase

The energy-transducing membranes of bacteria, chloroplasts, and mitochondria all harbor a proton-translocating ATPase which operates *in vivo* mainly in the ATP synthetase mode. This is the enzyme complex central to the coupling of respiration to phosphorylation. In all of these systems the complex is constructed on very similar lines. Each has a protruding particle containing the elements of the F_1ATPase connected to a membrane sector consisting of 2–4 peptides. The connecting peptides include the oligomycin-sensitivity-conferring protein. The precise number of separate peptides involved in and essential for the biologic function of this ATPase complex is not resolved, except perhaps in bacteria, where all peptides of the complex are specified by a single unc operon (Downie et al., 1979). The unc operon contains eight distinct structural genes coding for the five peptides of the F_1ATPase and three membrane peptides, including the membrane proteolipid. All of these eight peptides are similarly overproduced in E. coli when a lysogenic lambda phage including the unc operon is induced (Foster et al., 1980). All energy-transducing ATPases include a small molecular weight proteolipid, which appears to function as a protonophore (Criddle et al., 1977). When the proteolipid of yeast mitochondria was incorporated into artificial phospholipid vesicles, its proton-translocating activity could be inhibited by oligomycin. On the other hand, when the proteolipid was obtained from an oligomycin-resistant strain of yeast, the proton translocation in vesicles was much less sensitive to oligomycin inhibition. The relative insensitivity of the proteolipid to oligomycin was attributed to a single amino acid substitution in the peptide (Sebald et al., 1979). An acidic residue in the hydrophobic sector of the proteolipid is the site of interaction of low concentrations of dicyclohexylcarbodiimide with the ATPase complex (Tzagoloff et al., 1979). Oligomycin resistance in yeast has been assigned to mitochondrial genetic loci— oli-1, oli-2, and oli-3. Oli-1 and oli-3 mutants both affect the structure of the proteolipid, whose structure is unchanged in oli-2 mutants. The oli-2 locus is assigned to the structural gene on mitochondrial DNA for ATPase subunit 6 (or 3) (Macino and Tzagoloff, 1980). In contrast to yeast, the structural gene for the proteolipid of Neurospora crassa is in the nucleus (Sebald and Wachter, 1978). In mammalian cells such as the rat (DeJong et al., 1980) and man (Anderson et al., 1981) there appears to be no evidence for the mitochondrial specification of proteolipid structure. In the rat, the dicyclohexylcarbodiimide-binding protein appears to be made outside the mitochondria (DeJong et al., 1980). Human mitochondrial DNA does have a structural gene specifying subunit 6 of ATPase. One might therefore anticipate that oligomycin resistance in mammalian cells could be determined either in the mitochondria or the nucleus.

4. Oligomycin-Resistant Mutants in Mammalian Cells

A number of investigators have reported the isolation of oligomycin-resistant mutants. Kuhns and Eisenstadt (1979) found that oligomycin increased the doubling time by about twofold in their oligomycin-resistant L-cell mutants. The mitochondrial ATPase of these mutants was relatively resistant to oligomycin, the more so if the cells had been grown in the presence of the antibiotic. In some cases the ATPase of mutants was as much as 1000-fold more resistant to oligomycin. The ATPase was also relatively resistant to inhibition by dicyclohexylcarbodiimide and venturicidin. The locus of the mutant gene(s) was not established.

Chinese hamster cells resistant to oligomycin have been isolated by Breen and Scheffler (1979) and by Lagarde and Siminovitch (1979). Both mutations were apparently transmitted through the nucleus. The former mutant had a mitochondrial ATPase that was resistant to oligomycin. On the other hand, the latter mutant exhibited cross resistance to venturicidin and antimycin. However, respiration remained sensitive to all of these antibiotics even though the cell exhibited increased glycolysis.

Howell and Sager (1979) reported the isolation of an oligomycin-resistant mutant in cells that expressed chloramphenicol resistance in the absence of pyruvate. All three phenotypic markers, i.e., chloramphenicol resistance, oligomycin resistance, and pyruvate independence, appear to be cytoplasmically inherited as assessed by either direct cybrid transmission or by cotransmission of the three markers. According to these authors, wild-type cells were able to grow in both antimycin and oligomycin in the presence of pyruvate. In most other chloramphenicol-resistant mutants of mouse cells studied by these authors, pyruvate was required for the expression of the chloramphenicol resistance. Pyruvate in this case could be replaced by hypoxanthine or adenine and not nearly as effectively by α-ketoglutarate, uridine, or thymidine.

The human cell mutants resistant to rutamycin were found to have a mitochondrial ATPase that was fully sensitive to rutamycin inhibition (Wiseman and Attardi, 1979) and so were not studied further.

5. Rutamycin-Resistant Mouse Cells and Their ATPase

We have isolated a rutamycin-resistant mutant from clone 1D (C1D) by the irradiation of cells containing bromodeoxyuridine preferentially incorporated into the mitochondrial DNA (Lichtor and Getz, 1978). Rutamycin is a very close analog of oligomycin. Two subclones of this mutant were capable of growing in concentrations of rutamycin as high as 5 μg/ml, although most of the biochemical studies have been performed with cells grown in 0.1 μg rutamycin/ml. In all cases, growth occurred in medium lacking exogenous pyruvate. Growth in the presence of rutamycin was slower than that in the absence of the drug. Growth of this mutant was relatively resistant also to ossamycin, oligomycin, peliomycin (all analogs of rutamycin), leucinostatin, and efrapeptin. On the other hand, diuron and dicyclohexylcarbodiimide inhibited growth equally in clone 1D and the rutamycin-resistant mutant (TL-1). The ATPase of isolated mitochondria from TL-1 was relatively resistant to rutamycin and leucinostatin but not to efrapeptin, an inhibitor of the F_1 ATPase active site (Cross and Kohlbrenner, 1978). One possible basis for rutamycin resistance is reduced binding of the antibiotic to the sensitive protein. However, the rutamycin inhibition of oxidative phosphorylation by mitochondria from clone 1D and two rutamycin-resistant clones were indistinguishable (Fig. 1), suggesting that the affinity of the rutamycin-binding protein is unaltered at least in the intact ATP synthetase. This is reminiscent of the observations with isolated mitochondria from an *oli-2* mutant in yeast, which also exhibited a normal sensitivity of ATP synthetase to oligomycin despite a resistance of the ATPase to the same antibiotic (Somlo, 1977).

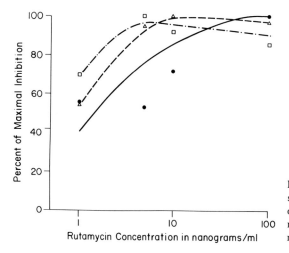

Figure 1. Percent of maximal inhibition of state III respiration versus rutamycin concentration. (●—●) Clone 1D; (▲- - -▲) rutamycin-resistant subclone 1; (☐- · -☐) rutamycin-resistant subclone 2.

With prolonged growth (3–4 years of TL-1 in 0.1 µg/ml of rutamycin) the mitochondrial ATPase appeared to be slightly less resistant to rutamycin than it had been. Growth of this mutant for about 100 generations in the absence of antibiotic did not further modify the response of ATPase to rutamycin. Many but not all cells cultivated in the absence of rutamycin grew readily upon re-exposure to the antibiotic.

The rutamycin-resistant phenotype was transferred to an 8-azaguanine-resistant mouse cell by the cytoplast of TL-1. The ATPase of these cybrids was highly resistant to rutamycin in cells grown continuously in 1 µg/ml of the antibiotic. In this case, too, the rutamycin resistance of mitochondrial ATPase was somewhat attenuated following many years growth under these conditions, although even then the resistance to the antibiotic remained much greater than the parent cells. In contrast to the situation with the original mutant, the ATPase of the cybrid grown for 50–100 generations in the absence of a selective agent was fully sensitive to rutamycin (Table I). The uninhibited ATPase specific activity of both the mutant and cybrid grown in the absence of rutamycin was higher than that of cells grown in the presence of the antibiotic. This raises the possibility that rutamycin remains bound to

Table I. ATPase Activities and Inhibition by Rutamycin

Cell line[a]	v_0 (µmPi/min/mg mt[b]/protein)	µg Rut/mg mt[b] protein required to cause 40% inhibition
C1D	0.180	0.12
TL-1	0.144	0.70
TL-1A	0.163	0.50
A9	0.245	0.03
enTL-1 × A9	0.174	40.00
enTL-1 × A9A	0.229	0.03

[a] enTL-1 × A9 denotes the cybrid. An A following a cell line name denotes growth in the absence of selective agents for >100 generations.
[b] Mitochondrial.

mitochondria of the latter cells, so that the ATPase of these mitochondria is already partially inhibited at the time of isolation. But this clearly cannot account for the rutamycin resistance of mutant TL-1 cells grown for many generations without the selective agent. Even in the cybrid the difference in rutamycin sensitivity of mitochondrial ATPase from cells grown in the presence and absence of rutamycin is not simply due to the removal of the inhibitor, as the ATPase of cybrid cells after a single transfer generation in the absence of the antibiotic retained a high though attenuated resistance to rutamycin. The cybrid presumably has at least two populations of mitochondria, one resistant to rutamyacin derived from TL-1 and one sensitive to rutamycin derived largely from the recipient cell. The selective conditions of growth probably determine the balance between these two mitochondrial populations, with the sensitive population predominating following prolonged growth without rutamycin selection. The ability of some cybrid cells to grow readily in rutamycin after prolonged growth in its absence suggests that the proportion of resistant to sensitive mitochondria can be promptly modified. Although some cells readily survive and multiply in rutamycin after long periods without selection pressure, some cells were promptly killed by rutamycin. These observations suggest that there is limited recombination between the two populations of mitochondrial genomes in the cybrid and that the two populations of mitochondria are not uniformly distributed throughout the cells of the culture. It is noteworthy that in contradistinction to the cybrid ATPase, the ATPase resistance of TL-1 mitochondria varies very little whether the cells are grown in the presence of selective agent or not.

6. Pleiotropic Characteristics of Rutamycin-Resistant Mutants

One of the advantages of the study of functionally deficient mutants of the mitochondria is the potential they provide for gaining insight into the normal functioning of the organelle as well as enabling understanding of the adaptations made in the physiology of the cell in response to such functional deficiencies. Among the metabolic alterations observed in the rutamycin-resistant mouse L cells were a very poor cloning efficiency, especially in the presence of rutamycin, an excessive lactic acid production, glucose dependence, a respiratory defect demonstrated mainly in complex I, and a defect in oxidative phosphorylation.

6.1. Glucose Dependence and Lactic Acid Production

One of the notable features of the rutamycin-resistant mutant was the very large amount of lactic acid produced during growth (Lichtor et al., 1979). At least five times as much lactic acid was produced by TL-1 than by clone 1D grown for 2 days in RPMI medium, despite the fact that the latter culture contained three times as many cells. Rutamycin-resistant cybrids also produced much more lactic acid than did A9 recipient cells. Less lactic acid

Table II. Growth Rates (GR) and Lactate Production (LP) as a Function of Hexose Source and Selective Agent

Cell line	Glucose		Glucose/Rut[b]		Galactose	
	GR[a]	LP	GR	LP	GR	LP
C1D	+	+	D	D	−	++
TL-1	−	+++	−	++++	D	D
TL-1A	+	++	+	+++	−	++
A9	−	++	D	D	−	+++
enTL-1 × A9	−	++++	−	++++	D	D
enTL-1 × A9A	+	++	−	+++	−	+++

[a] A plus sign denotes generation times <24 hr; a minus sign denotes generation times >24 hr. D denotes death.
[b] Rutamycin concentration for C1D, TL-1, and TL-1A was 0.1 µg/ml, and for A9, enTL-1 × A9, and enTL-1 × A9A was 1.0 µg/ml.

was produced by mutant cells and cybrid cells grown for long periods in the absence of rutamycin, though the acid produced in these cases still significantly exceeded the amount liberated by clone 1D and A9, respectively (Table II).

The lactic acid production by the mutant cells draws attention to the high usage of glycolysis with an impairment of respiration (see below). This was reinforced by the inability of mutant cells, both TL-1 and cybrids, to grow in galactose (plus pyruvate, i.e., Leibovitz medium); clone 1D and A9 were able to grow in this substrate very readily. Exclusion of rutamycin from Leibovitz medium did not permit growth of mutant or cybrid cells. However, supplementation of Leibovitz medium with glucose did support cell growth of both strains. When either mutant or cybrid was grown for several months in the absence of rutamycin, not only was less lactic acid produced, but the cells apparently regained the parental capacity to grow in Leibovitz medium. Addition of rutamycin to Leibovitz medium promptly killed these cells.

Others have observed glucose auxotrophy or dependence in respiration-deficient mutants. Thus Wiseman and Attardi (1979) observed the glucose dependence of their several respiration-deficient human cell mutants. These cytoplasmically inherited mutants were resistant to chloramphenicol and had reduced mitochondrial protein synthesis as well as reduced mitochondrial inner membrane enzymes, such as cytochrome oxidase, succinate-cytochrome c reductase, and oligomycin-sensitive ATPase. Howell and Sager (1979) have also reported that cytoplasmically inherited chloramphenicol-resistant mutants require glucose for the expression of their antibiotic resistance. Similar effects were observed in Chinese hamster cells by Ziegler and Davidson (1979). Respiratory-deficient mutants, determined by nuclear mutation, had earlier been reported to require not only glucose but also asparagine and carbon dioxide (Ditta et al., 1976). Indeed, the rutamycin-resistant mutant, TL-1, unlike clone 1D, was unable to grow in the absence of carbon dioxide (Lichtor, 1980). A series of Chinese hamster mutants have recently been reported to lack the ability to grow on galactose and to be deficient in complex I and complex III of the electron transport chain as well as having an uncoupled respiration (Whitfield et al., 1981).

6.2. Respiratory Deficiency

The rapid production of lactic acid suggested that the three carbon compounds produced by glycolysis could not be readily oxidized to carbon dioxide in vivo. To ascertain whether this was so, several features of the terminal respiratory pathway have been assessed (Lichtor et al., 1979). Endogenous respiration of whole cells of three of the mutant clones (TL-1, TL-2, and TL-3) was reduced by tenfold, but was substantially stimulated by the provision of exogenous succinate. The endogenous respiration of the cybrid was not seriously impaired. The respiration by isolated mitochondria revealed a marked deficiency of complex I, i.e., rotenone-sensitive NADH cytochrome c reductase, in the three mutants and the cybrid formed from one of these mutants. The oxidation of those substrates generating NADH was notably impaired, i.e., pyruvate, glutamate, and malate. Yet malate and glutamate dehydrogenase were either increased or unchanged. However, pyruvate decarboxylase was reduced by two-thirds in the mitochondria of all these cells. It is noteworthy that respiration supported by succinate was sensitive to inhibition by antimycin in the mitochondria from rutamycin-resistant mutants and the derived cybrid. The presence of fat droplets in the cytoplasm of TL-1 and the cybrid represents morphologic evidence of the defect in the oxidative capacity of the mitochondria.

The other complexes of the terminal respiratory pathway, succinate cytochrome c reductase and cytochrome c oxidase, were not significantly reduced, thus indicating that mitochondrial protein synthesis is probably unimpaired in these mutants. These cells were also sensitive to chloramphenicol inhibition, unlike the mitochondrial protein synthesis-deficient human mutants described by Wiseman and Attardi (1979), whose studies also revealed reduced succinate cytochrome c reductase and cytochrome c oxidase in their mutants.

The mitochondria from mutants TL-1 and TL-2 and from parent cell clone 1D all had similar state IV respiration (Table III), but this respiration in the mutants was insensitive to rutamycin inhibition. State IV respiration in clone 1D was inhibited about 20–25% by rutamycin. Respiration in clone 1D mitochondria with succinate as substrate was notably stimulated by the addition of ADP, yielding a respiratory control ratio of more than 2. ADP stimulated the respiration of rutamycin-resistant mitochondria to a much lesser extent, but in all cases the ADP-dependent respiration was equally sensitive to rutamycin inhibition (Fig. 1). Measurements of oxidative phosphorylation have not been made in the cybrid.

The respiratory behavior of the rutamycin-resistant mutant is very reminiscent of a nuclear-coded hamster mutant described by De Francesco et al. (1976). These mutants had a decreased rotenone-sensitive NADH coenzyme Q reductase as well as an impairment of pyruvate decarboxylase. Malate and glutamate dehydrogenase activities were normal in their mutants, which were, however, unable to grow in the absence of glucose and carbon dioxide. Similar Chinese hamster mutants were also described by Whitfield et al. (1981).

Table III. Effect of Rutamycin on Respiration of Mouse Mitochondria

	Oxygen consumed, nmole/min/mg protein	
Cell type	+3 mM Succinate	+3 mM Succinate + 0.3 mM ADP
Clone 1D		
No rutamycin	25.4	60.0
+0.1 g/ml rutamycin	20.9	19.8
+0.01 g/ml rutamycin	18.9	31.1
+0.005 g/ml rutamycin	19.7	38.8
+0.001 g/ml rutamycin	19.4	37.3
Mutant subclone 1		
No rutamycin	24.0	29.8
+0.1 g/ml rutamycin	24.0	21.6
+0.01 g/ml rutamycin	22.5	21.3
+0.005 g/ml rutamycin	22.0	22.0
+0.001 g/ml rutamycin	20.0	25.3
Mutant subclone 2		
No rutamycin	23.3	37.0
+0.1 g/ml rutamycin	25.7	24.6
+0.01 g/ml rutamycin	24.4	23.5
+0.005 g/ml rutamycin	23.3	22.4
+0.001 g/ml rutamycin	22.4	28.3

There appeared to be a discrepancy in the capacity of the whole cybrid cells to respire with succinate as substrate and the succinate oxidation by isolated cybrid mitochondria. While cybrid cells respired with succinate as substrate as well as the recipient A9 cells, the succinate oxidase specific activity of cybrid mitochondria was only about 40% of that of A9 cell mitochondria. This discrepancy appeared to be partially attributable to an increased mass of mitochondria (by 2- to 4-fold) in the cybrid grown in the presence of rutamycin. A similar inference was suggested by the calculation of mitochondrial mass based on cytochrome c oxidase activity measurements in isolated cybrid mitochondria and whole-cell homogenates (Table IV). Cybrid cells appeared to be larger than the cells from which they were formed. Larger amounts of "mitochondrial" protein were recovered from a given amount of cybrid cells than from a similar number of mutant or parent cells.

Table IV. Estimation of Mitochondrial Mass Based on Specific Activities for Succinate Cytochrome c Reductase and Cytochrome c Oxidase

	Mitochondrial mass, mg mt protein/10^{10} cells	
	Succinate cytochrome c reductase method	Cytochrome c oxidase method
C1D	357	379
TL-1	387	274
A9	294	—
enTL-1 × A9	1743	1316
enTL-1 × A9A	—	463

The calculation of mitochondrial mass on the basis of measurements of enzyme activities in the homogenate and mitochondria would be inconclusive if the enzyme employed for this calculation was in a different state of activation or inhibition in the homogenate compared to the mitochondria. However, the fact that qualitatively similar outcomes were achieved from several and separate experiments employing succinate cytochrome c reductase or cytochrome c oxidase markers gives confidence in the conclusions. When the cybrids were grown for many generations in the absence of selective agent (rutamycin), the proportionate cellular mitochondrial mass was decreased to the level of the original mutants. These results, taken together with the results of the cybrid ATPase resistance to rutamycin, suggest that when grown in the presence of antibiotic the cybrid cells are populated by two populations of mitochondria, one sensitive to the antibiotic and one resistant, with the latter predominating. On the other hand, when cybrid cells were grown for long periods in the absence of antibiotic, the sensitive mitochondrial population predominated. A small population of resistant mitochondria derived from TL-1 and capable of being rapidly expanded remained within the cybrid cell containing rutamycin-sensitive ATPase.

7. Conclusions

In a rutamycin-resistant mouse L cell whose antibiotic resistance was cytoplasmically inherited, a resistance of mitochondrial ATPase to rutamycin inhibition and the deficiency of complex I (NADH cytochrome c reductase) of the terminal respiratory pathway was documented. It is not clear whether these are the results of two independently determined cytoplasmic gene mutations or whether the one functional defect is the secondary consequence of the other. It is likely that the defect in pyruvate decarboxylase, on the other hand, is secondary to the complex I deficiency, since it is known that the accumulation of NADH such as may be anticipated with the reduced flux through complex I would result in a lesser proportion of pyruvate dehydrogenase complex in the active state (Pettit et al., 1975). Other biochemical dysfunctions may be attributed to either the altered ATPase or the imperfectly functioning NADH reductase. The latter respiratory problem may result in the inability to oxidize malate and glutamate in mutant cells, in the reduced production of ATP via oxidative phosphorylation, and in the reliance upon glycolysis, manifested by the high lactic acid production and the inability of mutant cells to grow on less efficient substrates of glycolysis than glucose. The altered resistance of the ATPase to rutamycin is probably related to a similar resistance of state IV but not state III respiration with a reduced efficiency of oxidative phosphorylation even with succinate as substrate, despite its capacity to be readily oxidized.

The precise genetic lesion remains unclear. Heretofore there has been no description of a mitochondrial genetic input into NADH coenzyme Q reductase, i.e., an NADH cytochrome c reductase defect unaccompanied by a succinate cytochrome c reductase defect. However, it must be borne in mind that eight reading frames remain unassigned in mitochondrial DNA (Anderson et al., 1981). On the other hand, mitochondrial ATPase does harbor subunits encoded in mitochondrial DNA; at least the gene for subunit six is present in human mitochondrial DNA (Anderson et al., 1981). The *oli-2* locus in the yeast mitochondrial genome is related to ATPase subunit six (Macino and Tzagoloff, 1980). Oligomycin resistance in yeast, mapping at the *oli-2* locus, is associated with a resistance of the ATPase to antibiotic but a sensitivity of the ATP synthetase (Somlo, 1977)—a situation very similar to that observed in this mouse cell mutant. There is no gene or reading frame in the human mitochondrial DNA corresponding to the yeast gene for subunit nine or proteolipid. In this connection the cytoplasmic mutant described here is sensitive to dicyclohexylcarbodiimide, which binds to the proteolipid.

As the growth environment of rutamycin-resistant mutants is modified by the inclusion of antibiotic in the medium, especially in the cybrid, indirect evidence has been developed for a modification of both the proportion and mass of mitochondria per cell. The mechanism by which these modifications are effected remains to be established. It is not clear that the presence of rutamycin in the medium simply aborts the replication of mitochondria containing normal genomes or how the accumulation of mitochondria with sensitive ATPase is favored in the absence of rutamycin.

References

Anderson, S., Bankier, A. T., Barrell, B. G., de Bruijn, M. H. L., Coulson, A. R., Drouin, J., Eperon, I. C., Nierlich, D. P., Roe, B. A., Sanger, F., Schreier, P. H., Smith, A. J. H., Staden, R., and Young, I. G., 1981, Sequence and organization of the human mitochondrial genome, *Nature* **290**:457–465.

Berk, A. J., and Clayton, D. A., 1974, Mechanism of mitochondrial DNA replication in mouse L-cells: Asynchronous replication of strands, segregation of circular daughter molecules, aspects of topology and turnover of initiation sequences, *J. Mol. Biol.* **86**:801–824.

Blanc, H., Wright, C. T., Bibb, M. J., Wallace, D. C., and Clayton, D. A., 1981, Mitochondrial DNA of chloramphenicol-resistant mouse cells contains a single nucleotide change in the region encoding the 3′ end of the large ribosomal RNA, *Proc. Natl. Acad. Sci. USA* **78**:3789–3793.

Bogenhagen, D., and Clayton, D. A., 1974, The number of mitochondrial deoxyribonucleic acid genomes in mouse L and human cells, *J. Biol. Chem.* **249**:7791–7795.

Bonitz, S. G., Coruzzi, G., Thalenfeld, B. E., and Tzagoloff, A., 1980, Assembly of the mitochondrial membrane system. Structure and nucleotide sequence of the gene coding for subunit 1 of yeast cytochrome oxidase, *J. Biol. Chem.* **255**:11927–11941.

Borst, P., and Grivell, L. A., 1978, The mitochondrial genome of yeast, *Cell* **15**:705–723.

Bos, J. L., Heyting, C., Borst, P., Arnberg, A. C., and Van Bruggen, E. F. J., 1978, An insert in the single gene for the large ribosomal RNA in yeast mitochondrial DNA, *Nature* **275**:336–338.

Breen, G. A. M., and Scheffler, I. E., 1979, Respiration-deficient Chinese hamster cell mutants: Biochemical characterization, *Somat. Cell Genet.* **5**:441–451.

Bunn, C., Wallace, D. C., and Eisenstadt, J., 1974, Cytoplasmic inheritance of chloramphenicol resistance in mouse tissue culture cells, *Proc. Natl. Acad. Sci. USA* **71**:1681–1685.

Clayton, D. A., and Teplitz, R. L., 1972, Intracellular mosaicism (nuclear$^-$/mitochondrial$^+$) for thymidine kinase in mouse L cells, *J. Cell Sci.* **10**:487.

Coen, D., Deutsch, J., Netler, P., Petrochilo, E., and Slonimski, P. P., 1970, Mitochondrial genetics.

I—Methodology and phenomenology, in: *Symposia of the Society for Experimental Biology*, No. 24 (P. L. Miller, ed.) Cambridge University Press, London, pp. 449–496.

Criddle, R. S., Packer, L., and Shieh, P., 1977, Oligomycin-dependent ionophoric protein subunit of mitochondrial adenosine triphosphatase, *Proc. Natl. Acad. Sci. USA* **74**:4306–4310, 1977.

Cross, R. L., and Kohlbrenner, W. E., 1978, The mode of inhibition of oxidative phosphorylation by efrapeptin (A23871). Evidence for an alternating site mechanism for ATP synthesis, *J. Biol. Chem.* **253**:4865–4873.

DeFrancesco, L., Scheffler, I. E., and Bissell, M. J., 1976, A respiratory-deficient Chinese hamster cell line with a defect in NADH-coenzyme Q reductase, *J. Biol. Chem.* **251**:4588–4595.

DeJong, L., Holtrop, M., and Kroon, A. M., 1980, The biogenesis of rat-liver mitochondrial ATPase: Evidence that the N,N'-dicyclohexyl carbodiimide-binding protein is synthesized outside the mitochondria, *Biochim. Biophys. Acta* **606**:331–337.

Ditta, G., Soderberg, K., Landy, F., and Scheffler, I. E., 1976, The selection of Chinese hamster cells deficient in oxidative energy metabolism, *Somat. Cell Genet.* **2**:331–344.

Downie, J. A., Gibson, F., and Cox, G. B., 1979, Membrane adenosine triphosphatases of prokaryotic cells, *Ann. Rev. Biochem.* **48**:103–131.

Dujon, B., 1980, Sequence of the intron and flanking exons of mitochondrial 21S rRNA gene of yeast strains having different alleles at the ω and rib-1 loci, *Cell* **20**:185–197.

Foster, D. L., Mosher, M. E., Futai, M., and Fillingame, R. H., 1980, Subunits of the H^+-ATPase of *Escherichia coli*. Overproduction of an eight-subunit F_1F_0-ATPase following induction of a lamda-transducing phage carrying the *unc* operon, *J. Biol. Chem.* **255**:12037–12041.

Giles, R. E., Stroynowski, I., and Wallace, D. C., 1980, Characterization of the mitochondrial DNA in chloramphenicol resistant interspecific hybrids and a cybrid, *Somat. Cell Genet.* **6**:543–554.

Haid, A., Schweyen, R. J., Beckmann, H., Kaudewitz, F., Solioz, M., and Schatz, G., 1979, The mitochondrial COB region in yeast codes for apocytochrome b and is mosaic, *Eur. J. Biochem.* **94**:451–464.

Howell, N., and Sager, R., 1979, Cytoplasmic genetics of mammalian cells: Conditional sensitivity to mitochondrial inhibitors and isolation of new mutant phenotypes, *Somat. Cell Genet.* **5**:833–845.

Kuhns, M. C., and Eisenstadt, J. M., 1979, Oligomycin-resistant mitochondrial ATPase from mouse fibroblasts, *Somat. Cell Genet.* **5**:821–832.

Lagarde, A. E., and Siminovitch, L., 1979, Studies on Chinese hamster ovary mutants showing multiple cross-resistance to oxidative phosphorylation inhibitors, *Somat. Cell Genet.* **5**:847–871.

Lewin, A. S., Morimoto, R., and Rabinowitz, M., 1979, Stable heterogeneity of mitochondrial DNA in grande and petite strains of *S. cerevisiae*, *Plasmid* **2**:155–181.

Lichtor, T. R., 1980, Cytoplasmic inheritance of rutamycin resistance and a respiratory deficiency in mouse fibroblasts, Ph.D. Thesis, University of Chicago.

Lichtor, T., and Getz, G. S., 1978, Cytoplasmic inheritance of rutamycin resistance in mouse fibroblasts, *Proc. Natl. Acad. Sci. USA* **75**:324–328.

Lichtor, T., Tung, B., and Getz, G. S., 1979, Cytoplasmically inherited respiratory deficiency of a mouse fibroblast line which is resistant to rutamycin, *Biochemistry* **18**:2582–2590.

Locker, J., and Rabinowitz, M., 1979, An overview of mitochondrial nucleic acids and biogenesis, in: *Methods in Enzymology*, Vol. 56, *Biomembranes, Part G—Biogenesis of Mitochondria, Organization and Transport* (S. Fleischer and L. Packer, eds.), Academic Press, New York, pp. 3–16.

Locker, J., Lewin, A. S., and Rabinowitz, M. 1979, The structure and organization of mitochondrial DNA from petite yeast, *Plasmid* **2**:155–181.

Macino, G., and Tzagoloff, A., 1980, Assembly of the mitochondrial membrane system: Sequence analysis of a yeast mitochondrial ATPase gene containing the oli-2 and oli-4 loci, *Cell* **20**:507–517.

Pettit, F. H., Pelley, J. W., and Reed, L. J., 1975, Regulation of pyruvate dehydrogenase kinase and phosphatase by acetyl-CoA/CoA and NADH/NAD ratios, *Biochem. Biophys. Res. Comun.* **65**:575–582.

Putrament, A., Baranowska, H., and Prazo, W., 1973, Induction by manganese of mitochondrial antibiotic resistance mutations in yeast, *Mol. Gen. Genet.* **126**:357–366.

Schatz, G., 1979, How mitochondria import proteins from the cytoplasm, *FEBS Lett.* **103**:203–211.

Schatz, G., and Mason, T. L., 1974, The biosynthesis of mitochondrial proteins, *Ann. Rev. Biochem.* **43**:51–87.

Sebald, W., and Wachter, E., 1978, Amino acid sequence of the putative protonophore of the energy-transducing ATPase complex, in: *29th Mossbacher Colloquium on Energy Conservation* (G. Schaefer and M. Klingenberg, eds.), Springer-Verlag, Berlin, pp. 228–236.

Sebald, W., Wachter, E., and Tzagoloff, A., 1979, Identification of amino acid substitutions in the dicyclohexylcarbodiimide-binding subunit of the mitochondrial ATPase complex from oligomycin resistant mutants of *Saccharomyces cerevisiae*, *Eur. J. Biochem.* **100**:599–607.

Shannon, C., Enns, R., Wheelis, L., Burchiel, K., and Criddle, R. S., 1973, Alterations in mitochondrial adenosine triphosphatase activity resulting from mutation of mitochondrial deoxyribonucleic acid, *J. Biol. Chem.* **248**:3004–3011.

Somlo, M., 1977, Effect of oligomycin on coupling in isolated mitochondria from oligomycin-resistant mutants of *Saccharomyces cerevisiae* carrying an oligomycin resistant ATP phosphohydrolase, *Arch. Biochem. Biophys.* **182**:518–524.

Tzagoloff, A., Akai, A., Needleman, R. B., and Zulch, G., 1975, Assembly of the mitochondrial membrane system. Cytoplasmic mutants of *Saccharomyces cerevisiae* with lesions in enzymes of the respiratory chain and in the mitochondrial ATPase, *J. Biol. Chem.* **250**:8236–8242.

Tzagoloff, A., Macino, G., and Sebald, W., 1979, Mitochondrial genes and translation products, *Ann. Rev. Biochem.* **48**:419–441.

Whitfield, C. D., Bostedor, R., Goodrum, D., Haak, M., and Chu, E. H. Y., 1981, Hamster cell mutants unable to grow on galactose and exhibiting an overlapping complementation pattern are defective in the electron transport chain, *J. Biol. Chem.* **256**:6651–6656.

Williamson, D. H., Johnston, L. H., Richmond, K. M. V., and Game, J. C., 1977, Two aspects of mitochondrial DNA structure: The occurrence of two types of mitochondrial DNA in rat liver and the isolation from rat liver of DNA complexes of high buoyant density, in: *Mitochondria 1977, Genetics and Biogenesis of Mitochondria* (W. Bandlow, R. J. Schweyen, K. Wolf, and F. Kaudewitz, eds.), Walter deGruyter, Berlin, pp. 1–24.

Wiseman, A., and Attardi, G., 1978, Reversible tenfold reduction in mitochondrial DNA content of human cells treated with ethidium bromide, *Mol. Gen. Genet.* **167**:51–63.

Wiseman, A., and Attardi, G., 1979, Cytoplasmically inherited mutations of a human cell line resulting in deficient mitochondrial protein synthesis, *Somat. Cell Genet.* **5**:241–262.

Ziegler, M. A., and Davidson, R. L., 1979, The effect of hexose on chloramphenicol sensitivity and resistance in Chinese hamster cells, *J. Cell Physiol.* **98**:627–635.

Chapter 11
Erythromycin Resistance in Human Cells

CLAUS-JENS DOERSEN and ERIC J. STANBRIDGE

1. Introduction

The biogenesis of mitochondria depends on the cooperation of the nuclear and mitochondrial genomes. The remarkable nature of the interaction is best seen in the assembly of the mitochondrial protein-synthesizing apparatus and the respiratory enzyme complexes of the inner mitochondrial membrane (Schatz and Mason, 1974; Borst and Grivell, 1978; Tzagoloff et al., 1979). Our understanding of the role of the mitochondrial genome in the biogenesis of mitochondria in lower eukaryotes has benefited from the vast number of mutants affecting mitochondrial function (Borst and Grivell, 1978; Tzagoloff et al., 1979). Many of these mutants were selected by virtue of their resistance to specific inhibitors of mitochondrial protein synthesis and respiration. A similar approach has been applied to somatic cells in recent years. Resistance to the mitochondrial protein synthesis inhibitors chloramphenicol (Bunn et al., 1974; Wallace et al., 1975; Mitchell and Attardi, 1978; Munro et al., 1978) and erythromycin (Doersen and Stanbridge, 1979) has been shown to be cytoplasmically inherited. Noncytoplasmically inherited resistance to carbomycin (Bunn and Eisenstadt, 1977a) and erythromycin (Molloy and Eisenstadt, 1979) has also been reported. Respiration deficiency resulting from defects in mitochondrial protein synthesis has been shown to be both cytoplasmically (Wiseman and Attardi, 1979) and nuclearly (Soderberg et al., 1979) inherited. Cytoplasmic inheritance of resistance to inhibition of respiration (Harris, 1978) and oxidative phosphorylation (Lichtor and Getz, 1977; Breen and Scheffler, 1980) has also been demonstrated. It is hoped that the continued characterization of these mutants and isolation of additional ones will enhance our understanding of the expression of the mammalian

---

CLAUS-JENS DOERSEN • Department of Microbiology, California College of Medicine, University of California—Irvine, Irvine, California 92717. Present address: Division of Biology, California Institute of Technology, Pasadena, California 91125. ERIC J. STANBRIDGE • Department of Microbiology, California College of Medicine, University of California—Irvine, Irvine, California 92717.

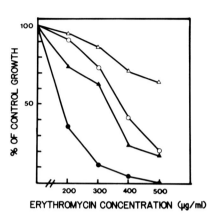

Figure 1. Effect of different buffering conditions on the growth sensitivity of HeLa and ERY2301 cells to increasing concentrations of ERY. After 6 days of culture the cells were harvested and counted using a hemocytometer. Each point represents the percent of growth in the absence of ERY and was calculated from the mean triplicate samples. HeLa (○) and ERY2301 (△) in media buffered by the bicarbonate–CO_2 system; HeLa (●) and ERY2301 (▲) in HEPES-buffered media.

mitochondrial genome, the organization and sequence of which has resulted from elegant molecular studies (Attardi, 1981).

We have focused our initial effort on isolating and characterizing mutants in human cell lines that are resistant to specific inhibitors of mitochondrial protein synthesis. Such mutants should prove to be useful tools for experimentally manipulating the expression of the mitochondrial genome and probing the complex interactions between nuclear and mitochondrial gene products in somatic cells. The inhibitory effects of erythromycin lactobionate (ERY*) on the growth of HeLa cells in culture and on cell-free mitochondrial protein synthesis have been reported (Doersen and Stanbridge, 1979). A cytoplasmically inherited mutant was isolated which displayed ERY-resistant mitochondrial protein synthesis activity. The initial characterization of two additional ERY-resistant mutants has revealed a new class of mitochondrial protein synthesis mutants in which the resistant phenotype is nuclearly encoded and inherited as a recessive trait in cell hybrids (Doersen and Stanbridge, 1982a).

2. Selection of Erythromycin-Resistant Cell Lines

2.1. Erythromycin Inhibition of Cell Proliferation Is pH Dependent

The most commonly used buffering system for cultured mammalian cells is the bicarbonate–CO_2-buffered medium. HeLa cells were relatively insensitive to the growth inhibitory effect of ERY in these culture conditions (Fig. 1). However, it can be seen that the growth of HeLa cells in sealed flasks containing medium buffered to pH 7.4 by the addition of 25 mM HEPES (N-2-hydroxyethylpiperazine-N'-2-ethanesulfonic acid) (Doersen and Stanbridge, 1979) was much more sensitive to the inhibitory effect of ERY. After several days of culture the bicarbonate–CO_2-buffered medium had a pH

---

*ERY refers specifically to erythromycin lactobionate. The term "erythromycin" is used to refer to all forms of erythromycin.

Table I. Buffer Combinations Used for Growth Media of Desired pH[a]

Buffer[b]	Buffer concentration, mM				
	pH = 7.0	7.2	7.4	7.6	7.8
PIPES	10	—	—	—	—
MOPS	10	15	10	10	—
HEPES	10	10	15	15	15
TES	—	—	10	10	15
NaH$_2$PO$_4$	10	2	2	—	—

[a] Final pH of growth medium was adjusted with 5 N NaOH or 5 N HCl. These formulations were adapted from Ceccarini and Eagle (1971) and Eagle (1971).
[b] PIPES, piperazine-N,N'-bis(2-ethanesulfonic acid). MOPS, morpholinopropanesulfonic acid. HEPES, N-2-hydroxyethylpiperazine-N'-2-ethanesulfonic acid. TES, N-tris(hydroxymethyl)methyl-2-aminoethanesulfonic acid.

of approximately pH 7.0–7.1, compared to the HEPES-buffered medium, which had a range of pH 7.3–7.4. That this difference in the pH of the media could account for the difference in the inhibitory effect of ERY was corroborated by employing media of specific pH and buffering capacity over a range of pH 7.0–7.8 (Table I) (Doersen and Stanbridge, 1982b). It should be noted that the greatest differential in ERY sensitivity between ERY2301, a cytoplasmically inherited ERY-resistant mutant, and HeLa was observed at pH 7.4, and demonstrates the appropriateness of the HEPES-buffered medium for the selection and maintenance of ERY-resistant cells (Doersen and Stanbridge, 1979).

2.2. Selection of ERY2301

The ERY-resistant cell line ERY2301 was isolated from HeLa cells treated with 0.5 µg/ml ethidium bromide for one cell generation (Doersen and Stanbridge, 1979). At this concentration of ethidium bromide the synthesis of mitochondrial DNA (Leibowitz, 1971; Smith et al., 1971; Nass, 1972; Wiseman and Attardi, 1978) and its transcription (Zylber et al., 1969) are preferentially inhibited in mammalian tissue culture cells. After allowing the cells to recover overnight, the ethidium bromide-treated cell population was exposed to 200 µg/ml ERY for approximately two population doublings. The concentration of ERY was then increased to 300 µg/ml. After an initial period of limited growth, the majority of the cells detached from the surface of the flask. These cells failed to proliferate when replated in ERY-free growth medium. The gradual loss of adherent cells continued until only a few colonies survived. These colonies, with the exception of ERY2301, were composed of large multinucleate cells which failed to proliferate upon subculture. ERY2301 retained the normal HeLa cell morphology and was serially cultured in the presence of 300 µg/ml ERY.

Although there is no direct evidence that the ethidium bromide was mutagenic in this selection system, similar treatments have been used in the selection of chloramphenicol-resistant cell lines (Spolsky and Eisenstadt,

1972; Bunn et al., 1974). It has been reported that ethidium bromide is not mutagenic in mammalian cells (Naum and Pious, 1971), because a quantitative induction of respiratory deficiency analogous to petite induction in yeast (Slonimski et al., 1968) was not observed. This may be due to the fact that mammalian cells contain from 1000 to 9000 mitochondrial DNA molecules (Bogenhagen and Clayton, 1974), or at least ten times the number of molecules per yeast cell (Williamson et al., 1977). Therefore, the expression of the mutant phenotype would be masked by the large excess of wild-type mitochondrial DNA molecules. Thus, only under an appropriate selective pressure would the mutant mitochondrial DNA molecule, by repopulating the cell to a sufficient level, be able to express its mutant phenotype at the cellular level.

2.3. Selection of ERY2305 and ERY2309

Several additional ERY-resistant clones were isolated by using a mutagenic procedure similar to the one employed in the selection of mitochondrial protein synthesis-deficient mutants (Wiseman and Attardi, 1979). HeLa cells were exposed to 50 ng/ml ethidium bromide for 2 days to reduce the mitochondrial target size. Wiseman and Attardi (1978) have shown that exposure to low concentrations of ethidium bromide results in a reversible reduction in the mitochondrial DNA content of human cells and that normal levels are quickly attained upon removal of the ethidium bromide. Therefore, after removal of the ethidium bromide the HeLa cells were immediately exposed to ethyl methane sulfonate at a concentration of 60 μg/ml for 24 hr. Following overnight recovery in growth medium the cells were exposed to 200 μg/ml ERY for 5 days, at which time the concentration of ERY was increased to 300 μg/ml. Approximately 1 week after the stepwise increase to 300 μg/ml ERY, colonies were readily apparent. Most of these colonies were composed of large multinucleate cells similar to those previously observed in the selection of ERY2301 (Doersen and Stanbridge, 1979). Several colonies with normal HeLa cell morphology were isolated 2–3 weeks later and the two most vigorously growing cell populations, ERY2305 and ERY2309, were chosen for further study. These ERY-resistant colonies were recovered with a frequency of approximately one per 10^6 cells originally exposed to the ethidium bromide. No colonies of cells capable of growing in the presence of ERY arose from untreated HeLa cell cultures or cells exposed only to ethyl methane sulfonate.

This mutagenic protocol was expressly used to obtain mutations in the mitochondrial DNA. It was therefore surprising that the ERY-resistant phenotype of ERY2305 and ERY2309 was found to be nuclearly encoded (see Section 4.2). However, it is clear that the ethidium bromide treatment combined with ethyl methane sulfonate greatly enhanced the recovery of ERY-resistant mutants. The synergistic effect of the ethidium bromide pretreatment may be due to the fact that concomitant with the reduction in mitochondrial DNA content, macromolecular complexes composed of both nuclear and mitochondrial gene products, e.g., the mitochondrial ribosomes and respiratory chain (Schatz and Mason, 1974; Attardi, 1981), are also depleted in ethidium bromide-treated cells (Wiseman and Attardi, 1978). Thus, an ethyl methane sulfonate-induced alteration in a nuclearly encoded gene product conferring ERY resistance at the level of mitochondrial protein synthesis would be phenotypically amplified during the period of mitochondrial recovery following the removal of the ethidium bromide.

2.4. Isolation of D98-ERY^r

D98-ERY^r represents a spontaneously arising ERY-resistant mutant which was isolated from a population of the hypoxanthine-guanine phosphoribosyl transferase-deficient HeLa cells line D98/AH-2 (Szybalski et al., 1962) exposed to 300 μg/ml ERY (Doersen and Stanbridge, 1979). In contrast to the selection of the other ERY-resistant cell lines, D98-ERY^r did not display an initial period of slow growth before expressing its final ERY-resistant phenotype.

3. Characterization of Erythromycin-Resistant Cell Lines

Several models can be proposed to explain the ability of the various ERY-resistant cell lines to continue to grow in the presence of the drug. The most intriguing possibility is that ERY resistance stems from an alteration in the mitochondrial protein-synthesizing apparatus. Alternatively, resistance may result from a change in either the plasma membrane or the mitochondrial membranes that affect the rate of transport of ERY. Finally, ERY resistance may be due to some unknown detoxification mechanism which inactivates the drug. Our preliminary characterizations of the mutant cells lines outlined below implicate the first two possibilities.

3.1. Cell Proliferation in the Presence of Erythromycin

As seen in Fig. 2, ERY2301, ERY2305, ERY2309 were similar in their relative growth sensitivity to increasing concentrations of ERY. The relative number of ERY2301, as well as ERY2305 and ERY2309, cells in the presence of 300 μg/ml reflects a growth rate that is somewhat less than observed in the absence of ERY. Only in the ERY concentrations greater than 400 μg/ml was any cell cytotoxicity noticed. In contrast, HeLa cells failed to proliferate at concentrations greater than 300 μg/ml and the few remaining adherent cells were no longer viable when replated in ERY-free medium (Doersen and Stanbridge, 1979). The ERY sensitivity of D98-ERY^r was intermediate, although at concentrations greater than 400 μg/ml ERY the level of resistance was comparable to the other ERY-resistant cell lines (Fig. 2). ERY2301,

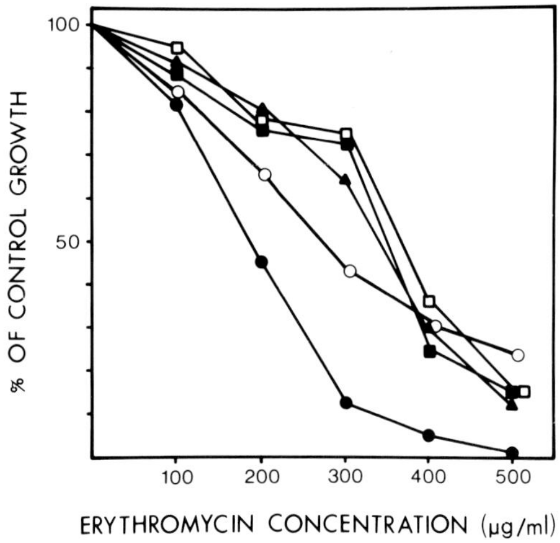

Figure 2. Effect of increasing concentrations of ERY on the proliferation of ERY-sensitive and ERY-resistant cell lines. After 6 days of culture the cells were harvested and counted using a hemocytometer. Each point represents the percent of growth in the absence of ERY and was calculated from the mean of triplicate samples. HeLa (●); D98-ERY^r (○); ERY2301 (▲); ERY2305 (■); ERY2309 (□).

ERY2305, ERY2309, and D98-ERY^r appear to be stably resistant to ERY, since all the resistant cells cultured in the absence of ERY for a number of population doublings grew to the same extent in ERY as parallel cultures maintained in the presence of ERY (Doersen and Stanbridge, 1979). All the ERY-resistant cell lines were similar to HeLa in that they failed to grow in the presence of chloramphenicol, suggesting the presence of an active chloramphenicol-sensitive mitochondrial protein-synthesizing apparatus.

3.2. *In Vivo* Mitochondrial Protein Synthesis

Mitochondrial protein synthesis can be assayed in cells growing in culture by using emetine or cycloheximide to selectively inhibit cytoplasmic protein synthesis (Mitchell et al., 1975). Briefly, equivalent numbers of cells (approximately 2×10^6 cells per 25-cm flask) were incubated at 37°C for 30 min in leucine-free growth medium supplemented with 5% dialyzed calf serum or the same medium containing ERY or chloramphenicol. Cycloheximide was then added to a final concentration of 300 µg/ml and 15 min later the cell cultures were exposed to [3,5-^{3}H]leucine (50 µCi/ml, 60 Ci/mmole) for 1 hr. The incorporation of [^{3}H]leucine was stopped by thoroughly washing the cells with ice-cold saline containing 100 mM cold leucine and hydrolyzing the cells in 1 N NaOH at 37°C for 30 min. After neutralizing with 1 N HCl, the samples were precipitated in ice-cold 10% trichloracetic acid and the radioactivity measured by liquid scintillation counting. The results in Table II indicate that ERY and chloramphenicol are able to inhibit mitochondrial protein synthesis in intact HeLa cells. A reasonable degree of inhibition by ERY and chloramphenicol, even at very high concentrations of the drug, was achieved only when the cells were preincubated for 30 min in the presence of the drug. Similar differences in the concentration of drug needed to inhibit cell growth and that needed to observe

Table II. *In Vivo* Mitochondrial Protein Synthesis

	Percent of cycloheximide-resistant incorporation[a]			
	Erythromycin		Chloramphenicol	
Cell line	300 µg/ml	800 µg/ml	50 µg/ml	200 µg/ml
HeLa	59 ± 5[b]	40 ± 2	29 ± 1	21 ± 4
ERY2301	89 ± 4	73 ± 1	28 ± 4	22 ± 1
ERY2305	85 ± 3	73 ± 6	19 ± 1	17 ± 6
ERY2309	84 ± 1	70 ± 8	19 ± 3	18 ± 4
D98-ERY[r]	86 ± 3	51 ± 8	25 ± 1	22 ± 3

[a] The incorporation of [^{3}H]leucine into acid-soluble material in the presence of 300 µg/ml cycloheximide was assayed for 1 hr at 37°C in HEPES-buffered minimal essential medium minus leucine supplemented with glutamine and 5% dialyzed calf serum.
[b] Values are the mean ± standard deviation of the percent of control incorporation for duplicate samples and were calculated from at least two separate experiments.

inhibition in *in vivo* mitochondrial protein synthesis assays have been reported for erythromycin (Molloy and Eisenstadt, 1979) and other mitochondrial protein synthesis inhibitors (Spolsky and Eisenstadt, 1972; Mitchell et al., 1975; Bunn and Eisenstadt, 1977a). Nevertheless, ERY2301, ERY2305, and ERY2309 all displayed significant ERY-resistant mitochondrial protein synthesis activity *in vivo*. D98-ERY[r] exhibited an intermediate ERY resistance; the incorporating activity of D98-ERY[r] was comparable to that observed for the other resistant cells at the lower concentration of ERY, but was significantly more sensitive to the higher concentration, although not to the extent observed for HeLa cells.

3.3. Cell-Free Mitochondrial Protein Synthesis

Cell-free mitochondrial protein synthesis assays have been essential in establishing the phenotype of mutants resistant to ERY (Doersen and Stanbridge, 1979) and chloramphenicol (Spolsky and Eisenstadt, 1972; Mitchell et al., 1975), as well as mutants defective in mitochondrial protein synthesis in respiratory-deficient cells (Ditta et al., 1977; Wiseman and Attardi, 1979). However, the cell-free characterization of the drug-resistant phenotype has been omitted from several recent reports dealing with putative mitochondrial protein synthesis mutants. The critical importance of the cell-free mitochondrial protein synthesis assays is exemplified by comparing the results obtained for D98-ERY[r] to those of the other ERY-resistant cell lines, ERY2301, ERY2305, and ERY2309 (Section 3.3.2).

Mitochondria were isolated in the 5000 g_{av} membrane fraction (Lederman and Attardi, 1970) by differential centrifugation of cell homogenates (Attardi et al., 1969) produced using a motor-driven A. H. Thomas homogenizer at approximately 1500 rpm. Endogenous mitochondrial protein synthesis was measured at 37°C by measuring the incorporation of [^{3}H]leucine into the cold trichloroacetic acid-insoluble material as previously described (Doersen and

Stanbridge, 1979, 1981). The reaction mixture contained, in a final volume of 250 μl, 50 mM Tricine-KOH, pH 7.8, 100 mM sucrose, 150 mM KCl, 10 mM $MgCl_2$, 5 mM Na_2HPO_4, 2 mM EDTA, 1 mM dithiothreitol, 2 mM ATP, 10 mM phosphenolpyruvate, 25 μg/ml pyruvate kinase, 100 μM of each amino acid minus leucine, 10 μCi/ml of [3,5-^{3}H]leucine (61 Ci/mmole), and the mitochondrial preparation at a final protein concentration of 0.5–1.5 mg/ml.

All mitochondrial preparations were refractory to the cytoplasmic protein synthesis inhibitor cycloheximide, but remained sensitive to chloramphenicol. Thus, the bulk of the incorporation of [^{3}H]leucine can be attributed to the mitochondrial protein-synthesizing apparatus and not due to contaminating rough endoplasmic reticulum. The use of the motor-driven A. H. Thomas homogenizer is essential in obtaining a membrane fraction particularly enriched in mitochondria.

3.3.1. Erythromycin Inhibition of Mitochondrial Protein Synthesis Is pH-Dependent

The pH of the reaction mixture is of critical importance in evaluating the inhibitory effect of ERY on HeLa mitochondrial protein synthesis (Doersen and Stanbridge, 1982b). The incorporation of [^{3}H]leucine into the cold acid-insoluble material by intact and Triton X-100 disrupted mitochondria was assayed in the previously described Tricine-KOH reaction mixture (Section 3.3) buffered to pH 7.4, 7.6, 7.8, or 8.0 with 1 N KOH. As seen in Table III, protein synthesis by intact mitochondria isolated from HeLa cells was relatively insensitive to ERY at pH 7.4 and 7.6, but at the higher pH's tested, pH 7.8 and 8.0 mitochondrial protein synthesis was inhibited approximately 50%. Increasing the concentration of ERY at a given pH did not result in a further increase in the level of ERY inhibition. When the mitochondrial

Table III. The Effect of pH on *In Vitro* Protein Synthesis by Intact and Triton X-100-Treated Mitochondria

				Percent of control incorporation[b]		
		Control incorporation,[a]	Cycloheximide	Erythromycin lactobionate		
Cell line	pH	cpm/mg	300 μg/ml	400 μg/ml	800 μg/ml	1200 μg/ml
HeLa	7.4	12,021	92 ± 4	79 ± 3	76 ± 4	68 ± 6
	7.6	11,768	93 ± 1	74 ± 7	72 ± 2	61 ± 1
	7.8	11,956	91 ± 2	57 ± 1	52 ± 1	49 ± 3
	8.0	10,075	90 ± 1	51 ± 4	44 ± 4	48 ± 1
HeLa + 0.01% Triton X-100	7.4	13,466	84 ± 5	70 ± 5	56 ± 6	45 ± 8
	7.6	12,640	89 ± 5	66 ± 2	58 ± 9	53 ± 4
	7.8	11,608	89 ± 4	54 ± 4	55 ± 1	48 ± 3
	8.0	10,934	89 ± 9	54 ± 6	49 ± 3	44 ± 6

[a] The incorporation of [^{3}H]leucine in the absence of inhibitors into the acid-insoluble fraction was assayed for 30 min and expressed as cpm/mg of mitochondrial protein.
[b] Values are the mean ± standard deviation of the percent of [^{3}H]leucine incorporated in the absence of inhibitors for each mitochondrial preparation and were calculated from at least two separate experiments.

membranes were disrupted by Triton X-100, ERY inhibition of HeLa mitochondrial protein synthesis was also pH dependent. In addition, at the lower pH's tested, pH 7.4 and 7.6, a greater degree of inhibition was observed with increased concentrations of ERY.

These results suggested that the pH of the reaction mixture influenced not only the permeability of the mitochondrial membranes, but also directly affected the ability of ERY to inhibit mitochondrial protein synthesis. Erythromycin, with a pK_a of 8.6, is also more inhibitory at higher pH's in bacterial *in vitro* protein-synthesizing systems (Mao and Wiegand, 1968) and it has been proposed that only the nonprotonated erythromycin molecule is inhibitory. Since the binding of erythromycin to bacterial ribosomes was not affected over a range of pH 6–9 (Mao and Putterman, 1969), it is reasonable to suggest that the maintenance of a sufficient inhibitory concentration of erythromycin is dependent on the pH of the reaction mixture. The data for HeLa mitochondrial protein synthesis in the presence of Triton X-100 tend to support this hypothesis.

It is interesting to note that the earlier experiments using intact rat liver mitochondria (Kroon and deVries, 1971; Towers et al., 1972; Ibrahim et al., 1973), which showed that erythromycin was ineffective in inhibiting mitochondrial protein synthesis, were carried out in buffers of pH 7.4 or 7.6. However, the experiments that demonstrated the ability of erythromycin to inhibit protein synthesis activity by isolated ribosomes were performed at the higher pH of 7.8 (Greco et al., 1973; Ibrahim and Beattie, 1973).

It is evident from our studies on mitochondrial protein synthesis in HeLa cells (Doersen and Stanbridge, 1979, 1982b) that, irrespective of the concentration of ERY, the presence of Triton X-100, or the pH used in the assay, the maximum inhibition observed was approximately 50% of control incorporation. This partial inhibition of protein synthesis by erythromycin in sensitive cells has also been reported in yeast mitochondrial systems (Linnane et al., 1968; Grivell et al., 1971, 1973) as well as bacterial *in vitro* systems (Mao and Wiegand, 1968). The reasons for this have not been elucidated in any of these systems, but may be related to the ability of erythromycin to inhibit the translocase step in protein synthesis rather than the peptidyl transferase reaction (Pestka, 1971).

3.3.2. Protein Synthesis by Intact Mitochondria

The results presented in Table IV show that protein synthesis in mitochondria isolated from HeLa D98/AH-2 cells was inhibited by ERY. ERY2301, ERY2305, and ERY2309 all displayed a comparable increased ERY-resistant mitochondrial protein synthesis activity. However, protein synthesis by mitochondria isolated from D98-ERY^r was as sensitive to ERY as the activity exhibited by mitochondria isolated from cell lines incapable of proliferating in the presence of ERY. Taking into consideration the data obtained from the *in vivo* mitochondrial protein synthesis assays (Table II), it seems most likely that ERY resistance in D98-ERY^r is due to an alteration in

Table IV. Protein Synthesis by Isolated Mitochondria

Source of mitochondria	Percent of control incorporation[a]		
	Cycloheximide 300 μg/ml	Erythromycin 800 μg/ml	Chloramphenicol 200 μg/ml
HeLa	89 ± 3	53 ± 5	15 ± 3
ERY2301	91 ± 3	70 ± 3	18 ± 3
ERY2305	90 ± 2	72 ± 5	18 ± 1
ERY2309	92 ± 2	70 ± 5	18 ± 6
D98/AH-2	89 ± 2	54 ± 3	16 ± 2
D98-ERY[r]	91 ± 3	50 ± 8	13 ± 1

[a] The incorporation of [³H]leucine into acid-insoluble fraction was followed for 30 min and expressed as cpm/mg mitochondrial protein. Values are the mean ± standard deviation of the percent of incorporation in the absence of inhibitors for each mitochondrial preparation and were calculated from at least two separate experiments. For further details see Doersen and Stanbridge (1979, 1982a).

the plasma membrane resulting in the reduced permeability of ERY. However, the inactivation of ERY by some detoxification mechanism cannot be ruled out.

3.3.3. Protein Synthesis by Triton X-100-Disrupted Mitochondria

In order to demonstrate that the ERY-resistant mitochondrial protein synthesis exhibited by ERY2301, ERY2305, and ERY2309 was not due to the impermeability of the mitochondrial membranes, protein synthesis assays were carried out in the presence of 0.01% Triton X-100 using mitochondrial preparations suspended in the same concentration of Triton X-100. As judged by electron microscopy, this concentration of Triton X-100 was sufficient to disrupt the integrity of the mitochondrial membranes (unpublished observation). The results in Table V reveal that ERY2301, ERY2305, and ERY2309 displayed the same degree of ERY-resistant protein synthesis activity as observed for intact mitochondria. Also, the level of incorporation in the presence or absence of ERY for HeLa mitochondrial preparations was similar to that observed for intact mitochondria.

Table V. Protein Synthesis by Triton X-100-Disrupted Mitochondria

Source of mitochondria	Percent of control incorporation[a]		
	Cycloheximide 300 μg/ml	Erythromycin 800 μg/ml	Chloramphenicol 200 μg/ml
HeLa	92 ± 5	51 ± 6	12 ± 2
ERY2301	91 ± 3	71 ± 9	15 ± 3
ERY2305	91 ± 4	72 ± 4	16 ± 6
ERY2309	89 ± 2	74 ± 6	15 ± 2

[a] The incorporation of [³H]leucine into acid-insoluble fraction was followed for 30 min in the presence of 0.01% Triton X-100 and expressed as cpm/mg mitochondrial protein. Values are the mean ± standard deviation of the percent of incorporation in the absence of inhibitors for each mitochondrial preparation and were calculated from at least two separate experiments. For further details see Doersen and Stanbridge (1979), 1982a).

3.4. Effects of Mycoplasma Contamination on the Erythromycin-Resistant Phenotype

Mycoplasmas are common contaminants of cell cultures (Stanbridge, 1971; Stanbridge and Doersen, 1978) and often remain undetected unless comprehensive cultural and biochemical detection methods are employed (Schneider et al., 1974; Russel et al., 1975). Even though these microorganisms reside in a predominantly extracellular environment in close association with the host plasma membrane (Manchee and Taylor-Robinson, 1969; Stanbridge and Weiss, 1978), they are able to affect the metabolic activities of their infected hosts in many ways (Stanbridge, 1971; Stanbridge and Doersen, 1978). Furthermore, because large numbers of organisms are usually present, mycoplasmal biological activities may be erroneously identified as being of host cell origin.

Because of the burgeoning interest in mitochondrial mutants in mammalian cells and a tendency by many investigators to disregard possible mycoplasma contamination, the effects of mycoplasma contamination on the phenotypic expression of ERY resistance as well as chloramphenicol resistance were studied (Doersen and Stanbridge, 1981). The effect of mycoplamas on the proliferation of drug-sensitive or drug-resistant cells ranged from no obvious effect to increased cell death. However, the protein synthesis activity of mitochondrial preparations from cell populations contaminated with mycoplasmas was clearly different from that observed for preparations that were free of mycoplasmas. The apparent drug-resistant or drug-sensitive phenotypes of the mycoplasma-infected cell lines, as measured by cell-free mitochondrial protein synthesis, were actually those of the contaminating mycoplasmas. The potential hazard for such erroneous interpretation of data derived from contaminated cell lines is further illustrated by the example of D98-ERY^r cells infected with the ERY-resistant *Mycoplasma orale*. Although D98-ERY^r grows in the presence of ERY (Fig. 2) and displays ERY-resistant mitochondrial protein synthesis activity *in vivo* (Table II), cell-free mitochondrial protein synthesis is sensitive to ERY (Table IV). However, when these cells were infected with the ERY-resistant *M. orale*, mitochondrial protein synthesis appeared to be resistant to ERY. This apparent resistance was due entirely to mycoplasmal protein synthesis (Doersen and Stanbridge, 1981).

4. Transfer of Erythromycin Resistance

The ability to transfer the ERY-resistant phenotype to a sensitive cell represents a potentially useful tool in the genetic analysis of mitochondrial biogenesis by selecting for the functional interaction between different nuclear and mitochondrial gene products. Furthermore, the method of transfer provides presumptive evidence regarding the cytoplasmic or nuclear origin of the resistant phenotype. Thus, although ERY2301, ERY2305, and ERY2309 have comparable ERY-resistant phenotypes in culture and in cell-free mitochondrial protein synthesis assays, these mutants fall into two distinguishable classes based on the transfer of ERY resistance.

4.1. Cytoplasmic Transfer of Erythromycin Resistance

The cytoplasmic transfer of the ERY-resistant phenotype of ERY2301 was demonstrated by enucleating ERY2301 cells, fusing the anucleate cytoplasts to sensitive cells, and selecting for ERY-resistant cybrids (Doersen and Stanbridge, 1979). The ERY-sensitive cell lines D98/AH-2 (Szybalski et al., 1962) and D98^{OR} (Weissman and Stanbridge, 1980) carry the recessive nuclear allele for HGPRT (hypoxanthine-guanine phosphoribosyl transferase) deficiency and are therefore resistant to 6-thioguanine, but will not grow in the presence of 300 µg/ml ERY. The cells ERY2301, ERY2305, and ERY2309 are wild type with respect to HGPRT activity and will not grow in the presence of 6-thioguanine. In selective medium containing ERY and 6-thioguanine, only fusions between the anucleate cytoplasts derived from ERY-resistant cells and D98/AH-2 or D98^{OR} cells will survive if ERY resistance is cytoplasmically inherited and is phenotypically dominant in cybrids. The results in Table VI show that no ERY-resistant or ERY-sensitive parental cells, with one exception, survived these selective conditions. The exceptional survivor was designated D98-ERY^r (see Section 2.4).

The frequency of cybrid formation using ERY2301 cytoplasts was between 10^{-3} and 10^{-4}, depending on the ratio of cytoplasts to cells, which ranged from 1:1 to 13:1 (Table VI). The frequencies are similar to those reported for chloramphenicol-resistant cybrid formation (Bunn et al., 1974; Wallace et al., 1975; Bunn and Eisenstadt, 1977b; Mitchell and Attardi, 1978; Munro et al., 1978) (see Bunn, this volume, Chapter 13). Chromosomes analysis of individual cybrid clones and the mass cybrid populations revealed

Table VI. Cytoplasmic Transfer of Erythromycin Resistance

Fusion parents[a]	Ratio of cytoplasts to cells	Number of cells plated	Number of colonies
enERY2301 × D98/AH-2	1:1	5×10^5	181
enERY2301 × D98/AH-2	6:1	5×10^5	508
enERY2301 × D98^{OR}	8:1	5×10^5	463
enERY2301 × D98^{OR}	13:1	5×10^5	519
D98/AH-2	—	2×10^6	1
D98^{OR}	—	4×10^6	0
ERY2301	—	3×10^6	0
enERY2305 × D98^{OR}	7:1	5×10^5	0
enERY2305 × D98^{OR}	11:1	1×10^6	0
enERY2309 × D98^{OR}	5:1	5×10^5	0
enERY2309 × D98^{OR}	9:1	1×10^6	0

[a] en designates the anucleate cytoplast preparation. For further details see Doersen and Stanbridge (1979, 1982a).

that cybrids displayed a chromosome complement comparable with that of the D98/AH-2 or D98^{OR} recipient (Doersen and Stanbridge, 1979). Furthermore, two cybrid clones displayed an ERY-resistant phenotype, both in culture and in cell-free mitochondrial protein synthesis assays, which was indistinguishable from ERY2301.

Attempts to demonstrate the cytoplasmic transfer of the ERY-resistant phenotype of ERY2305 and ERY2309 have been unsuccessful and clearly indicate that the resistant phenotype is not cytoplasmically inherited (Doersen and Stanbridge, 1982a).

4.2. Transfer of Erythromycin Resistance by Cell Hybridization

D98^{OR}, which, in addition to being HGPRT-deficient, is resistant to the plasma membrane Na$^+$/K$^+$ ATPase inhibitor ouabain, and has been used as a "universal fuser" to cells that are wild type with respect to HGPRT activity (Weissman and Stanbridge, 1981). Thus, hybrids from the fusion of D98^{OR} and ERY-resistant cells can be selected in the presence or absence of ERY.

The combined presence of ouabain and ERY greatly increases the ERY sensitivity of ouabain-resistant cells irrespective of their ERY phenotype. The optimum concentration of ERY necessary to select for ERY resistance in the presence of ouabain in these ouabain-resistant hybrids was determined to be 50 µg/ml. Whereas no hybrid colonies arose from the fusion of D98^{OR} and HeLa cells, D98^{OR} × ERY2301 hybrids were recovered in the presence of ERY with a frequency which was slightly less than that observed in the absence of ERY (Table VII).

The recovery of hybrids following the fusion of D98^{OR} and ERY2305 or ERY2309 cells was greatly influenced by the presence of ERY in the selective medium (Table VII). Hybrids selected in the absence of ERY arose with the expected frequency of approximately 10^{-5}. However, hybrids selected in the presence of ERY were recovered with a frequency about 20-fold less.

Subsequent ERY selection of those hybrid populations originally recovered in the absence of ERY was accompanied by extensive cell death. Only rare colonies arose from this ERY selection. These later hybrids proved to be

Table VII. Transfer of Erythromycin Resistance by Cell Hybridization

Parental cells	Selective medium^a	Total number of cells plated	Total number of colonies
D98^{OR} × HeLa	HAT + Oua	3 × 10^6	37
	HAT + Oua + ERY	6 × 10^6	0
D98^{OR} × ERY2301	HAT + Oua	1.2 × 10^7	123
	HAT + Oua + ERY	1.2 × 10^7	99
D98^{OR} × ERY2305	HAT + Oua	2.4 × 10^7	202
	HAT + Oua + ERY	2.4 × 10^7	6
D98^{OR} × ERY2309	HAT + Oua	2.4 × 10^7	235
	HAT + Oua + ERY	2.4 × 10^7	9

^{a} Abbreviations: HAT, hypoxanthine + aminopterin + thymidine; Oua, ouabain; ERY, erythromycin.

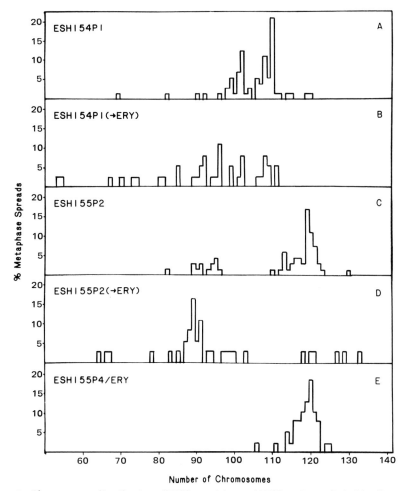

Figure 3. Chromosome distribution of ERY-sensitive and ERY-resistant hybrid cell populations. A minimum of 30 metaphase spreads was counted for each cell population. A. ESH154P1, a $D98^{OR} \times ERY2305$ hybrid population selected in HAT + ouabain. B. ESH154P1 (→ERY), an ESH154P1 population subsequently selected in ERY. C. ESH155P2, a $D98^{OR} \times ERY2309$ hybrid population selected in HAT + ouabain. D. ESH155P2 (→ERY), an ESH155P2 population subsequently selected in ERY. E. ESH155P4/ERY, a $D98^{OR} \times ERY2309$ hybrid population selected in HAT + ouabain + ERY. [Details may be found in Doersen and Stanbridge (1982a).]

as resistant to ERY as the hybrids originally selected directly in the presence of ERY, as well as the ERY2305 or ERY2309 parental cells. Furthermore, these ERY-resistant hybrids were shown to have lost a significant number of chromosomes compared to the hybrid population from which they were selected (Fig. 3). This reduction in chromosome number presumably reflects the loss of the ERY-sensitive determinant(s) due to chromosome segregation. These data are consistent with the interpretation that ERY resistance in ERY2305 and ERY2309 is nuclearly encoded and phenotypically recessive in cell hybrids. Thus, ERY resistance in these cell lines may be analogous to the recessive nature of emetine resistance in Chinese hamster cell hybrids (Gupta and Siminovitch, 1977; Wasmuth et al., 1980), suggesting that the ERY-sensitive phenotype of these cell hybrids results from the mixing of sensitive and resistant ribosomes in mitochondrial polysomes. This is in contrast to the cytoplasmically determined ERY-resistant phenotype of ERY2301, which is inherited as a dominant trait in both cybrid and hybrid cell fusions. The absence of significant phenotypic mixing in ERY2301-derived cybrids and hybrids suggests that the mitochondrial populations whose DNA presumably encode for the ERY-resistant determinant(s) do not interact to an appreciable extent with wild-type ERY-sensitive mitochondria.

5. Conclusion

It has been clearly documented in yeast that the macrolide antibiotic erythromycin inhibits mitochondrial protein synthesis and that resistance to this antibiotic is mitochondrially encoded and expressed at the level of mitochondrial protein synthesis (Linnane et al., 1968; Grivell et al., 1971, 1973). Information regarding the effects of erythromycin on mammalian cells is, however, more controversial.

Some investigators have speculated that mammalian cells are intrinsically resistant to the effect of erythromycin (Towers et al., 1972, 1973) whereas others have noted that this drug will inhibit rat liver mitochondrial protein synthesis (Kroon and deVries, 1971; Greco et al., 1973; Ibrahim and Beattie, 1973; Ibrahim et al., 1973). More recently we (Doersen and Stanbridge, 1979, 1982b) and others (Dixon et al., 1972; Molloy and Eisenstadt, 1979) have shown that erythromycin inhibits the proliferation of mammalian cells in culture. In our studies (Doersen and Stanbridge, 1979, 1982a,b) we have found that the ability of ERY to inhibit the growth of HeLa cells in culture is most likely due to the inhibitory action of the drug on the mitochondrial protein-synthesizing apparatus. We base this inference on the results obtained from both *in vivo* and cell-free mitochondrial protein synthesis assays (as discussed in Sections 3.2 and 3.3). Inhibition of mitochondrial protein synthesis would result in a depletion of functional respiratory complexes (Schatz and Mason, 1974) with a concomitant reduction in the respiratory capacity of the cells. Indeed, such a reduction in respiration has been observed in HeLa cells cultured for 24 hr in the presence of erythromycin at a concentration sufficient to inhibit cellular proliferation (Dixon et al., 1972). Furthermore, since even high concentrations of erythromycin do not directly affect mitochondrial respiration (Dixon et al., 1972), erythromycin appears to specifically act at the level of mitochondrial protein synthesis. This is in contrast to chloramphenicol, which also inhibits the NADH oxidase segment of the respiratory chain (Freeman, 1970).

It is apparent from the results outlined in this discussion that ERY resistance in human cells is expressed at the level of mitochondrial protein synthesis. Several different loci may influence this ERY phenotype. For example, it may be due to the inability of ERY to reach its mitochondrial site of action as a result of permeability changes in the plasma membrane, as appears to be the case for D98-ERY^r (although a detoxification mechanism cannot be discounted). Alternatively, the two classes of mitochondrial protein synthesis mutants, as represented by the cytoplasmically inherited ERY2301 and the nuclearly inherited ERY2305 and ERY2309, most likely result from alterations in the mitochondrial ribosome. The alterations are probably localized in the large ribosomal subunit, since erythromycin is thought to interact directly with the analogous subunit in prokaryotes (Pestka, 1971). Both the nuclear and mitochondrial genomes contribute structural gene products to the mitochondrial ribosome. The mitochondrial DNA codes for the two rRNA species which combine with ribosomal proteins encoded in the nuclear genome to form the large and small ribosomal subunits (Schatz and Mason, 1974; Attardi, 1981). These interact with mitochondrial mRNAs to form polysomes. Erythromycin-resistant loci in yeast map in the region of the mitochondrial genome coding for the large rRNA (Borst and Grivell, 1978; Tzagoloff et al., 1979). It is therefore reasonable to assume that the cytoplasmically inherited ERY-resistant phenotype of ERY2301 is due to a mutation resulting in an altered large rRNA. However, the possibility of an altered mitochondrially encoded ribosomal protein cannot at this time be excluded. The nuclearly encoded ERY-resistant phenotype of ERY2305 and ERY2309 is possibly due to an alteration in a ribosomal protein(s). The alternative possibility that ERY resistance results from the altered methylation of the rRNAs by a mutated methylase seems less likely, because the human mitochondrial rRNAs are not extensively methylated compared to their cytoplasmic counterparts (Attardi and Attardi, 1971).

The isolation and initial characterization of ERY-resistant cell lines, representing two distinct classes of protein synthesis mutants demonstrating cytoplasmic and nuclear inheritance, represent valuable additions to the growing number of somatic cell mutants affecting mitochondrial functions. Elucidation of the molecular nature of these mitochondrial protein synthesis mutants represents a formidable challenge and should provide valuable insights into the interaction between nuclearly and mitochondrially encoded gene products in the assembly and function of the mitochondrial ribosome.

References

Attardi, G., 1981, Organization and expression of the mammalian mitochondrial genome; A lesson in economy, *Trends Biochem. Sci.* **6**:86–89, 100–103

Attardi, B., and Attardi, G., 1971, Expression of the mitochondrial genome in HeLa cells. I. Properties of the discrete RNA components from the mitochondrial fraction, *J. Mol. Biol.* **55**:231–249.

Attardi, B., Cravioto, B., and Attardi, G., 1969, Membrane-bound ribosomes in HeLa cells. I. Their proportion to total cell ribosomes and their association with messenger RNA, *J. Mol. Biol.* **44**:47–70.

Bogenhagen, D., and Clayton, D. A., 1974, The number of mitochondrial deoxyribonucleic acid genomes in mouse L and human HeLa cells, *J. Biol. Chem.* **249**:7991–7995.

Borst, P., and Grivell, L. A., 1978, The mitochondrial genome of yeast, *Cell* **15**:705–723.

Breen, G. A. M., and Scheffler, I. E., 1980, Cytoplasmic inheritance of oligomycin resistance in Chinese hamster ovary cells, *J. Cell Biol.* **86**:723–729.

Bunn, C. L., and Eisenstadt, J. M., 1977a, Carbomycin resistance in mouse L-cells, *Somat. Cell Genet.* **3**:611–627.

Bunn, C. L., and Eisenstadt, J. M., 1977b, Cybrid formation in mouse L-cells: The influence of cytoplast-to-cell ratio. *Somat. Cell Genet.* **3**:331–341.

Bunn, C. L., Wallace, D. C., and Eisenstadt, J. M., 1974, Cytoplasmic inheritance of chloramphenicol resistance in mouse tissue culture cells, *Proc. Natl. Acad. Sci. USA* **71**:1681–1685.

Ceccarini, C., and Eagle, H., 1971, Induction and reversal of contact inhibition of growth by pH modification, *Nature New Biol.* **233**:271–273.

Ditta, G., Soderberg, K., and Scheffler, I. E., 1977, Chinese hamster cell mutant with defective mitochondrial protein synthesis, *Nature* **268**:64–67.

Dixon, H., Kellerman, G. M., and Linnane, A. W., 1972, Effect of mikamycin, carbomycin, spiramycin, erythromycin, and paromomycin, on growth and respiration of HeLa cells, *Arch. Biochem. Biophys.* **152**:869–875.

Doersen, C.-J., and Stanbridge, E. J., 1979, Cytoplasmic inheritance of erythromycin resistance in human cells, *Proc. Natl. Acad. Sci. USA* **76**:4549–4553.

Doersen, C.-J., and Stanbridge, E. J., 1981, Effects of mycoplasma contamination on phenotype expression of mitochondrial mutants in human cells, *Mol. Cell Biol.* **1**:321–329.

Doersen, C.-J., and Stanbridge, E. J., 1982a, Nuclear inheritance of erythromycin resistance in human cells: A new class of mitochondrial protein synthesis mutants, *Mol. Cell. Biol.*

Doersen, C.-J., and Stanbridge, E. J., 1982b, Erythromycin inhibition of cell proliferation and *in vitro* mitochondrial protein synthesis in human cells is pH dependent, *Biochim. Biophys. Acta,* submitted.

Eagle, H., 1971, Buffer combinations for mammalian cell culture, *Science* **174**:500–503.

Freeman, K. B., 1970, Effects of chloramphenicol and its isomers and analogues on the mitochondrial respiratory chain, *Can. J. Biochem.* **48**:469–478.

Greco, M., Pepe, G., and Saccone, C., 1973, Characterization of the monomer form of rat liver mitochondrial ribosome and its activity in poly U-directed polyphenylalanine synthesis, in: *The Biogenesis of Mitochondria* (A. M. Kroon and C. Saccone, eds.), Academic Press, New York, pp. 367–376.

Grivell, L. A., Reijnders, L., and deVries, H., 1971, Altered mitochondrial ribosomes in a cytoplasmic mutant of yeast, *FEBS Lett.* **16**:159–163.

Grivell, L. A., Netter, P., Borst, P., and Slonimski, P. P., 1973, Mitochondrial antibiotic resistance in yeast: Ribosomal mutants resistant to chloramphenicol, erythromycin and spiramycin, *Biochim. Biophys. Acta* **312**:358–367.

Gupta, R. S., and Siminovitch, L., 1977, The molecular basis of emetine resistance in Chinese hamster ovary cells: Alteration of the 40s ribosomal subunit, *Cell* **10**:61–66.

Harris, M., 1978, Cytoplasmic transfer of resistance to antimycin A Chinese hamster cells, *Proc. Natl. Acad. Sci. USA* **75**:5604–5608.

Ibrahim, N. G., and Beattie, D. S., 1973, Protein synthesis on ribosomes isolated from rat liver mitochondria: Sensitivity of erythromycin, *FEBS Lett.* **36**:102–104.

Ibrahim, N. G., Burke, J. P., and Beattie, D. S., 1973, Mitochondrial protein synthesis *in vitro* is not an artifact, *FEBS Lett.* **29**:73–76.

Kroon, A. M., and deVries, H., 1971, Mitochondriogenesis in animal cells: Studies with different inhibitors, in: *Autonomy and Biogenesis of Mitochondria and Chloroplasts* (N. K. Boardman, A. W. Linnane, and R. M. Smillie, eds.), Elsevier/North-Holland, Amsterdam, pp. 318–327.

Lederman, M., and Attardi, G., 1970, In vitro protein synthesis in a mitochondrial fraction from HeLa cells: Sensitivity to antibiotics and ethidium bromide, *Biochem. Biophys. Res. Commun.* **40**:1492–1500.

Leibowitz, R. D., 1971, The effect of ethidium bromide on mitochondrial DNA synthesis and mitochondrial DNA structure in HeLa cells, *J. Cell Biol.* **51**:116–122.

Lichtor, T., and Getz, G. S., 1977, Cytoplasmic inheritance of rutamycin resistance in mouse fibroblasts, *Proc. Natl. Acad. Sci. USA* **75**:324–328.

Linnane, A. W., Lamb, A. J., Christodoulou, C., and Lukins, H. B., 1968, The biogenesis of mitochondria. VI. Biochemical basis for the resistance of *Saccharomyces cerevisiae* towards antibiotics which specifically inhibit mitochondrial protein synthesis, *Proc. Natl. Acad. Sci. USA* **59**:1288–1293.

Manchee, R., and Taylor-Robinson, D., 1969, Studies on the nature of receptors involved in the attachment of tissue culture cells to mycoplasmas, *Br. J. Exp. Pathol.* **50**:66–75.

Mao, J. C.-H., and Putterman, M., 1969, The intramolecular complex of erythromycin and ribosome, *J. Mol. Biol.* **44**:347–361.

Mao, J. C.-H., and Wiegand, R. G., 1968, Mode of action of macrolides, *Biochim. Biophys., Acta* **157**:404–413.

Mitchell, C. H., and Attardi, G., 1978, Cytoplasmic transfer of chloramphenicol resistance in a human cell line, *Somat. Cell Genet.* **4**:737–744.

Mitchell, C. H., England, J. M., and Attardi, G., 1975, Isolation of chloramphenicol-resistant variants from human cell lines, *Somat. Cell. Genet.* **1**:215–234.

Molloy, P. L., and Eisenstadt, J. M., 1979, Erythromycin resistance in mouse L-cells, *Somat. Cell Genet.* **5**:585–595.

Munro, E., Siegle, R. L., Craig, I. W., and Sly, W. S., 1978, Cytoplasmic transfer of a determinant for chloramphenicol resistance between mammalian cell lines, *Proc. R. Soc. Lond.* B **201**:73–85.

Nass, M. M. K., 1972, Differential effects of ethidium bromide on mitochondrial and nuclear DNA synthesis *in vivo* in cultured mammalian cells, *Exp. Cell Res.* **72**:211–222.

Naum, Y., and Pious, D. A., 1971, Reversible inhibition of cytochrome oxidase accumulation in human cells by ethidium bromide, *Exp. Cell Res.* **65**:355–359.

Pestka, S., 1971, Inhibitors of ribosome functions, *Ann. Rev. Microbiol.* **25**:487–562.

Russell, W. C., Newman, C., and Williamson, D. H., 1975, A simple cytochemical technique for demonstration of DNA cells infected with mycoplasmas and viruses, *Nature* **253**:461–462.

Schatz, G., and Mason, T. L., 1974, The biosynthesis of mitochondrial proteins, *Ann. Rev. Biochem.* **43**:51–87.

Schneider, E. L., Stanbridge, E. J., and Epstein, C. J., 1974, Incorporation of ³H-uridine and ³H-uracil into RNA: A simple technique for the detection of mycoplasma contamination of cultured cells, *Exp. Cell Res.* **84**:311–318.

Slonimski, P. P., Perrodin, G., and Croft, J. H., 1968, Ethidium bromide induced mutation of yeast mitochondria: Complete transformation of cells into respiratory deficient non-chromosomal "petites," *Biochem. Biophys. Res. Commun.* **30**:232–239.

Smith, C. A., Jordan, J. M., and Vinograd, J., 1971, *In vivo* effects of intercalating drugs on the superhelix density of mitochondrial DNA isolated from human and mouse cells in culture, *J. Mol. Biol.* **59**:255–272.

Soderberg, K., Mascarello, J. T., Breen, G. A. M., and Scheffler, I. E. 1979, Respiration-deficient Chinese hamster cell mutants: Genetic characterization, *Somat. Cell Genet.* **5**:225–240.

Spolsky, C. M., and Eisenstadt, J. M., 1972, Chloramphenicol resistant mutants of human HeLa cells, *FEBS Lett.* **25**:319–324.

Stanbridge, E. J., 1971, Mycoplasmas and cell cultures, *Bacteriol. Rev.* **35**:206–227.

Stanbridge, E. J., and Doersen, C.-J., 1978, Some effects that mycoplasmas have upon their infected host, in: *Mycoplasma Infection of Cell Cultures* (G. J. McGarrity, D. G. Murphy, and W. W. Nichols, eds.), Plenum Press, New York, pp. 119–134.

Stanbridge, E. J., and Weiss, R. L., 1978, Mycoplasma capping on lymphocytes, *Nature* **276**:583–587.

Szybalski, W., Szybalska, E. H., and Ragni, G., 1962, Genetic studies with human cell lines. *Natl. Cancer Inst. Monogr.* **7**:75–89.

Towers, N. R., Dixon, H., Kellerman, G. M., and Linnane, A. W.,1972,Biogenesis of mitochondria 22. The sensitivity of rat liver mitochondria to antibiotics: a phylogenetic difference between a mammalian system and yeast, *Arch. Biochem. Biophys.* **151**:361–369.

Towers, N. R., Kellerman, G. M., and Linnane, A. W., 1973, Competition between non-inhibitory antibiotics and inhibitory antibiotics for binding by rat liver mitochondrial ribosomes, *Arch. Biochem. Biophys.* **155**:159–166.

Tzagoloff, A., Macino, G., and Sebald, W., 1979, Mitochondrial genes and translation products, *Ann. Rev. Biochem.* **48**:419–441.

Wallace, D. C., Bunn, C. L., and Eisenstadt, J. M., 1975, Cytoplasmic transfer of chloramphenicol resistance in human tissue culture cells, *J. Cell Biol.* **67**:174–188.

Wasmuth, J. J., Hill, J. M., and Vock, K. S., 1980, Biochemical and genetic evidence for a new class of emetine-resistant Chinese hamster cells with alterations in the protein biosynthetic machinery, *Somat. Cell Genet.* **6:**495–516.

Weissman, B. E., and Stanbridge, E. J., 1980, Characterization of ouabain resistant hypoxanthine guanine phosphoribosyl transferase deficient human cells and their usefulness as a general method for the production of human cell hybrids, *Cytogenet. Cell Genet.* **28:**227–239.

Williamson, D. H., Johnston, L. H., Richmond, K. M. V., and Games, J. C., 1977, Packaging and recombination of mitochondrial DNA in vegetatively growing yeast cells, in: *Mitochondria 1977, Genetics and Biogenesis of Mitochondria* (W. Bandlow, R. J. Schweyen, K. Wolf, and F. Kaudewitz, eds.), W. deGruyter, Berlin, pp. 1–24.

Wiseman, A. and Attardi, G., 1978, Reversible tenfold reduction in mitochondrial DNA content of human cells treated with ethidium bromide, *Mol. Gen. Genet.* **167:**51–63.

Wiseman, A., and Attardi, G., 1979, Cytoplasmically inherited mutations of a human cell line resulting in deficient mitochondrial protein synthesis, *Somat. Cell Genet.* **5:**241–262.

Zylber, E., Vesco, C., and Penman, S., 1969, Selective inhibition of the synthesis of mitochondria-associated RNA by ethidium bromide, *J. Mol. Biol.* **44:**195–204.

Chapter 12
Cytoplasmic Inheritance of Chloramphenicol Resistance in Mammalian Cells

DOUGLAS C. WALLACE

Chloramphenicol (CAP) resistance was the first cytoplasmic drug resistance marker to be described for mammalian cells. CAP and a variety of structurally related analogs (Fig. 1) are potent inhibitors of both bacteria and mammalian cells. Inhibition of bacterial growth is rapid, but inhibition of mammalian cell replication requires several generations (Spolsky and Eisenstadt, 1972; Bunn et al., 1974; D. C. Wallace et al., 1975).

1. CAP Inhibition of Bacterial and Mitochondrial Ribosomes

CAP blocks the peptidyl transferase reaction of the bacterial ribosome by binding to the 50S subunit. Puromycin and CpCpA trinucleotide cytosine cytidylate adenylate compete with CAP binding, indicating that CAP interacts with the a-site and inhibits the binding of the 3′ end of the aminoacyl-tRNA. Exclusion of the charged tRNA blocks growth of the nascent polypeptide at the p-site (Nierhaus and Nierhaus, 1973). Ribosome reconstitution experiments and monoiodoamphenicol affinity labeling studies have revealed that the *Escherichia coli* 50S ribosomal proteins L16 and L24 and the 30S subunit protein S6 are associated with the CAP binding site (Nierhaus and Nierhaus, 1973; Pongs et al., 1973; Pongs and Messer, 1976).

Eukaryotic-cell 80S cytosolic ribosomes are insensitive to CAP, though they are inhibited by cycloheximide (CHX) and emetine. Mitochondrial (mt) ribosomes, by contrast, are inhibited by CAP but not by CHX or emetine (Lamb et al., 1968; Perlman and Penman, 1970; Loeb and Hubby, 1968; Costantino and Attardi, 1975; Lansman and Clayton, 1975b; Ibrahim et al., 1974). Monoiodoamphenicol binds to the large mt ribosomal protein L2. The protein L2 has been implicated in the peptidyl transferase reaction (O'Brien et

---

DOUGLAS C. WALLACE • Department of Genetics, School of Medicine, Stanford University, Stanford, California 94305.

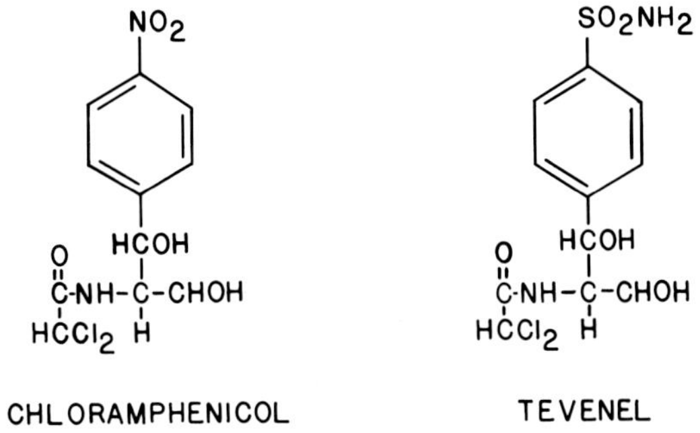

Figure 1. The structure of the biologically active D(−)-threo isomer of chloramphenicol (Das et al., 1966) and its sulfamoyl analog Tevenel (Freeman, 1970a).

al., 1980). CAP inhibition of mammalian cells thus results from the inhibition of mt protein synthesis. The delayed inhibition of growth presumably reflects the time necessary to deplete the preexisting mt proteins located in the several hundred cellular mitochondria (Nass, 1969; Posakony et al., 1975; Johnson et al. 1980; Spolsky, 1973).

2. Isolation of Cytoplasmic CAP-Resistant Mutants

Mammalian cell mutants can be isolated which are resistant to 50–100 µg/ml of CAP (Spolsky and Eisenstadt, 1972). Three factors must be considered when isolating such mutants: depletion of the resident CAP^S mtDNAs, specific mutagenesis of the mtDNA, and selection for the mutant mtDNA and mitochondrion. Tables I–IV list a number of mammalian CAP^R mutants isolated from different species and summarize the isolation procedures used. The most successful regimes have incorporated depletion of the cellular mtDNAs, followed by mtDNA mutagenesis, and then selection for the mutant mtDNAs during the repopulation of the cell with mitochondria.

Mitochondrial DNA depletion has been accomplished by ethidium bromide (EtBr) treatment (Spolsky and Eisenstadt, 1972; Wiseman and Attardi, 1978, 1979); removal of cytoplasm by enucleation (Kuhns and Eisenstadt, 1979) (see Chapter 9, this volume); and selectively labeling the mtDNA of thymidine kinease-deficient (TK^-) cells with 5-bromodeoxyuridine (BrdU) followed by visible light irradiation to destroy the photosensitized mtDNAs (Croizat and Attardi, 1975; Lansman and Clayton, 1975a; Lichtor and Getz, 1978) (see Chapter 10, this volume). Treatment with rhodamine-6G (R6G) should also prove effective in eliminating mtDNA (see below) (Ziegler and Davidson, 1981) (see Chapter 15, this volume).

A variety of mtDNA mutagenesis procedures have also been employed. Most promising is the inhibition of nuclear (n) DNA replication with CHX while simultaneously mutagenizing the replicating mtDNA with 2-aminopurine (Rosenberg et al., 1976; Shannon et al., 1973; Wiseman and Attardi, 1979) or nitrosoguanidine (Craig and Webb, 1979). The mtDNA of TK^- cells can be selectively mutagenized with BrdU (Clayton and Teplitz, 1972; R. B. Wallace and Freeman, 1975; Lichtor and Getz, 1978; Yatscoff et al., 1981). Mn^{++}, an effective mtDNA mutagen in yeast (Putrament et al., 1973, 1977), has

Table I. CAP-Resistant Mutants of Human and Rat Cells

Species, strain	Mutant	Mode of selection			Mutation[d]	Inherit[e]	Notes[f]	Ref.[g]
		Enrichment[a]	Mutagen[b]	Selection[c]				
Human, HeLa	296-1	EtBr	EtBr	CAP^{50}	$MtPS^R$	C (cyb, mtD, seq)	S, D	1, 4, 8
Human, HeLa	MC63, MC75	EtBr	MNNG	CAP^{50}	$MtPS^R$	C (cyb, seq)	S, MIK^R, CAR^R	3, 10–14 3, 7, 9
Human, HT1080	HT102W	—	Am	CAP^{50}, $HGlm$, HO_2	$MtPS^R$	C	Us, D	1, 8, 11
Human, WI-18-VA$_2$-B	CAP-23	EtBr	EMS	CAP^{40}, HG, HGlm	$MtPS^R$	C (cyb, mtD, seq)	S, MIK^R, CAR^R	5, 6
Human, WI-18-VA$_2$-B	mPS^- 1 to 8	EtBr	CHX + AP, EMS, Mn^{++}	RES^R RES^-	$MtPS^-$	C (cyb) C (cyb)	CAP^R	15
Rat, L6TG	$L6TGCAP^R$	Step	EMS	CAP^{100}	NS	C (cyb, re)		2

[a] Mutant mtDNA enrichment. EtBr, ethidium bromide. Step, selection by stepwise increases in CAP.
[b] Mutagenesis procedure. MNNG, N-methyl-N'-nitro-N-nitrosoguanidine. Am, aminopterin. EMS, ethyl methane sulfonate. CHX, cycloheximide. AP, 2-aminopurine. EMS, ethyl methane sulfonate.
[c] Selection conditions. Superscripts represent concentration in µg/ml. HGlm, ≥ 4 mM glutamine (2 mM glutamine is standard). HO_2, high O_2 tension. HG, $\geq 0.3\%$ glucose. RES^R, selected for resistance to respiration inhibitors rotenone$^{0.2}$, rutamycin^2, or antimycin^5 or^{10} in the absence of glucose, but with succinate and high glutamine. RES^-, mutants selected for respiration deficiency; selected first without glucose, but in high glutamine and succinate and treatment with BrdU plus light. Then, the BrdU and light were removed and the cells shifted back to glucose, no succinate, and high glutamine. In some instances, DMEM was used but the formulation was not specified. Though all DMEMs contain HGlm, different formulations can contain 0.1% glucose (Dulbecco and Freeman, 1959), 0.45% glucose, or 0.1% glucose + 1 mM pyruvate (Grand Island Biological Company, 1978/1979). In these instances, the conditions which seemed most likely to have been employed were given.
[d] Basis of CAP resistance. mtPS, mt protein synthesis. $MtPS^R$, CAP^R mtPS. $mtPS^-$, mtPS-deficient. NS, not stated.
[e] Location of mutant locus. Methods used to determine coding site listed within parentheses. C, cytoplasmic. N, nuclear. cyb, cybrid fusion. rc, reconstituted cells. mtD, linked to mtDNA restriction endonuclease site polymorphisms. seq, located by mtDNA sequencing. Rat-mouse cyb/re, cytoplasmic inheritance demonstrated using mouse cells or nuclei as recipients.
[f] S, stable. Us, unstable. D, dominant. MIK, mikamycin. CAR, carbomycin.
[g] References: 1. Blanc et al. (1981a). 2. Hayashi et al. (1980). 3. Kearsey and Craig (1981). 4. Kislev et al. (1973). 5. Mitchell et al. (1975). 7. Munro et al. (1978). 8. Oliver and Wallace (1982). 9. Siegel et al. (1976). 10. Spolsky and Eisenstadt (1972). 11. Wallace (1981). 12. Wallace et al. (1976). 14. Wallace et al. (1977). 15. Wiseman and Attardi (1979).

Table II. CAP-Resistant Mutants of Mouse Cells

Species, strain	Mutant	Mode of selection			Mutation[d]	Inherit[e]	Notes[f]	Ref.[g]
		Enrichment[a]	Mutagen[b]	Selection[c]				
Mouse, A9	501-1	EtBr	EtBr	CAP^{50}	MtPS^R	C (cyb, seq)	S, G-DEP	1, 2, 3, 11, 14
Mouse, A9	A9CAP^R	EtBr	EtBr	CAP^{50}, HG, HGlm	NS	C (rc)	Mouse human rc	8
Mouse, A9	A9B2	—	EMS	CAP^{50}	NS	C		4, 5, 6
Mouse, LM(TK$^-$)	T8, T22	LO$_2$ to HO$_2$	—	TEV^{100}	MtPS^R	(cyb, mtD, gt) NS	S	12
Mouse, EMT	EMTCH21	Step	—	CAP^{100}, HG, HGlm	NS	C (cyb)		13
Mouse, MK	MK920	Step	MNNG	CAP^{100}, HG, HGlm	NS	C (cyb)		13
Mouse, MT-29240	MT-0C-1 MT-0C-10	HCD	EMS	CAP^{50}, HG, HGlm	NS	C (cyb, mtD, gt)		4–7, 9, 10

[a] See Table I. LO$_2$ to HO$_2$, selection in suspension yielding low O$_2$ tension and then attached yielding high O$_2$ tension. HCD, selected at high cell density.
[b] See Table I.
[c] See Table I. TEV, Tevenel.
[d] See Table I.
[e] See Table I. gt, transferred via purified mtDNA.
[f] See Table I. G-DEP, requires glucose. Mouse-human rc, cytoplasmic inheritance demonstrated in reconstituted cells using human karyoplasts.
[g] References: 1. Blanc et al. (1981b). 2. Bunn et al. (1974). 3. Bunn et al. (1977). 4. Coon (1978). 5. Coon and Ho (1978). 6. Ho and Coon (1979). 7. Malech and Wivel (1976). 8. Nette et al. (1979). 9. Shay (1977). 10. Shay et al. (1978). 11. Wallace et al. (1976). 12. Wallace and Freeman (1975). 13. Ziegler (1978). 14. Ziegler and Davidson (1979).

Table III. CAP-Resistant Mutants of Mouse Cells

Species, Strain	Mutant	Enrichment[a]	Mutagen[b]	Selection[c]	Mutation[d]	Inherit[e]	Notes[f]	Ref.[g]
Mouse, 3T3	3T3CAP^R	NS	NS	CAP	NS	C (seq)	—	7
Mouse, 3T3	—	EtBr	—	CAP^{100}, HGlm	—	C (cyb)	—	3
Mouse, 3T3-A31	284-14 (CAP^R-PYA-DEP)	—	—	CAP^{50}, HG, PYR	NS	C (cyb)	S,G-DEP, P-DEP	5, 6
Mouse, SVT2	280-7 (CAP^R-PYR-DEP)	—	—	CAP^{50}, HG, PYR, HGlm	NS	C (cyb)	S,G-DEP, P-DEP	5, 6
Mouse, SVT2-107	107-6-4-1	—	—	CAP^{50}, HG, HGlm	NS	C (cyb)	G-DEP, P-IND	6
	107-6-4-2 (CAP^R-PYR-IND)							
Mouse, 280-7	280-7-3 (CAP^R-PYR-IND)	—	—	CAP^{50}, HG, HGlm	NS	C (cyb)	G-DEP, P-IND	6
Mouse, (M. molossinus) KYK	KYK-CAP	—	EMS	CAP^{50}	NS	C (cyb, mtD, gt)	—	1, 2, 4
Mouse, (M. caroli) MCH	MCH-C18	—	EMS	CAP^{50}	NS	C (cyb, mtD, gt)	—	1, 2, 4

[a] See Table I.
[b] See Table I.
[c] See Table I. PYR, 1 mM pyruvate.
[d] See Table I.
[e] See Table I.
[f] See Table I. P-DEP, requires pyruvate. P-IND, pyruvate not required.
[g] References: 1. Coon (1978). 2. Coon and Ho (1978). 3. Giguère and Morais (1981). 4. Ho and Coon (1979). 5. Howell and Sager (1978). 6. Howell and Sager (1979). 7. Kearsey and Craig (1981).

Table IV. CAP-Resistant Mutants of Hamster and Chick Cells

Species, strain[a]	Mutant	Mode of selection			Mutation[d]	Inherit[e]	Notes[f]	Ref.[g]
		Enrichment	Mutagen[b]	Selection[c]				
Chinese hamster V79	201, 204	—	—	CAP100, HG, HGlm	NS	C (cyb)	Vr	7
Chinese hamster, V79	5-3	—	—	CAP100, PYR	NS	C (cyb, R6G)	G-DEP	8, 9
Chinese hamster CHEF/18-1 205-30	294-7	—	MNNG	CAP50, HG, PYR	NS	C (cyb)	—	3
Chinese hamster CHEF/16-2	213-21-3	—	—	CAP50, PYR	NS	C (cyb)	—	3
Chinese hamster CHO	BT3	LO2 to HO2	BrdU	TEV150	NS	C (cyb)	NS	6
Chinese hamster V79	V79-G7	—	EMS	RES⁻	MtPS⁻	N (cyb)	R, G-DEP, ASP-DEP, HCO3⁻-DEP, EtBR^R	1, 2, 5
Chick, CEF	CEF	—	—	CAP100, PYR, UR, LGlm	MtPS^d			4

[a] See Table II.
[b] See Table I. BrdU, 25 μm/ml for 4 days.
[c] See Tables I–III. Ur, 2 μg/ml uridine. LG/m, 1 mm glutamine RES⁻, respiratory deficiency. Selected first in 0.36% glucose, high glutamine, hypoxanthine 10, uridine 10, no NaHCO3, and BrdU plus light. Then, the BrdU and light were removed and the cells grown at high glucose, 0.45% glutamine, high NaHCO3, and asparagine.
[d] See Table I. mPS^d, mtPS depressed in CAP.
[e] See Table I. R6G, eliminated by R6G treatment.
[f] See Table I–III. Vr, variable resistance in cybrids. R, respiratory deficient. ASP-DEP, asparagine dependent. HCO3⁻-DEP, bicarbonate dependent.
[g] References: 1. Ditta et al. (1976). 2. Ditta et al. (1977). 3. Howell and Sager (1978). 4. Morais et al. (1980). 5. Soderberg et al. (1979). 6. Yatscoff et al. (1981). 7. Yen and Harris (1978). 8. Ziegler and Davidson (1979). 9. Ziegler and Davidson (1981).

also been used for human cells (Wiseman and Attardi, 1979). The yeast petite-inducing agents EtBr (Bastos and Mahler, 1974) and the folate analogs (Wintersberger and Hirsch, 1973a,b) have been used to isolate mammalian CAP^R mutants (Spolsky and Eisenstadt, 1972; D. C. Wallace, 1981), but their effectiveness as mtDNA mutagens has not been established (Smith, 1977).

Designing an effective selection regime for CAP^R mutants is potentially complex. The nature of the cell line, the regime for CAP addition, and the composition of the medium must be considered. Though numerical data are lacking, the best results seem to be obtained when the factors affecting initial selective pressure are balanced to permit the cells to replicate for a period in CAP and then become blocked for replication without severe damage to the cells. Often, morphologically normal, but dormant, cells will remain in CAP for weeks or months. This maximizes the opportunity for the mutant mtDNAs to repopulate the cell during the prolonged, 2-week to several-month, selection period.

Cells of different origins often have different sensitivities to the drug and different propensities for yielding CAP^R mutants. Mouse 3T3 cells are more resistant to a variety of mt inhibitors than are their SV40-transformed counterparts (Howell and Sager, 1977). Established cell lines, in general, are readily killed by CAP, while chick embryo fibroblasts will grow indefinitely in CAP provided the medium is supplemented with uridine (Morais et al., 1980). Mouse LM(TK$^-$) cells generally give rise to CAP^R mutants at a high frequency and after relatively short periods of selection (R. B. Wallace and Freeman, 1975). In contrast, it has proven quite difficult to isolate CAP^R mutants in mouse 3T3 and SV40-transformed 3T3 cells. A two-stage selection is often necessary, with glucose and pyruvate being required during the initial selection to reduce the toxicity of CAP. These initial mutants require pyruvate to grow in CAP (PYR-DEP). A second round of selection is then required to isolate CAP^R but pyruvate-independent (PYR-IND) cells (Howell and Sager, 1979).

In different mutant selections CAP has been added at high initial concentrations or added at progressively increasing concentrations to prolong the replication period in CAP prior to cessation of growth. Tevenel has been used instead of CAP to isolate mutants in LM(TK$^-$) cells. CAP not only inhibits mt protein synthesis, but also mt NADH oxidase. At 1 mM, CAP causes a 50% inhibition of NADH oxidase activity (Freeman, 1970a). CAP selection is usually applied at 50–100 µg/ml (0.15–0.3 mM); hence, NADH oxidase inhibition may be an important second cause of toxicity. Tevenel is as effective as CAP at inhibiting mt protein synthesis, but has one-tenth the activity on NADH oxidase (Freeman, 1970a,b). Further, cells undergo five replications after addition of Tevenel as compared to only two for CAP (Fettes et al., 1972). Consequently, use of Tevenel both increases the specificity of selection and provides more time for mutant mtDNA enrichment. These factors may, in part, explain the high frequency of Tevenel^R mutants observed for LM(TK$^-$) (R. B. Wallace and Freeman, 1975).

The composition of the medium and, hence, the physiology of the cell also have a major effect on the toxicity of the drug. High concentrations of glucose reduce CAP toxicity, while less rapidly fermented sugars, such as mannose and galactose, increase toxicity (Ziegler and Davidson, 1979; Molloy and Eisenstadt, 1979; Howell and Sager, 1979). Pyruvate (1 mM), in the presence of high glucose, substantially reduces the toxicity of mt inhibitors, possibly by contributing essential Krebs cycle intermediates (Harris, 1980; Howell and Sager, 1979). Pyruvate or ethanol in low glucose medium increases the toxicity of mt ribosome inhibitors (Molloy and Eisenstadt, 1979), probably due to a shift to cellular reliance on oxidative metabolism. For human HT1080 cells, pyruvate is quite toxic above 1 mM (Wallace, unpublished results); hence, pyruvate toxicity should be checked prior to use. Toxicity to mt inhibitors has also been increased in low glucose medium by adding succinate (Wiseman and Attardi, 1979) and by using high glutamine and O_2 tension (Wallace, 1981). A shift from high to low glucose in the presence of glutamine has been shown to shift the proportion of HeLa energy produced by oxidative metabolism from more than 50% to 98% (Reitzer et al., 1979). Finally, Malech and Wivel (1976) have suggested that high cell density reduces CAP toxicity in high glucose medium, although I generally select mutants at low cell density to avoid cell killing due to intermittent medium depletion and excessive acid production.

Generally CAP^R mutants have been found to have CAP^R mt protein synthesis either by *in vitro* (Spolsky and Eisenstadt, 1972) or *in vivo* (Siegel et al., 1976) assays. Different mutants may have different levels of resistance (Oliver and Wallace, 1982) and some have been found to be cross-resistant to other mt ribosome inhibitors, such as carbomycin and mikamycin (Mitchell and Attardi, 1978; Mitchell et al., 1975; Munro et al.,, 1978; Siegel et al., 1976). In some instances, however, CAP resistance can result from a totally different mechanism. Mutations to mt protein synthesis deficiency obtained concomitantly with selections for respiratory deficiency have been observed in SV40-transformed human fibroblast WI-18-VA$_2$-G cells (Wiseman and Attardi, 1979) and for Chinese hamster V79 cells (Ditta et al., 1977). Though the hamster mutant is nuclear, the human mutant has been shown to be cytoplasmic and CAP^R.

3. Identification of Cytoplasmic CAP^R Mutants

Since mt genes are coded in both the nDNA and the mtDNA, it is essential that the cytoplasmic coding site of CAP^R mutants be confirmed. Currently, extrachromosomal CAP^R mutants can be identified in three ways: the transfer of CAP resistance via enucleated CAP^R cell fragments to CAP-sensitive (CAP^S) cells or nuclei, the segregation of the CAP resistance marker during mitotic replication, or the elimination of CAP^R determinnants with R6G prior to cell fusion. A fourth procedure, transfer of mouse cell resistance via purified mtDNA (Coon, 1978; Coon and Ho, 1977; Ho and Coon, 1979) has not yet been successfully applied to other systems. In contrast to mutants resistant to other mt ribosome inhibitors (Doersen and Stanbridge, 1979; Bunn and Eisenstadt, 1977b; Molloy and Eisenstadt, 1979), CAP^R mutants have generally been found to be cytoplasmic.

3.1. Cytoplasmic Transfer of CAP Resistance

In cytoplasmic transfer experiments, the nucleus of the CAP^R cell is physically removed and the membrane-bound cytoplasmic fragment containing the mitochondria (cytoplast) is fused either to a whole CAP^S cell or to its membrane-bound nuclear fragment (karyoplast). Successful cytoplast–cell fusions generate cybrids (Bunn et al., 1974; D. C. Wallace et al., 1975), while cytoplast–karyoplast fusions yield reconstituted cells (Shay, 1977). A typical cybrid fusion is diagrammed in Fig. 2 and compared to the classical hybrid fusion.

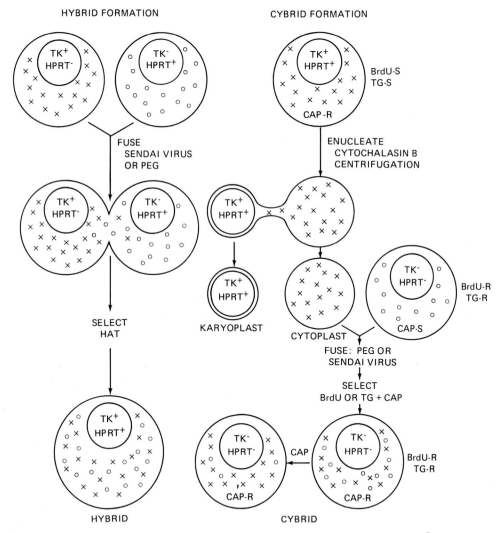

Figure 2. Diagrammatic representation of hybrid and cybrid fusions. (×) CAP^R mtDNA; (○) = CAP^S mt DNA. Abbreviations and terms are defined in the text. For HAT and CAP formulations, see Table V. TG is frequently used at 0.1 mM (Wallace, 1981).

The enucleation of cells is accomplished by disaggregating the cellular cytoskeleton with cytochalasin B (Carter, 1967) and removing the dense nucleus using a centrifugal field. Centrifugal force can be applied to the nucleus either by inverting attached cells in a centrifuge tube (Prescott et al., 1972; Veomett et al., 1976) or by suspending the cells in a Ficoll density gradient (Wigler and Weinstein, 1975) (see Chapter 6, this volume). Cytoplast and karyoplast preparations are generally contaminated with whole cells and cellular fragments. This can generally be ignored if the appropriate selective system is employed (Bunn et al., 1974; D. C. Wallace et al., 1975); however, in other instances, preparations of higher purity are required. Cytoplasts have been purified by passage of the preparation through 5.0-μm Unipore filters which retain the whole cells and karyoplasts but not the cytoplasts (Yatscoff et al., 1981) or by sorting out the cytoplasts from Hoechst 33258-stained cells using the fluorescent-activated cell sorter (FACS) (Schaap et al., 1979). Karyoplasts can be purified by differential attachment (Shay, 1977), by size using the FACS, or by density through the enucleation of cells loaded with tantalum particles (Clark et al., 1980) (see Chapter 20, this volume). The CAP^R cell cytoplasts are then fused to CAP^S recipient whole cells or karyoplasts using inactivated Sendai virus (Bunn et al., 1974) or polyethylene glycol (Shay, 1977; Yatscoff et al., 1981; Wilson et al., 1978).

The coding site of CAP resistance is determined by selecting the cybrids or reconstituted cells in CAP and for the recipient cell nucleus. If colonies appear, the mutation is coded in cytoplasm. Generally, 50–100 μg/ml of CAP is used in selection and colonies appear between 10 days and 3–4 weeks (Bunn et al., 1974; D. C. Wallace et al., 1975). Cytoplasmic CAP resistance is expressed immediately after cybrid or hybrid fusion, indicating that it is a dominant or codominant marker (Bunn et al., 1974, 1977; D. C. Wallace et al., 1975, 1977). Maximum cybrid yields have been obtained when CAP selection is applied immediately after fusion (Yatscoff et al., 1981).

A variety of drug selection regimes have been employed to select for the CAP^S cell nucleus. By using CAP^S recipient cells that are TK or hypoxanthine phosphoribosyl transferase$^-$ ($HPRT^-$) and CAP^R donor cells that are TK^+ or $HPRT^+$, the recipient nucleus can be selected for by addition of BrdU for TK^- recipient cells or 8-azaguanine (AG) and/or 6-thioguanine (TG) for $HPRT^-$ recipient cells. Use of a recessive marker in the recipient cell has the advantage that it not only selects against any unenucleated CAP^R parental cells, but also any hybrids in which such cells have participated (Bunn et al., 1974; D. C. Wallace et al., 1975). When selecting for the $HPRT^-$ marker, it is important to add the AG or TG to the medium at least 5 days after fusion, to permit decay of the HPRT transferred along with $HPRT^+$ cell cytoplast (Wallace et al., 1975). This does not seem to be a problem with the TK^- marker, possibly because of its rapid turnover (Littlefield, 1966).

Cybrid studies using recessive markers have revealed that the frequency of cybrid formation is generally much higher than the frequency of hybrid formation using the same cells. This effect can be greatly accentuated by increasing the cytoplast to cell ratio (Bunn and Eisenstadt, 1977a) (see

Chapter 13, this volume). This fact makes it possible to select for cybrids using dominant nuclear markers in the recipient cell. Since the majority of clones will be cybrids, these can be picked and confirmed using other nuclear characteristics. Cybrids have thus been successfully isolated by fusing the cytoplasts of CAP^R and TK^- or $HPRT^-$ cells to CAP^S but HAT^R recipient cells followed by selection in HAT + CAP (Wallace, 1981; Shay, 1977). Alternatively, cybrids have been isolated by fusing CAP^R CHO cytoplasts to CAP^S but 5,6-dichloro-1-β-D-ribofuranosyl-benzimidazole $(DRB)^R$ cells and selection in DRB + CAP (Yatscoff et al., 1981).

A typical set of hybrid and cybrid crosses between two human cell lines, HeLa and HT1080, is described in Table V. In the upper pair of cybrid and hybrid crosses, the HeLa parent was CAP^R, while in the lower pair, the HT1080 parent was CAP^R. Otherwise, the crosses were identical except for incidental differences in nuclear markers. All fusions were performed using a one-to-one ratio of the parents, and the cybrids were selected by either the dominant (HAT + CAP) or recessive (BrdU + CAP) nuclear marker strategy. In both pairs of crosses, the cybrid frequency was substantially higher than the hybrid frequency.

Once CAP^R clones have been isolated, it is essential that their nuclear origin be confirmed. Morphologic differences, unselected $drug^R$ markers, chromosome number, marker chromosomes, and isozyme variants (Bunn et al., 1974; D. C. Wallace et al., 1975) have all been used. In the HeLa–HT1080 crosses of Table V, the HT1080 cells were fibroblastic, near-euploid, had two marker chromosomes, had a Y chromosome, and were G6PD variant B. The HEH cybrids had exactly these traits. Similarly, HeLa BU25 cells were epithelial, aneuploid, had several marker chromosomes, lacked a Y chromosome, and were G6PD variant A. The WEH cybrids possessed these traits. These characteristics confirmed the cybrid origin of the HEHs and WEHs. They also clearly distinguished them from the TIR and RIB hybrids, which contained the sum of the HT1080 and BU25 nuclear characteristics (D. C. Wallace, 1981; Oliver and Wallace, 1982).

It is also important to confirm the CAP resistance phenotype of the cybrids. The resistance of their mt protein synthesis should be reaffirmed and

Table V. Cell Fusions Using CAP-Resistant Human Cells

Type of cross	Fusion[a] $CAP^R \times CAP^S$	Selection[b]	Frequency	Designation
Cybrid	en HEB7A × HT1080C	HAT + CAP	5.1×10^{-4}	HEH
Hybrid	HEB7A × HT1080-6TG1	HAT + CAP	6.5×10^{-5}	TIR
Cybrid	en HT102W × BU25	BrdU + CAP	1.3×10^{-3}	WEH
Hybrid	WER1A × BU25	HAT + CAP	4.2×10^{-5}	RIB

[a] en, enucleated parent. BU25 is a TK^- HeLa cell. HEB7A is a CAP^R derivative of BU25. HT1080C is a clone of HT1080, and HT102W is a CAP^R mutant of HT1080C. HT1080-6TG1 is an $HPRT^-$ derivative of HT1080. WER1A is a cybrid in which the HT102W CAP resistance was transferred to HT1080-6TG1.
[b] HAT, 0.1 mM hypoxanthine, 16 μM thymidine, 0.5 μM aminopterin, and 3 μM glycine. BrdU was used at 0.1 mM and CAP at 50 μg/ml.

they should be tested for other traits characteristic of the original mutant, such as cross-resistant to other mt ribosome inhibitors. In the HeLa–HT1080 fusion described in Table V, the HT1080 CAP^R mutant (HT102W) was less resistant to CAP than the HeLa mutant (296-1). The cybrids and hybrids acquired the level of resistance characteristic of the CAP^R parent (Oliver and Wallace, 1982).

3.2. Mitotic Segregation of CAP Resistance

It is often possible to confirm the cytoplasmic inheritance of CAP^R mutants by demonstrating the segregation of the phenotype from intraspecific hybrids and cybrids during mitotic replication and in the absence of chromosome segregation. The cytoplasms of CAP^R and CAP^S cells can be mixed in hybrid fusion by selection for nuclear gene complementation with HAT or in cybrid fusions by selection in CAP and for the recipient cell nucleus. In cybrid fusions, the CAP selection is removed as soon as the cybrid clones have been isolated. Both hybrids and cybrids are then propagated without CAP and periodically cloned with and without the drug to determine the proportion of cells that retain resistance (Bunn et al., 1977; D. C. Wallace et al., 1977).

In some crosses, CAP resistance is lost very rapidly. Hybrids and cybrids between CAP^R mouse L cells and CAP^S mouse RAG cells have been found to decline in the proportion of CAP^R cells by 90% within 12 doublings after fusion (Bunn et al., 1977). Other crosses are more balanced. Fusions of CAP^R HeLa cells to CAP^S HT1080 cells yield cell lines which retain their resistance for prolonged periods (D. C. Wallace, 1981, and unpublished results).

The molecular basis of segregation is unknown. It does not seem to be the product of gross differences in the mtDNA sequence. Comparison of the mtDNAs of mouse L and RAG cells by digestion with restriction enzymes *Taq*I, *Hha*I, *Hae*III, *Pst*I, and *Msp*I and by heteroduplex analysis (Ferguson and Davis, 1978) have revealed no restriction site differences or insertions or deletions of greater than 100 nucleotides (Stroynowski, Giles, and Wallace, unpublished results). One factor may be the fitness of the CAP^R mutant itself (Adoutte and Beisson, 1972). The HT1080 CAP^R mutant HT102W is less resistant than the HeLa CAP mutant 296-1, and HT102W and its cybrids and hybrids seem to be generally less stable than comparable cell lines derived from 296-1 (D. C. Wallace et al., 1977; D. C. Wallace, 1981).

3.3. Elimination of CAP^R Determinants with R6G

R6G provides a final procedure for differentiating between cytoplasmic- and nuclear-coded CAP^R mutants. A 3-day treatment of cells with 1 µg/ml R6G is highly toxic, yet such cells can be rescued by fusion with cytoplasts from untreated cells. R6G must then damage cytoplasmic rather than nuclear genetic determinants (Ziegler and Davidson, 1981).

When cells carrying a cytoplasmic CAP^R marker are treated with R6G and then fused to CAP^S cells or cytoplasts, the resulting cell lines are generally found to be CAP^S (Ziegler and Davidson, 1981). Thus, this procedure provides a rapid method for screening for cytoplasmically inherited CAP^R mutants (see Chapter 15, this volume.)

4. Assignment of CAP Resistance to the Mitochondrial DNA

The CAP resistance locus has been assigned to the mammalian mtDNA by correlating its inheritance to strain-specific (Wallace, 1981) and species-specific (Ho and Coon, 1979) mtDNA restriction-endonuclease-site polymorphisms. Intraspecific restriction site polymorphisms are relatively common in human cells (Brown, 1980; Giles et al., 1980a; Case and Wallace, 1981), but are rare or absent in mouse cell lines derived from most major inbred lines (Giles and Wallace, unpublished results; Ferris et al., 1982). Differences in restriction patterns have been observed for wild mice, some inbred mouse strains (Ferris et al., 1982), and some inbred strains of rat (Hayashi et al., 1978; Kroon et al., 1978; Francisco et al., 1979). Little is published about mtDNA polymorphisms in hamster.

At present, mtDNA polymorphisms between cell lines can only be detected by screening the mtDNAs with a variety of enzymes, either using purified mtDNA (D. C. Wallace, 1981) or whole-cell DNA and Southern transfers (Case and Wallace, 1981). Because of problems with species incompatibility, when possible such assignments should be performed in intraspecific systems. The assignment of the human CAP^R locus to the mtDNA will be used as an example (D. C. Wallace, 1981).

HeLa- and HT1080-cell mtDNAs were screened with nine restriction endonucleases and found to differ in cleavage patterns for three of these: *Hae*II, *Hha*I, and *Hae*III. The results for *Hae*II will be used for illustration. *Hae*II cleaves HT1080 mtDNA into six fragments and HeLa mtDNA into eight (Fig. 3). In HeLa, the 8.6-kilobase (kb) fragment of HT1080 contains two additional sites which result in its cleavage into 4.5-, 2.1-, and 1.9-kb fragments. By digestion of the mtDNAs of the HeLa–HT1080 hybrids and cybrids (Table V) with *Hae*II and noting the presence of the distinctive 8.6-kb HT1080 fragment and 2.1-, and 1.9-kb HeLa fragments, the parental origin of the mtDNAs in each cell line can be determined. The results from an analysis of four hybrids is shown in Fig. 3. The two TIR hybrids expressed HeLa CAP resistance and contained primarily HeLa mtDNA, as shown by the 2.1- and 1.9-kb fragments. RIB21 expressed HT1080 CAP resistance and contained predominantly HT1080 mtDNA, as revealed by the 8.6-kb fragment. RIB22 was discordant and will be discussed below.

The proportion of the parental mtDNAs was quantitated by determining the relative number of molecules (Ni) in each polymorphic band (i in kb) and dividing these values by the Ni of nonpolymeric bands. Nonpolymorphic bands, such as the 1.4- and 1.3-kb fragments in this example, provided a value for the sum of the two parental mtDNAs when compared to the polymorphic fragments of the same preparation. The Ni values were obtained by dividing

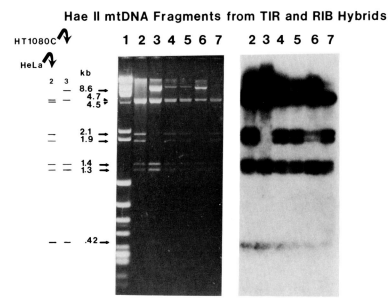

Figure 3. Representative gel of restriction endonuclease analysis of mtDNAs from CAP^R hybrids. Strain origins are defined in the text and in Table V. Closed circular mtDNAs were purified using CsCl–EtBr gradients. All DNAs, except for channel 1, contained 0.5 µg mtDNA digested with HaeII (Bethesda Research Laboratories, Rockville, MD). DNA fragments were separated on a 1% agarose gel containing 0.01 µg/ml EtBr and photographed using EtBr fluorescence (left panel). The DNA was then transferred to diazobenzyloxymethyl paper, hybridized to ^{32}P-labeled mtDNA–cRNA transcribed with E. coli RNA polymerase, and autoradiographed, right panel (Giles et al., 1980b; Wallace, 1981). Channel 1: λDNA digested with HindIII + φX174 DNA digested with HincII for size standards, 1.5 µg. Channel 2: S3. Channel 3: HT1080C. Channel 4: TIR21. Channel 5: TIR22. Channel 6: RIB21. Channel 7: RIB22.

the EtBr fluorescence of each band (peak area obtained from densitometry of photographic negatives) by the molecular size of the fragment in kb as determined by external size standards (Fig. 3; D. C. Wallace, 1981). The number of molecules of HT1080 and HeLa origin were then calculated by the formulas $N_{8.6}/[(N_{1.3} + N_{1.4})/2]$ and $[(N_{2.2} + N_{1.9})/2]/[(N_{1.3} + N_{1.4})/2]$, respectively. The results from four cybrids and five hybrids are summarized in Fig. 4 and are consistent with CAP resistance being coded on the mtDNA. The HEH cybrids and TIR hybrids expressed HeLa CAP resistance and retained predominantly HeLa mtDNA. The WEH1A cybrid and RIB21 and RIB24 hybrids expressed HT1080 CAP resistance and contained primarily HT1080 mtDNA. The WEH5 cybrid was allowed to segregate the HT1080 CAP resistance and then rechallenged for a brief time with CAP. It contained a roughly equal mixture of the two mtDNAs. RIB22 was the only discordant clone.

In addition to assigning CAP resistance to the mtDNA, these studies led to an important discovery. All of the CAP^R cybrids and hybrids retained some CAP^S mtDNA even after selection in CAP for up to 50 doublings. Apparently, CAP selection biased the proportion of mtDNA molecules toward the CAP^R parent but did not eliminate the CAP^S mtDNAs. This discovery indicates that

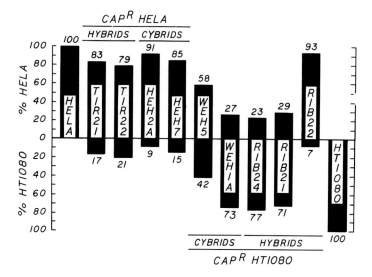

Figure 4. Quantitation of HeLa and HT1080 mtDNAs in CAP^R hybrids and cybrids, using *Hae*II restriction endonuclease fragments. See text and Wallace (1981) for procedures and details on strain origin and passage history.

the cellular mt phenotype is an average of the mtDNA genotypes and suggests that semiquantitative data may generally be required to assign genes to the mtDNA. It also explains why cybrids selected in CAP can still revert to CAP^S in the absence of selection.

5. Molecular Basis of CAP Resistance

Mammalian mtDNA, like yeast mtDNA, codes for one copy each of the large and small rRNA genes (Anderson et al., 1981). In yeast, CAP resistance has been found to correlate with single base changes in the large rRNA gene in the region coding for the 3' end of the rRNA (Dujon et al., 1977; Dujon, 1980). Comparable results have now been obtained for two human and two mouse CAP^R mutants (Blanc et al., 1981a,b; Kearsey and Craig, 1981).

The region of the mouse mtDNA large rRNA gene corresponding to the 3' end of the rRNA has been analyzed using two approaches. In one, the region was cloned into pBR325 using the 2.1-kb *Eco*RI fragment and sequenced by the Maxam–Gilbert (1980) procedure (Blanc et al., 1981b). In the other, it was cloned into pBR322 using the large 5.4-kb *Bam*Hl fragment and the 1.3-kb *Eco*RI–*Bam*Hl subfragment cut out and purified. This fragment was digested with *Sau*3A and subcloned into the *Sau*3A site of bacteriophage M13mp2. The resulting phage were used for sequencing by the dideoxy procedure (Sanger et al., 1980; Kearsey and Craig, 1981).

The same region of the human mtDNA was also analyzed in two ways. In one, the 1.5-kb *Kpn*I–*Eco*RI fragment was cloned into a specifically constructed derivative of pBR325 containing the herpes simplex TK gene. This insert contains a *Kpn*I site and permits cloning of *Eco*RI–*Kpn*I fragments

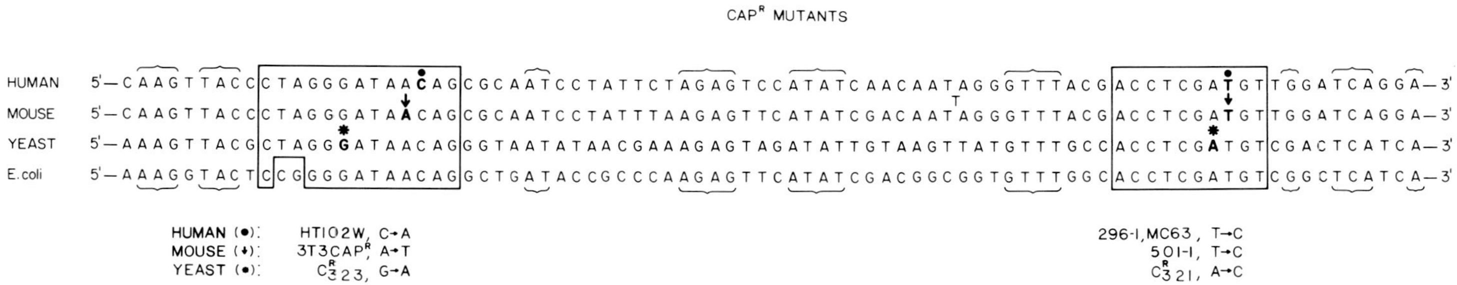

CAP^R MUTANTS

```
HUMAN  5'-CAAGTTACC CTAGGGATAAC AG CGCAATCCTATTCTAGAGTCCAATCAACAATAGGG TTTACG ACCTCGAT GT TGGATCAGGA -3'
                                                          T                        ••  →
MOUSE  5'-CAAGTTACC CTAGGGATAA→A CAG CGCAATCCTATTTAAGAGTTCATATCGACAATAGGGTTTACG ACCTCGAT GT TGGATCAGGA -3'
                            *                                                        *
YEAST  5'-AAAGTTACG CTAGG GATAACAGG GTAATAACGAAAGATAGATATTGTAAGTTATGTTTGCC ACCTCGA A TGT CGACTCATCA -3'
                                                                                      •
E.coli 5'-AAAGGTACT CC GGGATAACAGG CTGATACCGCCCAAGAGTTCATATCGACGGGGTGTTTGG CACCTCGATG T CGGCTCATCA -3'
```

HUMAN (●):   HT102W, C→A   296-I,MC63, T→C
MOUSE (→):   3T3CAP^R, A→T   501-I,   T→C
YEAST (*):   C^R_323, G→A   C^R_321,   A→C

Figure 5. Comparison of the nucleotide sequences of the genes corresponding to the 3' ends of the large rRNAs. The strand corresponding to the rRNA sequence is given and reproduces nucleotides 2920–3003 of the human mtDNA (Anderson et al., 1981); nucleotides 2486–2570 of the mouse LA9 mtDNA (Van Etten et al., 1980); nucleotides −77 to +7 of the yeast mtDNA (Dujon, 1980); and nucleotides 2433–2516 of the *E. coli* 23S rRNA (Brosius et al., 1980). Highly conserved regions are boxed and the base changes associated with CAP resistance located within them: HT102W and 296-1 from Blanc et al. (1981a); MC63 and 3T3CAP^R from Kearsey and Craig (1981); 501-1 from Blanc et al. (1981b); and C^R_{323} and C^R_{321} from Dujon (1980).

using the pBR325 CAP gene EcoRI site. The rRNA region was then sequenced using the Maxam–Gilbert method (Blanc et al., 1981a). In the second procedure, the CAP^R human mtDNA was shotgun-cloned into the M13mp2 Sau3A site and the appropriate fragment identified by hybridization with the appropriate mouse M13mp2 phage. The human phage insert was then sequenced by the dideoxy procedure (Kearsey and Craig, 1981).

All of the sequence changes associated with CAP resistance have been located in the same region of the large rRNA presented in Fig. 5. The upper three lines report the appropriate sequence for human, mouse, and yeast mtDNAs, while the lower line presents the E. coli sequence. Two highly conserved regions were found, indicated by boxes; one region of 13 nucleotides on the left, and the other of ten nucleotides on the right. All base changes associated with CAP resistance were located in these two regions. Three of the five mammalian mutants were found to have exactly the same change, a T to C transition one nucleotide 3' to the yeast C_{321}^R mutation (Blanc et al., 1981a,b; Kearsey and Craig, 1981). The remaining two mutants, one mouse and one human, were discovered in the left-hand boxed sequence and together with the yeast mutant C_{321}^R define a six-nucleotide region associated with CAP binding (Blanc et al., 1981a; Kearsey and Craig, 1981).

Though the two conserved regions are separated by 42 nucleotides of less conserved sequence, they can be brought together by the formation of an intervening stem and loop structure (Fig. 6; Blanc et al., 1981b; Baer and

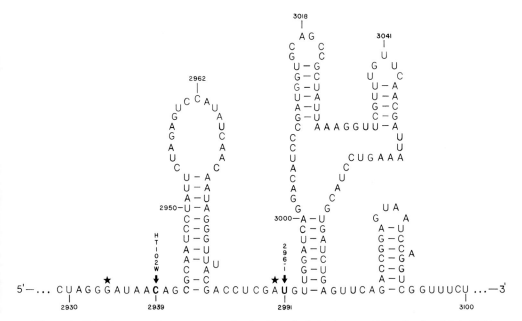

Figure 6. Putative secondary structure in the 3' end of the human 16S mitochondrial rRNA adapted from Baer and Dubin (1981) and Blanc et al. (1981b). The RNA regions corresponding to mtDNA nucleotides 2929–3100 is represented. The CAP^R mutations of 296-1 and HT102W are shown. The star indicates the position of the yeast CAP^R mutations.

Dubin, 1981). Since CAP is known to block the peptidyl transferase reaction by binding to E. coli ribosomal proteins L16, L24, and S6 and bovine mt ribosomal protein L2, this catalytic site and these proteins must be associated with this region of the large rRNA.

6. The Biochemistry of Cytoplasmic CAP Resistance

Genetic studies have shown that CAP resistance is dominant, but it was unclear if this was due to the independent expression of separate CAP^R and CAP^S mitochondria within the cell or to the translation of the mRNA from the CAP^S mtDNAs by the ribosomes of the CAP^R mtDNA. To distinguish between these two alternatives, the expression of a pair of mitochondrially synthe-

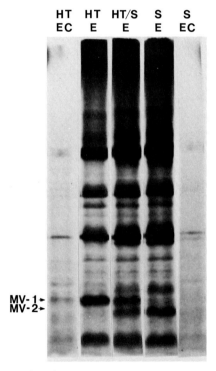

Figure 7. Mitochondrially synthesized marker polypeptides MV-1 and MV-2. Cells were labeled with [^{35}S] methionine in the presence of 100 µg/ml emetine (E) or E plus 100 µg/ml CAP (EC). The mitochondria were isolated, solubilized, and separated on a 13% polyacylamide gel with a 4.75% stacking gel. The gel was dried and fluorographed (Oliver and Wallace, 1982). S, HeLa S3; HT, HT1080C; HT/S, an equal mixture of the radioactivity from the two parents. E channels contained (1.4–1.5) × 10^4 cpm, while EC channels contained equal protein but (1.2–4.5) × 10^3 cpm. [Wallace et al. (1982), reproduced by permission of Cold Spring Harbor Laboratories.]

sized marker polypeptides present in HeLa and HT1080 cells was examined by labeling the HeLa–HT1080 hybrids and cybrids (Table V) in emetine with and without CAP. HT1080 has a 15-kilodalton (kD) polypeptide designated MV-1, while HeLa has a faster 14-kD variant designated MV-2 (Fig. 7; Yatscoff et al., 1978; Oliver and Wallace, 1982). Quantitation of the relative expression in emetine of MV-1 and MV-2 in HeLa–HT1080 cybrids and hybrids by densitometry revealed that their synthesis was directly proportional to the ratio of the parental mtDNAs. Thus, MV-1 and MV-2 were assigned to the mtDNA and shown to be expressed constitutively (Oliver and Wallace, 1982).

The level of expression of MV-1 and MV-2 was then examined in the HeLa–HT1080 cybrids and hybrids in the presence of emetine + CAP. If the CAP^R and the CAP^S mitochondria were metabolically separate, then labeling in CAP should eliminate the expression of the marker polypeptide linked to the CAP^S mtDNA and yield a pattern equivalent to that of the CAP^R parent. Alternatively, if the two mtDNAs cooperate and complement each other, then the CAP^R ribosomes should translate the mRNA of the CAP^S mtDNA and both marker polypeptides should continue to be synthesized. The latter alternative proved to be the case. Figure 8 shows that labeling in CAP only partially shifted the ratio of expression of the two marker polypeptides toward that of the CAP^R parent (Oliver and Wallace, 1981; D. C. Wallace et al., 1982).

These observations suggest that the mitochondria must fuse within the cell and that the CAP^R ribosomes can translate the mRNA of the CAP^S mtDNA in CAP and in the presence of the CAP^S ribosomes. This result explains the dominance of CAP resistance, the prolonged retention of CAP^S mtDNAs in CAP^R hybrids and cybrids, and the reason why a single mutant CAP^R mtDNA can, on selection, repopulate a previously CAP^S cell.

Analysis of the expression of MV-1 and MV-2 in RIB22 revealed one additional point. RIB22, though able to grow slowly in CAP, retained predominantly CAP^S mt protein synthesis, consistent with 95% of its mtDNA

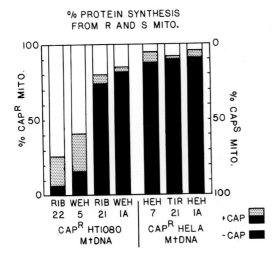

Figure 8. Effect of CAP on expression of polypeptides linked to CAP^R and CAP^S mtDNAs. Proportion of mt protein synthesis contributed by the CAP^R mtDNA in 100 μg/ml emetine (filled bar) and 100 μg/ml emetine + 100 μg/ml CAP (filled plus stippled bar). [Oliver and Wallace (1982); figure from Wallace et al. (1982), by permission of Cold Spring Harbor Laboratory.]

being derived from the CAP^S parent. Hence, the CAP resistance of RIB22 was not due to the acquisition of the HeLa CAP^R locus through recombination. Rather, it is more likely that the 5% CAP^R HT1080 mtDNA was sufficient to permit slow growth of the cell in CAP or, alternatively, that the cell was simply respiratory-deficient and in a physiologic state comparable to that of chick embryo fibroblasts when grown in CAP and uridine (Morais et al. 1980). The existence of such cell lines potentially complicates the assignment of the CAP resistance locus to the mtDNA and also means that it is essential that restriction site polymorphisms be used together with the CAP resistance locus to assign genes to the mtDNA.

7. Cytoplasmic CAP Resistance in Interspecific Crosses

Cytoplasmic CAP resistance has also proven to be a powerful tool in interspecific somatic cell genetic studies. It has permitted studies on the evolution of the mammalian mitochondrion and on nucleocytoplasmic interactions and now offers the opportunity for controlling the direction of chromosome and mtDNA segregation in interspecific somatic cell hybrids.

The evolutionary divergence of mammalian mt genes has been examined by using CAP selection to determine if the mtDNA of one species can be incorporated into the cell of another. Initial studies examined the exchange of CAP^R mtDNAs between human and mouse cells. Enucleated CAP^R HeLa cells were fused to CAP^S mouse RAG and L cells and enucleated CAP^R mouse L cells were fused to CAP^S HeLa cells. Numerous fusions were performed, but only a few colonies were obtained. These were isolated from the fusion of CAP^R HeLa cytoplasts to mouse cells (D. C. Wallace et al., 1976; Munro et al., 1978; D. C. Wallace and Eisenstadt, 1979).

The parental origin of the mtDNAs from various of these cell lines was examined by three different methods. In the first, the buoyant densities of the mtDNAs were determined and compared with those of the human and mouse parents (Attardi and Attardi, 1972; Clayton et al., 1971; D. C. Wallace, 1975). In the second, whole-cell DNAs bound to filters were hybridized to species-specific ^{32}P-labeled mtDNA–cRNA probes (Coon et al., 1973; Wallace, Pollack, and Eisenstadt, unpublished results). In the last, and most rigorous method, purified mtDNAs were digested with BamHl or HhaI (enzymes which yield dramatically different human and mouse mtDNA cleavage patterns) and the restriction fragments transferred to diazobenzyloxymethyl paper. The bound fragments were then hybridized to ^{32}P-human mtDNA–cRNA and the human mtDNA fragments identified by autoradiography. This probe was then removed and the fragments rehybridized to ^{32}P-mouse mtDNA–cRNA and the mouse fragments identified by autoradiography. In this way, the parental origin of the sequence in each restriction fragment could be determined (Giles et al., 1980b). Within the limits of the techniques used, every line examined was found to have only mouse mtDNA (D. C. Wallace, 1975; Wallace, Pollack, and Eisenstadt, unpublished results; Giles et al., 1980b).

Thus, these experiments failed to establish human mtDNA within mouse cells by CAP selection.

To date, only one cell line has been reported to have a human nucleus and mouse mtDNA. The chromosomes of this reconstituted cell were found to be those of the SV40-transformed human fetal lung cell parent, but the mtDNA was observed to have the same buoyant density as mouse mtDNA (Nette et al., 1979).

Comparable fusions have been performed between cytoplasts of CAP^R mouse L cells and CAP^S Syrian hamster cells. Mouse–Syrian hamster hybrids have been observed to have a more balanced chromosome and mtDNA segregation pattern than mouse–human hybrids (Eliceiri, 1973). Initially, a high frequency of CAP^R cybrid colonies was obtained. However, these stopped growing prior to being transferred and most of them degenerated (D. C. Wallace et al., 1976; D. C. Wallace and Eisenstadt, 1979). Analysis of the mtDNAs of two of the surviving cybrid lines by hybridization of species-specific probes to disk-bound cellular DNA revealed that, within the limits of the method, these lines contained hamster mtDNA (Wallace, Pollack, and Eisenstadt, unpublished results). Thus, like the mouse–human experiments, it was not possible to establish mouse mtDNA in hamster cells.

Efforts to establish rat mtDNA in mouse cells met with comparable results. CAP^R rat cytoplasts were fused to mouse whole cells or karyoplasts. Though cell lines were obtained having only mouse chromosomes, restriction enzyme analysis revealed only mouse mtDNA (Hayashi et al., 1980).

The only clearly successful interspecific transfer of mtDNA using the CAP resistance marker has been reported for transfers between species and subspecies of mice. CAP resistance was transferred by cybrid fusions from Mus caroli to Mus musculus musculus. The resulting cybrids contained only M. caroli mtDNA as determined by restriction endonuclease cleavage pattern. CAP resistance was also transferred from M. musculus molossinus to M. musculus and these cybrids retained only M. m. molossinus mtDNA. Interestingly, in the reciprocal transfer of M. m. musculus CAP resistance to M. m. molossinus, cybrids were found to contain both musculus and molossinus mtDNAs. Growth of these cells without CAP resulted in the coordinate loss of CAP resistance and the musculus mtDNA (Ho and Coon, 1979).

With the exception of the single report of a human–mouse reconstituted cell, these results suggest a clear trend. Beyond selected, closely related species, the nucleus of one mammalian species is unable to sustain the replication and expression of the mtDNA of another mammalian species. Those cell lines that are obtained are best explained as the product of new CAP^r mutations occurring in the recipient cell mtDNA. Mitochondrial DNAs can be exchanged between some species of the same genera, but even in these cases, transfers between other strains do not succeed. In addition, successful transfers often result in the complete replacement of the mtDNA of the CAP^S recipient, a result different from that found in the human HeLa–HT1080 cell experiments (D. C. Wallace, 1981). This suggests that some interspecific incompatibility exists between the mtDNAs of even closely related species.

Since nuclear-coded genes of even widely different organisms have been found to complement each other in somatic cell hybrids, it is surprising that mt incompatibility was found between the mitochondria of even closely related species. This observation is consistent with the view that mtDNA genes are diverging much faster than nuclear genes (Brown et al., 1979) and thus may be important in speciation. However, the mitochondrion is a complex and highly integrated organelle. Hence, the observed incompatibility might merely reflect the fact that a far greater number of genetic interactions must succeed for the interspecific complementation of mtDNA genes than is required for the interspecific complementation of single nuclear genes.

The high degree of species incompatibility between nuclei and the mtDNAs made it possible to determine which chromosomal elements were necessary for the retention of the mtDNA. Fusions between human cells and established mouse cells are known to result in hybrids which have lost all of the human mtDNA and segregated the human chromosomes but which retained a complete set of mouse chromosomes and pure mouse mtDNA (Clayton et al., 1971; Attardi and Attardi, 1972). By fusing CAP^R human HeLa cells to CAP^S mouse cells and selection in CAP, it was hoped that the human mtDNA would be retained along with the chromosomes necessary for its expression. Direct selection of CAP^R HeLa–CAP^S mouse RAG cell hybrids or CAP^R mouse L cell–CAP^S HeLa cell hybrids in HAT + CAP proved impossible. CAP^R HeLa–CAP^S RAG hybrids were obtained by selection first in HAT and then by removing HAT and challenging the hybrids with CAP for extended periods. All of the resulting cell lines were found to contain a few human chromosomes but a complete set of mouse chromosomes and pure mouse mtDNA. One hybrid has a single RAG chromosome complement and grew rapidly, while the others had 1½ RAG chromosome complements and grew very poorly (D. C. Wallace and Eisenstadt, 1979; Giles et al., 1980b).

The direct selection of a CAP^R HeLa–CAP^S RAG hybrid was achieved by selection in CAP + ouabain. This line was found to have pure HeLa mtDNA. However, it also contained 1½ human chromosomal complements and only a partial mouse chromosome complement (D. C. Wallace and Eisenstadt, 1979; Giles et al., 1980b).

Fusion of CAP^R mouse L cells to CAP^S Syrian hamster cells and selection directly in HAT + CAP invariably yielded hybrids with a complete mouse chromosome set and a partial hamster chromosome complement (D. C. Wallace and Eisenstadt, 1979). Analysis of the mtDNA from three of these hybrids using species-specific probes hybridized to disk-bound whole-cell DNA revealed that they all contained predominantly, if not exclusively, mouse mtDNA (Wallace, Pollack, and Eisenstadt, unpublished data). Similarly, fusion of CAP^R Chinese hamster cells to CAP^S mouse 3T3 cells and selection in HAT + CAP resulted in hybrids with two hamster chromosome complements and one mouse complement (Ziegler and Davidson, 1981).

The overwhelming conclusion of these results is that the mtDNA retained in an interspecific hybrid will invariably be that of the parent for which a complete chromosome complement was retained. In cases where the mtDNA of one of the parents can be selected directly in CAP at the time of fusion, then the hybrids will generally retain the CAP^R mtDNA and also a complete complement of chromosomes from the CAP^R parent. This will occur even if the direction of chromosome and mtDNA segregation is contrary to that normally found for the cells used. In cases where the hybrid has already become committed to segregate the mtDNA and chromosomes of the CAP^R parent, either by initial HAT selection or by removal of the CAP^R parent's nucleus, then selection in CAP will not rescue the CAP^R mtDNA. Rather, the cells will either die, acquire resistance by secondary mutation in the CAP^S mtDNA, or initiate anaerobic growth in CAP and in the absence of mt protein synthesis. Secondary mutations would explain the interspecific cybrid results and the actively growing HeLa–RAG hybrid selected first in HAT and then in CAP. Anaerobic growth might explain the slowly growing HeLa–RAG HAT and then CAP-selected hybrids as well as the CAP^R human–mouse reconstituted cell and possibly the intraspecific RIB22 hybrid.

It is now well established that hybrids lose all of the mtDNA of the parent whose chromosomes are segregated (De Francesco et al., 1980; Clayton et al., 1971; Attardi and Attardi, 1972). Results using the mtDNA-linked CAP resistance marker demonstrate that the direction of mtDNA segregation can be reversed by selection in CAP and that invariably this also results in a reversal of the direction of chromosome segregation. These results clearly demonstrate that a complete chromosomal complement is essential for retaining the mtDNA of one of the parents, but they also raise the possibility that the direction of chromosome segregation is determined by the direction of mtDNA segregation. This seems even more reasonable, since it has recently been observed that pretreatment of Chinese hamster cells with R6G prior to their fusion to mouse cells greatly increases the viability and stability of the resulting hamster–mouse hybrids (Ziegler and Davidson, 1981). This suggests that the interspecific incompatibility of mtDNAs of even these closely related species greatly decrease hybrid viability. Mitochondrial DNA incompatibility between distantly related species, such as mouse and human, might then be expected to be lethal unless one of the parental mtDNAs was rapidly and completely eliminated. Those hybrids that did lose one parental mtDNA would survive, but the loss of that parent's mtDNA would in turn dictate that the chromosomes of that parent would also segregate.

It remains to be determined what the molecular basis of the mtDNA incompatibility is. Is it of the same nature as the factors responsible for the segregation of CAP resistance in intraspecific hybrids and cybrids (D. C. Wallace, 1975)? Could it be related to a species-specific restriction-modification systems (Sager and Kitchin, 1975)?

Regardless of the causal relationship between mtDNA and chromosome segregation in interspecific hybrids, it is clear that selection for mtDNA-coded CAP resistance permits selection for those hybrids that have retained the CAP^R parent's genome and segregated the mtDNA and chromosomes of the CAP^S parent. This fact should prove very useful in experiments on comparative mapping of chromosomal genes of different species using somatic cell hybrids.

Acknowledgments

The author would like to express his appreciation to Drs. R. Giles and N. Oliver for assistance in preparation of Figs. 3 and 7 and for helpful discussions. This research was supported by National Institutes of Health Grants GM 24285 and GM 28428, National Science Foundation Grant PCM 80-21871, and March of Dimes Basic Research Grant 1-788.

References

Adoutte, A., and Beisson, J., 1972, Evolution of mixed populations of genetically different mitochondria in *Paramecium aurelia*, *Nature* **235**:393–396.

Anderson, S., Bankier, A. T., Barrell, B. G., de Bruijn, M. H. L., Coulson, A. R., Drouin, J., Eperon, I. C., Nierlich, D. P., Roe, B. A., Sanger, F., Schreier, P. H., Smith, A. J. H., Staden, R., and Young, I. G., 1981, Sequence and organization of the human mitochondrial genome, *Nature* **290**:457–465.

Attardi, B., and Attardi, G., 1972, Fate of mitochondrial DNA in human–mouse somatic cell hybrids, *Proc. Natl. Acad. Sci. USA* **69**:129–133.

Baer, R. J., and Dubin, D. T., 1981, Methylated regions of hamster mitochondrial ribosomal RNA: Structural and functional correlates, *Nucleic Acids Res.* **9**:323–337.

Bastos, R. de N., and Mahler, H. R., 1974, Molecular mechanisms of mitochondrial genetic activity. Effects of ethidium bromide on the deoxyribonucleic acid and energetics of isolated mitochondria, *J. Biol. Chem.* **249**:6617–6627.

Blanc, H., Adams, C. A., and Wallace, D. C., 1981a, Different nucleotide changes in the large rRNA gene of the mitochondrial DNA confer chloramphenicol resistance on two human cell lines, *Nucleic Acids Res.* **9**:5785–5795.

Blanc, H., Wright, C. T., Bibb, M. J., Wallace, D. C., and Clayton, D. A., 1981b, Mitochondrial DNA of chloramphenicol-resistant mouse cells contains a single nucleotide change in the region encoding the 3′ end of the large ribosomal RNA, *Proc. Natl. Acad. Sci. USA* **78**:3789–3793.

Brosius, J., Dull, T. J., and Noller, H. F., 1980, Complete nucleotide sequence of a 23S ribosomal RNA gene from *Escherichia coli*, *Proc. Natl. Acad. Sci. USA* **77**:201–204.

Brown, W. M., 1980, Polymorphism in mitochondrial DNA of humans as revealed by restriction endonuclease analysis, *Proc. Natl. Acad. Sci. USA* **77**:3605–3609.

Brown, W. M., George, M., Jr., and Wilson, A. C., 1979, Rapid evolution of animal mitochondrial DNA, *Proc. Natl. Acad. Sci. USA* **76**:1967–1971.

Bunn, C. L., and Eisenstadt, J. M., 1977a, Cybrid formation in mouse L cells: the influence of cytoplast-to-cell ratio, *Somat. Cell Genet.* **3**:335–341.

Bunn, C. L., and Eisenstadt, J. M., 1977b, Carbomycin resistance in mouse L cells, *Somat. Cell Genet.* **3**:611–627.

Bunn, C. L., Wallace, D. C., and Eisenstadt, J. M., 1974, Cytoplasmic inheritance of chloramphenicol resistance in mouse tissue culture cells, *Proc. Natl. Acad. Sci. USA* **71**:1681–1685.

Bunn, C. L., Wallace, D. C., and Eisenstadt, J. M., 1977, Mitotic segregation of cytoplasmic determinants for chloramphenicol resistance in mammalian cells I: Fusions with mouse cell lines, *Somat. Cell Genet.* **3**:71–92.

Carter, S. B., 1967, Effects of cytochalasins on mammalian cells, *Nature* **213**:261–264.

Case, J. T., and Wallace, D. C., 1981, Maternal inheritance of mitochondrial DNA polymorphisms in cultured human fibroblasts, *Somat. Cell Genet.* **7**:103–108.

Clark, M. A., Shay, J. W., and Goldstein, L., 1980, Techniques for purifying L-cell karyoplasts with minimal amounts of cytoplasm. *Somat. Cell Genet.* **6**:455–464.

Clayton, D. A., and Teplitz, R. L., 1972, Intracellular mosaicism (nuclear$^-$/mitochondrial$^+$) for thymidine kinase in mouse L cells, *J. Cell Sci.* **10**:487–493.

Clayton, D. A., Teplitz, R. L., Nabholz, M., Dovey, H., and Bodmer, W., 1971, Mitochondrial DNA of human–mouse cell hybrids, *Nature* **234**:560–562.

Coon, H. G., 1978, The genetics of the mitochondrial DNA of mammalian somatic cells, their hybrids and cybrids, *Natl. Cancer Inst. Monogr.* **48**:45–55.

Coon, H. G., and Ho, C., 1978, Transformation of cultured cells to chloramphenicol resistance by purified mammalian mitochondrial DNA, in: *Genetic Interaction and Gene Transfer*, Brookhaven Symposium, Volume 29 (C. W. Anderson, ed.), Brookhaven Natl. Lab., Upton, New York, pp. 166–177.

Coon, H. G., Horak, I., and Dawid, I. B., 1973, Propagation of both parental mitochondrial DNAs in rat–human and mouse–human hybrid cells, *J. Mol. Biol.* **81**:285–298.

Costantino, P., and Attardi, G., 1975, Identification of discrete electrophoretic components among the products of mitochondrial protein synthesis in HeLa cells, *J. Mol. Biol.* **96**:291–306.

Craig, I., and Webb, M., 1979, Resistance to antimycin A: A new extrachromosomal genetic marker in mammalian cells, *J. Supramol. Struct.* **10–12**(Suppl. 3):396 Abt.

Croizat, B., and Attardi, G., 1975, Selective *in vivo* damage by 'visible' light of BrdU-containing mitochondrial DNA in a thymidine kinase-deficient mouse cell line with persistent mitochondrial enzyme activity, *J. Cell Sci.* **19**:69–84.

De Francesco, L., Attardi, G., and Croce, C. M., 1980, Uniparental propagation of mitochondrial DNA in mouse–human cell hybrids, *Proc. Natl. Acad. Sci. USA* **77**:4079–4083.

Das, H. K., Goldstein, A., and Kanner, L. C., 1966, Inhibition by chloramphenicol of the growth of nascent protein chains in *Escherichia coli*, *Mol. Pharmacol.* **2**:158–170.

Ditta, G., Soderberg, K., Landy, F., and Scheffler, I. E., 1976, The selection of Chinese hamster cells deficient in oxidative energy metabolism, *Somat. Cell Genet.* **2**:331–344.

Ditta, G., Soderberg, K., and Scheffler, I. E., 1977, Chinese hamster cell mutant with defective mitochondrial protein synthesis, *Nature* **268**:64–67.

Doersen, C.-J., and Stanbridge, E. J., 1979, Cytoplasmic inheritance of erythromycin resistance in human cells, *Proc. Natl. Acad. Sci. USA* **76**:4549–4553.

Dujon, B., 1980, Sequence of the intron and flanking exons of the mitochondrial 21S rRNA gene of yeast strains having different alleles at the ω and rib-1 loci, *Cell* **20**:185–197.

Dujon, B., Colson, A. M., and Slonimski, P. P., 1977, The mitochondrial genetic map of *Saccharomyces cerevisiae*: Compilation of mutants, genes, genetic and physical maps, in: *Mitochondria 1977: Genetics and Biogenesis of Mitochondria* (W. Bandlow, R. J. Schweyen, K. Wolf, and F. Kaudewitz, eds.), W. de Gruyter, Berlin, pp. 579–669.

Dulbecco, R., and Freeman, G., 1959, Plaque production by the polyoma virus, *Virology* **8**:396–397.

Eliceiri, G. L., 1973, The mitochondrial DNA of hamster–mouse hybrid cells, *FEBS Lett.* **36**:232–234.

Ferguson, J., and Davis, R. W., 1978, Quantitative electron microscopy of nucleic acids, in: *Advanced Techniques in Biological Electron Microscopy II* (J. K. Koehler, ed.), Springer-Verlag, Berlin, pp. 123–171.

Ferris, S. D., Sage, R. D., and Wilson, A. C., 1982, Evidence from mtDNA sequences that common laboratory strains of inbred mice are descended from a single female, *Nature* **295**:163–165.

Fettes, I. M., Haldar, D., and Freeman, K. B., 1972, Effect of chloramphenicol on enzyme synthesis and growth of mammalian cells, *Can. J. Biochem.* **50**:200–209.

Francisco, J. F., Brown, G. G., and Simpson, M. V., 1979, Further studies on types A and B rat mtDNAs: Cleavage maps and evidence for cytoplasmic inheritance in mammals, *Plasmid* **2**:426–436.

Freeman, K. B., 1970a, Effects of chloramphenicol and its isomers and analogues on the mitochondrial respiratory chain, *Can. J. Biochem.* **48**:469–478.

Freeman, K. B., 1970b, Inhibition of mitochondrial and bacterial protein synthesis by chloramphenicol, *Can. J. Biochem.* **48**:479–485.

Giguere, L., and Morais, R., 1981, On suppression of tumorigenicity in hybrid and cybrid mouse cells, *Somat. Cell Genet.* **7**:457–471.

Giles, R. E., Blanc, H., Cann, H. M., and Wallace, D. C., 1980a, Maternal inheritance of human mitochondrial DNA, *Proc. Natl. Acad. Sci. USA* **77**:6715–6719.

Giles, R. E., Stroynowski, I., and Wallace, D. C., 1980b, Characterization of mitochondrial DNA in chloramphenicol-resistant intraspecific hybrids and a cybrid, *Somat. Cell Genet.* **6**:543–554.

Grand Island Biological Company, 1978/1979, *Gibco Catalogue*, Gibco, Grand Island, New York.

Harris, M., 1980, Pyruvate blocks expression of sensitivity to antimycin A and chloramphenicol, *Somat. Cell Genet.* **6**:699–708.

Hayashi, J.-I., Yonekawa, H., Gotoh, O., Watanabe, J., and Tagashira, Y., 1978, Strictly maternal inheritance of rat mitochondrial DNA, *Biochem. Biophys. Res. Commun.* **83:**1032–1038.

Hayashi, J.-I., Gotoh, O., Tagashira, Y., Tostu, M., and Sekiguchi, T., 1980, Identification of mitochondrial DNA species in interspecific cybrids and reconstituted cells using restriction endonuclease, *FEBS Lett.* **117:**59–62.

Hightower, M. J., Fairfield, F. R., and Lucas, J. J., 1981, A staining procedure for identifying viable cell hybrids constructed by somatic cell fusion, cybridization, or nuclear transplantation, *Somat. Cell Genet.* **7:**321–329.

Ho, C., and Coon, H. G., 1979, Restricted mitochondrial DNA fragments as genetic markers in cytoplasmic hybrids, in: *Extrachromosomal DNA, ICN–UCLA Symposium on Molecular and Cellular Biology*, Volume XV (D. J. Cummings, P. Borst, I. B. Dawid, S. M. Weissman, and C. F. Fox, eds.), Academic Press, New York, pp. 501–514.

Howell, N., and Sager, R., 1977, Differential effects of mitochondrial inhibitors on normal and tumorigenic mouse cells, *Fed. Proc.* **36:**356 Abt.

Howell, A. N., Sager, R., 1978, Tumorigenicity and its suppression in cybrids of mouse and Chinese hamster cell lines, *Proc. Natl. Acad. Sci. USA* **75:**2358–2362.

Howell, N., and Sager, R., 1979, Cytoplasmic genetics of mammalian cells: Conditional sensitivity to mitochondrial inhibitors and isolation of new mutant phenotypes, *Somat. Cell Genet.* **5:**833–845.

Ibrahim, N. G., Burke, J. P., and Beattie, D. S., 1974, The sensitivity of rat liver and yeast mitochondrial ribosomes to inhibitors of protein synthesis, *J. Biol. Chem.* **249:**6806–6811.

Johnson, L. V., Walsh, M. L., and Chen, L. B., 1980, Localization of mitochondria in living cells with rhodamine 123, *Proc. Natl. Acad. Sci. USA* **77:**990–994.

Jongkind, J. F., Verkerk, A., and Tanke, H., 1979, Isolation of human fibroblast heterokaryons with two-color flow sorting (FACS II), *Exp. Cell Res.* **120:**444–448.

Kearsey, S. E., and Craig, I. W., 1981, Altered ribosomal RNA genes in mitochondria from mammalian cells with chloramphenicol resistance. *Nature* **290:**607–608.

Kislev, N., Spolsky, C. M., and Eisenstadt, J. M., 1973, Effect of chloramphenicol on the ultrastructure of mitochondria in sensitive and resistant strains of HeLa, *J. Cell Biol.* **57:**571–579.

Kroon, A. M., de Vos, W. M., and Bakker, H., 1978, The heterogeneity of rat-liver mitochondrial DNA, *Biochim. Biophys. Acta* **519:**269–273.

Kuhns, M. C., and Eisenstadt, J. M., 1979, Oligomycin-resistant mitochondrial ATPase from mouse fibroblasts, *Somat. Cell Genet.* **5:**821–832.

Lamb, A. J., Clark-Walker, G. D., and Linnane, A. W., 1968, The biogenesis of mitochondria 4: The differentiation of mitochondrial and cytoplasmic protein synthesizing systems *in vitro* by antibiotics, *Biochim. Biophys. Acta* **161:**415–427.

Lansman, R. A., and Clayton, D. A., 1975a, Selective nicking of mammalian mitochondrial DNA *in vivo*: Photosensitization by incorporation of 5-bromodeoxyuridine, *J. Mol. Biol.* **99:**761–776.

Lansman, R. A., and Clayton, D. A., 1975b, Mitochondrial protein synthesis in mouse L-cells: Effect of selective nicking on mitochondrial DNA, *J. Mol. Biol.* **99:**777–793.

Lichtor, T., and Getz, G. S., 1978, Cytoplasmic inheritance of rutamycin resistance in mouse fibroblasts, *Proc. Natl. Acad. Sci. USA* **75:**324–328.

Littlefield, J. W., 1966, The periodic synthesis of thymidine kinase in mouse fibroblasts, *Biochim. Biophys. Acta* **114:**398–403.

Loeb, J. N., and Hubby, B. G., 1968, Amino acid incorporation by isolated mitochondria in the presence of cycloheximide, *Biochim. Biophys. Acta* **166:**745–748.

Malech, H. L., and Wivel, N. A., 1976, Transfer of murine intracisternal A particle phenotype in chloramphenicol-resistant cytoplasts, *Cell* **9:**383–391

Maxam, A. M., and Gilbert, W., 1980, Sequencing end-labeled DNA with base-specific chemical cleavages, in: *Methods in Enzymology*, Volume 65 (L. Grossman and K. Moldave, eds.), Academic Press, New York, pp. 499–560.

Mitchell, C. H., and Attardi, G., 1978, Cytoplasmic transfer of chloramphenicol resistance in a human cell line, *Somat. Cell Genet.* **4:**737–744.

Mitchell, C. H., England, J. M., and Attardi, G., 1975, Isolation of chloramphenicol-resistant variants from a human cell line, *Somat. Cell Genet.* **1:**215–234.

Molloy, P. L., and Eisenstadt, J. M., 1979, Erythromycin resistance in mouse L cells, *Somat. Cell Genet.* **5**:585–595.

Morais, R., Gregoire, M., Jennotee, L., and Gravel, D., 1980, Chick embryo cells rendered respiration-deficient by chloramphenicol and ethidium bromide are auxotrophic for pyrimidines, *Biochem. Biophys. Res. Commun.* **94**:71–77.

Munro, E., Siegel, R. L., Craig, I. W., and Sly, W. S., 1978, Cytoplasmic transfer of a determinant for chloramphenicol resistance between mammalian cell lines, *Proc. R. Soc. Lond. (Biol.)* **201**:73–85.

Nass, M. M. K., 1969, Mitochondrial DNA I: Intramolecular distribution and structural relations of single- and double-length circular DNA, *J. Mol. Biol.* **42**:521–528.

Nette, E. G., Sit, H. L., Clavey, W., and King, D. W., 1979, Isolation of viable reconstituted cells from human karyoplasts fused to mouse cytoplasts, *Exp. Cell Res.* **121**:143–151.

Nierhaus, D., and Nierhaus, K. H., 1973, Identification of the chloramphenicol-binding protein in *Escherichia coli* ribosomes by partial reconstitution, *Proc. Natl. Acad. Sci. USA* **70**:2224–2228.

O'Brien, T. W., Denslow, N. D., Harville, T. O., Hessler, R. A., and Matthews, D. E., 1980, Functional and structural roles of proteins in mammalian mitochondrial ribosomes, in: *The Organization and Expression of the Mitochondrial Genome* (A. M. Kroon and C. Saccone, eds.), Elsevier/North-Holland Biomedical Press, Amsterdam, pp. 301–305.

Oliver, N. A., and Wallace, D. C., 1982, Assignment of two mitochondrially-synthesized polypeptides to the human mitocondrial DNA and their use in the study of ıntracellular mitochondrial interaction, *Mol. Cell Biol.* **2**.

Perlman, S., and Penman, S., 1970, Mitochondrial protein synthesis: Resistance to emetine and response to RNA synthesis inhibitors, *Biochem. Biophys. Res. Commun.* **40**:941–948.

Pongs, O., and Messer, W., 1976, The chloramphenicol receptor site in *Escherichia coli in vivo* affinity labeling by monoiodoamphenicol, *J. Mol. Biol.* **101**:171–184.

Pongs, O., Bald, R., and Erdmann, V. A., 1973, Identification of chloramphenicol-binding protein in *Escherichia coli* ribosomes by affinity labeling, *Proc. Natl. Acad. Sci. USA* **70**:2229–2233.

Posakony, J. W., England, J. M., and Attardi, G., 1975, Morphological heterogeneity of HeLa cell mitochondria visualized by a modified diaminobenzidine staining technique, *J. Cell Sci.* **19**:315–329.

Prescott, D. M., Myerson, D., and Wallace, J., 1972, Enucleation of mammalian cells with cytochalasin B, *Exp. Cell Res.* **71**:480–485.

Putrament, A., Baranowska, H., and Prazmo, W., 1973, Induction by manganese of mitochondrial antibiotic resistance mutations in yeast, *Mol. Gen. Genet.* **126**:357–366.

Putrament, A., Baranowska, H., Ejchart, A., and Jachymczyk, W., 1977, Manganese mutagenesis in yeast. VI. Mn^{2+} uptake, mitDNA replication and E^R induction. Comparison with other divalent cations, *Mol. Gen. Genet.* **151**:69–76.

Reitzer, L. J., Wice, B. M., and Kennell, D., 1979, Evidence that glutamine, not sugar, is the major energy source for cultured HeLa cells, *J. Biol. Chem.* **254**:2669–2676.

Ringertz, N. R., 1978, Reconstruction of cells by Sendai virus-induced fusion of cell fragments, *Natl. Cancer Inst. Monogr.* **48**:31–36.

Rosenberg, E., Mora, C., and Edwards, D. L., 1976, Selection of extranuclear mutants of *Neurospora crassa*, *Genetics* **83**:11–24.

Sager, R., and Kitchen, R., 1975, Selective silencing of eukaryotic DNA, *Science* **189**:426–433.

Sanger, F., Coulson, A. R., Barrell, B. G., Smith, A. J. H., and Roe, B. A., 1980, Cloning in single-stranded bacteriophage as an aid to rapid DNA sequencing, *J. Mol. Biol.* **143**:161–178.

Schaap, G. H., Van der Kamp, A. W. M., Öry, F. G., and Jongkind, J. F., 1979, Isolation of enucleate cells using a fluorescence activated cell sorter (FACS II), *Exp. Cell Res.* **122**:422–426.

Shannon, C., Enns, R., Wheelis, L., Burchiel, K., and Criddle, R. S., 1973, Alterations in mitochondrial adenosine triphosphatase activity resulting from mutation of mitochondrial deoxyribonucleic acid, *J. Biol. Chem.* **248**:3004–3011.

Shay, J. W., 1977, Selection of reconstituted cells from karyoplasts fused to chloramphenicol-resistant cytoplasts, *Proc. Natl. Acad. Sci. USA* **74**:2461–2464.

Shay, J. W., Peters, T. T., and Fuseler, J. W., 1978, Cytoplasmic transfer of microtubule organizing centers in mouse tissue culture cells, *Cell* **14**:835–842.

Siegel, R. L., Jeffreys, A. J., Sly, W., and Craig, I. W., 1976, Isolation and detailed characterization of human cell lines resistant to D-threochloramphenicol, *Exp. Cell Res.* **102**:298–310.

Smith, C. A., 1977, Absence of ethidium bromide induced nicking and degradation of mitochondrial DNA in mouse L-cells, *Nucleic Acids Res.* **4**:1419–1427.

Soderberg, K., Mascarello, J. T., Breen, G. A. M., and Scheffler, I. E., 1979, Respiratory-deficient Chinese hamster cell mutants: Genetic characterization, *Somat. Cell Genet.* **5**:225–240.

Spolsky, C. M., 1973, Chloramphenicol resistant mutants of HeLa cells, Ph. D. thesis, Yale University.

Spolsky, C. M., and Eisenstadt, J. M., 1972, Chloramphenicol-resistant mutants of human HeLa cells, *FEBS Lett.* **25**:319–324.

Van Etten, R. A., Walberg, M. W., and Clayton, D. A., 1980, Precise localization and nucleotide sequence of the two mouse mitochondrial rRNA genes and three immediately adjacent novel tRNA genes, *Cell* **22**:157–170.

Veomett, G., Shay, J., Hough, P. V. C., and Prescott, D. M., 1976, Large-scale enucleation of mammalian cells, in: *Methods in Cell Biology*, Volume XIII (D. M. Prescott, ed.), Academic Press, New York, pp. 1–6.

Wallace, D. C., 1975, Cytoplasmic genetics in mammalian tissue culture cells, Ph.D. Thesis, Yale University.

Wallace, D. C., 1981, Assignment of the chloramphenicol resistance gene to mitochondrial deoxyribonucleic acid and analysis of its expression in cultured human cells, *Mol. Cell. Biol.* **1**:697–710.

Wallace, D. C., and Eisenstadt, J. M., 1979, Expression of cytoplasmically inherited genes for chloramphenicol resistance in interspecific somatic cell hybrids and cybrids, *Somat. Cell Genet.* **5**:373–396.

Wallace, D. C., Bunn, C. L., and Eisenstadt, J. M., 1975, Cytoplasmic transfer of chloramphenicol resistance in human tissue culture cells, *J. Cell Biol.* **67**:174–188.

Wallace, D. C., Pollack, Y., Bunn, C. L., and Eisenstadt, J. M., 1976, Cytoplasmic inheritance in mammalian tissue culture cells, *In Vitro* **12**:758–776.

Wallace, D. C., Bunn, C. L., and Eisenstadt, J. M., 1977, Mitotic segregation of cytoplasmic determinants for chloramphenicol resistance in mammalian cells II: Fusions with human cell lines, *Somat. Cell Genet.* **3**:93–119.

Wallace, D. C., Oliver, N. A., Blanc, H., and Adams, C. W., 1982, A system to study human mitochondrial genes: Application to chloramphenicol resistance, in: *Mitochondrial Genes* (P. Slonimski, P. Borst, and G. Attardi, eds.), Cold Spring Harbor Lab., Cold Spring Harbor, New York.

Wallace, R. B., and Freeman, K. B., 1975, Selection of mammalian cells resistant to a chloramphenicol analog, *J. Cell Biol.* **65**:492–498.

Wigler, M. H., and Weinstein, I. B., 1975, A preparative method for obtaining enucleated mammalian cells, *Biochem. Biophys. Res. Commun.* **63**:669–674.

Wilson, J. M., Howell, N., Sager, R., and Davidson, R. L., 1978, Polyethylene-glycol-mediated cybrid formation: High-efficiency techniques and cybrid formation without enucleation, *Somat. Cell Genet.* **4**:745–752.

Wintersberger, U., and Hirsch, J., 1973a, Induction of cytopalsmmic respiratory deficient mutants in yeast by the folic acid analogue, methotrexate I: Studies on the mechanism of petite induction, *Mol. Gen. Genet.* **126**:61–70.

Wintersberger, U., and Hirsch, J., 1973b, Induction of cytoplasmic respiratory deficient mutants in yeast by the folic acid analogue, methotrexate II: genetic analysis of the methotrexate-induced petites, *Mol. Gen. Genet.* **126**:71–74.

Wiseman, A., and Attardi, G., 1978, Reversible tenfold reduction in mitochondrial DNA content of human cells treated with ethidium bromide, *Med. Gen. Genet.* **167**:51–63.

Wiseman, A., and Attardi, G., 1979, Cytoplasmically inherited mutations of a human cell line resulting in deficient mitochondrial protein synthesis, *Somat. Cell Genet.* **5**:241–262.

Yatscoff, R. W., Goldstein, S., and Freeman, K. B., 1978, Conservation of genes coding for proteins synthesized in human mitochondria, *Somat. Cell Genet.* **4**:633–645.

Yatscoff, R. W., Mason, J. R., Patel, H. V., and Freeman, K. B., 1981, Cybrid formation with recipient cell lines containing dominant phenotypes, *Somat. Cell Genet.* **7**:1–9

Yen, R. C. K., and Harris, M., 1978, Cytoplasmic transfer of chloramphenicol resistance in Chinese hamster cells, *Cell Structure and Function* **3**:79–88.
Ziegler, M. L., 1978, Phenotypic expression of malignancy in hybrid and cybrid mouse cells, *Somat. Cell Genet.* **4**:477–489.
Ziegler, M. L., and Davidson, R. L., 1979, The effect of hexose on chloramphenicol sensitivity and resistance in Chinese hamster cells, *J. Cell. Physiol.* **98**:627–635.
Ziegler, M. L., and Davidson, R. L., 1981, Elimination of mitochondrial elements and improved viability in hybrid cells, *Somat. Cell Genet.* **7**:73–88.

Chapter 13

The Influence of Cytoplast-to-Cell Ratio on Cybrid Formation

CLIVE L. BUNN

1. Principles

1.1. Uses of Cytoplast Fusions

Fusions in which one of the partners is a cytoplasmic fragment of a cell, or cytoplast, have been extensively used in recent years to investigate a variety of problems in biology. For example, cytoplasmic inheritance can be observed. Properties of a cell that can be transferred and maintained through its cytoplasm to a recipient cell are presumed to be controlled by cytoplasmic hereditary factors. Thus, cytoplasmic transfer of chloramphenicol (CAP) resistance in mouse and human cells suggested that mitochondrial DNA (mtDNA) controlled CAP resistance (Bunn et al., 1974; Wallace et al., 1975), and this suggestion has proved correct (Clark and Shay, 1980; Blanc et al., 1981). Similarly, the cytoplasmic transfer of murine intracisternal A particles (Malech and Wivel, 1976) and of microtubule-organizing centers (Shay et al., 1978) suggest at least a partial replicative autonomy for these particles.

Nuclear transplantation experiments in amphibian oocytes have shown that cytoplasmic substances can control nuclear gene activities (Gurdon, 1974). The second use of cytoplast fusions is in the search for cytoplasmic regulatory substances in mammalian cultured cells which may be involved in the activation or repression of specific genes or sets of genes (Gopalakrishnan et al., 1977; Gopalakrishnan and Anderson, 1979; Lipsich et al., 1979; Halaban et al., 1980). The role of cytoplasmic regulators in differentiating systems has been similarly investigated in reactivation of the chick red cell nucleus *in vitro* (Ege et al., 1975; Rao, 1976; Lipsich et al., 1978), myogenic differentiation (Ringertz et al., 1978), and in differentiation of pluripotent murine teratocarcinoma cells *in vitro* (Linder et al., 1979; McBurney and Strutt, 1979; Linder, 1980) and *in vivo* (Watanabe et al., 1978).

Cytoplast fusions have also been used to examine the role of cytoplasmic

---

CLIVE L. BUNN • Department of Biology, University of South Carolina, Columbia, South Carolina 29208.

factors in the expression of complex phenotypes such as tumorigenicity and its suppression (Howell and Sager, 1978; Ziegler, 1978; Coon, 1979; Clark et al., 1980a; Halaban et al., 1980; Shay and Clark, 1980), senescence in human diploid fibroblasts (Wright and Hayflick, 1975a; Muggleton-Harris and Hayflick, 1976; Bunn and Tarrant, 1980), and in the control of the G_1 phase of the cell cycle (Jonak and Baserga, 1979). Recently, cytoplast fusions have been used in studies on genetic complementation groups in human lysosomal enzyme diseases (DeWit-Verbeek et al., 1978). The purpose of this paper is to consider the cytoplasmic input or "dosage," that is, the contribution of the donor cytoplasm to the resultant cell, and how this input may be varied to alter the properties of the resultant cell.

1.2. Properties of Cytoplasts

Cytoplasts are prepared by treating cells with cytochalasin B (Aldrich Chemical Co., Milwaukee, Wisconsin; Serva, Heidelberg, West Germany; Imperial Chemical Industries, Cheshire, U.K.), followed by centrifugation to remove nuclei, a process referred to as enucleation. The resultant cytoplasmic fragments are termed *cytoplasts* and the nuclear fragments termed *karyoplasts* (Shay et al., 1973) or alternatively *cytoplasts* and *minicells* (Ege and Ringertz, 1975; Krondahl et al., 1977). I shall use the terms cytoplast and karyoplast in this paper.

Cells may be enucleated from coverslips, that is, when attached to substrates (Prescott et al., 1972), or in suspension by centrifugation through Ficoll gradients (Wigler and Weinstein, 1975). The attached method usually produces a cytoplast preparation (after removal from the coverslip with trypsin solutions) which is less contaminated with unenucleated whole cells than in the suspension method. However, much larger quantities of cytoplasts can be prepared by the suspension method, and a second centrifugation can "clean up" the cytoplast preparation to 95–99% purity with most cell lines (see G. Veomett, this volume, Chapter 6, for methods of enucleation).

Cytoplasts remain morphologically intact for at least 24 hr, and can support synthesis of poliovirus and vaccinia virus for at least 12 hr (Prescott et al., 1971; Pollack and Goldman, 1973). Cytoplasmic protein synthesis generally declines exponentially with a half-life of 4–5 or up to 12 hr, although mitochondrial protein synthesis only declines by 20–30% in 20 hr (Ege et al., 1975; Wigler and Weinstein, 1975; England et al., 1978). Interferon production, if started before enucleation, can continue for 16 hr (Burke and Veomett, 1977), and of about 500 polypeptides detected by two-dimensional gel electrophoresis, about 10% show quantitative differences when compared with whole cells 10 hr after enucleation (Linder, 1980).

In general, about 90% of original cells' cytoplasm is contained in the cytoplast, with the remaining 10% containing ribosomes, occasional mitochondria, and some endoplasmic reticulum surrounding the nucleus in the karyoplast (Wise and Prescott, 1973; Krondahl et al., 1977). A recent paper describes a method of labeling cytoplasms with the heavy metal tantalum, and thereby selecting for karyoplasts with only 2–4% of the original cell cytoplasm (Clark et al., 1980b). Cells, cytoplasts, and karyoplasts can also be separated by size using filtration through Unipore filters (Yatscoff et al., 1981). Most karyoplasts cannot survive, except for some which may contain large amounts of cytoplasm (Krondahl et al., 1977; Clark et al., 1980b).

It should be appreciated that a cytoplast preparation, while metabolically active, is probably extremely heterogeneous because of the trauma of its isolation (Poste, 1973; Prescott and Kirkpatrick, 1973; Wright, 1973). The initial cell population, and therefore the cytoplast population also, is almost certainly heterogeneous with respect to stage in the cell cycle, and it has been shown that cytoplasts at differing stages of the cell cycle exert different influences on recipient cells (Lipsich et al., 1978). When enucleating HeLa cells, two bands are commonly observed in Ficoll gradients, consisting of small and large cytoplast populations (Bunn, unpublished). Another consideration when fusing cytoplasts to cells is that cells vary widely in nuclear/ cytoplasmic ratios. The cytoplast of a human HeLa cell, for example, is much smaller than that of a recipient human diploid fibroblast cell (Bunn and Tarrant, 1980), and similarly a rat myoblast cytoplast is substantially larger than its fusion partner mouse teratocarcinoma cell (Linder et al., 1979).

1.3. Selection of Fusion Products

Methods of cell fusion are discussed in detail in this volume and will be briefly mentioned below (Section 2.3). The fusion product of a cytoplast with a whole cell which is a proliferating, viable cell (presumably with a mixed cytoplasm) is called a *cybrid*. The transient product of such a fusion, that is, an initial fusion product which may or may not divide, has been called a *fusion cybrid* (as opposed to a *viable* cybrid) by Bunn and Eisenstadt (1977), and such products that only traverse one cell cycle have been called *cybridoids* by Jonak and Baserga (1979).

A cell produced from the fusion of a karyoplast with a cytoplast is called a *reconstituted* cell (Veomett et al., 1974; Krondahl et al., 1977). The fusion product of a karyoplast with a whole cell has been termed a *karyobrid* (McBurney and Strutt, 1979) or nuclear hybrid (Weide et al., Chapter 21, this volume). Reconstituted cells have been produced from intraspecies fusions with mouse cells (Veomett et al., 1974; Burke and Veomett, 1977; Shay, 1977; Halaban et al., 1980; Shay and Clark, 1980) and with human cells (Muggleton-Harris and Hayflick, 1976). Interspecies reconstituted cells have been constructed between rat and mouse (Ege and Ringertz, 1975; Krondahl et al., 1977; Lipsich et al., 1979) and between mouse and human cells (Nette et al., 1978; Hightower and Lucas, 1980). For this chapter, the important difference between reconstituted cell fusions and cell–cytoplast fusions is that the former does not involve (much) dilution of components of the donor cytoplasm by those of the recipient cytoplasm.

There are three methods of selecting cybrids from a mixed population of parental cells and fusion products. The first uses selectable genetic markers preferably in both nuclei, and also in the cytoplasm to be transferred.

Selection is *for* the nucleus of the recipient cell (or karyoplast) and *for* the "donor" cytoplasm, and *against* the nucleus of the enucleated cell. The last marker is particularly useful, as it selects against any cells contaminating the cytoplast preparation. As nuclear markers are present in both parental cells, they should both be recessive (such as thymidine kinase deficiency or hypoxanthine phosphoribosyl transferase deficiency) or both dominant. The cytoplasmic marker must be dominant (Bunn et al., 1974; Wallace et al., 1975). This method is used if the investigation aims to establish cytoplasmic components with hereditary continuity. Any such proliferating cybrids must be stable for the cytoplasmic component in the absence of selection. It is also the "cleanest" method, as contaminants can only appear at the frequency of spontaneous mutation of each of the markers involved. This method has also been used in the production of reconstituted cells (Shay, 1977; Nette et al., 1978; Shay and Clark, 1980).

However, there are only a few cytoplasmic markers available (see Chapters 9–12 this volume), and it is tedious to induce them in cell lines. Furthermore, genetic selection is slow and not apparent until several cell divisions after fusion. Events occurring soon after fusion of unmarked cells or karyoplasts with cytoplasts have been observed by differential labeling of the parental cell lines with polystyrene (latex) beads (Veomett et al., 1974; Ege and Ringertz, 1975; Krondahl et al., 1977; Levine and Cox, 1978; Jonak and Baserga, 1979; Linder et al., 1979; Jongkind et al., 1980; Linder, 1980; James and Veomett, 1981; Rabinovitch and Norwood, 1981). Resultant cybrids are then directly observed *in situ*. This method is described below (Section 2.1).

Another method of cybrid selection has been described by Wright and Hayflick (1975b) and Wright (1978). In the absence of genetic markers, the whole cell can be poisoned with certain doses of iodoacetate and rotenone together. This cell can be rescued by fusion with a noninhibitor-treated cytoplast. The dosage of each inhibitor for the desired toxicity level should be carefully calibrated, and appropriate controls performed (Wright, this volume, Chapter 5).

2. Procedures

The aim of these experiments is to investigate the influence of variation in the ratio of cytoplasts to cells in fusion mixtures on (a) the number of fusion events occurring between cells and cytoplasts, and (b) the yield of proliferating cybrids. As explained above, bead labeling is a means by which early fusion events are observed, and stable proliferating cybrids are best detected by selection in inhibitors for the appropriate genetic markers. The equipment used for these experiments should be available in any laboratory that routinely cultures mammalian cells, and I shall confine the descriptions of cell growth, enucleation, and fusion only to those points that are of special relevance to these experiments. These techniques are explained in other sections of this volume. Variations with other published methods will be noted.

2.1. Polystyrene Bead Labeling

The cells that are to be enucleated and fused to recipient cells must be labeled with polystyrene beads. Beads are available in a range of sizes from Warrington Polysciences, Warrington, Pennsylvania, Dow Diagnostics, Indianapolis, Indiana, or E. F. Fullam, Schenectady, New York. Beads of approximate diameter 0.5, 1.0, and 2.0 μm are most commonly used, and all are readily phagocytosed by a wide variety of attached mammalian cells grown in culture. No data rigorously comparing uptake by different cell types in different growth states are available, but I prefer to add the beads to an exponentially growing culture at about 10^4 cells/cm^2 or $(2-3) \times 10^5$/ml of medium.

Beads come in sealed vials. If desired, they can be sterilized by adding one part of bead suspension to two parts absolute ethanol. Store solutions at 4°C (Levine and Cox, 1978). Beads are added to exponentially growing cell cultures at a final concentration of 2×10^7 beads/ml of medium. Veomett et al., (1974) used 10^6/ml and 5×10^6/ml, Jonak and Baserga (1979) used 10^7/ml and 2×10^7/ml, and Levine and Cox (1978) used $(5-8) \times 10^6$/ml. The time of exposure of cells to the beads varies between 18 and 48 hr to obtain maximum uptake, and is probably influenced by the growth rate of the cells. I have obtained good uptake from an 18-hr exposure, about one population doubling, for mouse A9 cells and beads of 1.01 μm.

These conditions result in over 90% of the mouse A9 cells containing beads, and the number of beads per cell varies from 1 to 30. Beads can be visualized by phase contrast microscopy at magnifications of 400× to 500×. At the end of the time of exposure to beads, the medium and bead suspension is removed, and the cells washed with three changes of an equal volume (to the original medium volume) of phosphate-buffered saline.

At this stage, the cell monolayer is removed by trypsinization and prepared for enucleation (see Section 2.2). Cytoplasts lose very few beads during enucleation, and it is more convenient to label them before enucleation. Also, my experience is that the trypsinization serves the useful purpose of removing beads that adhere to the cell surface, as shown by microscopic inspection before and after trypsinization. Furthermore, this (gentle) trypsinization aids enucleation and also subsequent fusion, in my experience from controlled experiments, probably by slightly damaging the cell membrane. Cells grown in suspension are labeled in the same way, washed by three changes of phosphate-buffered saline and centrifugation at 500g for 5 min each, and similarly treated with Viokase (Grand Island Biological Company, Grand Island, New York) or trypsin-EDTA (0.05% w/v trypsin, 0.02% EDTA, Flow Laboratories, McLean, Virginia) prior to enucleation and fusion.

No beads are seen in the nuclei of cells, at least for the diameter of beads used here (Veomett et al., 1974; Bunn and Eisenstadt, 1977; Krondahl et al., 1977). Similar conditions of bead concentration, bead size, and time of exposure seem to result in a wide range of beads per cell, from a few to 100–200, apparently according to cell type. Experiments in which a smaller number of beads per cell are obtained generally result in a higher proportion of unlabeled cells. A 90% labeling of cells with 5–30 beads per cell or cytoplast should be aimed for.

2.2. Enucleation

Bead-labeled cells are suspended in a solution of 10% Ficoll (Pharmacia, Piscataway, New Jersey), and this suspension is gently layered over a step gradient of Ficoll concentrations 12.5%, 15.0%, 16.0%, and 17.0%, with 25.0% at the bottom of the centrifuge tube. The volumes of each layer, from 10% to 25%, for tubes for the SW41 Beckman rotor are 3, 2, 0.5, 0.5, 2, and 2 ml, respectively, and for the SW27 Beckman rotor are 3, 3, 1, 1, 3, and 3 ml. All Ficoll solutions are made in calcium-free minimal essential medium (Earle's salts) without serum for suspension cultures, (MEM-S, Flow Laboratories) and contain 10 μg cytochalasin B/ml (Bunn and Eisenstadt, 1977). These gradients are centrifuged at 20,000g for 1 hr at 30°C (Wigler and Weinstein, 1975). Bands rich in cytoplasts appear at the interface of the 10% and 12.5% Ficoll layers. This procedure was used for experiments with mouse A9 fibroblasts. Human HeLa cell enucleation sometimes produced two bands, one of small cytoplasmic fragments at the 10%/12.5% interface, and one of larger, more "intact-looking" cytoplasts at the 12.5%/15% interface. The larger cytoplasts were used in fusions (Bunn and Tarrant, 1980).

After centrifugation, the bands are collected by removing each layer from the top with a long-tipped Pasteur pipette, and the cytoplast-rich bands added to 20 ml nonselective medium with serum and centrifuged at 500g for 5 min to remove residual Ficoll and cytochalasin B. The supernatant is removed, the cytoplasts are resuspended in a small volume (2–5 ml) of nonselective medium, and the suspension is incubated at 37°C for 30 min to allow the cytoplasts to recover and resume spherical morphology. During this time, a drop is removed, placed on a microscope slide, and mixed with a drop of lactopropionic orcein stain (0.5 g orcein, 23 ml lactic acid, 23 ml propionic acid, water to 100 ml). The percent enucleation (the number of cytoplasts to the *total* number of cytoplasts, cells, and karyoplasts) is calculated at this point by counting under phase contrast microscopy at 400✕ magnification. The number of beads per cell is also counted. Ninety percent or more enucleation should be achieved.

2.3. Cell–Cytoplast Fusion

Cells and cytoplasts to be fused are combined in a tube and centrifuged at 500g for 5 min, after which time the supernatant is removed. Both polyethylene glycol (PEG) and inactivated Sendai virus produce fusion products in good yields, with PEG generally giving better and much less variable results. Wilson et al. (1978) have reported PEG fusion yields tenfold higher than Sendai virus-produced yields, and have described cybrids produced by PEG alone, without cytochalasin B treatment.

In this experiment, the efficiency of fusion is being measured, and hence it is important to keep the total number of particles to be fused and the amount of fusogen constant. In mouse cell experiments (Bunn and Eisenstadt, 1977) cytoplasts and cells were combined in ratios of 1:5, 1:1, 2:1, 3:1, 5:1, 10:1, and 20:1 (cytoplast:cell), each fusion mixture containing a total of 4×10^6 particles (cytoplasts + cells) and 500 hemagglutinating units of inactivated Sendai virus. I have used PEG 1000 at 40% w/v for 1 min exposure to the fusion mixture followed by three washes in 20 ml MEM-S, (minimal essential medium, suspension) for similar fusions between human HeLa cytoplasts and human Lesch-Nyhan diploid fibroblasts, with good results (Bunn and Tarrant, 1980). Detailed cell fusion protocols are described by Kennett (1979).

Immediately after fusion, aliquots are removed for assay of fusion efficiency (see Section 2.4). In a parallel experiment, *identical except for the bead labeling* (see Section 2.4), the various fusion mixtures are inoculated into growth medium containing inhibitors that select for proliferating cybrids. For the mouse cell lines enucleated 501-1 and whole cell LMTK⁻, the selective medium contained 50 μg CAP/ml and 30 μg 5-bromo-2'-deoxyuridine (BrdU)/ml. The BrdU selects for the LMTK⁻ nucleus which has a lesion in its thymidine kinase gene and against any unenucleated 501-1 cells (TK⁺), and the CAP selects for the CAP resistance of the 501-1 cytoplasm (Bunn and Eisenstadt, 1977; Kennett, 1979). Control cultures of identical numbers of parental cells, similarly treated with fusogen, should be inoculated alone into selective medium.

2.4. Efficiency of Cybrid Fusion and Proliferating Cybrid Formation with Varying Cytoplast:Cell Ratios

In the experiment to measure proliferating cybrids, colonies arising in selective medium are counted at an appropriate time after fusion. For clearly visible colonies of 30–100 cells, this is 12 days for fusions involving mouse cells, and 14–18 days for fusions involving human cells. In the experiment to determine fusion efficiency (i.e., initial fusion cybrid formation) with bead-labeled cytoplasts, the fusion mixture, if inoculated into selective medium after counting fusion events, produces a lower number of colonies than the fusion mixture with unlabeled cytoplasts. Apparently, the polystyrene beads interfere with cybrid viability. Therefore comparisons between fusion cybrids and viable (colony-forming) cybrids are made in separate experiments.

In comparing separate experiments, the assumption is made that beads do not affect fusion efficiency. This assumption is reasonable, as fusion itself is a cell-surface phenomenon, and no beads remain on the cell surface after trypsinization. Levine and Cox (1978) demonstrated that beads did not effect hybridization frequency, and the bead-labeled reconstituted cells of Krondahl et al. (1977) could proliferate. However, Levine and Cox (1978) also report that beads had no effect on the short-term growth of human diploid fibroblasts or on hamster kidney cell plating efficiency, in contrast to our observation. There may be more subtle long-term effects on hybrid and cybrid viability than those apparent in 6 or 8 days. Indeed, James and Veomett (1981) have reported that labeling human diploid fibroblasts in midlife with 2.0-μm beads decreased their lifespan by about eight population doublings, even though by six population doublings earlier the number of beads had been diluted by growth to one to five per cell. Further, James and Veomett (1981) reported that similar conditions had no effect on human HeLa cells. It is possible that toxic effects by beads depend on the number and size of the beads phagocytosed.

2.4.1. Formation of Proliferating Cybrids

In a typical experiment, colony counts of proliferating mouse cell cybrids showed that, over the range of cytoplast–cell ratios of 1:5 through 20:1, the number of viable cybrids per 10^5 recipient cells increased from 14 to 284, or to almost one viable cybrid per 300 recipient cells (Bunn and Eisenstadt, 1977). The increase in cybrid yield with increase in cytoplast–cell ratio constitutes further evidence for the cytoplasmic inheritance of the selected cytoplasmic marker, and should be demonstrated, in my view, in all cases of cytoplasmic inheritance where possible. Various fusion protocols for cybrid formation have been compared by Wilson et al. (1978), and maximum yields are reported of 243–542 cybrid colonies per 10^4 recipient cells, or about one per 20 or 40 recipient cells. The numbers clearly vary with cell lines and fusion protocols, but the increase with increasing cytoplast to cell ratio under constant conditions is clear.

2.4.2. Initial Formation of Fusion Cybrids

Examination of lactopropionic orcein-stained aliquots of each of the various fusion mixtures reveals cytoplasts (mostly with beads), cells containing a nucleus and beads in the cytoplasm, karyoplasts, and cells with a nucleus but no beads. All categories except karyoplasts are counted. Unlabeled, nucleated cells are almost certainly unfused recipient cells. Cells with nuclei and beads could be either cybrids or "donor" cells that escaped enucleation. The proportion of these unenucleated donor cells expected to accompany a given number of cytoplasts can be calculated from the enucleation efficiency measured above (Section 2.2). This number can then be subtracted from the cells with nuclei + beads count to yield the number of fusion cybrids formed.

For example [data from Bunn and Eisenstadt (1977)]:

(a) Cells without beads, 47 counts.
(b) Cytoplasts, 192 counts.
(c) Cells with nuclei and beads, 61 counts.
(d) Previously determined enucleation efficiency, 90%.
(e) Therefore, with 192 cytoplasts, expect $192 \times 10/90 = 21$ contaminating cells.
(f) Steps (c)–(e) give calculated total cybrids, 40.

The formation of initial cybrids is then expressed as the number of cybrids divided by the total number of cybrids and unfused cells, or $40/(47 + 40) = 0.46$ in this case. Thus 46% of the recipient cells actually fused with a cytoplast at this cytoplast–cell ratio, and this can be compared with the number of viable proliferating cybrid colonies at that ratio. (Karyoplasts are not counted, as they do not affect the results. Also, for the purposes of the experiment, apparent cytoplast–cytoplast and cell–cell fusions are counted as one cytoplast and one cell, respectively.)

3. Comments: Cytoplasmic Dosage

It has been shown in the experiments described above that the number of initial fusion cybrids per total number of recipient cells increased over the cytoplast–cell range of 1:5 to 5:1, and then reached a plateau with a small decline at 20:1 (Bunn and Eisenstadt, 1977). This result can be explained by a relatively higher number of cytoplasts at each ratio being involved in fusions with cells, with a possible excess of cytoplast–cytoplast fusions at 20:1. At the plateau, approximately 50% of the recipient cells had fused with a cytoplast. The wide variation in the number of beads per cytoplast made it impossible to quantify the number of fusions involving more than two parents (i.e., multiple fusions). However, from the sizes and shapes of the cells, it appeared that more multiple fusions between a cell and several cytoplasts occurred at higher cytoplast–cell ratios.

Comparing these data on the number of fusion events at each ratio with the numbers of viable cybrid colonies at each ratio, it is apparent that, at ratios of one cytoplast to five cells, about one in 500 cybrid fusions produces a viable cybrid, whereas at 10:1, one in 224 fusions produces a viable cybrid, and at 20:1, one in 60 fusions is "successful" (Bunn and Eisenstadt, 1977). Thus, increasing the cytoplast–cell ratio appears to increase the success of cytoplast–cell fusion events.

The reasons for this result are not clear, but all of the possibilities are relevant to the interpretation of results obtained in other cytoplast–cell fusion experiments. For example, the increased success of fusion events at higher fusion ratios may come from the higher number of fusions between large cytoplasts with minimal damage from the enucleation process and maximum metabolic activity and recipient cells. Thus a "selection of the fittest" cytoplast–cell fusion may occur. The heterogeneity of cytoplast preparations has been noted above (Section 1.2).

Alternatively, at higher fusion ratios there may be more multiple fusions of cytoplasts with cells, increasing the cytoplasmic input into the recipient cell. Unfortunately, at the moment it is not possible to quantify cytoplasmic input by counting the number of beads per cybrid, because of the variation in beads per cell. If this could be standardized, precise cytoplasmic dosage experiments could be performed.

In both cases there is an *effective* increase in cytoplasmic input into the cybrid. An increase in cytoplasmic input may be necessary for the expression of any cytoplasmic characteristic for two reasons. First, a critical concentration of a substance may have to be reached for the effects of the substance to be seen, and by increasing the input the *passive* dilution of the substance by the recipient cytoplasm is reduced. This dilution is minimized in cytoplast–karyoplast fusions. Second, the recipient cytoplasm may *actively* inhibit the incoming cytoplasmic factor, and therefore an increase in cytoplasmic input may be needed to overcome the inhibition. This could be a kind of cytoplasmic incompatibility mechanism.

These two processes can be illustrated in the case of the CAP-resistance transfer described here. The number of mtDNA molecules per cell can vary widely. Human Hela cells, for instance, contain about 9000 mtDNA molecules, while mouse L cells contain 900–1100 (Bogenhagen and Clayton, 1974). The number of mtDNA molecules necessary to phenotypically express resistance is unknown, but clearly dilution of the CAP-resistant mtDNA pool by sensitive recipient mtDNA molecules would not help the expression of CAP resistance.

An example of the second process would be an active segregation of CAP-resistant mtDNA by elimination and/or nonreciprocal recombination resulting in gene conversion of the CAP-resistant allele. Segregation of CAP-resistant determinants has been described in mouse and human cells (Bunn et al., 1977; Wallace et al., 1977), and a gene conversion process has been described for mitochondrial genes in yeast (Dujon et al., 1974).

Differences in results obtained from similar cell–cytoplast and karyoplast–cytoplast fusions would support the idea of the importance of cytoplasmic dosage and/or cytoplasmic incompatibility. Wallace et al. (1976) reported great difficulty in selecting CAP-resistant hybrids or cybrids from mouse–human fusions. However, Nette et al. (1978) have reported the isolation of reconstituted human–mouse cells using CAP resistance as a cytoplasmic genetic marker. Also, Wright and Hayflick (1975a) reported the absence of cytoplasmic effects on senescence in human diploid fibroblasts using cell–cytoplast fusions and selective chemical toxicity (see Section 2.3), while Muggleton-Harris and Hayflick (1976) reported some cytoplasmic influence on senescence in similar experiments using karyoplast–cytoplast fusions between young and old cells.

Considerable variation in results has occurred in experiments investigating the suppression of tumorigenicity. Normal cytoplasts can apparently suppress tumorigenicity when fused to recipient hamster (Howell and Sager, 1978), rat (Coon, 1979), or mouse cells (Giguere and Morais, 1981), but other mouse cytoplast–cell fusions do not show the suppression of tumorigenicity (Howell and Sager, 1978; Ziegler, 1978; Halaban et al., 1980). Differences in the stability of cytoplasmic elements or in nuclear gene dosage (as reflected in the ploidy of the recipient cell) have been suggested to account for this variation (Halaban et al., 1980; Giguere and Morias, 1981). The critical factor may be the ratio of cytoplasmic substance(s) to nuclear target genes.

Similar contradictions appear to occur in fusions studying gene expression and differentiation. Cytoplasmic regulators can apparently extinguish hemoglobin inducibility (Gopalakrishnan et al., 1977) and activate phenylalanine hydroxylase and (transiently) tyrosine aminotransferase (Gopalakrishnan and Anderson, 1979; Lipsich et al., 1979). However, there is no evidence for cytoplasmic regulators in similar fusions investigating melanocytic functions (Halaban et al., 1980), myogenic differentiation (Ringertz et al., 1978), or teratocarcinoma cell differentiation (Watanabe et al., 1978; Linder et al., 1979; McBurney and Strutt, 1979; Linder 1980). The reports of Ringertz et al. (1978), Lipsich et al. (1979), McBurney and Strutt (1979), and Halaban et al. (1980) all used karyoplast–cytoplast fusions, and Linder et al. (1979) and Linder (1980) used donor cytoplasts that were much larger than recipient cells and fused with a 5:1 cytoplast–cell ratio. Consequently, these different results may reflect fundamental differences in gene control mechanisms rather than in dosage of postulated cytoplasmic regulators.

References

Blanc, H., Wright, C. T., Bibb, M. J., Wallace, D. C., and Clayton, D. A., 1981, Mitochondrial DNA of chloramphenicol-resistant mouse cells contains a single nucleotide change in the region encoding the 3′ end of the large ribosomal RNA, *Proc. Natl. Acad. Sci. USA* **78:**3789–3793.

Bogenhagen, D., and Clayton, D. A., 1974, The number of mitochondrial DNA genomes in mouse L and human Hela cells, *J. Biol. Chem.* **249:**7991–7995.

Bunn, C. L., and Eisenstadt, J. M., 1977, Cybrid formation in mouse L cells: The influence of cytoplast-to-cell ratio, *Somat. Cell Genet.* **3:**335–341.

Bunn, C. L., and Tarrant, G. M., 1980, Limited lifespan in somatic cell hybrids and cybrids, *Exp. Cell Res.* **127:**385–396.

Bunn, C. L., Wallace, D. C., and Eisenstadt, J. M., 1974, Cytoplasmic inheritance of chloramphenicol resistance in mouse tissue culture cells, *Proc. Natl. Acad. Sci. USA* **71:**1681–1685.

Bunn, C. L., Wallace, D. C., and Eisenstadt, J. M., 1977, Mitotic segregation of cytoplasmic determinants for chloramphenicol resistance in mammalian cells I: Fusions with mouse cell lines, *Somat. Cell Genet.* **3:**71–92.

Burke, D. C., and Veomett, G., 1977, Enucleation and reconstruction of interferon-producing cells, *Proc. Natl. Acad. Sci. USA* **74:**3391–3395.

Clark, M. A., and Shay, J. W., 1980, A method for producing mitochondrial chimeras, *J. Cell Biol.* **87:**293a.

Clark, M. A., Lorkowski, G., and Shay, J. W., 1980a, Suppression of tumorigenicity in cybrids. *J. Cell Biol.* **87:**292a.

Clark, M. A., Shay, J. W., and Goldstein, L., 1980b, Techniques for purifying L-cell karyoplasts with minimal amounts of cytoplasm. *Somat. Cell. Genet.* **6:**455–464.

Coon, H. G., 1979, Tumorgenicity of cybrids, *J. Cell Biol.* **83:**449a.

DeWit-Verbeek, H. A., Hoogeveen, A., and Galjaard, H., 1978, Complementation studies with enucleated fibroblasts from different variants of B-galactosidase deficiency, *Exp. Cell Res.* **113:**215–218.

Dujon, B., Slonimski, P. P., and Weill, L., 1974, Mitochondrial genetics IX: A model for recombination and segregation of mitochondrial genomes in *Saccharomyces cerevisiae*, *Genetics* **78:**415–437.

Ege, T., and Ringertz, N., 1975, Viability of cells reconstituted by virus-induced fusion of minicells with anucleate cells, *Exp. Cell Res.* **94:**469–473.

Ege, T., Zeuthen, J., and Ringertz, N. R., 1975, Reactivation of chick erythrocyte nuclei after fusion with enucleated cells, *Somat. Cell Genet.* **1:**65–80.

England, J. M., Costantino, P., and Attardi, G., 1978, Mitochondrial RNA and protein synthesis in enucleated African Green Monkey cells, *J. Mol. Biol.* **119:**455–462.

Giguere, L., and Morais, R., 1981, On suppression of tumorigenicity in hybrid and cybrid mouse cells, *Somat. Cell Genet.* **7:**457–471.

Gopalakrishnan, T. V., and Anderson, W. F., 1979, Epigenetic activation of phenylalanine hydroxylase in mouse erythroleukemia cells by the cytoplast of rat hepatoma cells, *Proc. Natl. Acad. Sci. USA* **76:**3932–3936.

Gopalakrishnan, T. V., Thompson, E. B., and Anderson, W. F., 1977, Extinction of hemoglobin inducibility in Friend erythroleukemia cells by fusion with cytoplasm of enucleated mouse neuroblastoma or fibroblast cells, *Proc. Natl. Acad. Sci. USA* **74:**1642–1646.

Gurdon, J. B., 1974, *The Control of Gene Expression in Animal Development*, Clarendon Press, Oxford.

Halaban, R., Moellman, G., Godawska, E., and Eisenstadt, J. M., 1980, Pigmentation and tumorigenicity of reconstituted, cybrid, and hybrid mouse cells, *Exp. Cell Res.* **130:**427–435.

Hightower, M. J., and Lucas, J. J., 1980, Construction of viable mouse–human hybrid cells by nuclear transplantation, *J. Cell. Physiol.* **105:**93–103.

Howell, A. N., and Sager, R., 1978, Tumorigenicity and its suppression in cybrids of mouse and chinese hamster cell lines, *Proc. Natl. Acad. Sci. USA* **75:**2358–2362.

James, L., and Veomett, G. E., 1981, Decreased longevity of human diploid cells after incorporation of latex spheres within their cytoplasm, *Exp. Cell Res.* **132:**468–473.

Jonak, G. J., and Baserga, R., 1979, Cytoplasmic regulation of two G_1-specific temperature sensitive functions, *Cell* **18:**117–123.

Jongkind, J. F., Verkerk, A., Schaap, G. H., and Galjaard, H., 1980, Non-selective isolation of fibroblast cybrids by flow sorting, *Exp. Cell Res.* **130:**481–484.

Kennett, R. H., 1979, Cell fusion, in: *Methods in Enzymology*, Volume LVIII (W. B. Jakoby and I. B. Pastan, eds.), Academic Press, New York, pp. 345–359.

Krondahl, U., Bols, N., Ege, T., Linder, S., and Ringertz, N. R., 1977, Cells reconstituted from cell fragments of two different species multiply and form colonies, *Proc. Natl. Acad. Sci. USA* **74:**606–609.

Levine, M. P., and Cox, R. P., 1978, Use of latex particles for analysis of heterokaryon formation and cell fusion, *Somat. Cell Genet.* **4:**507–512.

Linder, S., 1980, Teratoma cybrids. An analysis of the post-fusion effects of myoblast cytoplasms on embryonal carcinoma cells, *Exp. Cell Res.* **130:**159–167.

Linder, S., Brzeski, H., and Ringertz, N. R., 1979, Phenotypic expression in cybrids derived from teratocarcinoma cells fused with myoblast cytoplasms, *Exp. Cell Res.* **120:**1–14.

Lipsich, L. A., Lucas, J. J., and Kates, J. R., 1978, Cell cycle dependence of the reactivation of chick erythrocyte nuclei after transplantation into mouse L929 cell cytoplasts, *J. Cell Physiol.* **97:**199–208.

Lipsich, L. A., Kates, J. R., and Lucas, J. J., 1979, Expression of a liver specific function by mouse fibroblast nuclei transplanted into rat hepatoma cytoplasts, *Nature* **281:**74–76.

Malech, H. L., and Wivel, N. A., 1976, Transfer of murine intracisternal A particle phenotype in chloramphenicol resistant cytoplasts, *Cell* **9:**383–391.

McBurney, M. W., and Strutt, B., 1979, Fusion of embryonal carcinoma cells to fibroblast cells, cytoplasts, and karyoplasts, *Exp. Cell Res.* **124:**171–180.

Muggleton-Harris, A. L., and Hayflick, L., 1976, Cellular aging studied by the reconstruction of replicating cells from nuclei and cytoplasms isolated from normal human diploid cells, *Exp. Cell Res.* **103:**321–330.

Nette, E. G., Sit, H. L., Clavey, W., and King, D. W., 1978, The use of nuclear and cytoplasmic genetic markers in the selection of human–mouse reconstituted cells, *J. Cell Biol.* **79:**392a.

Pollack, R., and Goldman, R., 1973, Synthesis of infective poliovirus in BSC-1 monkey cells enucleated with cytochalasin B, *Science* **179:**915–916.

Poste, G., 1973, Anucleate mammalian cells: Applications in cell biology and virology, in: *Methods in Cell Biology*, Volume VII (D. M. Prescott, ed.), Academic Press, New York, pp. 211–249.

Prescott, D. M., and Kirkpatrick, J. B., 1973, Mass enucleation of cultured animal cells, in: *Methods in Cell Biology*, Volume VII (D. M. Prescott, ed.), Academic Press, New York, pp. 189–202.

Prescott, D. M., Kates, J., and Kirkpatrick, J. B., 1971, Replication of vaccinia virus DNA in enucleated L-cells, *J. Mol. Biol.* **59:**505–508.

Prescott, D. M., Meyerson, D., and Wallace, J., 1972, Enucleation of mammalian cells with cytochalasin B, *Exp. Cell Res.* **71:**480–485.

Rabinovitch, P. S., and Norwood, T. H., 1981, Rapid kinetics of polyethylene glycol-mediated cell fusion, *Somat. Cell Genet.* **7**:281–287.

Rao, M. V. N., 1976, Reactivation of chick erythrocyte nuclei in young and senescent cells, *Exp. Cell Res.* **102**:25–30.

Ringertz, N. R., Krondahl, U., and Coleman, J. R., 1978, Myogenic expression after fusion of minicells from rat myoblasts (L6) with mouse fibroblast (A9) cytoplasms, *Exp. Cell Res.* **113**:233–246.

Shay, J. W., 1977, Selection of reconstituted cells from karyoplasts fused to chloramphenicol resistant cytoplasts, *Proc. Natl. Acad. Sci. USA* **74**:2461–2464.

Shay, J. W., and Clark, M. A., 1980, Alternative method of identifying reconstituted cells, *Proc. Natl. Acad. Sci. USA* **77**:381–384.

Shay, J. W., Porter, K. R., and Prescott, D. M., 1973, Observations on the nuclear and cytoplasmic portions of CHO cells enucleated with cytochalasin B, *J. Cell Biol.* **59**:311a.

Shay, J. W., Peters, T. T., and Fuseler, J. W. 1978, Cytoplasmic transfer of microtubule organizing centers in mouse tissue culture cells, *Cell* **14**:835–842.

Veomett, G., Prescott, D. M., Shay, J., and Porter, K. R., 1974, Reconstruction of mammalian cells from nuclear and cytoplasmic components separated by treatment with cytochalasin B, *Proc. Natl. Acad. Sci. USA* **71**:1999–2002.

Wallace, D. C., Bunn, C. L., and Eisenstadt, J. M., 1975, Cytoplasmic transfer of chloramphenicol resistance in human tissue culture cells, *J. Cell Biol.* **67**:174–188.

Wallace, D. C., Pollack, Y., Bunn, C. L., and Eisenstadt, J. M., 1976, Cytoplasmic inheritance in mammalian tissue culture cells, *In Vitro* **12**:758–776.

Wallace, D. C., Bunn, C. L., and Eisenstadt, J. M., 1977, Mitotic segregation of cytoplasmic determinants for chloramphenicol resistance in mammalian cells. II: Fusions with human cell lines, *Somat. Cell Genet.* **3**:93–119.

Watanabe, T., Dewey, M. J., and Mintz, B., 1978, Teratocarcinoma cells as vehicles for introducing specific mutant mitochondrial genes into mice, *Proc. Natl. Acad. Sci. USA* **75**:5113–5117.

Wigler, M. H., and Weinstein, I. B., 1975, A preparative method for obtaining enucleated mammalian cells, *Biochem. Biophys. Res. Commun.* **63**:669–674.

Wilson, J. N., Howell, N., Sager, R., and Davidson, R. L., 1978, Polyethylene glycol-mediated cybrid formation: High efficiency techniques and cybrid formation without enucleation, *Somat. Cell Genet.* **4**:745–752.

Wise, G. E., and Prescott, D. M., 1973, Ultrastructure of enucleated mammalian cells in culture, *Exp. Cell Res.* **81**:65–72.

Wright, W. E., 1973, The production of mass populations of anucleate cytoplasms, in: *Methods in Cell Biology*, Volume VII (D. M. Prescott, ed.), Academic Press, New York, pp. 203–210.

Wright, W. E., 1978, The isolation of heterokaryons and hybrids by a selective system using irreversible biochemical inhibitors, *Exp. Cell Res.* **112**:395–407.

Wright, W. E., and Hayflick, L., 1975a, Nuclear control of cellular aging demonstrated by hybridization of anucleated and whole cultured normal human fibroblasts, *Exp. Cell Res.* **96**:113–121.

Wright, W. E., and Hayflick, L., 1975b, Use of biochemical lesions for selection of human cells with hybrid cytoplasms, *Proc. Natl. Acad. Sci. USA* **72**:1812–1816.

Yatscoff, R. W., Mason, J. R., Patel, H. V., and Freeman, K. B., 1981, Cybrid formation with recipient cells containing dominant phenotypes, *Somat. Cell Genet.* **7**:1–10.

Zeigler, M. L., 1978, Phenotypic expression of malignancy in hybrid and cybrid mouse cells, *Somat. Cell Genet.* **4**:477–489.

Chapter 14

Transformation of Mitochondrially Coded Genes into Mammalian Cells Using Intact Mitochondria and Mitochondrial DNA

MIKE A. CLARK, TIM L. REUDELHUBER, and JERRY W. SHAY

1. Introduction

The antibiotics chloramphenicol (CAP) and efrapeptin (EF) inhibit mitochondrial protein synthesis and ATPase, respectively, and thus kill sensitive (wild-type) mammalian cells. However, a variety of mutant cell lines have been isolated and characterized which are resistant to these otherwise lethal antibiotics. We have utilized two mutant murine cell lines, the AMT-BU-A1 (Malech and Wivel, 1976) and TL-1 (Lichtor and Getz, 1978), to investigate the possibility of isolating mitochondria from the resistant cells and using them to confer antibiotic resistance on sensitive cells. In addition, experiments were conducted comparing the ability of intact mitochondria or cloned mitochondrial DNA to confer antibiotic resistance on sensitive cells. It is expected that these techniques will be of use not only to those studying mammalian mitochondrial genetics and the mechanisms of DNA-mediated gene transfer, but also to those involved in elucidating the mechanisms by which cytoplasts exert effects on nuclear-coded functions.

2. Materials and Methods

2.1. Cell Lines

The murine cells used in these experiments as sources of mitochondria were the CAP^r EF^s AMT-BU-A1 (AMT) cell line (obtained from D. N. Wivel,

NIH, Bethesda, Maryland) and the MRR CAP^r EF^r TL-1 cell line (obtained from Dr. G. Getz, University of Chicago, Chicago, Illinois). The CAP^s EF^s P815 cell line was used as a source of antibiotic-sensitive mitochondria.

The CAP^s EF^s recipient cell lines included the murine Y-1 cell line (obtained from the American Type Culture Collection, Rockville Maryland) and PCC4 Aza 1 (obtained from Dr. W. Wright, University of Texas Health Science Center, Dallas, Texas), as well as the human HT1080 (Dr. C. Croce, The Wistar Institute, Philadelphia, Pennsylvania).

All cells used in these experiments, with the exception of the Y-1 cells, were grown in Dulbecco's medium supplemented with 10% fetal calf serum. The Y-1 cells were grown in Ham's F12 medium supplemented with 15% horse and 2.5% fetal calf serum. All cells were grown antibiotic-free in a water-saturated atmosphere containing 95% air and 5% CO_2. The cells were routinely tested and found to be free of mycoplasma contamination by [^{3}H]uridine and [^{3}H]uracil (Schneider et al., 1974) and scanning electron microscopy (Tai and Quinn, 1977).

2.2. Mitochondrial Isolation

The mitochondrial isolation procedure used in these experiments was that previously described by Bogenhagen and Clayton (1974) and consists in the following: Approximately 10^9 cells were detached from the substrate using a rubber policeman and quickly cooled to 4° for the remainder of the isolation. The cells were then washed twice in 134 mM NaCl, 5 mM KCl, 0.7 mM Na_2NPO_4, and 2.5 mM Tris pH 7.5 and allowed to swell in 10 mM NaCl, 1.5 mM $CaCl_2$, and 10 mM Tris pH 7.5 for 10 min before being homogenized in a Dounce homogenizer. After homogenization, 0.7 vol of 0.5 M manitol, 0.17 M sucrose, 7 mM Tris, and 7 mM EDTA was quickly added. The nuclei and whole-cell contaminants were removed by two centrifugations of 400g for 4 min and the mitochondria were then obtained by centrifugation of the supernatant at 26×10^3g for 20 min. The mitochondria were then purified using a 1–1.5 M discontinuous sucrose gradient made up in 5 mM EDTA, 10 mM buffer Tris pH 7.5. The mitochondria were removed from the interface and washed once in growth medium prior to being added to the cells. All solutions, with the exception of the final wash, were sterilized 24 hr prior to use with 0.04% diethylpyrocarbonate.

2.3. Procedures for Transformation with Intact Mitochondria

Purified mitochondria were concentrated by centrifuging 2×10^4g for 20 min and then resuspending the pellet in complete growth medium before serially diluting it to the desired concentration. Cells to which the mitochondria were to be added were grown to confluency in 96-well Falcon dishes ($\sim 3 \times 10^4$ cells/well). The medium was separated from the cells and replaced with 0.2 ml of complete medium containing mitochondria. After ~12 hr of incubation, the monolayers of recipient cells were washed twice, trypsinized, and replated in 10-cm petri dishes containing either 50 µg/ml of CAP (Sigma, St. Louis, Missouri) or 0.1 µg/ml of EF (Lilly, Indianapolis, Indiana). After two weeks of growth the dishes were either fixed in ethanol: acetic acid (3:1) and Giemsa-stained in order to determine the efficiency of transformation, or the clones were removed and saved for further analysis.

2.4 Procedures for Transformation with Purified DNA

Purified DNA was transformed into recipient cells by the method of Linnenbach et al. (1980). These studies were performed using either purified mitochondrial DNA (Bogenhagen and Clayton, 1974) or recombinant plasmid PSL1180, which contains the mitochondrial genome of an oligomycin–chloramphenicol-resistant mouse cell line (Lansman, personal communication). Cloned mitochondrial DNA was either transformed alone or was cotransformed with recombinant plasmid PSV2gpt, which carries an SV40 promoter linked to the *Escherichia coli* xanthine phosphoribosyl transferase (XGPRT) gene (Mulligan and Berg, 1981).

3. Results

Y-1 and PCC4 wild-type (CAP^s and EF^s) cells, plated out in medium containing 50 µg/ml CAP or 0.1 µg/ml EF, were able to attach to the substrate. After 4–5 days of incubation, however, the cells became vacuolated, developed pycnotic nuclei, and eventually detached from the substrate (Figs. 1B and 1C). Repeated attempts to isolate spontaneous mutants resistant to these drugs were unsuccessful and we estimated that the incidence must be $<10^{-8}$. When CAP^s EF^s Y-1 cells were incubated with mitochondria obtained from the CAP^r EF^r TL-1 cell line many colonies resistant to both CAP and EF were observed (Figs. 1G and 1H). One possible explanation for these results was that the mitochondrial preparation contained intact viable cells. This possibility was excluded by incubating the mitochondrial preparation in growth medium for 2 weeks (Fig. 1A) and by using recipient cells which contained additional markers. The Y-1 cell origin of the antibiotic-resistant cells was demonstrated by treating the cells with 100 milliunits of ACTH for 30 min (Clark and Shay, 1979). A mean of 49 out of 50 CAP^r colonies and 47 out of 50 EF^r colonies in three separate experiments were responsive to ACTH by morphologic analysis. Similar results were obtained with the PCC4 cell line. In these experiments the CAP^r and EF^r cells were demonstrated to be PCC4 cells by their resistance to 8-azaguanine. This is a nuclear mutation which allows the PCC4 cells but not AMT or TL-1 cells to proliferate in the presence of 8-azaguanine. The specificity of this phenomenon was demonstrated by isolating mitochondria from CAP^s EF^s P815 cells. These mitochondria were unable to confer either CAP^r or EF^r on Y-1 or PCC4 cells (Fig. 1F and 1I).

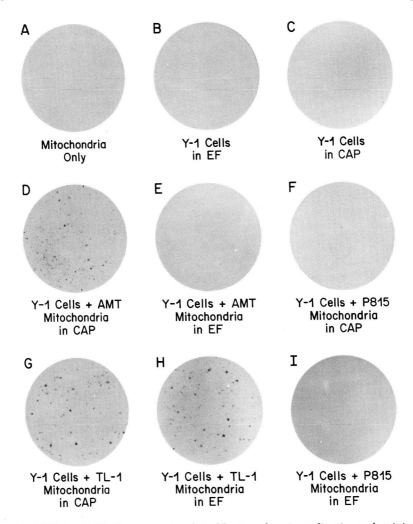

Figure 1. All dishes in this figure were incubated for 2 weeks prior to fixation and staining. Dish A shows the result of incubating isolated mitochondria, which indicates that this fraction did not contain any viable cells. Y-1 cells are EF^s and CAP^s and fail to grow when challenged with 0.1 µg/ml EF or 50 µg/ml CAP (dishes B and C). When Y-1 cells were incubated with CAP^r EF^s AMT mitochondria, many CAP^r clones of Y-1 cells appeared (dish D). Although the AMT mitochondria were able to confer CAP resistance, they were unable to confer EF resistance (dish E). In control experiments CAP^s EF^s mitochondria obtained from P815 mastocytoma cells were unable to confer either CAP or EF resistance (dishes F and I). If CAP^r EF^r TL-1 mitochondria were used, it was observed that both CAP-resistant (dish G) and EF-resistant (dish H) clones of Y-1 cells grew. Similar results were obtained using PCC4 Aza-1 wild-type recipient cells.

The incidence of CAP^r and EF^r was shown to be dependent on the number of mitochondria used in the transformation (Table I). Supernatants of cell extracts, depleted of mitochondria by centrifugation at $2 \times 10^4 g$ for 20 min, were ~80-fold reduced in the ability to confer CAP^r EF^r. Furthermore, exposure of mitochondria to UV light resulted in a reduction in their ability to confer antibiotic resistance (Table II).

Table I. Number of Antibiotic-Resistant Colonies Observed 2 Weeks after Transformation[a]

Mitochondrial equivalents	AMT (CAP^r EF^s)				TL-1 (CAP^r EF^r)				P815 (CAP^s EF^s)			
	CAP		EF		CAP		EF		CAP		EF	
	Y-1	PCC$_4$	Y-1	PCC$_4$	Y-1	PCC$_4$	Y-1	PCC$_4$	Y-1	PCC$_4$	Y-1	PCC$_4$
4×10^4	1	0	0	0	0	0	0	0	0	0	0	0
1.5×10^5	7	110	0	0	2	49	11	30	0	0	0	0
6×10^5	18	318	0	0	31	178	19	157	0	0	0	0
2.5×10^6	112	307	0	0	106	276	143	241	0	0	0	0

[a] Mitochondria were isolated from the CAP^r EF^s AMT, CAP^r EF^r TL-1, and CAP^s EF^s P815 cell lines as described. This table shows the number (average of three experiments) of CAP^r or EF^r clones observed after 2 weeks of growth in either CAP or EF when varying numbers of mitochondria (one mitochondrial equivalent = the number of mitochondria obtained from one cell) were incubated with a constant number (3×10^4) of Y-1 or PCC$_4$ cells. It was observed that the AMT mitochondria could confer CAP resistance but not EF resistance on both the Y-1 and PCC$_4$ cell lines, while the TL-1 mitochondria could confer both CAP and EF resistance with similar frequencies. In control experiments, mitochondria obtained from CAP^s EF^s P815 cells could not confer either CAP or EF resistance.

Table II[a]

	UV inactivation of mitochondria	
	Number CAP^r clones	Transformation frequency
Control	47	1.5×10^{-3}
30 min UV	17	5.66×10^{-4}
60 min UV	3	1.00×10^{-4}

[a] Results obtained (average of three experiments) when a constant number of mitochondria (obtained from 10^6 AMT cells) were incubated with 3×10^4 Y-1 cells. Control mitochondria, which were not exposed to ultraviolet light (UV), were able to confer CAP^r at a frequency of approximately 1.5×10^{-3}. However, if the same number of mitochondria were exposed to UV [Sylvania germicidal UV light (#G15T8) at a distance of 10 cm] then the transformation frequency was greatly reduced.

The stability of transformation was tested by isolating 12 antibiotic-resistant clones (six from dishes containing CAP and six from dishes containing EF). These clones were isolated after 2 weeks growth in antibiotics, grown for 1 week in the absence of antibiotics, then rechallenged with CAP or EF. It was observed that all 12 clones retained their antibiotic resistance. Furthermore, these clones were maintained for an additional 17 weeks in the presence of the antibiotic and all clones appeared to be stably transformed.

Additional genetic material in the mitochondrial genome was transferred to the recipient cells in addition to the antibiotic resistance originally selected for. After incubation of CAP^s EF^s Y-1 cells with CAP^r EF^r TL-1 mitochondria, one-half of the cells were plated in dishes containing CAP and the other half in EF. A total of five clones of CAP^r Y-1 cells and six clones of EF^r Y-1 cells were picked at random, isolated, and then tested for the ability to grow in the reciprocal antibiotic. All 11 clones were resistant to both antibiotics. This is in agreement with the observations of Wigler et al. (1980), who demonstrated a high incidence of cotransfer of genetic material using cloned DNA.

If the mitochondria were responsible for the stable transfer of antibiotic resistance, then mitochondrial transformants could be used as antibiotic-resistant donors in enucleation and cybrid experiments. CAP^r and EF^r mitochondrial transformants were enucleated using cytochalasin B (Shay and Clark, 1979) and then fused to CAP^s and EF^s Y-1 cells. The cytoplasms of two of the clonal cell lines tested were able to transfer antibiotic resistance, indicating that the TL-1 and AMT transforming material was incorporated into the cytoplasm.

Not all cell lines were equally capable of acting as recipients for transformation using purified mitochondria. Murine CAP^r and EF^r mitochondria could not confer either CAP or EF resistance on human Lesch-Nyhan cells. This result was consistent with previously published cybrid data which demonstrated that CAP resistance could not be transferred across species (Wallace et. al., 1975, 1977; Bunn et al., 1974, 1977; Wallace, 1981). In addition, not all murine cell lines were equally capable of transformation. After repeated attempts we have been unsuccessful in conferring CAP or EF resistance to the murine C17-S1-D-T984 cell line, which was originally derived from a teratoma. It is not yet understood why there was variability in the efficiency of mitochondrial transformation. However, this observation is not unlike previous findings which indicated that murine cell lines differed widely in their ability to incorporate isolated nuclear DNA (Wigler et al., 1980).

To date, we have been unable to transform Y-1 and PCC4 cells using calcium phosphate-precipitated mitochondrial DNA or cloned mitochondrial DNA from oligomycin-resistant cells. In addition we have cotransformed plasmid PSV2gpt (Mulligan and Berg, 1981) with PSL1180-cloned mitochondrial DNA and selected for cells that had taken up the XGPRT gene, using mycophenolic acid. To date we have isolated six clones that are XGPRT$^+$; however, all of these clones have failed to grow in oligomycin or chloramphenicol.

4. Discussion

In summary, isolated mitochondria were used to confer CAP and EF resistance with high efficiency. There was a high cotransfer frequency of genes not originally selected for, and such cells were stably transformed even in the absence of selection pressure. Some of these transformed cells were able to, in turn, transfer antibiotic resistance cytoplasmically, a result which suggests that intact mitochondria or mitochondrial genes coded are incorporated in the recipient cells. In contrast to previous reports (Coon, 1978; Coon and Ho, 1978), we have been unable to confer antibiotic resistance to sensitive cells using isolating mitochondrial DNA even though these cells could be transformed using a recombinant plasmid carrying the E. coli XPRT gene. The present study provides additional support for the hypothesis that mitochondria are responsible for conferring chloramphenicol and efrapeptin resistance.

Acknowledgments

This work was supported by grants GM-29261 and PCM-8023070 to J. W. Shay and GM-22201 to W. T. Garrard.

We would like to thank Drs. Woodring Wright and William T. Garrard for their helpful suggestions. We would also like to thank Dr. Paul Berg (Stanford University) and Dr. Robert Lansman (University of Georgia) for providing us with the cloned DNA used in this study.

References

Bogenhagen, D., and Clayton, D. A., 1974, The number of mitochondrial deoxyribonucleic acid genomes in mouse L and human Hela cells. Quantitative isolation of mitochondrial deoxyribunucleic acid, *J. Biol. Chem.* **249:**7991–7995.

Bunn, C. L., Wallace, D. C., and Eisenstadt, J. M., 1974, Cytoplasmic inheritance of chloramphenicol resistance in mouse tissue culture, *Proc. Natl. Acad. Sci. USA* **71:**1681–1685.

Bunn, C. L., Wallace, D. C., and Eisenstadt, J. M., 1977, Mitotic segregation of cytoplasmic determinants for chloramphenicol resistance in mammalian cells, *Somat. Cell Genet.* **3:**71–92.

Clark, M. A., and Shay, J. W., 1979, The response of whole and enucleated adrenal cortical tumor cells (Y-1) to ACTH treatment, *Scanning Electron Microscopy* **3:**527–535.

Coon, H. G., 1978, Genetics of the mitochondrial DNA of mammalian somatic cells, their hybrids and cybrids, *Natl. Cancer Inst. Monogr.* **48:**45–55.

Coon, H. G., and Ho, C., 1978, Transformation of cultured cells to chloramphenicol resistance by purified mammalian mitochondrial DNA, in: *Genetic Interactions and Gene Transfer*, Brookhaven Symposium, Volume 29 (C. W. Anderson, ed.), Brookhaven Natl. Lab., Upton, New York, pp. 166–177.

Lichtor, T., and Getz, G., 1978, Cytoplasmic inheritance of rutamycin resistance in mouse fibroblasts, *Proc. Natl. Acad. Sci. USA* **75:**324–329.

Linnenbach, A., Huebner, K., and Croce, C. M., 1980, DNA-transformed murine teratocarcinoma cells: Regulation of expression of simian virus 40 tumor antigen in stem versus differentiated cells, *Proc. Natl. Acad. Sci. USA* **77:**4875–4879.

Malech, H. L., and Wivel, N. A., 1976, Transfer of murine intracisternal A particle phenotype in chloramphenicol-resistant cytoplasts, *Cell* **9:**383–391.

Mulligan, R. C., and Berg, P., 1981, Selection for animal cells that express the *Escherichia coli* gene coding for xanthine-guanine phosphoribosyl transferase, *Proc. Natl. Acad. Sci. USA* **78:**2072–2076.

Schneider, E. L., Stanbridge, E. J., and Epstein, C. J., 1974, Incorporation of 3H-uridine and 3H-uracil into RNA: A simple technique for the detection of mycoplasma contamination of cultured cells, *Exp. Cell Res.* **84:**311–318.

Shay, J. W., and Clark, M. A. (1979). Nuclear control of tumorigenicity in reconstructed cell by PEG-induced fusion of cell fragments, *J. Supramol. Struct.* **11:**33–49.

Tai, Y. H., and Quinn, P. A., 1977, Rapid detection of mycoplasma contamination in tissue culture by SEM, *Scanning Electron Microscopy* **2:**291–299.

Wallace, D. C., 1981, Assignment of the chloramphenicol resistance gene to mitochondrial deoxybonucleic acid and analysis of its expression in cultured human cells, *Mol. Cell. Biol.* **1:**697–710.

Wallace, D. C., Bunn, C. L., and Eisenstadt, J. W., 1975, Cytoplasmic transfer of chloramphenicol resistance in human tissue culture cells, *J. Cell Biol.* **67:**174–188.

Wallace, D. C., Bunn, C. L., and Eisenstadt, J. M., 1977, Mitotic segregation of cytoplasmic determinants for chloramphenicol resistance in mammalian cells. II. Fusions with human cell lines, *Somat. Cell Genet.* **3:**93–119.

Wigler, M., Perucho, M., Kurtz, D., Dana, S., Pellicer, A., Axel, R., and Silverstein, S., 1980, Transformation of mammalian cells with an amplifiable dominant-acting gene, *Proc. Natl. Acad. Sci. USA* **77:**3567–3570.

Chapter 15
Mitochondrial Influences in Hybrid Cells

MICHAEL L. ZIEGLER

"Although the law, that the substances giving the definite and hereditary characters to the cell are entirely contained in the nucleus, is at times spoken of as a very probable hypothesis . . . ."

"Simple reflection shows . . . that the determination whether or not this Theory of Inheritance (*Vererbungs-Theorie*) is true, can be settled in one way alone, viz., to take two different sorts of cells, utilizing the nucleus of one and the protoplasm of the other, to form a new cell. If the nucleus and protoplasm are so constituted that they can exist together, then will the properties arising from this cell, made artificially, answer our question. For then either the exclusive qualities of that cell will develop which had held the nucleus, or those will arise that come from the protoplasm, or lastly, from a mixture of both . . . ." (Boveri, 1893).

1. Introduction

The analysis of genetic mechanisms underlying phenotypic expression in somatic cells depends on techniques which can demonstrate the presence of particular heritable components. For genetic elements located in the nucleus of the cell, numerous chromosomal, drug resistance, and biochemical markers are now available which allow study of specific loci. Similarly, for the analysis of mitochondrial inheritance in somatic cells there has been increased availability of drug resistance markers which can be transmitted by cytoplasmic transfer (Bunn et al., 1974; Lictor and Getz, 1978; Harris, 1978). Specific differences in buoyant density and restriction endonuclease digest patterns of mitochondrial DNA also have served as markers for the identification of cytoplasmic genetic determinants (Dawid et al., 1974; Case and Wallace, 1981).

In addition, there is a unique aspect to the analysis of mitochondrial genetics. Large populations of genomes reside in and are coordinated in their expression within single cells. In order to understand mitochondrial influences on phenotypic expression, techniques are needed to manipulate these

---

MICHAEL L. ZIEGLER • Department of Pathology and College of Dentistry, University of Kentucky Medical Center, Lexington, Kentucky 40536.

genome populations. In single-celled eukaryotes, ultraviolet irradiation and ethidium bromide have been shown to affect cytoplasmic loci and decrease transmission of a treated cell's determinants during mating (Sager and Ramanis, 1967, 1973; Dujon et al., 1975; Gillham et al., 1974). Until recently there have been no similar methods for altering input of cytoplasmic determinants during hybridization of mammalian cells in culture. However, our studies indicate that rhodamine 6G (R6G), a toxic mitochondria-specific fluorescent compound, can be used to decrease transmission of the mitochondrial chloramphenicol (CAP) resistance marker in somatic cell hybrids (Strugger, 1938; Gear, 1974; Johnson et al., 1980; Ziegler and Davidson, 1981). In addition, the use of R6G has revealed that interactions involving cytoplasmic determinants can influence viability and growth potential of hybrid cells (Ziegler and Davidson, 1981). The techniques and results of these experiments are summarized briefly in the present chapter.

2. Experimental Methods and Treatments with R6G

Chinese hamster and mouse lines were used in experimental assays. Line 5-3tg was isolated from V79 hamster lung fibroblasts as resistant to 100 μg/ml CAP and 10 μg/ml thioguanine (Ziegler and Davidson, 1979). The thymidine kinase-deficient hamster peritoneal line $B_{14}I_{50}$ (BI) was used as a complementary fusion partner for 5-3tg (Humphrey and Hsu, 1965). Mouse cells of the thymidine kinase-deficient line 3T3-4E were fused with 5-3tg for production of interspecific hybrids (Matsuya and Green, 1969). Hybrids between the above cell types could be isolated conveniently using a standard HAT (hypoxanthine, aminopterin, thymidine) selection system subsequent to polyethylene glycol-induced fusion. Fusion mixtures contained 2×10^6 5-3tg cells and either 2×10^5 3T3-4E or BI (or BI cytoplasts) cells.

Medium composition is critical for maintenance of cells treated with R6G prior to fusions. Ditta and co-workers (1976) have reported that cells with impaired mitochondrial energy function become dependent on high glucose concentrations, CO_2, and asparagine for survival. In addition, cells incubated in mitochondrial inhibitors display better survival when pyruvate is added to the culture medium (Harris, 1980). Since mitochondrial determinants become nonfunctional or are eliminated in the presence of R6G (as shown by loss of the CAP^R marker in subsequent hybrids), treated cells may become increasingly dependent on these particular medium constituents. Specifically, in the experiments described, Eagle's minimum essential medium (with 10% fetal calf serum) was used and contained 11.0 mM glucose, 0.3 mM asparagine, 26 mM sodium bicarbonate, and 1.0 mM sodium pyruvate (designated HG medium).

The protocol for treatment of V79 derivatives was standardized to obtain a large drop in plating efficiency which was not accompanied by decreases in subsequent hybrid frequency. Best results were obtained when 3×10^6 5-3tg cells were plated in 75-cm² flasks with 30 ml HG medium which contained 1 μg/ml R6G. After incubation for 3 days in R6G, cells were rinsed twice with

Table I. Rescue of R6G-Treated 5-3tg Cells by BI Cells and Cytoplasts[a]

Cells	R6G pretreatment[b]	Plating efficiency, %	Number of colonies in selective medium[c]	
			HAT	AZG
Parental lines				
5-3tg	−	88.0	0	NT
5-3tg-R6G	+	0.042	0	39.0
BI	−	57.0	0	36.0
Fusion mixtures				
5-3tg × BI	−	NT	89.0	NT
5-3tg-R6G × BI	+	NT	70.7	60.5
5-3tg-R6G × BI cytoplasts	+	NT	4.2	284.3

[a] Table reproduced from Ziegler and Davidson (1981). NT, not tested.
[b] Line 5-3tg was treated for 3 days with 1.0 μg/ml R6G prior to fusion.
[c] Numbers for hybrid mixtures represent the number of colonies per 10^5 5-3tg cells plated. Parental cells were assayed using 10^5 cells/dish. All numbers represent averages for at least three cultures. Line 5-3tg has a relative plating efficiency of 100% in 20 μg/ml AZG. Only colonies with more than 100 cells were counted.

HG medium and allowed to incubate for 1–2 hr in drug-free medium prior to fusion with an untreated cell type. This treatment reduced plating efficiency by up to 2000-fold without marked effect on hybrid frequency (Table I).

Thus far, published studies utilizing R6G treatments have employed only isolates of the Chinese hamster lung fibroblast V79 (Ziegler and Davidson, 1981). Although other cell types have been treated successfully, derivatives of V79 appeared to work best in the assay systems reported. It thus seems appropriate to describe briefly the morphologic and growth response of V79 cells to R6G treatment. Although 1 μg/ml R6G was highly toxic to V79 cells, this concentration of drug had little or no effect on gross cellular morphology. Cells appeared flattened and remained attached to flasks for up to 5 days without deterioration in the presence of R6G. In other cell lines (3T3-4E, BI, HeLa) degenerative alterations in morphology more closely paralleled decreases in viability as measured in plating efficiency assays. Even though the morphology of V79 cells was not adversely affected, cell growth was severely inhibited by 1 μg/ml R6G. During a typical 3-day incubation untreated cells underwent four population doublings, while treated cells completed just over one doubling. These observations concerning medium composition, treatment, and cellular response may be of use in choosing appropriate culture conditions and cell lines for experimental studies utilizing R6G.

3. Induced Segregation of Mitochondrial Determinants by R6G

A number of experiments were performed in which Chinese hamster cells were pretreated with R6G and then fused with untreated cells of either hamster or mouse origin. The results of such fusions were analyzed in terms of hybrid frequency, viability, and expression of CAP resistance. First, experiments were carried out to determine if pretreatment of one parental cell type with R6G affects subsequent hybridization frequencies. Doses of R6G that reduced the plating efficiency of 5-3tg by 2000-fold had little effect on the appearance of hybrid colonies when compared with untreated controls (see Table I). The number of colonies formed in control fusions between untreated partners was 89.0 per 100,000 cells plated, while fusions performed with one partner pretreated with R6G yielded 70.7 colonies per 100,000 cells plated. In addition, pretreatment of one partner did not alter the chromosome composition of hybrids subsequently formed (Table II). These data showed that whole untreated cells could effectively rescue cells lethally treated with R6G, and that R6G killing was not accompanied by gross chromosomal damage.

Since R6G binds preferentially to mitochondrial membranes, it seemed possible that mitochondria were the site of cell killing. If this were the case, healthy cytoplasm alone donated by a cell enucleated with cytochalasin B should be able to rescue cells lethally treated with R6G. Second, a decrease in effective input of mitochondrial markers from the treated parent might be expected. Experimental tests of these predictions were made. For this purpose enucleated cells (cytoplasts) derived from untreated Chinese hamster BI cells were fused with lethally treated 5-3tg cells and the fused cells were exposed to azaguanine (AZG) to select cytoplasmic hybrids (cybrids). As seen in Table I, these cultures yielded nearly five times the number of colonies found in control whole-cell hybrid platings. In addition, chromosome compositions of cells isolated from putative cybrid cultures were indistinguishable from those of 5-3tg. Thus, the data showed that BI cytoplasm alone could effectively rescue R6G-treated 5-3tg cells, and suggested that R6G killing was based on alterations in cytoplasmic systems.

In order to test whether R6G killing was based on elimination of viable mitochondrial determinants, levels of resistance to CAP were determined in both hybrid and cybrid combinations. Fusion products of line 5-3tg can serve

Table II. Properties of Intraspecific Hybrid Cells[a]

Cells	Total number of chromosomes[b]	Relative plating efficiency in CAP,[c] %
Parental lines		
5-3tg	22.3 (20–24)	100.5
BI	22.9 (21–24)	0 (<1.0)
Hybrids		
5-3tg × BI	43.9 (42–47)	76.0
5-3tg-R6G × BI	42.6 (37–46)	5.3
Cybrids: 5-3tg-R6G × BI cytoplasts		
Clone 1	22.3 (22–24)	0 (<1.0)
Clone 2	22.6 (19–24)	3.4
Clone 7	22.8 (22–24)	7.4

[a] Table reproduced from Ziegler and Davidson (1981).
[b] Numbers represent the mean of at least 20 chromosome spreads. Numbers in parentheses represent the range.
[c] Dishes were inoculated with 100 cells in either drug-free or CAP-containing HG medium. Relative plating efficiency is the ratio of the number of colonies in CAP-containing medium to the number formed in drug-free medium (×100). Values represent averages of three cultures.

in this assay, as 5-3tg is resistant to 100 μg/ml CAP (Ziegler and Davidson, 1979). Table II shows the relative plating efficiencies of parental and hybrid cells in CAP. It is clear from these data that R6G pretreatment of the CAP^R line 5-3tg markedly decreases expression of CAP resistance in both hybrid and cybrid combinations, and can be used to control the input of cytoplasmic determinants in fused cells.

4. Elimination of Incompatibility in Interspecific Crosses

R6G has also been employed in interspecific crosses between mouse and Chinese hamster cells (Ziegler and Davidson, 1981). As in intraspecific crosses, R6G pretreatment of the hamster line decreased transmission of the CAP^R mitochondrial marker. In addition, no adverse effect was found on initial chromosome constitutions of hybrids. These data suggested that the input of mitochondrial determinants also could be manipulated in interspecific crosses.

In the particular cell combination used, there appeared an incompatibility such that hybrid colonies arose and grew rapidly for several generations, but then growth rate slowed and the majority of clones (approximately 80%) lost viability entirely. Such incompatibility, although poorly understood, is rather common in interspecific hybrids (Jami and Grandchamp, 1971; Graves, 1972; Graves and Koschel, 1980; Mascarello et al., 1980). For instance, Jami and Grandchamp (1971) made the observation that there were large variations in growth potential of mouse–human hybrids which depended on the particular cell types used. In one cross fewer than 10% of the hybrid colonies originally isolated remained viable.

A potential insight into hybrid incompatibility was made when it was found that R6G pretreatment of the hamster parent in a hamster–mouse cross entirely eliminated the growth difficulties of resultant hybrids (see Fig. 1). Since the initial chromosome composition (5 days after fusion) of untreated and treated hybrids was similar, unbalanced chromosome contributions did not appear to be involved in the incompatibility (Ziegler and Davidson, 1981). Another explanation seemed more likely. R6G eliminated hamster mitochondrial determinants, which are incompatible with mouse nuclear or cytoplasmic (or both) factors. Although the mechanism causing hybrid inviability is not known, it is clear from these studies that cytoplasmic determinants can influence growth parameters in hybrid cells.

5. R6G and the Direction of Chromosome Loss

Several groups have reported a correlation between the loss of mitochondrial determinants and the direction of chromosome loss in interspecific hybrids (Attardi and Attardi, 1972; Clayton et al., 1971; Coon et al., 1973; Eliceiri, 1973). For example, assays of mouse–human hybrids revealed that

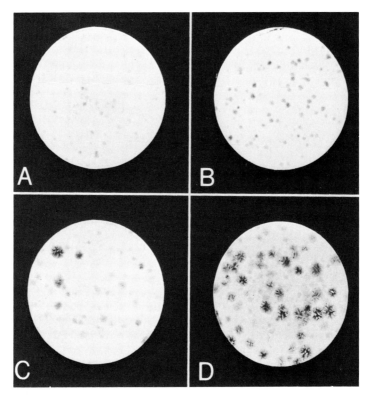

Figure 1. Growth of untreated and R6G-pretreated 5-3tg × 3T3-4E-hybrids in HAT medium. Cell lines 5-3tg and 3T3-4E were mixed, fused, and selected in HAT medium. Cultures received equivalent numbers of cells and were stained after 7 or 13 days. (A) 5-3tg × 3T3-4E, 7 days; (B) 5-3tg-R6G × 3T3-4E, 7 days; (C) 5-3tg × 3T3-4E, 13 days; (D) 5-3tg-R6G × 3T3-4E, 13 days. [Figure reproduced from Ziegler and Davidson (1981).]

the mitochondrial DNA of the parent that was segregating chromosomes was undetectable or present in marginal amounts (De Francesco et al., 1980). Although speculation as to mechanisms have been made, cause and effect in this relationship have been difficult to distinguish (Wallace and Eisenstadt, 1979).

The ability to manipulate cytoplasmic determinants prior to fusion with R6G allows investigation of associations between mitochondria and chromosome stability in hybrid cells. Chromosome distributions of 3T3-4E mouse × 5-3tg hamster hybrid crosses assayed about 3 weeks after fusion are shown in Fig. 2. As can be seen, parental Chinese hamster cells contain primarily metacentric chromosomes, while mouse cells contain exclusively telocentric chromosomes. Untreated hybrid crosses show a large variation in chromosome numbers. However, this variation arises secondarily, as early hybrids show an expected one mouse–one hamster chromosome constitution (Ziegler and Davidson, 1981). There appears to be a selective advantage in the untreated cross for hybrids containing multiple genomes from one parent. Similar observations have been made in other Chinese hamster–mouse

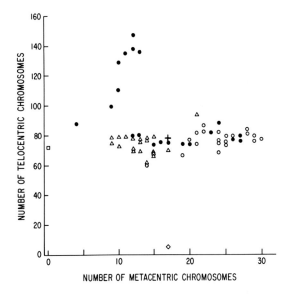

Figure 2. Chromosome composition of 5-3tg × 3T3-4E hybrids at 3 weeks post-fusion. 5-3tg (◇); 3T3-4E (□); 5-3tg-R6G × 3T3-4E (△); 5-3tg × 3T3-4E (●); 5-3tg × 3T3-4E fusion mixture selected in HAT + 100 µg/ml CAP (○); summation of 5-3tg and 3T3-4E chromosome compositions (+).

crosses (Lanfranchi and Marin, 1981). In contrast, pretreatment of the hamster parent with R6G allows rapid growth of hybrids containing one genome from each parent. Also shown in Fig. 2 are chromosome distributions of an untreated cross selected in HAT + 100 µg/ml CAP. The retention of the hamster CAP resistance appears correlated with the presence of multiple hamster genomes.

Amid the various hybrid chromosome complements, a pattern arises with respect to the direction of chromosome loss. Regardless of parental cell treatment or hybrid selection, approximately ¼ to ½ of the hamster metacentric chromosomes have been lost from hybrid cells. Mouse chromosomes have not been lost substantially except in the case of untreated hybrids which contain two mouse–one hamster genome. It is clear that in this hybrid combination the speed and direction of chromosome loss are unaffected by either selection (by incubation in HAT plus CAP) or elimination (by pretreatment with R6G) of hamster mitochondrial determinants.

6. Summary

A technique is now available for the control of transmission of mitochondrial determinants in somatic cell hybrids. Simple incubation of one parental cell type in R6G prior to hybridization significantly reduces the expression of mitochondrial markers contributed by the treated cell. In addition, the use of R6G has revealed that interactions involving mitochondrial determinants can influence the growth and viability of interspecific somatic cell hybrids. Thus, R6G may prove valuable for exploring both specific aspects of mitochondrial genetics as well as general relationships between cytoplasmic and nuclear genetic systems.

Acknowledgments

I thank Karen Taylor for assistance in the preparation of this chapter.

References

Attardi, B., and Attardi, G., 1972, Fate of mitochondrial DNA in human–mouse somatic cell hybrids, *Proc. Natl. Acad. Sci. USA* **69**:129–133.
Boveri, T., 1893, An organism produced sexually without characteristics of the mother, *Am. Naturalist* **27**:222–232.
Bunn, C. L., Wallace, D. C., and Eisenstadt, J. M., 1974, Cytoplasmic inheritance of chloramphenicol resistance in mouse tissue culture cells, *Proc. Natl. Acad. Sci. USA* **71**:1681–1685.
Case, J. T., and Wallace, D. C., 1981, Maternal inheritance of mitochondrial DNA polymorphisms in cultured human fibroblasts, *Somat. Cell Genet.* **7**:103–108.
Clayton, D. A., Teplitz, R. L., Nabholz, M., Dovey, H., and Bodmer, W., 1971, Mitochondrial DNA of human–mouse cell hybrids, *Nature* **234**:560–562.
Coon, H. G., Horak, I., and Dawid, I. B., 1973, Propagation of both parental mitochondrial DNAs in rat–human and mouse–human hybrid cells, *J. Mol. Biol.* **81**:285–298.
Dawid, I. G., Horak, I., and Coon, H. G., 1974, The use of somatic cells as an approach to mitochondrial genetics in animals, *Genetics* **78**:459–479.
De Francesco, L., Attardi, G., and Croce, C. M., 1980, Uniparental propagation of mitochondrial DNA in mouse–human cell hybrids, *Proc. Natl. Acad. Sci. USA* **77**:4079–4083.
Ditta, G., Soderberg, K., Landy, F., and Scheffler, I. E., 1976, The selection of Chinese hamster cells deficient in oxidative energy metabolism, *Somat. Cell Genet.* **2**:331–344.
Doersen, C., and Stanbridge, E. J., 1979, Cytoplasmic inheritance of erythromycin resistance in human cells, *Proc. Natl. Acad. Sci. USA* **76**:4549–4553.
Dujon, B., Kruszewska, A., Slonimski, P. P., Bolotin-Fukuhara, M., Coen, D., Deutsch, J., Netter, P., and Weill, L., 1975, Mitochondrial Genetics X: Effects of UV irradiation on transmission and recombination of mitochondrial genes in *Saccharomyces cerevisiae*, *Mol. Gen. Genet.* **137**:29–72.
Eliceiri, G. L., 1973, The mitochondrial DNA of hamster–mouse hybrid cells, *FEBS Lett.* **36**:232–234.
Gear, A. R. L., 1974, Rhodamine 6G: A potent inhibitor of mitochondrial oxidative phosphorylation, *J. Biol. Chem.* **249**:3628–3637.
Gillham, N. W., Boynton, J. E., and Lee, R. W., 1974, Segregation and recombination of non-Mendelian genes in *Chlamydomonas*, *Genetics* **78**:439–457.
Graves, J. A. M., 1972, Cell cycles and chromosome replication patterns in interspecific somatic hybrids, *Exp. Cell Res.* **73**:81–94.
Graves, J. A. M., and Koschel, K. W., 1980, Changes in the cell cycle during culture of mouse–chinese hamster cell hybrids, *J. Cell. Physiol.* **102**:209–216.
Harris, M., 1978, Cytoplasmic transfer of resistance to antimycin A in Chinese hamster cells, *Proc. Natl. Acad. Sci. USA* **75**:5604–5608.
Harris, M., 1980, Pyruvate blocks expression of sensitivity to antimycin A and chloramphenicol, *Somat. Cell Genet.* **6**:699–708.
Humphrey, R. M., and Hsu, T. C., 1965, Further studies on biological properties of mammalian cell lines resistant to 5-bromodeoxyuridine, *Texas Rep. Biol. Med.* **23**:321.
Jami, J., and Grandchamp, S., 1971, Karyological properties of human–mouse somatic hybrids, *Proc. Natl. Acad. Sci. USA* **68**:3097–3101.
Johnson, L. V., Walsh, M. L., and Chen, L. B., 1980, Localization of mitochondria in living cells with rhodamine 123, *Proc. Natl. Acad. Sci. USA* **77**:990–994.
Lanfranchi, G., and Marin, G., 1981, Evidence for the derivation of mammalian somatic hybrids from polykaryocytes, *Exp. Cell Res.* **133**:255–260.
Lichtor, T., and Getz, G. S., 1978, Cytoplasmic inheritance of rutamycin resistance in mouse fibroblasts, *Proc. Natl. Acad. Sci. USA* **75**:324–328.

Mascarello, J. T., Soderberg, K., and Scheffler, I. E., 1980, Assignment of a gene for succinate dehydrogenase to human chromosome 1 by somatic cell hybridization, *Cytogenet. Cell Genet.* **28**:121–135.

Matsuya, H., and Green, H., 1969, Somatic cell hybrid between the established human line D98 (presumptive HeLa) and 3T3, *Science* **163**:697–698.

Sager, R., and Ramanis, Z., 1967, Biparental inheritance of nonchromosomal genes induced by ultraviolet irradiation, *Proc. Natl. Acad. Sci. USA* **58**:931–937.

Sager, R., and Ramanis, Z., 1973, The mechanism of maternal inheritance in *Chlamydomonas*: Biochemical and genetic studies, *Theor. Appl. Genet.* **43**:101–108.

Strugger, S., 1938, Die Vitalfärbung des Protoplasmas mit Rhodamin B und 6G, *Protoplasma* **30**:85–100.

Wallace, D. C., and Eisenstadt, J. M., 1979, Expression of cytoplasmically inherited genes for chloramphenicol resistance in interspecific somatic cell hybrids and cybrids, *Somat. Cell Genet.* **5**:373–396.

Ziegler, M. L., and Davidson, R. L., 1979, The effect of hexose on chloramphenicol sensitivity and resistance in Chinese hamster cells, *J. Cell. Physiol.* **98**:627–636.

Ziegler, M. L., and Davidson, R. L., 1981, Elimination of mitochondrial elements and improved viability in hybrid cells, *Somat. Cell Genet.* **7**:73–88.

Chapter 16
Shedding of Tumor Cell Surface Membranes

ANDREJS LIEPINS

1. Introduction

Various antigenic determinants of the cell surface membrane not only have been shown to be modulated in the plane of the cell surface by various physiologic and experimental conditions (Nowotny et al., 1974; Poskitt et al., 1976; Raz et al., 1978; Van Blitterswijk et al., 1979), but have also been transferred to recipient cells by membrane vesicle–cell hybridization techniques (Poste and Nicolson, 1980; Volsky et al., 1981). These lines of evidence have provided the necessary impetus for the development of methods for the procurement of cell surface membranes without cell disruption or the use of drugs that may alter their biologic properties.

Cell disruption methods, which have been generally used for the isolation and characterization of the biologic properties associated with the cell surface membrane, have the inherent disadvantage of producing mixed populations of "right-side-out" and "inside-out" membrane vesicles (Walsh et al., 1976). Alternatively, a variety of chemicals or drugs have been used to induce shedding of cell surface membranes in the form of closed membrane vesicles. Among the reportedly effective agents are formaldehyde, N-ethylmaleimide, dithiothreitol, and iodoacetate (Scott, 1976; Hoerl and Scott, 1978). The membrane vesicles obtained in conjunction with these agents were reported to be enriched seven to tenfold with respect to 5'-nucleotidase activity, with a decrease in the activity of glucose-6-phosphatase and the complete absence of NADH–cytochrome c reductase (Scott, 1976). Furthermore, by means of immunologic studies, the shed membrane vesicles were shown to bear antigenic determinants of the parent cell line from which they were derived (Pearson and Scott, 1977). Agents such as cytochalasin B and D (Godman and

---

ANDREJS LIEPINS • Memorial University, Faculty of Medicine, St. John's, Newfoundland, Canada A1B 3V6.

Miranda, 1978; Hart et al., 1980), the local anesthetics dibucaine, tetracaine, and procaine (Nicolson et al., 1976), and the toxin phalloidin (Virtanen and Miettinen, 1980) have well-known effects on the function and organization of the cellular cytoskeletal elements and have also been reported to be effective inducers of cell surface membrane vesiculation and in some cases shedding. It should be noted that the shedding of cell surface membranes via membrane vesicles is not an event limited to *in vitro* experimental conditions, but appears to occur *in vivo* during tumor growth (Nowotny et al., 1974; Raz et al., 1978; De Broe et al., 1977; Van Blitterswijk et al., 1979; Petitou et al., 1978) and in hyperplastic thyroid epithelium (Zeligs and Wollman, 1977) as well as during the normal differentiation of rabbit thymocytes (Roozemond and Urli, 1981). Another well-documented example of highly specialized physiologic shedding occurs in the rod photoreceptor cells of the retina, in which the outer segment of these cells, comprised of disks or flattened membrane vesicles containing the light-sensitive pigment rhodopsin, is continuously formed and shed during exposure to light [reviewed by Black (1980a)].

From the functional point of view, membranes shed by tumor cells have been implicated in a number of biologic events that favor tumor growth and metastasis. Various studies have provided evidence that the shedding of tumor cell membranes that bear antigenic properties the same as those of the parent tumor cells may function as a means of escape from the host's immune defense mechanisms (Boyse et al., 1965; Kim et al., 1975; Davey et al., 1976). Platelet aggregation (Gasic et al., 1978) and plasminogen activator (Jones et al., 1975; Quigley, 1976; Jaken and Black, 1979) activities have also been shown to be associated with membrane vesicles shed by transformed cells. These lines of evidence have led Black (1980b) to suggest an alternative, novel approach for the treatment of cancer, which would involve the prevention of shedding of cell surface membranes that is prevalent in malignant cells. Evidently, the understanding of the physiologic mechanisms underlying the shedding of cell surface membranes with their associated biologic activities could contribute to the understanding of the malignant behavior of transformed cells.

In this chapter, evidence obtained in this laboratory regarding the induction by low temperature of membrane vesicle (MV) formation and shedding by tumor cells is presented and discussed in some detail (Liepins and Hillman, 1981). These data show that the MV-shedding process is energy dependent, in that it can be effectively inhibited by metabolic inhibitors as well as by agents that stabilize cytoskeletal components. The shedding process results in a significant reduction in the average cell diameter of the cell population, suggesting that membrane synthesis and turnover is sufficient to account for the membranes exfoliated via the membrane vesicles. The membrane permeability to trypan blue is not altered by the low-temperature-induced MV shedding. It remains to be established, however, whether the MV shedding process is related to possible subsequent changes in cell viability or is a manifestation of cell injury as suggested by some studies (Liepins et al., 1978; Hoerl and Scott, 1978; Trump et al., 1980).

2. Materials and Methods

2.1. Cell Lines

Mastocytoma P815 cells were maintained in tissue culture RPMI medium supplemented with 8% fetal bovine serum (Flow Laboratories, Mississauga, Ontario, Canada) and containing the antiobiotics penicillin and streptomycin, 100 units/ml and 100 µg/ml, respectively (Difco Laboratories, Detroit, Michigan) at 37°C in a humidified atmosphere of 5% CO_2–95% air. Cells were subcultured by dilution on alternate days using tissue culture flasks (from any supplier) of various sizes, depending on need. Large cell numbers when required were obtained by growing cells in 1-liter spinner flasks.

2.2. Induction of Membrane Vesicle Shedding

Logarithmically growing P815 cells were washed once in RPMI medium (without fetal bovine serum) by centrifugation at 200g for 5 min and the pelleted cells were resuspended in fresh RPMI medium to a concentration of $1.0–2.0 \times 10^6$ cells/ml. Suspensions of washed P815 cells were placed in an ice bath (0°C) for 1 hr and subsequently allowed to return to room temperature (22°C). Formation and shedding of membrane vesicles from the cell surfaces were monitored by observation of cell samples in a hemocytometer. In general 85–95% of the cell population was found to have formed and shed membrane vesicles between 1–2 hr at 22°C. Control cell samples were maintained (a) continuously at 22°C without a low-temperature period and (b) continuously at 0°C.

2.3. Harvesting of Membrane Vesicles

Cells were separated from the shed membrane vesicles by layering 10 ml of the cell and vesicle containing medium over a step gradient of 5% (15 ml) and 10% (10 ml) Ficoll 400 (Pharmacia, Dorval, Quebec, Canada) in 50-ml Oak Ridge-type polycarbonate tubes (Fisher Sci. Co., Ottawa, Ontario, Canada) and centrifuged at 225g for 12 min (1000 rpm, CRU-5000 centrifuge, DAMON/JEC Div.). Under these conditions most of the membrane vesicles band at the 5–10% Ficoll interface, whereas the cells pellet at the bottom of the tube. The membrane vesicles were harvested by removing the upper layer of culture medium and 5% Ficoll including the interface region. The vesicles were pelleted at 27,000g for 10 min (Beckman J21B centrifuge; JA20 angle rotor). The membrane vesicle pellet was resuspended and washed in 0.1 M phosphate or any other desired isotonic buffer.

2.4. Cell Size Distribution

The cell size distribution was determined with a Coulter counter equipped with a Channelyzer and an X-Y Recorder II (Coulter Electronics, Hialeah, Florida), using a probe of 100 μm aperture with an aperture current setting of 1.0 and an amplification factor setting of 2.0.

2.5. Effect of Nucleotides on the Shedding of Cell Surface Membranes

All nucleotides were obtained from Sigma Chemical Co. (St. Louis, Missouri). Stock solutions of adenosine-3':5'-cyclic monophosphoric acid (cAMP), $N^6,O^{2'}$-dibutyryladenosine-3':5'-cyclic monophosphoric acid (dbcAMP), guanosine 5'-triphosphate (GTP), and guanosine-3':5'-cyclic monophosphoric acid (cGMP) were prepared at 10 mM in RPMI tissue culture medium. Theophylline was prepared at 10 μg/ml of RPMI medium and in culture medium containing 5 mM dbcAMP. Cells from logarithmically growing cultures were collected by centrifugation, washed once in RPMI medium, and resuspended in medium containing the appropriate nucleotide and allowed to remain at room temperature (22°C) for 30 min before being submitted to low-temperature (0°C) for the induction of shedding of cell surface membranes. The effects of various nucleotides and theophylline were assessed microscopically after having allowed the cold-treated cells to return

Table I. Effects of Cyclic Nucleotides on the Formation and Shedding of Membrane Vesicles (MV) by P815 Mastocytoma Cells

Concentration	Percent cells shedding MV[a]	Percent control cells shedding MV[d]	Percent inhibition[b]	Percent viability[c]
cAMP 5mM	69.6	88.3	21.2	≥95
10mM	52.3	87.6	40.3	≥95
dbcAMP 5mM	69.5	90.0	22.8	≥95
10mM	70.0	89.0	21.3	≥95
5 μg/ml Theophylline + dbcAMP 5mM	65.3	89.0	26.6	≥95
10 μg/ml Theophylline + dbcAMP 5 mM	59.6	89.3	33.3	≥95
Theophylline 5 μg/ml	64.0	90.6	29.4	≥95
10 μg/ml	59.2	88.8	33.4	≥95
GTP 5mM	58.0	86.7	33.1	≥95
10mM	54.0	90.6	40.4	≥95
cGMP 5mM	62.0	91.5	32.2	≥95
10mM	53.5	90.0	40.6	≥95

[a] Percent average of P815 cells shedding MV in medium containing various nucleotides.
[b] Percent inhibition is given by Eq. (1) in the text.
[c] Viability: based on trypan blue exclusion.
[d] Percent average of P815 cells shedding MV in medium without various nucleotides.

to 22°C for 1 hr (Table I). Cell viability was monitored by the Trypan Blue exclusion method.

2.6. Effects of Deuterium Oxide on the Shedding of Cell Surface Membranes

Various concentrations of deuterium oxide were prepared (Sigma Chemical Co.) using 10× RPMI medium to obtain the desired proportion of D_2O in the medium (Table II). Cell pellets were resuspended in the D_2O-containing medium at concentrations of $1–2 \times 10^6$ cells/ml and incubated in an ice bath (0°C) for 1 hr before being returned to room temperature (22°C). The effects of various D_2O concentrations were assessed microscopically after 1 hr at 22°C and expressed as % inhibition versus control cell samples by the formula

$$\% \text{ inhibition} = 100 - \frac{\% \text{ cells shedding MV in the presence of } D_2O \text{ or drugs}}{\% \text{ cells shedding MV in control medium}} \quad (1)$$

2.7. Electron Microscopy

For transmission electron microscopy, pellets of cells and membrane vesicles were fixed in 2.4% glutaraldehyde in 0.1 M Na-cacodylate buffer (JBEM Services, Pointe Claire, Dorval, Quebec, Canada) containing 0.01% $CaCl_2$ for 2 hr at room temperature, washed overnight in buffer, and postosmicated for 1 hr in buffered 2.0% OsO_4 followed by buffered washes, dehydration, and embedding in Epon 812 by routine procedures. Thin sections stained with uranyl acetate and lead citrate were examined and photographed in a Phillips 300 electron microscope.

3. Results and Discussion

3.1. Inductive Stimuli for MV Formation and Shedding

As mentioned in the introduction, there are a variety of agents capable of inducing vesiculation and shedding of cell surface membranes. The simplest

Table II. Effects of Deuterium Oxide on the Formation and Shedding of Membrane Vesicles (MV) by P815 Mastocytoma Cells

Percent D_2O	Percent cells shedding MV	Percent control (cold-shocked) cells shedding MV	Percent inhibition	Percent viability
90	2.5	90	97	≥95
68	7.5	85	91	≥95
45	11.4	91	87	≥95
23	25	91	72	≥95
11	59	92	36	≥95

and most effective appears to be a low-temperature period (1–2 hr at 0°C) followed by an equivalent time period at room temperature (22°C). Under these conditions 85–95% of the mastocytoma cell population formed and shed MV from their surfaces (Figs. 3–5; Tables I, II). Since the primary requirement for the MV shedding process to occur appears to be low temperature, a condition which is known to disrupt cytoplasmic microtubules (Ostlund et al., 1980), we have previously postulated that the disruption of these cytoskeletal structures might be the essential requirement for the MV shedding process to be initiated (Liepins and Hillman, 1981). This interpretation is strengthened when viewed in conjunction with the results obtained by Meek and Puck (1979), who have reported a correlation between bleb formation, i.e., vesiculation of the cell surface membrane, and the disorganization of the cytoplasmic microtubule network. Actin-containing filaments have been shown to be associated with the vesiculating regions of the cell surface membrane (Virtanen and Mietinen, 1980; Britch and Allen, 1981), suggesting that these fibrils may be involved in providing the contractile forces required for the vesiculation and MV shedding process. Not surprisingly, the MV shedding process requires metabolic energy, since it can be prevented by various metabolic inhibitors (Liepins and Hillman, 1981). The low-temperature induction of MV shedding has been found to be effective for a variety of ascites tumor lines, i.e., P815 mastocytoma, L1210 murine lymphocytic leukemia, and S194 myeloma, as well as anchorage-dependent cells such as 3T3 and 3T6 cell lines (Liepins, unpublished).

3.2. Morphologic Aspects of MV Shedding

After the P815 cells incubated on ice for 1–2 hr were returned to room temperature (22°C) the formation of cell surface blebs was initiated within 30 min. These cell surface blebs led to the formation of round and turgid membrane vesicles which separated from the cell surface and appeared to be highly mobile while in the vicinity of the cell surface (Fig. 3–5). The MV and cells remained trypan blue-excluding throughout the shedding process. The shedding process, when completed after 1–2 hr at room temperature, resulted in a significant reduction in the average cell diameter (Fig. 1), suggesting that cell surface membrane turnover is sufficient to account for the membranes exfoliated with the MV. Control cells, maintained either at room temperature or continuously on ice, displayed neither cell surface MV formation nor shedding (Figs. 2 and 6).

At the ultrastructural level, thin sections through the MV shedding region showed vesicles containing primarily granular material suggestive of ribonucleoprotein particles and which may be free ribosomes (Figs. 7–9). After the separation of the MV from the cells on a Ficoll step gradient, the contents of the larger MV included membrane and mitochondrial profiles (Fig. 8). At high magnification, the membranes enclosing the vesicles displayed the classical tripartite structure of unit membranes (Fig. 9).

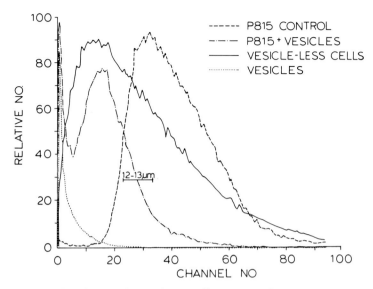

Figure 1. Cell size distribution of control P815 cells maintained at room temperature (- - -); P815 cells and shed MV in medium before Ficoll separation (- · -); P815 cells separated from shed MV through Ficoll gradient (—); separated MV fraction (· · ·). Bar indicates size distribution of paper mulberry pollen grains (12–13 μm).

Cells separated after the MV shedding was completed were found to have a narrow cytoplasm containing electron-dense ribonucleoprotein granules with few mitochondria and endoplasmic reticulum profiles occupying a small region of the narrow cytoplasm (Figs. 10 and 11). The surface of these cells was devoid of microvilli, which were abundant in control cells. The nucleus showed peripheral condensation of the chromatin, with occasional invagination of the inner membrane of the nuclear envelope (Fig. 10).

3.3. Inhibition of MV Shedding

As previously reported, the MV shedding process can be effectively inhibited (90%) by 10 mM 2-deoxy-D-glucose, 10 mM sodium azide, and 2 mM 2,4-dinitrophenol when added to the cell-containing medium before cooling the cells on ice (Liepins and Hillman, 1981), clearly demonstrating the energy requirements of this process. These results are in agreement with those of Hoerl and Scott (1978) and Godman et al. (1975), who have reported that the formaldehyde- and cytochalasin-induced vesiculation processes were also inhibited by metabolic inhibitors. It has been suggested that the plasma membrane vesiculation induced by various chemicals or anoxia is the result of cell injury (Hoerl and Scott, 1978; Trump et al., 1980). At the present time, based on the available data, it is difficult to unequivocally equate cell injury with shedding of MV, since the low-temperature-induced shedding does not alter the viability of those cells with respect to trypan blue and ruthenium red

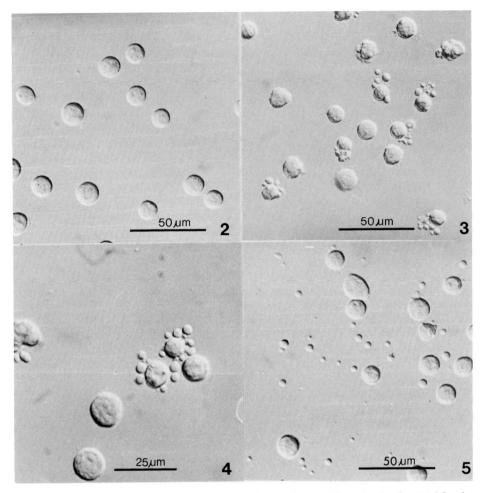

Figure 2. Interference contrast photomicrograph of control P815 cells maintained at 22°C for the duration of the experiment.

Figure 3. Photomicrograph of P815 cells incubated at 0°C for 1 hr followed by 30 min at 22°C, illustrating the membrane vesicle (MV) formation and shedding.

Figure 4. Higher magnification of P815 cells with shed MV still in their immediate vicinity after 1 hr at 22°C.

Figure 5. After 1–1.5 hr at 22°C the shed MV appear free in the medium.

exclusion or their subsequent susceptibility to T-lymphocyte-mediated lysis (Liepins, unpublished), unless cellular injury is considered to be a sequence of events leading to a "point of no return" as conceptualized by Trump et al. (1980). In such a case, low-temperature-induced MV shedding could be considered to be a state of injury preceding the "point of no return" at which cells can still maintain viability. Subculturing of the P815 cells after separation from the shed MV has shown that these cells can recover their

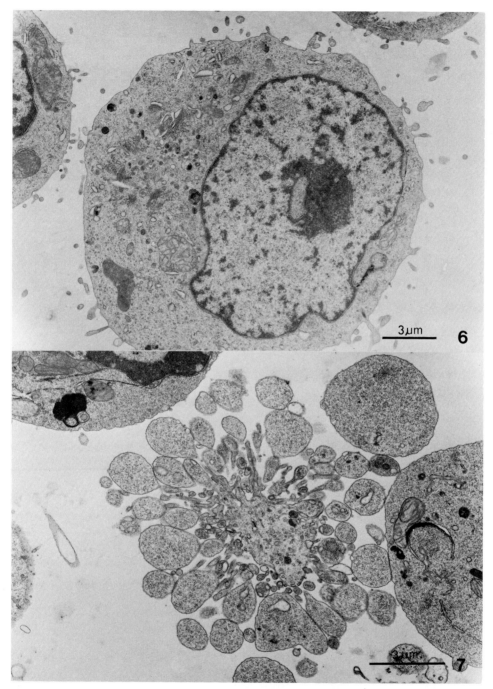

Figure 6. Transmission electron micrograph of P815 mastocytoma cells from the room-temperature (22°C) controls. Cell surface villi, cytoplasmic organelles, and euchromatic nuclei are characteristic for this ascites tumor line.

Figure 7. Section through the membrane vesicle shedding region of a P815 cell incubated at 0°C for 1 hr and followed by 1 hr at 22°C. The MV are heterogeneous in size and contain primarily electron-dense, ribosome-like particles with few membrane inclusions.

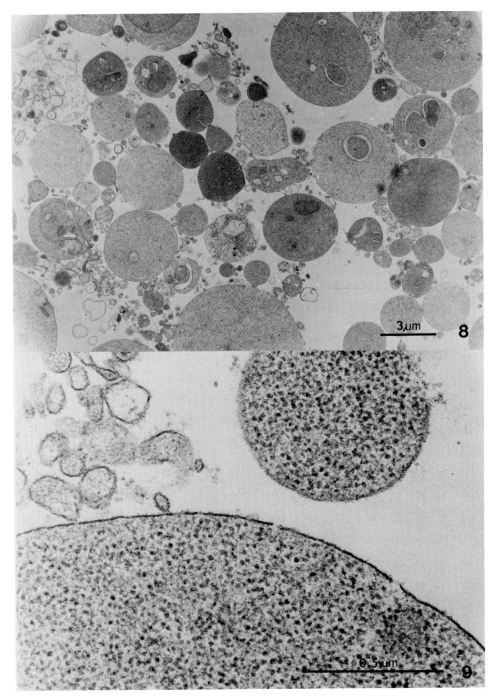

Figure 8. MV fraction separated through a Ficoll gradient. Membrane inclusions and occasional mitochondrial profiles can be found in the larger MV.

Figure 9. High magnification reveals that membranes enveloping the vesicles have typical unit membrane structure.

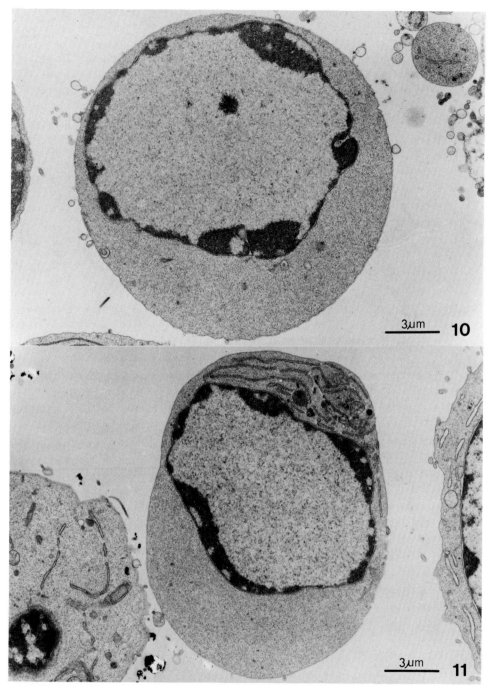

Figure 10. P815 cell after the MV shedding process was completed (1–2 hr at 22°C) and separated from the shed vesicles through a Ficoll gradient. Note the smooth cell surface devoid of microvilli, narrow cytoplasm devoid of organelles, and the nucleus displaying peripheral chromatin condensation interrupted by occasional invaginations of inner nuclear envelope membrane.

Figure 11. P815 cell after MV shedding showing endoplasmic reticulum and few mitochondria polarized within a small region of the cytoplasm. Perinuclear heterochromatin is characteristic of cells exposed to low temperature (0°C) for the induction of the MV shedding process.

logarithmic growth pattern after a lag period of 48 hr (Liepins, unpublished). It is evident that cell cloning studies will be required to determine the extent of cell injury caused by the incubation of cells at low temperature.

Since disruption of microtubules appears to be an essential requirement for the MV shedding process to occur, various agents reported to stabilize these cytoskeletal components were investigated with regard to their possible inhibitory effects on the shedding process. Cyclic nucleotides and their derivatives have been reported to stabilize microtubules against disruption by low temperature (Brinkley et al., 1975; Kirkland and Burton, 1972) as well as by stimulating tubulin polymerization (Porter et al., 1974; Brinkley et al., 1975; Borisy et al., 1972). Results summarized in Table I show that cAMP, GTP, and cGMP, when present at 10 mM in the culture medium before exposure of the cells to low temperature, caused a 40% inhibition of MV shedding. The cAMP derivative dbc-AMP had an inhibitory effect of 22.8% at 5 mM and showed no dose-dependent increase at 10 mM. However, in the presence of the phosphodiesterase inhibitor theophylline, the inhibitory effect of dbcAMP rose to 33.3%. Interestingly, theophylline alone at 10 μg/ml gave the same level of inhibition as it did in conjunction with dbcAMP. This suggests that inhibition of phosphodiesterase activity is sufficient to attain intracellular nucleotide concentrations capable of preventing a significant fraction of the cell population from forming and shedding MV. Higher concentrations of cyclic nucleotides were also tested but gave no higher inhibitory effects than at 10 mM (data not shown). The fact that cAMP has been shown in many instances to cause pronounced microtubule-related morphologic and behavioral changes in cells, i.e., reverse transformation, ruffling, substrate adhesion, blebbing, and locomotion (Puck, 1977; Dedman et al., 1979), strongly suggests a key role for microtubule depolymerization in the initiation of the shedding process.

Deuterium oxide, one of the most effective microtubule-stabilizing agents (Katz and Crespi, 1970), when present in the culture medium at a concentration of 90% inhibited an average of 97% of the cell population from forming and shedding MV (Table II). Lower concentrations gave proportionally less inhibition. These data further support the postulated role of microtubule disruption as being an essential requirement for the MV shedding process to occur.

3.4. General Considerations of the MV Shedding Process

An increasing body of evidence has accumulated indicating that the modulation of cell surface antigens is not exclusively dependent on the physicochemical properties of the cell surface membranes, but also on a highly organized cytoskeletal system of microfilaments and microtubules which appear to be involved in the anchorage and distribution of the cell surface antigenic determinants (Sundqvist and Ehrnst, 1976; Koch and Smith, 1978; Edelman, 1976; Nordquist et al., 1977). The transmembrane modulation of tumor cell surface antigens has been postulated to play a significant role in tumor cell susceptibility and/or escape from the host's immune defense mechanisms (Boyse et al., 1965; Kim et al., 1975; Alexander, 1974; Davey et al., 1976). Indeed, extracellular membrane vesicles bearing a high density of antigenic determinants in common with the tumor cells have been isolated from body fluids of tumor-bearing hosts (Van Blitterswijck et al., 1979; Nowotny et al., 1974; Davey et al., 1976; Raz et al., 1978).

Current work in this laboratory has shown that the low-temperature-induced MV shedding also induces the sequestration of concanavalin A receptors at the MV-forming areas of the cells. Likewise, immunofluorescence studies indicate that the murine mammary tumor virus envelope-associated glycoprotein-52 expressed on L1210 lymphocytic leukemia cells is also sequestered at the vesiculating regions of the cell surface and is shed in conjunction with the MV (Liepins, unpublished). These data suggest that the membranes enveloping the vesicles may be enriched in cell surface antigenic determinants associated with the cell's malignant phenotype and may provide a source for the isolation and characterization of these cell surface components. The specific activity of the plasma membrane enzyme marker 5'-nucleotidase is about two times higher in membranes isolated from MV than in control plasma membrane preparations obtained from control whole-cell homogenates (Liepins and Patrick, unpublished).

The recent development of membrane bioengineering methods has allowed the implantation via MV of various cell surface antigenic determinants into recipient cells, thus altering their metastatic and antigenic properties (Poste and Nicolson, 1980; Prujansky-Jakobovits et al., 1980; Volsky et al., 1981). These lines of evidence have provided provocative new information concerning the biologic properties associated with the cell surface membrane and focus on the importance of understanding the physiologic mechanisms that govern the anchorage, modulation, and shedding of cell surface molecules.

Acknowledgments

The work reported in this chapter was supported by a grant from the Medical Research Council of Canada (MA-7249). The electron microscopy work was kindly contributed by P. Hyam. Alison Hillman provided continuous and enthusiastic research efforts. Skillful and prompt secretarial help was provided by Susan Mouland.

References

Alexander, P., 1974, Escape from immune destruction by the host through shedding of surface antigens: Is this a characteristic shared by malignant and embryonic cells? *Cancer Res.* **34**:2077–2882.

Black, P. H., 1980a, Shedding from the cell surface of normal and cancer cells, *Adv. Cancer Res.* **32**:75–199.

Black, P. H., 1980b, Shedding from normal and cancer-cell surfaces, *N. Engl. J. Med.* **303**:1415–1416.

Borisy, G. G., Olmstead, J. B., and Klugman, R. A., 1972, In vitro aggregation of cytoplasmic microtubule subunits, *Proc. Natl. Acad. Sci. USA* **69**:2890–2894.

Boyse, E. A., Old, L. J., and Stockert, E., 1965, The TL (Thymus leukemia) antigen: A review, in: *Immunopathology*, 4th Int. Symp. Monte Carlo (P. Grabar and P. A. Miescher, eds.), pp. 23–40, Grune and Stratton, Inc. N.Y.

Brinkley, B. R., Fuller, G. M., and Highfield, D. P., 1975, Cytoplasmic microtubules and transformed cells in culture: Analysis by tubulin antibody immunofluorescence, *Proc. Natl. Acad. Sci. USA* **72**:4981–4985.

Britch, M., and Allen, T. D., 1981, The effects of cytochalasin B on the cytoplasmic contractile network revealed by whole-cell transmission electron microscopy, *Exp. Cell Res.* **131**:161–172.

Davey, G. C., Currie, G. A., and Alexander, P., 1976, Spontaneous shedding and antibody induced modulation of histocompatibility antigens on murine lymphomata: Correlation with metastatic capacity, *Br. J. Cancer* **33**:9–14.

De Broe, M. E., Wieme, R. J., Logghe, G. N., and Roels, F., 1977, Spontaneous shedding of plasma membrane fragments by human cells in vivo and in vitro, *Clin. Chim. Acta* **81**:237–245.

Dedman, J. R., Brinkley, B. R., and Means, A. R., 1979, Regulation of microfilaments and microtubules by calcium and cyclic AMP, in: *Advances in Cyclic Nucleotide Research*, Volume II (P. Greengard and G. A. Robison, eds.), Raven Press, New York, pp. 131–174.

Edelman, G. M., 1976, Surface modulation in cell recognition and cell growth, *Science* **192**:218–226.

Gasic, G. J., Boettiger, D., Catalfamo, J. L., Gasic, T. B., and Stewart, G. J., 1978, Aggregation of platelets and cell membrane vesiculation by rat cells transformed in vitro by Rous Sarcoma Virus, *Cancer Res.* **38**:2950–2955.

Godman, G. C., and Miranda, A. F., 1978, Cellular contractibility and the visible effects of cytochalasins, in: *Cytochalasins—Biochemical and Cell Biological Aspects*, North-Holland Research Monographs: Frontiers of Biology, Volume 46 (S. W. Tanenbaum, ed.), Elsevier/North-Holland Biomedical Press, New York, pp. 278–429.

Godman, G. C., Miranda, A. F., Deitch, A. D., and Tanenbaum, S. W., 1975, Action of cytochalasin-D on cells of established lines, *J. Cell Biol.* **64**:644–667.

Hart, I. R., Raz, A., and Fidler, I. J., 1980, Effects of cytoskeleton-disrupting agents on the metastatic behavior of melanoma cells, *J. Natl. Cancer Inst.* **64**:891–900.

Hoerl, B. J., and Scott, R. E., 1978, Plasma membrane vesiculation: A cellular response to injury, *Virchows Arch. B Cell Path.* **27**:335–345.

Jaken, S., and Black, P. H., 1979, Differences in intracellular distribution of plasminogen activator in growing, confluent, and transformed 3T3 cells, *Proc. Natl. Acad. Sci. USA* **76**:246–250.

Jones, P. A., Laug, W. E., and Benedict, W. F., 1975, Fibrinolytic activity in a human fibrosarcoma cell line and evidence for the induction of plasminogen activator secretion during tumor formation, *Cell* **6**:245–252.

Katz, J. J., and Crespi, H. L., 1970, Isotope effects in biological systems, in: *Isotope Effects in Chemical Reactions*, ACS Monograph 167 (C. J. Collins and N. S. Bowman, eds.), Van Nostrand Reinhold, New York, pp. 266–363.

Kim, U., Baumler, A., Carruthers, C., and Bielat, K., 1975, Immunological escape mechanism in spontaneously metastasizing mammary tumors, *Proc. Natl. Adac. Sci. USA* **72**:1012–1016.

Kirkland, W. L., and Burton, P. R., 1972, Dibutyryl cyclic AMP-mediated stabilization of mouse neuroblastoma cell neurite microtubules exposed to low temperature, *Nature New. Biol.* **240**:205–207.

Koch, G. L. E., and Smith, M. J., 1978, An association between actin and the major histocompatibility antigen H-2, *Nature* **273**:274–278.

Liepins, A., and Hillman, A. J., 1981, Shedding of tumor cell surface membranes, *Cell Biol. Internatl. Reports* **5**:15–26.

Liepins, A., Faanes, R. B., Choi, Y. S., and de Harven, E., 1978, T-lymphocyte mediated lysis of tumor cells in the presence of alloantiserum, *Cell. Immunol.* **36**:331–344.

Meek, W. D., and Puck, T. T., 1979, Role of the microfibrilar system in knob action of transformed cells, *J. Supramol. Struct.* **12**:335–354.

Nicolson, G. L., Smith, J. R., and Poste, G., 1976, Effects of local anesthetics on cell morphology and membrane associated cytoskeletal organization in BALB/3T3 cells, *J. Cell Biol.* **68:**395–402.
Nordquist, R. E., Anglin, J. H, and Lerner, M. P., 1977, Antibody induced antigen redistribution and shedding from human breast cancer cells, *Science* **197:**366–367.
Nowotny, A., Groshman, J., Abdelnoor, A., Rote, N., Cynara Yang, Waltersdorff, R., 1974, Escape of TA3 tumors from allogeneic immune rejection: Theory and experiments, *Eur. J. Immunol.* **4:**73–78.
Ostlund, R. E., Leung, J. T., and Hajek, S. V., 1980, Regulation of microtubule assembly in cultured fibroblasts, *J. Cell Biol.* **85:**386–391.
Pearson, G. R., and Scott, R. E., 1977, Isolation of virus-free Herpes virus saimiri antigen-positive plasma membrane vesicles, *Proc. Natl. Acad. Sci. USA* **74:**2546–2550.
Petitou, M., Tuy, F., Rosenfeld, C., Mishal, Z., Paintrand, M., Jasnin, C., Mathe, G., and Inbar, M., 1978, Decreased microviscosity of membrane lipids in leukemic cells: Two possible mechanisms, *Proc. Natl. Acad. Sci. USA* **75:**2306–2310.
Porter, K. R., Puck, T. T., Hsie, A. W., and Kelly, D., 1974, An electron microscope study of the effects of dibutyryl cyclic AMP on Chinese hamster ovary cells, *Cell* **2:**145–162.
Poskitt, P. K. F., Poskitt, T. R., and Wallace, J. H., 1976, Release into culture medium of membrane-associated, tumor-specific antigen by B-16 melanoma cells (39332), *Proc. R. Soc. Exp. Biol. Med.* **152:**76–80.
Poste, G., and Nicolson, G. L., 1980, Arrest and metastasis of blood-borne tumor cells are modified by fusion of plasma membrane vesicles from highly metastatic cells, *Proc. Natl. Acad. Sci. USA* **77:**399–403.
Prujansky-Jakobovits, A., Volsky, D. J., Loyter, A., and Sharon, N., 1980, Alteration of lymphocyte surface properties by insertion of foreign functional components of plasma membrane, *Proc. Natl. Acad. Sci. USA* **77:**7247–7251.
Puck, T. T., 1977, Cyclic AMP, the microtubule–microfilament system, and cancer, *Proc. Natl. Acad. Sci. USA* **74:**4491–4495.
Quigley, J. P., 1976, Association of a protease (plasminogen activator) with a specific membrane fraction isolated from transformed cells, *J. Cell Biol.* **71:**472–486.
Raz, A., Barzilai, R., Spira, G., and Inbar, M., 1978, Oncogenicity and immunogenicity associated with membranes from cell-free ascites fluid of lymphoma-bearing mice, *Cancer Res.* **38:**2480–2485.
Roozemond, R. C., and Urli, D. C., 1981, Fluorescence polarization studies and biochemical properties of membranes exfoliated from the cell surface of rabbit thymocytes in situ, *Biochim. Biophys. Acta* **643:**327–338.
Scott, R. E., 1976, Plasma membrane vesiculation: A new technique for the isolation of plasma membranes, *Science* **194:**743–745.
Sundqvist, K.-G., and Ehrnst, A. 1976, Cytoskeletal control of surface membrane mobility, *Nature* **264:**226–231.
Trump, B. F., Berezesky, J. K., Laiho, K. U., Osormio, A. R., Mergner, W. J., and Smith, M. W., 1980, The role of calcium in cell injury. A review, in: *Scanning Electron Microscopy II*, SEM Inc., AMF O'Hare (Chicago), Illinois, pp. 437–462.
Van Blitterswijk, W. J., Emmelot, P., Hilkmann, M., Hilgers, J., and Feltkamp, C. A., 1979, Rigid plasma-mambrane derived vesicles, enriched in tumor-associated surface antigens (MLr), occurring in the ascites fluid of a murine leukemia (GRSL), *Int. J. Cancer* **23:**62–70.
Virtanen, J., and Mietinen, A., 1980, The role of actin in the surface integrity of cultured rat liver parenchymal cells, *Cell Biol. International Reports* **4:**29–36.
Volsky, D. J., Ahrlund-Richter, L., Dalianis, T., and Klein, G., 1981, Implantation of mouse histocompatibility antigens into membranes of cultured tumor cells, *Eur. J. Immunol.* **11:**341–344.
Walsh, F. S., Barber, B. H., and Crumpton, M. J., 1976, Preparation of inside-out vesicles of pig lymphocyte plasma membrane, *Biochemistry* **15:**3557–3563.
Zeligs, J. D., and Wollman, S. H., 1977, Ultrastructure of blebbing and phagocytosis of blebs by hyperplastic thyroid epithelial cells in vivo, *J. Cell. Biol.* **72:**584–594.

Chapter 17
Production of Microcytospheres

GERD G. MAUL and JOSEF WEIBEL

1. Introduction

Cytoplasmic vesicles can be isolated by several methods which utilize agents with cross-linking capabilities, formaldehyde or glutaraldehyde (Scott et al., 1979). These partially fixed vesicles are unable to attach to a substrate. Since, in certain types of experiments, cytoplasmic vesicles containing only the active ribosomal components and a normal cell membrane are desirable, a method was needed to ensure minimal contamination of vesicles with other organelles. Cytochalasin B has been used in the past to break down the microfilaments associated with the cell membrane (Carter, 1967) and to enucleate cells (Prescott et al., 1971; Veomett et al., 1976; Wigler and Weinstein, 1975; Gopalakrishnan and Tompson, 1975). The resulting cytoplasts can then be fused with cells or karyoplasts of different origin (Shay, 1977). However, these cytoplasts contain all of the cellular organelles. We describe here a procedure that is speedy, sterile, and applicable to a variety of experiments in which it is essential that cytoplasmic parts of one cell type containing only defined components are fused with other cells. These types of cytoplasmic vesicles are named microcytospheres.

2. Materials and Methods

Microcytospheres were generated from suspension cultures of mouse L cells or HeLa S_3 cells. High cytochalasin B concentrations were used to induce a blebbing of the cell membrane. These blebs or vesicles can be separated from the cell by agitation and then collected by low-g-force centrifugation.

Cells at any concentration begin to bleb immediately if cytochalasin B is added to a concentration of 10 µg/ml and if the temperature remains above 25°C. Cold cells will not respond. To conserve cytochalasin B, washed cells are pelleted and resuspended in about three to four times their packed volume in prewarmed complete medium or medium without serum to which cytocha-

---

GERD G. MAUL and JOSEF WEIBEL • Wistar Institute of Anatomy and Biology, Philadelphia, Pennsylvania 19104.

lasin B is added. After 1–2 min, the cells look like mulberries as viewed in phase contrast microscopy. The cells are then agitated with a Vortex for about 2 min to remove the vesicles. Gentle homogenization in a Dounce homogenizer results in cell damage and contamination of the vesicles with cell debris. Manual shaking of the test tube with the cells results in a low yield, but the microcytospheres are very small and have the least amount of membrane-bound cytoplasmic organelles.

The cell and microcytosphere suspension is then diluted in medium and centrifuged at setting 4 in a tabletop International centrifuge for 4 min. This will remove whole cells and any large microcytospheres. The supernatant is then placed in pointed test tubes and centrifuged at 2000 rpm for 10 min in an International centrifuge (800g). The pellet will contain all but the smallest microcytospheres.

Mitotic HeLa cells were collected by selectively shaking off the rounded cells; mitotic inhibitors were not used (Maul et al., 1972). Chromosomes were then removed from the spindle by cold treatment.

Cells to be examined by transmission electron microscopy were stabilized in 0.3% glutaraldehyde in medium without serum, pelleted, fixed with 3% glutaraldehyde in 0.1 M PIPES [piperazine-N,N'-bis(2-ethanesulfonic acid)] buffer at pH 7.4, and post-fixed in osmium tetroxide in the same buffer at room temperature. After rapid dehydration, the pellets were embedded in Epon, stained with uranyl acetate and lead citrate, and observed in a Zeiss 10A transmission electron microscope. For scanning electron microscopy, 0.3% glutaraldehyde-stabilized cells and microcytospheres were placed on poly L-lysine-covered plastic dishes and critical point-dried. They were observed in a JEOL U3 scanning electron microscope.

3. Results

HeLa cells are densely covered with microvilli in untreated cells (Fig. 1). After the addition of cytochalasin B, the cell membrane produces the vesicles. The process, which should be more or less completed in 2–3 min, can be viewed in a phase contrast microscope. The temperature should be well above 20°C. For cells growing at 37°C, the yield of microcytospheres drops precipitously as room temperature is approached. Amphibian cells, however, yield optimal amounts of microcytospheres at ~20°C. One may have to determine the optimal temperature for each mammalian cell line, because they do not react equally. There was little difference in yield for mouse L and HeLa cells as judged by pellet size at 27–37°C. All cells in a population also do not produce microcytospheres equally well; at the bottom left on Fig. 2, there are three cells with no vesicles on their surfaces. Microvilli do not disappear during the initial 5–10 min, but if incubated for longer periods (30 min) in the presence of cytochalasin B, they begin to diminish, and the vesicles not removed by agitation begin to cap (Fig. 3). ATP seems necessary for the formation of microcytospheres, since inhibitors of oxidative phosphorylation

Production of Microcytospheres

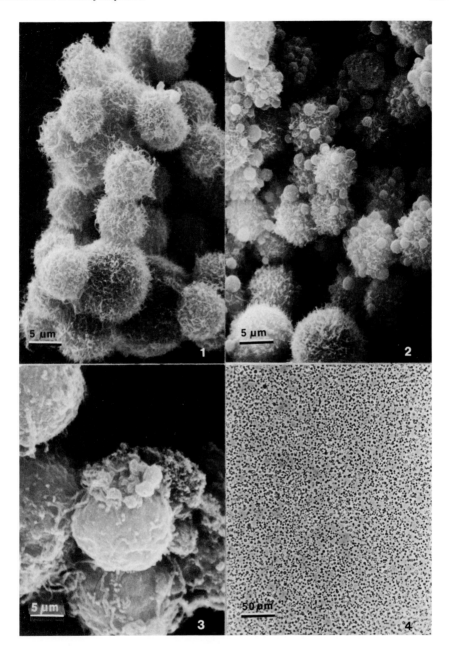

Figure 1. Control HeLa S_3 cells with many microvilli. Size marker, 5 μm.

Figure 2. Cytochalasin B-treated cells fixed 60 sec after the addition of the drug. Vesicles have formed, but microvilli are still present. Size marker, 5 μm.

Figure 3. Cells after vortexing and centrifugation were fixed 10 min after the addition of the drug. Few microvilli are present and the remaining vesicles have capped. Size marker, 5 μm.

Figure 4. Phase contrast micrograph of isolated microcytospheres. Size marker, 50 μm.

(antimycin A, 12.5 µg/ml) prevented vesicle formation if added to the cell suspension 10 min before the addition of cytochalasin B.

After only one cycle of centrifugation, essentially no whole cells remain as observed in phase contrast microscopy (Fig. 4). The pellets were tested for potential low-level contamination with whole cells by plating microcytospheres from 10^8 cells in a petri dish. As a control, approximately 100 cells, pretreated with cytochalasin B, were added to microcytospheres in an extra plate. After 1 week, many colonies could be seen in the control, but none in the dish containing microcytospheres only. We recovered about 27 vesicles per cell (HeLa) as counted by phase microscopy. This is a third more than one can count on scanning micrographs, assuming one sees half on the vesicles. This discrepancy may come about by loss of cells during preparation. An estimate of surface membranes recovered from the total is not possible, due to the large error in estimating the microvilli-covered surface of the HeLa cells. A comparison of one-dimensional sodium dodecyl sulfate (SDS) gel electrophoretic patterns shows no difference in the Coomassi Blue recognizable bands despite the absence of a nuclear component in the microcytospheres (data not shown).

The microcytospheres contain mostly ribosomes (Fig. 5). There are few mitochondria and lipid droplets in any of the small spheres, but some contain single, rough-surfaced endoplasmic reticulum cisternae. About 10% of the microcytospheres appear less dense than normal cytoplasm. They are considered damaged, but must have the ability to reseal, since ferritin was excluded if mixed with the microcytospheres before fixation. After 24 hr in fresh medium with serum, only 25% have retained their original density, nearly half appear damaged as judged by their reduced density, and the remainder appear denser than the control. If the microcytospheres are left to settle on a plastic dish, they will attach like a cell. This was determined by examining sections cut perpendicular to the plane of the plastic dish (data not shown). But many microcytospheres can be washed away with a gentle stream of medium. No microfilaments are detected in the attached microcytospheres. Mitotic cells, if pretreated to remove the chromosomes from the spindle, yield cell fragments with single chromosomes (Fig. 6).

If HeLa cells infected with adenovirus are treated with high concentrations of cytochalasin B, the microcytospheres obtained will contain the virus (Fig. 7).

Each preparation should be examined in the phase contrast microscope. Whole cells may remain after the first slow-speed spin due to accidental resuspension of the cell pellet during removal of the microcytosphere suspension, and additional centrifugation at slow and high speed should be done. Because some vesicles are lost during the slow-speed spin, the preparation should be underlaid with medium containing just enough sterile glycerol or sucrose (~5%) to facilitate the separation of top and bottom layers.

Larger and smaller microcytospheres can be fractionated by resuspending the pellet in 5% sucrose in complete medium and underlaying it with 10% sucrose in medium. Centrifugation at 1000 rpm for 10 min leaves

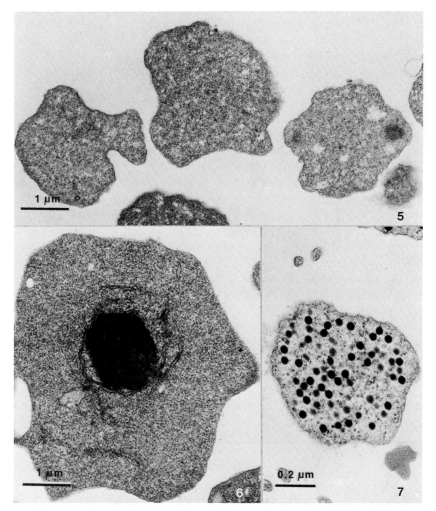

Figure 5. Cross section of microcytospheres fixed 30 min after the addition of the drug. Size marker, 1 μm.

Figure 6. Microcytosphere derived from a mitotic cell suspension. Size marker, 1 μm.

Figure 7. Microcytosphere produced from HeLa cells infected with adenovirus 5. Size marker, 0.2 μm.

microcytospheres in the 5% sucrose-containing medium which are essentially free of membrane-bound cell organelles.

4. Discussion

Cytochalasin B has been used to divide cells into a nuclear and a cytoplasmic component (Carter, 1967). Initially, this was possible only for cells growing as monolayers (Prescott et al., 1971; Veomett et al., 1976), but suspension culture cells have also been successfully enucleated (Wigler and Weinstein, 1975; Gopalakrishnan and Tompson, 1975) and chromosomes removed from mitotic cells (Sunkara et al., 1977). Ficoll density gradients and high-speed centrifugation are necessary for these procedures. Our objective was to obtain, under mild conditions, large quantities of cytoplasts containing little or no membrane-bound cytoplasmic organelles. The use of various aldehydes to induce plasma membrane vesiculation has been described (Scott et al., 1979; Scott and Maercklein, 1979). Aside from their fixation properties, those agents did not give any satisfactory results in suspension cultures.

Vesiculation on the surface of cells was observed at high (10 μg/ml) concentrations of cytochalasin B. This vesiculation was energy and temperature dependent and resulted in the production of large quantities of microcytospheres. They can be fractionated into different size classes, smaller ones nearly devoid of membrane-bound cell organelles, and larger ones containing only single rough endoplasmic reticulum (RER) cisternae and an occasional mitochondrion. Advantages of this technique are the speed (about 15 min), elimination of high-speed centrifugation, and the low number of transfers and solutions necessary. Most importantly, the active reagent, cytochalasin B, can induce vesicle formation in suspension cultures, unlike the aldehydes, and its effects are reversible. These microcytospheres can therefore be used like cytoplasts for fusion experiments [see Clark and Shay (1982) for review]. Other potential applications in fusion research are suggested by showing that the adenovirus 5-infected HeLa cells produce vesicles filled with the virions, and that microcytospheres containing single chromosomes can be generated.

Acknowledgments

The expert technical assistance of Maria Obrocka and Lynn Zukosky is acknowledged.

This work was supported by U. S. Public Health Service research grant GM-21615 from the National Institute of General Medical Sciences and CA-10815 from the National Cancer Institute.

References

Carter, S. B., 1967, Effects of cytochalasins on mammalian cells, *Nature* **213**:261–264.
Clark, M. A., and Shay, J. W., 1982, Long lived cytoplasmic factors that suppress adrenal steroidogenesis, *Proc. Natl. Acad. Sci USA* **79**:1144–1148.
Gopalakrishnan, T. V., and Tompson, E. B., 1975, A method for enucleating cultured mammalian cells, *Exp. Cell Res.* **96**:435–439.
Maul, G. G., Maul, H. M., Scogna, J. E., Lieberman, M. W., Stein, G. S., Hsu, B. Y., and Borun, T. W., 1972, Time sequence of nuclear pore formation in phytohemagglutinin-stimulated lymphocytes and in HeLa cells during the cell cycle, *J. Cell. Biol.* **55**:433–447.
Prescott, D. M., Myerson, D., and Wallace, J., 1971, Enucleation of mammalian cells with cytochalasin B, *Exp. Cell Res.* **71**:480–485.

Scott, R. E., and Maercklein, P. B., 1979, Plasma membrane vesiculation in 3T3 and SV3T3 cells: Factors affecting the process of vesiculation, *J. Cell Sci.* **35**:245–252.

Scott, R. E., Perkins, R. G., Zschunke, M. A., Hoerl, B. J., and Maercklein, P. B., 1979, Plasma membrane vesiculation in 3T3 and SV3T3 cells: Morphological and biochemical characterization, *J. Cell Sci.* **35**:229–243.

Shay, J. W., 1977, Selection of reconstituted cells from karyoplasts fused to chloramphenicol-resistant cytoplasts, *Proc. Natl. Acad. Sci. USA* **74**:2461–2464.

Sunkara, P. S., Al-Bader, A. A., and Rao, P. N., 1977, Mitoplasts: Mitotic cells minus the chromosomes, *Exp. Cell Res.* **107**:444–447.

Veomett, G., Shay, J. W., Hough, P. V., and Prescott, D. M., 1976, Large-scale enucleation of mammalian cells, in: *Methods in Cell Biology*, Volume XIII (D. M. Prescott, ed.), Academic Press, New York, pp. 1–6.

Wigler, M. H., and Weinstein, J. B., 1975, A preparative method for obtaining enucleated mammalian cells, *Biochem. Biophys. Res. Commun.* **63**:669–674.

Chapter 18
Isolation and Characterization of Mitoplasts

POTU N. RAO, PRASAD S. SUNKARA, and
ABDULLATIF A. AL-BADER

1. Introduction

The discovery by Carter (1967) that cytochalasin B (CB), a metabolite produced by the fungus *Helminthosporium dematiodeum*, causes nuclear extrusion and spontaneous enucleation in a small proportion of treated cells has led to the development of a number of techniques to study nucleocytoplasmic interactions, cytoplasmic inheritance, and the creation of reconstituted cells. CB has a profound effect on the morphology of mammalian cells in culture and is best known for blocking cytokinesis without affecting the karyokinesis or the nuclear division that usually leads to the formation of binucleate cells (Carter, 1967; Ladda and Estensen, 1970). However, the unique feature of CB, its induction of spontaneous nuclear extrusion, has been exploited by the cell biologist to the maximum extent.

The frequency of enucleation in mammalian cells has been increased to almost 100% by combining CB treatment with high-speed centrifugation (Prescott et al., 1971; Veomett et al., 1976). By this procedure, it is possible to separate a monolayer culture into two fractions: (1) the cytoplasts (the enucleated cells), which remain attached to the bottom of the dish; (2) the karyoplasts (the nuclei surrounded by a thin shell of cytoplasm enclosed within an outer cell membrane), which collect at the bottom of the centrifuge tube. The cytoplasmic content of a karyoplast may be less than 10% of that of a whole cell. Special methods have been developed to enucleate cells that do not naturally attach to the culture dish (Wigler and Weinstein, 1975; Gopalakrishnan and Thompson, 1975). In all these studies, only interphase cells were used for enucleation studies.

For the present study, we were especially interested in enucleating

---

POTU N. RAO • Department of Developmental Therapeutics, University of Texas System Cancer Center, M. D. Anderson Hospital and Tumor Institute, Houston, Texas 77030. PRASAD S. SUNKARA • Merrell Research Center, Cincinnati, Ohio 45215. ABDULLATIF A. AL-BADER • Department of Pathology, Faculty of Medicine, Kuwait University, Kuwait.

mitotic cells, i.e., removing the mitotic apparatus consisting of the spindle and the chromosomes. Cytoplasms from mitotic cells (mitoplasts) that are not contaminated with chromosomes may be useful in studying the nature of the mitotic factors involved in chromosome condensation.

Usually chromosomes in eukaryotic cells are visualized either during mitosis or meiosis. However, cell fusion between mitotic and interphase cells induced by either UV-inactivated Sendai virus or polyethylene glycol (PEG) results in the transformation of the interphase nucleus into discrete chromosomes under the influence of the factors (proteins) present in the mitotic cell (Johnson and Rao, 1970). The chromosomes resulting from the interphase nucleus are called prematurely condensed chromosomes (PCC). The morphology of the PCC depends on the position of the interphase cell in the cell cycle at the time of fusion with the mitotic cell. G1-PCC have single chromatids, S-PCC exhibit a pulverized appearance, and G2-PCC consist of two chromatids and appear very similar to prophase chromosomes (Johnson and Rao, 1970).

Rao and Johnson (1974) have demonstrated that, during the induction of premature chromosome condensation, proteins from the mitotic cell that were prelabeled with ^{3}H-amino acids become associated with the PCC of the unlabeled interphase cell. The nature of these factors is not yet clearly understood, because of the difficulties in obtaining pure preparations of PCC without contaminating mitotic chromosomes. Since the method of PCC induction involves fusion of interphase cells with mitotic cells, it is almost impossible to separate the PCC from mitotic chromosomes completely. Differential gradient centrifugation of the mixtures of PCC and metaphase chromosomes failed to give satisfactory results.

We believed that the alternative approach was to induce PCC by using mitoplasts instead of mitotic cells as the inducer cell. Such a fusion between mitoplasts and interphase cells would yield PCC and nuclei from the uninduced and unfused interphase cells. Separation of PCC from the nuclei by standard methods should be relatively easier and thus we could obtain PCC in reasonably pure and sufficiently large quantities for biochemical analysis of proteins associated with the PCC. We could then compare them with those found in interphase nuclei. Such a comparative study might help us understand the nature of the factors involved in chromosome condensation and cell division.

Hence the object of this study was to obtain mitoplasts in large quantities and study their characteristics, particularly their ability to induce premature chromosome condensation when fused with interphase cells.

2. Isolation of Mitoplasts

2.1. Cells and Cell Synchrony

HeLa cells were routinely grown as a suspension culture at 37°C in Eagle's minimal essential medium (MEM) supplemented with nonessential amino acids, sodium pyruvate, and glutamine (1% each) and 10% heat-inactivated fetal calf serum. The cells have a cell cycle time of 22 hr with a G1 period of 10.5 hr, S period of 7 hr, G2 period of 3.5 hr, and mitosis of 1.0 hr duration (Rao and Engelberg, 1966).

HeLa cells were synchronized into mitosis by the following method. An exponentially growing suspension culture was partially synchronized by a single excess (2.5 mM) thymidine (TdR) block of 16-hr duration. Then the TdR block was removed by centrifuging the cells and resuspending them in regular culture medium. The resuspended cells were plated in 150-mm plastic culture dishes and incubated in humidified CO_2 (5% in air) incubator at 37°C for 4 hr, during which time the cells attached to the dish. At the end of this incubation period, the unattached and floating cells were removed by aspiration and fresh medium was added to the dishes. The dishes were then kept in a pressure chamber filled with N_2O at 80 psi at 37°C. Ten hours later, N_2O was released and the rounded and loosely attached mitotic cells were harvested by selective detachment, i.e., either by gentle tapping or pipetting (Rao, 1968; Rao and Johnson, 1972). Cells thus obtained had a mitotic index of about 98%.

The N_2O-blocked mitotic HeLa cells displayed a typical colchicine metaphase (C-metaphase) configuration when examined by light microscopy. When the cells were returned to a 37°C incubator, mitosis resumed as the cells progressed from metaphase to anaphase and telophase within 1.5 hr after the reversal of the N_2O block. Ultrastructural studies of the N_2O-blocked cells revealed a bipolar spindle with centriole pairs at each pole (Brinkley and Rao, 1973).

We have also collected mitotic cells by using Colcemid (0.05 µg/ml) instead of N_2O block. However, Colcemid-arrested mitotic cells were devoid of mitotic spindles and were irreversibly blocked in C-metaphase.

2.2. Chemicals

A stock solution of CB (Aldrich Chemical Co.) was prepared by dissolving it in dimethyl sulfoxide (4.0 mg/ml), and stored at 4°C. Ficoll (Pharmacia Fine Chemicals) was dissolved by stirring in Eagle's MEM without serum to a final concentration of 25% (w/v), sterilized by autoclaving, and stored at 4°C. [³H]uridine (5–15 Ci/mmole) was obtained from New England Nuclear (Boston, Massachusetts).

2.3. Extrusion of Chromosomes from Mitotic Cells

We followed with slight modification the method of Wigler and Weinstein (1975) for the enucleation of cells in suspension using Ficoll density gradients (see also Veomett, Chapter 6, this volume). The density gradients were prepared as follows: cellulose nitrate tubes (38 ml) were sterilized by UV light and then carefully filled with the following layers of Ficoll: 4 ml of 25%, 4 ml of 17%, 1 ml of 16%, 1 ml of 15%, and 4 ml of 12.5%, all in complete MEM containing 10 µg/ml of CB and preequilibrated with CO_2 to pH 7.0–7.2. Mitotic cells were collected, centrifuged at 1000g for 10 min, and resuspended in 12.5% Ficoll in complete MEM containing 10 µg/ml of CB to a final concentration of 1×10^7 cells in 3 ml. This cell suspension, essentially free of cell clumps, was applied on the already prepared gradients and overlayed with 20 ml of MEM containing CB. The gradients were then centrifuged in a Beckman SW 27 swinging bucket rotor in a Beckman L2 ultracentrifuge for 1 hr at 25,000 rpm and 36°C. At the end of centrifugation, cell fractions at appropriate banding interfaces were carefully aspirated with a Pasteur pipette from the top of the tube. The banding fractions were diluted with 15 ml of McCoy's 5A modified medium supplemented with 16% heat-inactivated fetal calf serum and harvested by centrifugation at 1000g for 10 min. The cell pellets were washed twice, then resuspended in McCoy's medium, plated in petri dishes, and incubated at 37°C in a humidified CO_2 incubator. The purity of the mitoplasts was determined by cytocentrifuge preparations. The cytocentrifuge preparations were fixed in methanol–acetic acid (3:1), air-dried, and stained with Giemsa (Sunkara et al., 1977).

At the end of centrifugation of mitotic HeLa cells in a Ficoll gradient, three major bands of cell fractions were observed. The band at 0–12.5% Ficoll region contained mainly cell debris. The 15–17% Ficoll interface contained the mitoplasts (Figs. 1A and 1B), and the 17–25% region contained primarily chromosome clusters surrounded by a thin film of cytoplasm (Fig. 1C). Routinely, we recovered 50–60% of cells layered on the gradient as mitoplasts, with a purity of 90–95%. The optimal cell load on the Ficoll gradient was found to be 1×10^7 cells. At low cell densities, recovery of the mitoplasts was poor, whereas higher densities yielded mitoplasts of low purity. Furthermore, cells blocked in mitosis by N_2O yielded mitoplasts of greater purity than those blocked by Colcemid. By using all six buckets of the SW 27 rotor, 3×10^7 mitoplasts could be obtained in a single preparation.

3. Characterization of the Mitoplasts

3.1. Morphologic Features

Like mitotic cells, the mitoplasts, upon plating in culture dishes, remained spherical and unattached even after a prolonged period (20 hr) of incubation. They remained intact and viable, as shown by the dye (trypan blue) exclusion method. The mitoplasts obtained from Colcemid-arrested mitotic cells contained occasionally one or two chromosomes, whereas the mitoplasts from the N_2O-blocked mitotic cells remained relatively free of chromosomes (Figs. 1A and 1B).

The CB treatment and centrifugation of both Colcemid- and N_2O-arrested mitotic cells resulted in the extrusion of chromosomes in clusters. The two mitotic inhibitors, Colcemid and N_2O, differ in their effects on the mitotic apparatus. Colcemid causes the dissolution of the mitotic spindle, leading to the formation of C-mitosis (Eigsti and Dustin, 1955). The mitotic spindle is present in N_2O-blocked mitotic cells, but it is slightly deformed (Brinkley and

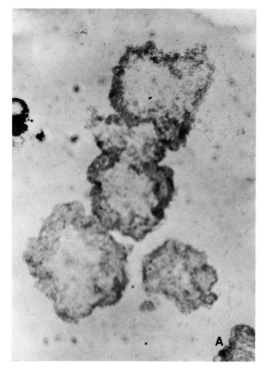

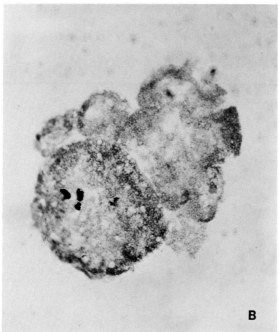

Figure 1. (A) Mitoplasts obtained from N_2O-blocked mitotic HeLa cells. (B) Mitoplasts of Colcemid-blocked mitotic cells. Two chromosomes have been retained in one of the mitoplasts. (C) The extruded chromosome clusters from Colcemid-blocked mitotic cells. (D) Mitotic cells synchronized by selective detachment after Colcemid treatment. ×630. [From Sunkara et al. (1977).]

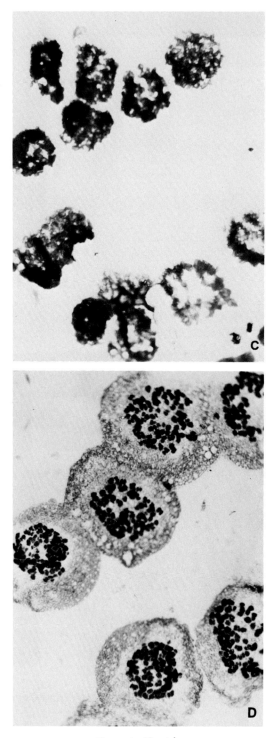

Figure 1. (Cont.)

Rao, 1973). The fact that chromosomes are extruded from the Colcemid-blocked cells as clusters rather than separate entities suggests that some residual ultrastructure holds the C-mitotic chromosomes together. These results suggest that this technique could be used to isolate relatively pure and intact mitotic apparatus in large quantities for biochemical studies.

3.2. Metabolic Activity of the Mitoplasts

Protein and RNA synthesis of mitoplasts were determined by measuring the incorporation of ^{3}H-labeled leucine and uridine (1 μCi/ml), respectively, for 2 hr. The N_2O-arrested mitotic cells used in these labeling studies had undergone ultracentrifugation in Ficoll density gradients similar to mitoplasts, except that there was no CB in the medium. This treatment appeared to delay the division of these cells by 3 hr. The amount of label incorporated into acid-precipitable material was determined according to the method of Mans and Novelli (1961).

The results of this study indicate that the mitoplasts are able to synthesize RNA and protein, as indicated by the incorporation of [^{3}H]uridine and [^{3}H]leucine (Table I). These results indicate little difference between the mitoplasts and the mitotic cells. Since the rates of RNA and protein synthesis are at a minimum during mitosis (Prescott, 1976), the extrusion of chromosomes does not seem to affect these processes. Hence the RNA and protein synthesis observed with the mitoplasts might reflect the mitochondrial activity. However, we have no explanation for the increased incorporation of uridine by the Colcemid-blocked cells in comparison with those blocked by N_2O.

4. Induction of Premature Chromosome Condensation (PCC) with Mitoplasts

Our initial attempts to induce PCC by fusing mitoplasts with interphase cells with the help of UV-inactivated Sendai virus were not successful. It appeared that the membranes of the mitoplasts were easily disrupted during the fusion procedures.

Table I. RNA and Protein Synthesis in Mitoplasts and Mitotic Cells[a] (cpm/10^6 cells)

Incorporation of	Mitotic cells blocked by		Mitoplasts blocked by	
	N_2O	Colcemid	N_2O	Colcemid
[^{3}H]Leucine	283	293	433	283
[^{3}H]Uridine	1861	3016	1604	2993

[a] Average of two experiments [From Sunkara et al. (1977)].

Subsequently, we stabilized the mitoplasts with spermine, a naturally occurring polyamine known to strengthen bacterial membranes (Tabor, 1962). After the removal of chromosomes, the mitoplasts were resuspended in MEM plus spermine (25 μM), plated in 35-mm plastic culture dishes, and incubated for 1 hr in a humidified CO_2 incubator at 37°C. The spermine-treated mitoplasts were then fused with interphase cells using UV-inactivated Sendai virus as described earlier (Rao and Johnson, 1970).

The chromosome preparations were made as follows: The fusion mixture was suspended in 0.075 M potassium chloride and left at room temperature for 10 min. At the end of incubation, the cell suspension was centrifuged and the cell pellet was resuspended in Carnoy's fixative (three parts absolute methanol and one part glacial acetic acid) for 5 min and kept at room temperature. The cells were centrifuged and the cell pellet was again resuspended in 5 ml of Carnoy's solution and then spun down immediately. Two to three drops of Carnoy's solution was added to the pellet, the cells resuspended by gentle tapping, and then the cells were dropped on wet slides. The chromosome preparations were stained with Giemsa and scored for PCC.

Conventional fusion procedures involving 15-min incubation at 4°C and 45 min at 37°C tended to break the mitoplasts. Hence, we standardized the fusion procedure to get better yields of PCC with mitoplasts (Sunkara et al., 1980). This involved the following: The mitoplasts and the interphase cells (2×10^6 each) were spun together at 800 rpm for 5 min and washed in MEM without serum. The supernatant was removed and 0.5 ml of UV-inactivated Sendai virus was added. It was kept at 4°C for 5 min, followed by a 10-min incubation at 37°C. After this, two drops of heat-inactivated fetal calf sera were added to each fusion mixture. We were able to obtain pure PCC by fusion between mitoplasts and interphase cells (Fig. 2). About 5% of the fusion products of the mitoplasts and interphase cells exhibited PCC.

5. Conclusions

The results of this study indicate the following:

1. Mitotic apparatus and chromosomes could be extruded from mitotic HeLa cells arrested by N_2O or Colcemid by combining CB treatment with centrifugation on a Ficoll gradient.

2. The chromosomes were extruded as a cluster but not as single chromosomal entities even in Colcemid-blocked mitotic cells. This suggests that even though the mitotic apparatus was dissolved, some infrastructure held the chromosomes together.

3. Mitoplasts can be obtained in relatively large amounts (3×10^7 mitoplasts in a single preparation) for biochemical studies.

4. Like mitotic cells, mitoplasts do not attach to the dish even after 20 hr incubation.

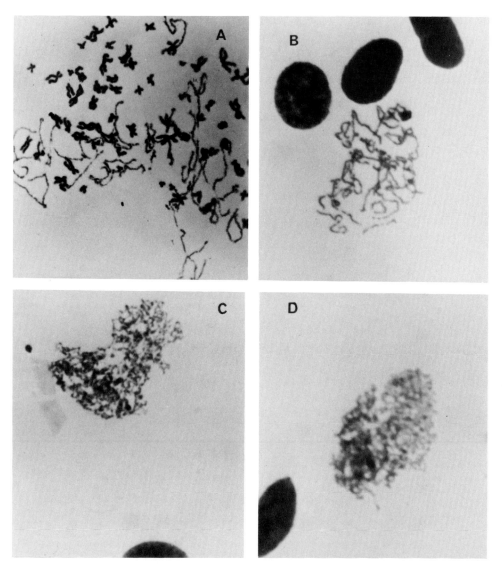

Figure 2. (A) G1 PCC induced by fusion between a mitotic cell and a G1 cell. The darkly stained chromosomes with two chromatids are mitotic chromosomes. The G1 PCC are thin, long chromosomes with one chromatid. (B) Early, (C) mid, and (D) late G1 PCC were obtained by a fusion between HeLa mitoplasts and HeLa G1 cells. [Sunkara et al. (1980).]

5. Premature chromosome condensation can be induced in interphase cells by fusing them to mitoplasts. This observation suggests that the factors responsible for chromosome condensation are present in the cytoplasm of the mitotic cells. By this technique it should be possible to obtain large quantities of PCC, not contaminated with mitotic chromosomes, to study the nature of the factors involved in chromosome condensation and cell division.

References

Brinkley, B. R., and Rao, P. N., 1973, Nitrous oxide: Effects on the mitotic apparatus and chromosome movement in HeLa cells, *J. Cell Biol.* **58**:96–106.

Carter, S. B., 1967, Effect of cytochalasins on mammalian cells, *Nature* **213**:261–264.

Eigsti, O. J., and Dustin, P., 1955, *Colchicine in Agriculture, Medicine, Biology and Chemistry,* Iowa State University Press, Ames, Iowa.

Gopalakrishnan, T. V., and Thompson, E. B., 1975, A method for enucleating cultured mammalian cells, *Exp. Cell Res.* **96**:435–439.

Johnson, R. T., and Rao, P. N., 1970, Mammalian cell fusion. II. Induction of premature chromosome condensation in interphase nuclei, *Nature* **226**:717–722.

Ladda, R. L., and Estensen, R. D., 1970, Introduction of a heterologous nucleus into enucleated cytoplasms of cultured mouse L-cells, *Proc. Natl. Acad. Sci. USA* **67**:1528–1533.

Mans, R. J., and Novelli, G. D., 1961, Measurement of the incorporation of radio-active amino acids into protein by a filter-paper disk method, *Arch. Biochem. Biophys.* **94**:48–53.

Prescott, D. M., 1976, The cell cycle and the control of cellular reproduction, *Adv. Genet.* **18**:99–177.

Prescott, D. M., Myerson, D., and Wallace, J., 1971, Enucleation of mammalian cells with cytochalasin B, *Exp. Cell Res.* **71**:480–485.

Rao, P. N., 1968, Mitotic synchrony in mammalian cells treated with nitrous oxide at high pressure, *Science* **160**:774–776.

Rao, P. N., and Engelberg, J., 1966, Effects of temperature on the mitotic cycle of normal and synchronized mammalian cells, in: *Cell Synchrony—Biosynthetic Regulation,* (I. L. Cameron and G. M. Padilla, eds.), Academic Press, New York, pp. 332–352.

Rao, P. N., and Johnson, R. T., 1972, Cell fusion and its application to studies on the regulation of cell cycle, in: *Methods in Cell Physiology,* Volume V (D. M. Prescott, ed.), Academic Press, New York, pp. 75–126.

Rao, P. N., and Johnson, R. T., 1974, Regulation of cell cycle in hybrid cells, in: *Control of Proliferation in Animal Cells,* Cold Spring Harbor Conferences on Cell Proliferation, Volume 1, Cold Spring Harbor Laboratories, Cold Spring Harbor, pp. 785–800.

Sunkara, P. S., Al-Bader, A. A., and Rao, P. N., 1977, Mitoplasts: Mitotic cells minus the chromosomes, *Exp. Cell Res.* **107**:444–448.

Sunkara, P. S., Al-Bader, A. A., Riker, M. A., and Rao, P. N., 1980, Induction of prematurely condensed chromosomes by mitoplasts, *Cell Biol. Internatl. Rep.* **4**:1025–1029.

Tabor, C. W., 1962, Stabilization of protoplasts and spheroplasts by spermine and other polyamines, *J. Bacteriol.* **83**:1101–1111.

Veomett, G., Shay, J., Hough, P. V. C., and Prescott, D. M., 1976, Large-scale enucleation of mammalian cells, in: *Methods in Cell Biology,* Volume XIII (D. M. Prescott, ed.), Academic Press, New York, pp. 1–14.

Wigler, M. H., and Weinstein, I. B., 1975, A preparative method for obtaining enucleated mammalian cells, *Biochem. Biophys. Res. Commun.* **63**:669–674.

Chapter 19

Nuclear Transplantation with Mammalian Cells

MARGARET J. HIGHTOWER and JOSEPH J. LUCAS

1. Introduction

Early microsurgical nuclear transplantation experiments with nonmammalian systems suggested that cytoplasmic elements participate in the regulation of nuclear gene expression and replication. With cells from *Rana pipiens* (Briggs and King, 1960), *Xenopus leavis* (Gurdon, 1962), and *Drosophila melanogaster* (Okada et al., 1974), it was shown that egg cell cytoplasm could redirect the differentiative pathway of nuclei from cells at much later stages of development. Moreover, in *Stentor coeruleus*, for example, nuclear DNA synthesis likewise appeared to be regulated, at least in part, by cytoplasmic factors (deTerra, 1967). With mammalian cells, numerous somatic cell hybridization experiments demonstrated that the patterns of gene expression of two parental cell types could be stably altered when a hybrid cell containing a mixed genome was constructed [reviewed by Ringertz and Savage (1976); Lucas (1982)]. For example, rat hepatoma cells that secreted albumin were fused to mouse fibroblasts that did not. Some hybrid clones secreted only mouse *or* rat albumin, while others secreted both rat *and* mouse albumin (Peterson and Weiss, 1972). Results of this and other similar experiments showed that a complex array of interactions is possible when two very different cell types are fused. They also suggested the existence, in animal cells, of elements which can interact functionally with a foreign nucleus and either positively or negatively regulate the expression of certain genes. However, as a perusal of the literature will illustrate, use of somatic cell fusion techniques has resulted in neither the definitive identification nor determination of the molecular nature or modes of action of these putative genetic regulators. Perhaps the primary problem hindering progress in this area is the complexity of hybrid cells. They contain a mixed genome,

---

MARGARET J. HIGHTOWER • Department of Microbiology, State University of New York at Stony Brook, Stony Brook, New York 11794. Present address: Cold Spring Harbor Laboratory. Cold Spring Harbor, New York 11724.   JOSEPH J. LUCAS • Department of Microbiology, State University of New York at Stony Brook, Stony Brook, New York 11794.

often suffering massive chromosomal losses, rearrangements, and other aberrations, within a mixed cytoplasm. It was surmised that analysis of cells constructed by introduction of single nuclei into foreign cytoplasms might yield useful information.

The first major advance in developing a technique for nuclear transplantation with mammalian cells was the discovery by Carter (1967) of the unique effects of a group of fungal metabolites, called cytochalasins, on cells in culture. Cytochalasin B, perhaps by disrupting microfilament structure (Wessels et al., 1971), induces extrusion of the cell's nucleus on a thin stalk of cytoplasm. Soon after Carter's observations, it was found that application of a centrifugal force to cytochalasin-treated cultures induced most of the stalks to break, thus yielding cultures of enucleated cells, or cytoplasts (Prescott et al., 1972; Wright and Hayflick, 1972). In addition, it produced karyoplasts, nuclei surrounded by thin shells of cytoplasm and plasma membrane (Ege et al., 1974; Shay et al., 1974; Zorn et al., 1980). Veomett and colleagues (1974) showed that cytoplasts and karyoplasts could be fused together with Sendai virus to reform whole viable cells. This chapter describes modifications of these techniques by which large cultures of nuclear transplants that are free of contamination by cytoplast or karyoplast donor whole cells can be routinely prepared. In addition, methods for identification of the hybrids are described.

2. Preparation of Cytoplasts

2.1. Method of Enucleation

For most purposes, cytoplasts were prepared as monolayers on 60-mm tissue culture dishes (Falcon, Oxnard, California). However, when fluorescence microscopy was to be done on the hybrids, cytoplasts were prepared on glass disks (14-mm in diameter) that were cut from microscope slides. Enucleation of Detroit 532 cells (a human diploid fibroblast cell strain) was inefficient (50–80%) if cells were enucleated at or near confluence or senescence of the culture. Therefore, only subconfluent, low-passage cultures were used. Some cell types did not show such a dependence on density or age of culture. With the human fibroblasts, the plastic dishes or glass disks were first coated by wetting them with a solution of collagen (Ethicon Research, Somerville, New Jersey; diluted 1:25 in 0.5% acetic acid) and allowing the residual collagen to air-dry. They were then sterilized by irradiation with ultraviolet light for 15 min. With some cell types, collagen coating was necessary to increase their adherance to the substrate. Other cell types (e.g., mouse L929) did not require coated surfaces. The enucleation technique (Lipsich et al., 1978; Hightower and Lucas, 1980) is based on the procedure of Follett (1974). The dishes were placed upside down in sterile cylinders (cut from plastic centrifuge bottles) which contained 150 ml each of normal growth medium with 10 μg/ml cytochalasin B (Aldrich Chem. Co., Milwaukee, Wisconsin). After a 30 min incubation at 37°C, the bottles were transferred to

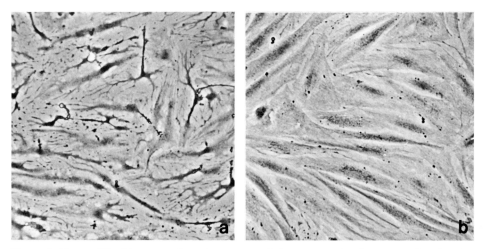

Figure 1. Human fibroblasts enucleated using cytochalasin B. (a) A culture immediately after enucleation; (b) the culture after incubation for 1 hr in normal growth medium. During this period, the cytoplasts assumed the normal morphology and cytoarchitecture of the parental whole cells.

a warmed Sorvall GSA rotor and centrifuged at 10,500 rpm (18,000g) for 50 min at 34°C. Care must be taken that the rotor temperature does not exceed 37°C. After centrifugation, the plates were removed from the bottles and the cytoplasts were permitted to recover in normal growth medium for 1-2 hr in a 37°C, 5% CO_2 incubator. After this period, one plate was stained (for example, with Giemsa stain) to determine the efficiency of enucleation. Using this procedure with Detroit 532 cells, we find that 99% of the cells are routinely enucleated. Cytoplasts were prepared on glass disks by placing the disks, cell side down, in sterilized centrifuge tubes containing 5 ml of cytochalasin-B containing medium, incubating these for 30 min at 37°C, and centrifuging in a Sorvall SS-34 rotor. Then the disks were transferred to tissue culture dishes, submerged in growth medium, and allowed to recover at 37°C for 1-2 hr. Cytoplast cultures immediately after preparation and after a brief recovery period are shown in Fig. 1.

Each different cell strain or line required a unique set of conditions—cytochalasin concentration, incubation time, centrifugation time and speed—to achieve optimal enucleation. These conditions must be empirically determined for each new cell type used.

2.2. Separation of Cytoplasts and Whole Cells

Despite repeated attempts, some cell lines could not be enucleated with high efficiency. In these cases, it was often possible to obtain pure populations of cytoplasts by separating them from remaining whole cells on density gradients of renograffin (Lipsich et al., 1979a). For example, it was found that monolayer cultures of Hepa-2 cells (a mouse hepatoma cell line) could be enucleated with an efficiency of only 50%. Cells and cytoplasts were removed from the dishes by treatment with trypsin, and a suspension of 10^6 bodies in 1 ml of phosphate-buffered saline (PBS) was prepared. This was layered onto a 14-ml, 15–30% (v/v) linear gradient of Squibb renograffin-76. The tube was centrifuged at 1000g for 5 min at 25°C. The visible, separated bands of cytoplasts and whole cells were easily removed from the tubes. Cytoplasts prepared in this way excluded the dye trypan blue, would readhere to a surface, and could be used to construct, by nuclear transplantation, whole viable cells. The particular method described here was applicable to some cell types. Altering the range of the gradient, the velocity of centrifugation, or the composition of the gradient solvent (for example, to normal growth medium with serum) aided in the preparation of other types of cytoplasts.

2.3. Some Characteristics of Cytoplasts

Many physical and biochemical properties of cytoplasts have been well studied by several laboratories. They are able to synthesize proteins (Prescott et al., 1972; Croce and Koprowski, 1973) and to support the replication of vesicular stomatitis virus (Follett et al., 1974) and polio virus (Pollack and Goldman, 1973), to support vaccinia virus-directed DNA synthesis (Prescott et al., 1971) and rabies virus-directed RNA and protein synthesis (Wiktor and Koprowski, 1974), and to rescue simian virus 40 from transformed cells (Croce and Koprowski, 1973). Goldman et al. (1973) showed that cytoplasts are capable of substrate adhesion, membrane ruffling, pinocytosis, and locomotion. In addition, they contain all of the types of organelles normally found in the cytoplasm, including centrioles (Shay et al., 1974; Zorn et al., 1980).

3. Preparation of Karyoplasts

3.1. Method of Enucleation

The method of enucleation for preparing karyoplasts differed from that used for preparing cytoplasts (see Sections 2.1 and 2.2). The procedure described here optimized the formation of a discrete, recoverable pellet of karyoplasts and apparently minimized damage to them (Lucas et al., 1976).

Two days before enucleation, cells were plated onto plastic sheets that were cut from tissue culture flasks. These sheets, called bullets, were made such that two of them could fit back-to-back (cell sides out) into a Sorvall 50-ml centrifuge tube. For enucleation of A9 cells (a mouse fibroblast line), the cell-covered bullets were first placed in tubes containing normal growth medium and centrifuged for 15 min at 7000 rpm (9500g) in a Sorvall HS-4 rotor at 35°C. This removed loosely adherent cells and minimized subsequent contamination of the karyoplast pellet. After this centrifugation the bullets were transferred to tubes containing normal growth medium plus 10 μg/ml cytochalasin B, incubated for 15 min at 37°C, and then centrifuged for 45 min at 7000 rpm in a Sorvall HS-4 rotor at 35°C. The bullets were then removed from each tube, the cytochalasin medium was decanted, and the pellets were resuspended in normal growth medium. It should be noted that freshly prepared karyoplasts were very fragile and too vigorous "pipetting" invariably killed them.

Bullets, like 60-mm dishes and glass disks, could also be coated with collagen to enhance cell adherance. After each use the bullets were rinsed thoroughly and dried, treated with concentrated H_2SO_4 for 5 min, rinsed extensively with running tap water, rinsed with glass distilled water, air-dried, and sterilized before reuse. As with cytoplast preparation, the conditions for optimal enucleation of bullet cultures varied for each cell type used.

3.2. Purification of Karyoplasts

Crude karyoplast preparations were contaminated with cytoplasmic fragments, a small number of whole cells, and, as the time after enucleation increased, an increasing number of dead karyoplasts. A procedure for the complete purification of karyoplasts was devised (Lucas et al., 1976; Zorn et al., 1980) (see also Chapter 20, this volume). For purifying karyoplasts from mouse L929 or A9 cells the following procedure was performed.

The karyoplasts and cytoplasmic fragments were separated on a 1–6% Ficoll gradient at 1g for 90 min. The gradient was maintained at 37°C in a humid, 5% CO_2 atmosphere. After 90 min, two distinct bands were identified in the gradient. The upper band was removed by aspiration and consisted of cytoplasmic fragments. The lower band containing the karyoplasts was removed and diluted with growth medium. The karyoplasts were then pelleted and resuspended in fresh growth medium.

Whole cells were eliminated from the karyoplast preparation by two successive 90-min incubations in tissue culture dishes. The karyoplasts lost the ability to attach to and spread upon tissue culture dishes, while the whole cells retained this ability. Whole-cell contamination has been estimated by several methods to be between 0.4% and 4% of the karyoplast preparation before any purification (Lucas et al., 1976; G. A. Zorn, unpublished results). After the 3 hr of incubation, the whole-cell contamination was greatly reduced.

Living and dead karyoplasts were separated using a solution developed for the isolation of monocytes and other nucleated cells from whole blood (Boyum, 1968). The solution contains 5% Ficoll and 9% diatrizoate sodium and is marketed under the name Ficoll-paque (Pharmacia, Uppsala, Sweden). The karyoplasts were pelleted and resuspended in medium at a concentration of 1×10^7 karyoplasts per ml and layered on top of 5 ml of Ficoll-paque. Next, 1 ml of medium was layered on top of the karyoplast suspension. The medium and the karyoplast suspension remained in two distinct layers. The tubes were centrifuged at 800 rpm (130g) at room temperature for 15 min. After centrifugation, a band could be identified at the interface between the medium and the Ficoll-paque. Examination of the karyoplasts in the band revealed that 98% of the karyoplasts were viable (trypan blue-excluding). Pelleted on the bottom of the tube and in suspension in the rest of the Ficoll-paque were the dead karyoplasts. These karyoplasts were 99% nonviable.

For some experiments, it was not necessary to remove cytoplasmic fragments and dead karyoplasts. But in *all* nuclear transplantation experiments, karyoplasts were first placed in tissue culture dishes in order to minimize subsequent parental whole-cell contamination of the hybrid cultures.

3.3. Some Characteristics of Karyoplasts

Karyoplasts, too, have been extensively characterized (see Chapter 20). They are surrounded by a plasma membrane and contain approximately 10% of the whole-cell volume of cytoplasm. They contain some ribosomes, endoplasmic reticulum, and a few mitochondria, but do not contain centrioles (Ege et al., 1974; Shay et al., 1974; Zorn et al., 1980). When prepared and cultured properly, a portion (10%) of the karyoplasts from some cell lines (mouse fibroblasts A9, L929, 3T3, neuroblastoma N18TG2, and monkey epithelial CV-1) but apparently not others (human HeLa, potoroo PTK-2) will regenerate their lost cytoplasm to reform whole viable cells (Lucas et al., 1976; Brown et al., 1980; Zorn et al., 1980; J. D. White, J. Bruno and J. J. Lucas, submitted for publication). A detailed characterization of this process and further definition of the conditions required to optimize regeneration frequencies are described in the articles cited above.

4. Nuclear Transplantation

4.1. Preparation of Sendai Virus

The Ender's strain of Sendai virus was prepared and assayed using modifications of standard procedures [Watkins (1971)]. Ten-day-old fertilized hen's eggs were inoculated with 0.1 ml of Earle's balanced salt solution (EBS) containing 0.01–0.02 hemagglutinating units of virus. After 3 days of incubation at 37°C, the eggs were chilled and clear allantoic fluid was harvested and pooled. The fluid was heated at 37°C for 15 min and then centrifuged (300g) for 10 min at 4°C in the Sorvall SS34 rotor. The supernatant was removed and centrifuged (20,000g) for 30 min at 4°C in the SS34 rotor. The pellet was suspended in Earle's balanced salt solution (EBS) and assayed for its ability to agglutinate sheep erythrocytes. Virus stocks, having titers of 2000–10,000 hemagglutinating units per milliliter, were stored frozen at −70°C. Before use in fusion experiments, virus was exposed for 5 min, at a distance of 15 cm, to a General Electric G8T5 ultraviolet light bulb.

4.2. Fusion of Cytoplasts and Karyoplasts

This procedure (Lucas and Kates, 1976; Lipsich et al., 1978; Hightower and Lucas, 1980) was devised to optimize the fusion of one cytoplast to one karyoplast, so that the hybrid cells were fairly homogeneous with respect to both cytoplasmic and nuclear contributions.

After their 1–2-hr recovery period, cytoplast monolayers were placed on a tray resting on ice and washed twice with cold EBS. Then, 0.5 ml of cold Sendai virus in EBS (approximately 400 HAU/ml) was added to each 60-mm dish of cytoplasts and the virus was allowed to adsorb to the cytoplast monolayers for 15 min. Meanwhile, the karyoplasts were pooled in one tube; centrifuged at 800 rpm for 6 min in an IEC PR-6000 centrifuge, gently resuspended in a small volume of either EBS or, in some cases, in complete growth medium with serum, and placed on ice. After the 15-min virus-adsorption period, excess virus was removed by aspiration and 0.25 ml (1–5 × 10^7 karyoplasts) of the karyoplast suspension was added to each dish. The plates were kept for 15 min on ice to allow karyoplasts to adsorb to the virus-covered cytoplasts. To aid adsorption, the plates were gently rocked every 3–5 min. Then they were transferred to a 37°C incubator to allow fusion to occur. After 45 min at 37°C, the plates were washed vigorously several times with EBS or serum-free medium. At this point, the plates were examined using a phase contrast microscope; if any residual unfused karyoplasts were seen, the monolayers were washed several more times with EBS or medium. The normal growth medium in which the nuclear donor cell type was maintained was then placed in the plates.

One plate from each experiment was always stained and the percentage

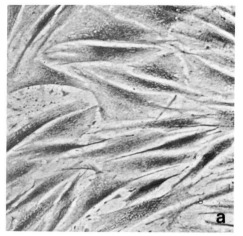 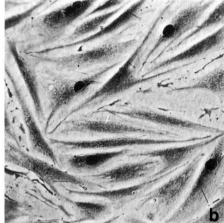

Figure 2. Nuclear Transplantation with mammalian cells. (a) A culture of enucleated human Detroit 532 fibroblasts. (b) The culture after fusion to karyoplasts prepared from mouse A9 fibroblasts. About 20% of the cytoplasts were successfully renucleated. The cultures were stained to emphasize the appearance of the nuclei.

of cytoplasts that were "renucleated" was estimated. Pictures of a control culture of cytoplasts and of hybrid cells are shown in Fig. 2. The remaining cytoplasts, which had not been renucleated, died and peeled from the vessel surface within about 2 days. If cultures free of contaminating cytoplasts are needed for analysis prior to this time, the hybrid cells can be purified by centrifugation through a density gradient of renograffin as described above.

This procedure has been used in the construction of hybrid cells using a variety of cell types. Also, it should be noted that rather extensive attempts were made to use polyethylene glycol (PEG) as the fusing agent in nuclear transplantation. Regardless of the conditions used, it proved to be much more toxic to cytoplasts and karyoplasts than Sendai virus (L. A. Lipsich and J. J. Lucas, unpublished results).

5. Identification of Hybrid Cells

5.1. Drug and Toxin Sensitivities

If appropriate mutant cell lines are available, their use permits definitive identification of hybrid cells. As described elsewhere in this volume, a variety of nuclear and cytoplasmically determined drug resistance markers can be employed. One can also take advantage of natural differences in the parental cell types, such as the high resistance of rodent cells to diphtheria toxin, a potent killer of human cells in culture [see, for example, Hightower and Lucas (1980)]. However, although these procedures are useful for establishing the frequency of successful transplantation for a particular cell system, their use as selective agents has a serious disadvantage: elimination of the contaminating cells takes several days to several weeks. By then, most of the interesting cytoplasmic–nuclear interactions which occur in the hybrid cells have probably been completed.

5.2. Immunofluorescent Staining of Fixed Cells

The efficiency of nuclear transplantation, and also the degree of contamination by parental whole cells, can be accurately quantitated by examining the hybrids for traits characteristic of the cytoplasm and nucleus of the two parental whole cells. For example, it was noted that staining of cells with the fluorochrome Hoechst 33258 permitted distinction of mouse and human nuclei (Moser et al., 1975). We combined this technique for distinguishing nuclei with indirect immunofluorescence techniques developed either for (1) studying the structure of the actin- and myosin-containing cytoskeleton of cells; or for (2) localizing fibronectin on the cell surface, to definitively identify and quantitate mouse A9 nuclear–human Detroit 532 cytoplasmic hybrids (Hightower and Lucas, 1980). Since only the human parent cell type had a highly ordered array of microfilament bundles throughout its cytoplasm and large fibronectin deposits on its surface, hybrids could be easily detected as cells having these traits, yet containing mouse nuclei, as determined by Hoechst staining. Detailed descriptions of these immunofluorescence techniques were described previously (Hightower and Lucas, 1980).

In brief, the cytoplast donors were grown and enucleated on glass disks, fused to karyoplasts, and then, after a brief recovery period, fixed with 3.7% formaldehyde and cold acetone. The fixed cells were then treated with rabbit antibodies prepared against either myosin or fibronectin and stained with rhodamine-conjugated goat anti-rabbit IgG. Finally, their nuclei were stained by incubation for 1 min with Hoechst 33258 (0.5 μg/ml in PBS). The disks were rinsed in deionized water and mounted, cell side up, on microscope slides. The disks were then covered with a #1 thickness coverslip. The mounting medium used was Aquamount (Lerner Laboratories, Stamford, Connecticut). Some results of these procedures are shown in Fig. 3a–3d. Mouse nuclei, exhibiting the characteristic fluorescent polka dot pattern, were seen within stained human cytoplasts.

5.3. Hoechst–Rhodamine Staining of Living Cells

For certain applications, the techniques described in the preceding section have two serious disadvantages: they can only be used with cells possessing such distinguishing characteristics and they result in the death of the cells. An alternative method which has occasionally been used involves the marking of cells with ingested latex spheres of different sizes or colors (Veomett et al., 1974; Zorn and Anderson, 1981). However, application of this technique, too, often has numerous disadvantages, which have been discussed in detail previously (Hightower and Lucas, 1980).

A recently developed method (Hightower et al., 1981), modifications of which can also be used to identify whole-cell hybrids and cybrids, employs a combination of the Kodak laser dye rhodamine 123 and the Hoechst fluorochrome 33258, which accumulate in the mitochondria and nuclei, respectively, of living cells (Johnson et al., 1980; Arndt-Jovin and Jovin, 1977). When used as described here, they appeared to be nontoxic and eventually lost from the cells. A hybrid cell stained by this procedure is shown in Figs. 3e and 3f.

In this method, the cytoplasts and karyoplasts were labeled before fusion with rhodamine 123 and Hoechst 33258, respectively. Transplantation was then performed and hybrids were identified as cells containing both fluorescent mitochondria and nuclei. In this experiment (shown in Fig. 3), the cytoplasts were labeled by incubating them at 37°C for 30 min in normal growth medium containing 10 μg/ml of rhodamine 123. They were then washed twice with warm EBS or serum-free medium. Karyoplasts were labeled by incubating them at 37°C for 90 min in normal growth medium containing 5 μg/ml Hoechst 33258. They were washed twice with warm EBS or serum-free medium and then fused to cytoplasts as described in Section 4.2. As both dyes are light-sensitive, these manipulations were performed in subdued light. The rhodamine dye was prepared as a 1 mg/ml (100×) stock

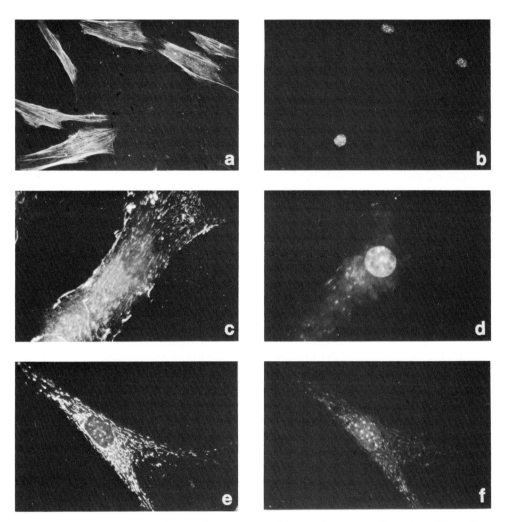

Figure 3. Mouse–human hybrid cells constructed by nuclear transplantation. The cells contained mouse nuclei, identified by staining with the Hoechst fluorochrome 33258 (b,d,f), within human cytoplasts containing an ordered array of microfilament bundles (a), a surface studded with deposits of fibronectin (c), or mitochondria stained with the Kodak dye rhodamine 123. For Figs. a–d, the cells were fixed and stained after nuclear transplantation. Figures e and f show a living cell formed from a cytoplast labeled before fusion with rhodamine and a karyoplast labeled before fusion with Hoechst 33258. The techniques used, including indirect immunofluorescence using antisera against myosin (a,b) or fibronectin (c,d), are described in the text. Note that in Figs. a and b (which are at a lower magnification than c–f), three out of the six cytoplasts in this field received mouse nuclei.

solution in distilled water and was stored at 4°C in a foil-wrapped container. Under these conditions, it was found to be stable for 2–3 wk. The solution was sterilized by passage through a Millipore 0.2-μm pore size filter. Hoechst 33258 was prepared and stored as a 0.5 mg/ml (100×) stock solution in PBS. It was found that the dye was apparently altered by autoclaving and also bound tightly to sterilizing filters. Therefore, stock solutions were sterilized by adding a drop of chloroform.

6. Analysis of Hybrid Cells

Using the techniques described here, hybrid cells suitable for many types of biologic and biochemical analyses can be produced in large numbers. An average of 20% of enucleated human Detroit 532 cytoplasts were renucleated by fusion to A9 karyoplasts. No detectable contamination by the A9 parent cell type was found, while the degree of contamination by the cytoplasmic donor cell type depended only on the initial efficiency of enucleation. Approximately 90% of the hybrids were viable, that is, were capable of continued growth and division. They began to change shape almost immediately after construction and, within a few days, resembled almost exactly the nuclear parent cell type. A detailed analysis of (1) the nuclear-directed alterations of cytoskeletal structure occurring soon after transplantation, and (2) the overall patterns of polypeptide synthesis in these hybrid cells, as detected by two-dimensional gel electrophoresis, will be described in a forthcoming publication (M. J. Hightower, J. Bruno, and J. J. Lucas).

The efficiency of transplantation varied for each particular cell system studied. For example, about 10% of enucleated rat HTC cells were renucleated using mouse A9 karyoplasts (Lipsich et al., 1979b), while greater than 90% of enucleated chick embryo fibroblasts were renucleated with dormant avian erythrocyte nuclei (Bruno et al., 1981). The high efficiency of fusion in this latter system permitted a detailed examination of gene expression in these hybrid cells. Of particular interest was the finding that only a subset of the globin polypeptides normally expressed in mature erythrocytes was synthesized in the hybrid cells. Results such as these suggest that hybrid cell systems constructed by nuclear transplantation will indeed be useful in analyzing the interaction of cytoplasm and nucleus in determining patterns of gene expression in eukaryotic cells.

Acknowledgments

The techniques described here were developed by a group which included, in addition to the authors, Arthur Brings, John Bruno, Eric Fairfield, Joseph R. Kates, Leah Ann Lipsich, Elizabeth Szekely, Jeffrey D. White, and Glenn A. Zorn. The research was supported by grants from the National Cancer Institute, DHHS, and the American Cancer Society. J. J. Lucas is an Established Investigator of the American Heart Association.

References

Arndt-Jovin, D. J., and Jovin, T. M., 1977, Analysis and sorting of living cells according to deoxyribonucleic acid content, *J. Histochem. Cytochem.* **25**:585–589.

Boyum, A., 1968, Isolation of mononuclear cells and granulocytes from human blood, *Scand. J. Clin. Lab Invest.* (*Suppl. 47*) **21**:770.

Briggs, R., and King, T. J., 1960, Nucleocytoplasmic interactions in eggs and embryos, in: *The Cell*, Volume 1 (J. Brackett and A. E. Mirsky, eds.), Academic Press, New York, pp. 537–617.

Brown, R. L., Wible, L. J., and Brinkley, B. R., 1980, Cytoplasmic microtubule assembly-disassembly in enucleated cells and regenerating karyoplasts, *Cell Biol. Internatl. Rep.* **4**:453–458.

Bruno, J., Reich, N. R., and Lucas, J. J., 1981, Synthesis of globin in hybrid cells constructed by transplantation of dormant avian erythrocyte nuclei into enucleated fibroblasts, *Mol. Cell Biol.* **1**:1163–1176.

Carter, S. B., 1967, Effects of cytochalasins on mammalian cells, *Nature* **213**:261–264.

Croce, C. M., and Koprowski, H., 1973, Enucleation of cells made simple and rescue of SV-40 by enucleated cells made even simpler, *Virology* **51**:227–229.

deTerra, N., 1967, Macronuclear DNA synthesis in *Stentor*: Regulation by a cytoplasmic initiator, *Proc. Natl. Acad. Sci. USA* **57**:607–614.

Ege, T., Hamberg, H., Krondahl, U., Ericson, J., and Ringertz, N. R., 1974, Characterization of minicells (nuclei) obtained by cytochalasin enucleation, *Exp. Cell Res.* **87**:365–377.

Follett, E. A. C., 1974, A convenient method for enucleating cells in quantity, *Exp. Cell Res.* **84**:72–78.

Follett, E. A. C., Pringle, C. R., Wunner, W. H., and Skehel, J. J., 1974, Virus replication in enucleate cells: Vesicular stomatitis virus and influenza virus, *J. Virol.* **13**:394–399.

Goldman, R. D., Pollack, R., and Hopkins, N. H., 1973, Preservation of normal behavior by enucleated cells in culture, *Proc. Natl. Acad. Sci. USA* **70**:750–754.

Gurdon, J. B., 1962, Adult frogs derived from the nuclei of single somatic cells, *Develop. Biol.* **4**:256–273.

Hightower, M. J., and Lucas, J. J., 1980, Construction of viable mouse–human hybrid cells by nuclear transplantation, *J. Cell. Physiol.* **105**:93–103.

Hightower, M. J., Fairfield, E., and Lucas, J. J., 1981, A staining procedure for identifying viable cell hybrids constructed by somatic cell fusion, cybridization or nuclear transplantation, *Somat. Cell Genet.* **7**:321–329.

Johnson, L. V, Walsh, M. L., and Chen, L. B., 1980, Localization of mitochondria in living cells with rhodamine 123, *Proc. Natl. Acad. Sci. USA* **77**:990–994.

Lipsich, L. A., Lucas, J. J., and Kates, J. R., 1978, Cell cycle dependence of the reactivation of chick erythrocyte nuclei after transplantation into mouse L929 cell cytoplasts, *J. Cell. Physiol.* **97**:199–208.

Lipsich, L. A., Lucas, J. J., and Kates, J. R., 1979a, Separation of cytoplasts and whole cells using density gradients of renografin, *J. Cell. Physiol.* **98**:637–642.

Lipsich, L. A., Kates, J. R., and Lucas, J. J., 1979b, Expression of a liver-specific function by mouse fibroblast nuclei transplanted into rat hepatoma cytoplasts, *Nature* **281**:74–76.

Lucas, J. J., 1982, Somatic cell hybridization, in: *Eukaryotic Genes: Their Structure, Activity and Regulation* (N. Maclean, S. P. Gregory, and R. A. Flavell, eds.), Butterworth, London.

Lucas, J. J., and Kates, J. R., 1976, The construction of viable nuclear-cytoplasmic hybrid cells by nuclear transplantation, *Cell* **7**:397–405.

Lucas, J. J., and Kates, J. R., 1977, Nuclear transplantation with mammalian cells, in: *Methods in Cell Biology*, Volume XV (D. M. Prescott, ed.), Academic Press, New York, pp. 359–370.

Lucas, J. J., Szekely, E., and Kates, J. R., 1976, The regeneration and division of mouse L-cell karyoplasts, *Cell* **7**:115–122.

Moser, G., Dorman, B. P., and Ruddle, F. H., 1975, Mouse–human heterokaryon analysis with a 33258 Hoechst–Giemsa technique, *J. Cell Biol.* **66**:376–405.

Okada, M., Kleinman, A., and Schneiderman, H. A., 1974, Chimeric *Drosophila* adults produced by transplantation of nuclei into specific regions of fertilized eggs, *Develop. Biol.* **39**:286–294.

Peterson, J. A., and Weiss, M. C., 1972, Expression of differentiated functions in hepatoma cell hybrids: Induction of mouse albumin production in rat hepatoma–mouse fibroblast hybrids, *Proc. Natl. Acad. Sci. USA* **69:**571–575.

Pollack, R., and Goldman, R., 1973, Synthesis of infective poliovirus in BSC-1 monkey cells enucleated with cytochalasin B, *Science* **179:**915–916.

Prescott. D. M., Kates, J., and Kirkpatrick, J. B., 1971. Replication of vaccinia virus DNA in enucleated L-cells, *J. Mol. Biol.* **59:**505–508.

Prescott, D. M., Myerson, D., and Wallace, J., 1972, Enucleation of mammalian cells with cytochalasin B, *Exp. Cell Res.* **71:**480–485.

Ringertz, N. R., and Savage, R. E., 1976, *Cell Hybrids*, Academic Press, New York.

Shay, J. W., Porter, K. R., and Prescott, D. M., 1974, The surface morphology and fine structure of CHO (Chinese hamster ovary) cells following enucleation, *Proc. Natl. Acad. Sci. USA* **71:**3059–3063.

Veomett, G., Prescott, D. M., Shay, J., and Porter, K. R., 1974, Reconstruction of mammalian cells from nuclear and cytoplasmic components separated by treatment with cytochalasin B, *Proc. Natl. Acad. Sci. USA* **71:**1999–2002.

Watkins, J. F., 1971, Fusion of cells for virus studies and production of cell hybrids, in: *Methods in Virology*, Volume 5 (T. K. Maramorasch and H. Koprowski, eds.), Academic Press, New York, pp. 1–32.

Wessels, N. K., Spooner, B. S., Ash, J. F., Bradley, M. O., Luduena, M. A., Wrenn, E. L., and Yamada, K. M., 1971, Microfilaments in cellular and developmental processes: Contractile microfilament machinery of many cell types is reversibly inhibited by cytochalasin B, *Science* **171:**135–143.

Wiktor, T. J., and Koprowski, H., 1974, Rhabdovirus replication in enucleated host cells, *J. Virol.* **14:**300–306.

Wright, W. E., and Hayflick, L., 1972, Formation of anucleate and multinucleate cells in normal and SV-40 transformed WI-38 by cytochalasin B, *Exp. Cell Res.* **74:**187–194.

Zorn, G. A., and Anderson, C. W., 1981, Adenovirus type 2 expresses fiber in monkey–human hybrids and reconstructed cells, *J. Virol.* **37:**759–769.

Zorn, G. A., Lucas, J. J., and Kates, J. R., 1980, Purification and characterization of regnerating mouse L929 karyoplasts, *Cell* **18:**659–672.

Chapter 20

Techniques for Purifying L-Cell Karyoplasts with Minimal Amounts of Cytoplasm

MIKE A. CLARK and JERRY W. SHAY

1. Introduction

Cells may be enucleated by centrifuging monolayers of cells in the presence of the drug cytochalasin B (CB) (Prescott et al., 1972; Wright and Hayflick, 1972). These procedures allow one to obtain large numbers of enucleated cells (cytoplasts) and nuclei surrounded by a thin layer of cytoplasm containing an intact outer plasma membrane. The nuclear fragments were originally termed karyoplasts and described as being incapable of regeneration. Later it was reported that cytoplasts could be fused to karyoplasts using inactivated Sendai virus (Veomett et al., 1974). The resulting "reconstructed" cells are of potential use in many aspects of somatic cell studies in areas such as gene regulation, differentiation, and transformation (Shay, 1977; Shay and Clark 1979) (also see Chapter 19, this volume). A major difficulty in these experimental procedures has been the presence of a low percentage of cells in the cytoplast population which fail to enucleate and the more serious problem of a small fraction (approximately 6%) of the whole cells which detach during enucleation and thus contaminate the karyoplast population.

Even though methods have been reported for circumventing this latter problem, these approaches are not able to successfully remove all of the whole cells from the karyoplast fraction. We first attempted to purify karyoplasts on the basis of sedimentation velocity in a serum gradient, but this method has not yet achieved desirable levels of purity (>99.9%) (Shay and Clark, 1977). A second method using differential adhesion (Lucas et al., 1976) requires a long time and thus lowers the viability of the purified karyoplasts.

We report here two additional methods for obtaining purified karyoplasts that satisfy the requirements for use in cell reconstruction experiments. With these methods we obtain pure karyoplasts with a yield of 24–50% of the

---

MIKE A. CLARK and JERRY W. SHAY • Department of Cell Biology, University of Texas Health Science Center, Dallas, Texas 75235. Present address for Dr. Clark: Wistar Institute of Anatomy and Biology, Philadelphia, Pennsylvania 19104.

original whole-cell population, depending on the technique chosen, with whole-cell contaminants (and karyoplast regeneration) accounting for as little as one clone of cells arising from 10^6 karyoplasts.

The first method takes advantage of the ability of the cells to ingest nontoxic, heavy tantalum (Ta) particles which are 1–3 μm in size. The Ta-containing cells are centrifuged in CB and the karyoplast fraction is then placed on a discontinuous Ficoll gradient. The whole cells (and karyoplasts that contain more than the minimal amount of cytoplasm) have a greatly increased density because of the Ta particles and are readily separated from the smallest karyoplasts, which contain few or no particles. The second method consists in purifying the karyoplasts using a fluorescence-activated cell sorter (FACS). The FACS is able to separate whole cells from the smaller karyoplasts on the basis of size (light scatter) at the rate of up to 5×10^3 particles/sec.

In addition, data are included on the metabolic decay of karyoplasts as determined by trypan blue dye exclusion and by uridine and leucine incorporation. These studies show that small, highly purified karyoplasts do not regenerate and that the purified karyoplasts may be fused to cytoplasts using polyethylene glycol (PEG) or inactivated Sendai virus to produce reconstructed cells that are viable and able to form clones of cells which then may be used for studying nuclear–cytoplasmic interactions (see also Chapters 19 and 21).

2. Materials and Methods

2.1. Cell Culture

Mouse L-929 cells were grown in Dulbecco's medium supplemented with 10% fetal calf serum (Flow Laboratories, Rockville, Maryland) without antibiotics. Cells to be enucleated were plated at a density of 4×10^5 cells per T25 flask (Falcon Plastics, Oxnard, California) and incubated for 48 hr at 37°C in an atmosphere containing 5% CO_2 plus 95% air. In addition, these cells were found to be free from mycoplasma contamination by three different assays: growth in selection medium (Hayflick, 1965), uridine/uracil incorporation assay (Schneider et al., 1974), and scanning electron microscopic observations (Tai and Quinn, 1977).

2.2. Tantalum Preparation

Tantalum particles (Norton Co., Newton, Massachusetts), 1–3 μm in size, were autoclaved in a small beaker containing a magnetic stirring bar. Sterile Earle's balanced salt solution (EBSS) was added to the Ta particles to give a final concentration of 0.04 g of Ta/ml. The solution was stirred continuously while 0.2 ml was removed and added to each flask of cells 24 hr prior to enucleation. This procedure ensured that constant amounts of well-dispersed

Ta particles were added to each flask of cells. This procedure resulted in virtually all cells (>99.9%) containing >12 Ta particles.

2.3. Cell Enucleation Procedure

Karyoplasts were obtained using the enucleation procedure of Shay and Clark (1979). The pellet resulting from enucleation, which contained karyoplasts and whole cells, was resuspended in EBSS for spinner culture (SEMEM) (GIBCO, Grand Island, New York) for further use. Cytoplasts that were used for cell reconstruction experiments were obtained using an alternate method (Prescott and Myerson, 1972). The cytochalasin B used for the enucleation was obtained from Aldrich Chemical Co. (Milwaukee, Wisconsin).

2.4. Karyoplast Purification Using Ta

The karyoplast preparation was purified in a chamber (Fig. 5) similar to one previously described (Shall, 1975). In advance of the purification procedure, the chamber and tubing were coated with Siliclad (Becton Dickinson Co., Parsippany, New Jersey) and autoclaved. An enucleation pellet suspended in 1.5 ml of SEMEM was placed in the bottom of the chamber, and 1 ml of 3% Ficoll (mol. wt. 400,000; Sigma Chemical Co., St. Louis, Missouri) in SEMEM solution was pumped in below the enucleation

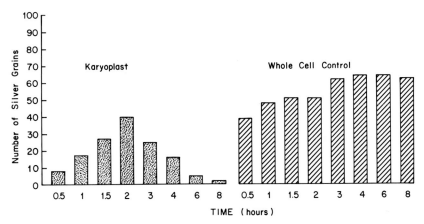

Figure 1. After enucleation, the CB was removed from the karyoplasts and fresh medium was added at time 0. The karyoplasts were then incubated at 37°C and pulsed with [^{3}H]uridine 30 min prior to fixation. The karyoplasts were then prepared for autoradiography and metabolic activity was measured by counting the silver grains. The whole-cell controls were exposed to CB but not centrifuged. They were then treated the same as karyoplasts. The numbers indicated on the Y axis are the average number of grains found over 100 karyoplasts or whole cells. Whole cells contaminating the karyoplast preparations were excluded from the analysis.

suspension at the rate of 0.5 ml/min. The suspension was allowed to settle for 1 hr at room temperature (~21°C), during which time various components of the enucleation pellet descended into the Ficoll under the force of unit gravity.

Since Ta has a density of >15 times that of a cell, those entities that contained Ta particles sedimented much faster than did the Ta-free karyoplasts. Fractions of the gradient were then collected from the top by displacing the chamber contents from the bottom by pumping in a solution of 30% Ficoll in SEMEM at a rate of 0.5 ml/min.

2.5. Karyoplast Purification Using the Cell Sorter

The Becton-Dickinson fluorescence-activated cell sorter (FACS III) was set up to the following specifications: Wavelength 488nm; light output 300 MW; PMT voltage 500 V; gain of 4; pulse height analyses divided into 256 channels. Unpurified karyoplasts were washed once with Puck's saline solution A, pH 6.5, then resuspended in Puck's A at a concentration of 10^7/ml and sorted. Purified karyoplasts were collected in Siliclad glass tubes and used for cell reconstruction experiments or placed in culture to study possible regeneration.

2.6. Karyoplast Viability

The viability of karyoplasts was determined by the exclusion of trypan blue dissolved in PBS to give a final concentration of 0.04 g/ml. The percent of cells that were viable at various time intervals after enucleation was determined by counting ~400 karyoplasts.

2.7. Metabolic Activity Assays

Karyoplast RNA and protein synthesis rates were evaluated by autoradiography. Karyoplasts were plated out in petri dishes containing glass coverslips and incubated at 37°C. Prior to fixation (30 min) [^{3}H]uridine or [^{3}H]leucine (New England Nuclear, Boston, Massachusetts) was added to give a final concentration of 10 μCi/ml. The coverslips were fixed in ethanol and acetic acid (3:1) at the following times post-enucleation: 0.5, 1.0, 1.5, 2.0, 4.0, 6.0, 8.0 hr. The coverslips were air-dried and coated with Kodak NTB3 emulsion diluted 1:1 with water. After 7 days of exposure, the coverslips were developed and stained with Giemsa dye. Silver grains over individual karyoplasts were counted to quantitate RNA and protein synthesis. Whole-cell control rates were established by growing cells on glass coverslips until 50% confluent. The coverslips containing whole cells were placed in 10 μg CB for 30 min and then rinsed and placed in growth medium. Whole cells were pulsed with radioactive precursors and fixed at the same intervals after CB treatment as were karyoplasts.

2.8. Cell Reconstruction

To reconstruct cells using Sendai virus, the monolayer procedure (Veomett et al., 1974) was used with minor modifications. After cell enucleation, coverslips containing attached cytoplasts were incubated for 1 hr at 37°C in growth medium and then cooled to 4°C for 20 min. The monolayers of cytoplasts were then washed twice with EBSS, pH 8.0, at 4°C followed by the addition of 200 HAU of UV-inactivated Sendai virus in 1.0 ml of EBSS, pH 8.0, 4°C for 20 min. The cytoplasts were washed twice with EBSS, pH 8.0, 4°C and the EBSS was then removed. The karyoplasts at 100 times the number of cytoplasts were suspended in a saline solution (PBS) and added to the cytoplasts for 45 min at 4°C. The coverslips were then warmed to 37°C for 20 min and then washed twice with growth medium.

An alternative fusion procedure was to use polyethylene glycol (PEG) (400 ml. wt.; Fisher Pittsburgh, Pennsylvania) and a modification of the procedure described by Clark et al. (1977).(See also Chapter 22, this volume.) Cytoplasts were allowed to recover for 30 min after enucleation in growth medium at 37°C. The cytoplasts were then trypsinized from the coverslips and suspended in 0.25 ml of growth medium. Karyoplasts at 100 times the number of cytoplasts were suspended in 0.25 ml of growth medium and then added to the cytoplasts. Next, 0.5 ml of PEG was added, gently mixed (1 min), diluted with 9 ml of growth medium plus serum, and then centrifuged at 500g for 3 min. The cells were then removed and plated out in petri dishes containing growth medium.

2.9. Electron Microscopy

Cells and karyoplasts to be used for TEM were sorted directly into 1.5-ml polypropylene tubes (Brinkman, Westburg, New York). The cells and karyoplasts were fixed in 0.3 M Cacodylate-buffered 3% glutaraldehyde, post-fixed in 1% OsO_4, dehydrated in acetone, and then embedded in Epon. The cells were then thin-sectioned, stained in uranyl acetate and lead citrate, and finally observed using a JEOL 100B transmission electron microscope.

3. Results

The centrifugation of L-929 cells in the presence of cytochalasin B results in >95% enucleation of the cells remaining attached to the substrate. The pellet formed in the bottom of the flask by such treatment contains mostly karyoplasts and ~5% whole cells.

The incorporation of radioactive precursors by the karyoplasts was studied autoradiographically in order to exclude contaminating whole cells from the analysis. The rate of incorporation of [^{3}H]uridine (Fig. 1) increased during the first 2 hr post-enucleation, but then sharply decreased until it became indistinguishable from the background level at 8 hr post-enucleation. The rate of incorporation of [^{3}H]leucine followed a similar pattern, increasing

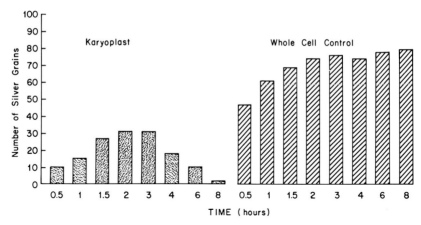

Figure 2. As in Fig. 1, but for karyoplasts pulsed with [³H]leucine.

during the first 2 hr post-enucleation and to background by 8 hr post-enucleation (Fig. 2). When the viability of karyoplasts was examined using trypan blue dye exclusion it was observed that at 12-hr post-enucleation only 8% continued to exclude the dye (Fig. 3). However, those entities that did exclude the dye were well spread out and identified as being whole-cell contaminants.

3.1. Results from the Ta Purification Procedure

Essentially all of the L-929 cells exposed to Ta particles for 4 hr engulfed ⩾12 Ta particles (Fig. 4) and most of these were found near the periphery of the nucleus but never in the nucleus. The viability of the Ta-containing cells, as determined by trypan blue exclusion, was ~99.5%. Of the small number of cells that did not take up Ta particles, greater than 50% also failed to exclude

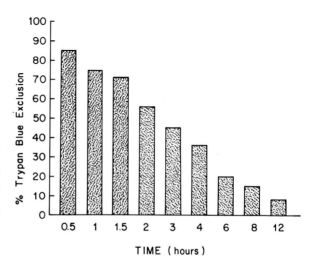

Figure 3. Karyoplast viability determined using trypan blue dye. The percent found excluding the dye was calculated from a sample size of ~400 unpurified karyoplasts at each time period. The 8% still excluding the dye 12 hr post-enucleation were found to be contaminating whole cells.

Purifying L-Cell Karyoplasts

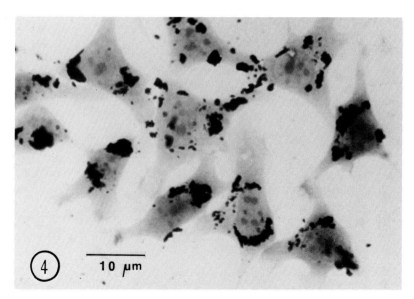

Figure 4. A light micrograph of L-929 cells which had ingested Ta. This metal, which is physiologically inert, was not observed to have any detrimental affects on the cells. Because of the high density of Ta (16) as compared to cell density (1.05–1.10), whole cells that contained Ta sediment faster than Ta-free karyoplasts in a Ficoll gradient.

the dye. Other than a possible slight reduction in growth rate, the cells did not appear adversely affected by the Ta.

Following the enucleation of cells that had been exposed to Ta particles, 95–98% of the cells that remained attached to the growth substrate were found to be enucleated and had retained many Ta particles. The enucleation pellet contained karyoplasts, whole cells, cytoplasmic fragments, and free Ta particles. Most of the cellular material was in the form of karyoplasts, with ~5% of the material being whole cells.

After placing the enucleation pellet in the separation chamber (Fig. 5), the material separated into several fractions. The top fraction was found to

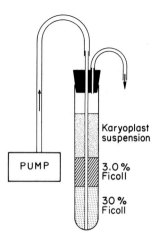

Figure 5. The separation chamber used to purify karyoplasts from Ta-containing whole-cell contaminants. After enucleation the enucleation pellet was placed in the bottom of the tube and the Ficoll solutions were pumped in from below at the rate of 0.5 ml/min. The apparatus was then allowed to stand undisturbed at unit gravity for 1 hr while the karyoplasts separated from the whole cells. The purified karyoplasts were then collected at the top of the 0.3% Ficoll layer by displacing the contents of the chamber at the rate of 0.5 ml/min.

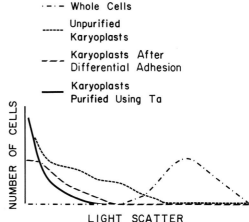

Figure 6. The results obtained using the cell sorter to analyze four populations of karyoplasts and whole cells according to light scatter. Each population consisted of at least 5×10^6 cells or karyoplasts.

contain 50% of all karyoplasts produced by the enucleation. These karyoplasts contained 0–1 Ta particles. Approximately 3% of the unpurified karyoplasts were able to form a viable colony of cells in 2 weeks. However, 10^6 purified karyoplasts plated in T75 flasks gave an average of one colony after 2 weeks of incubation. Ten intact L-929 control cells and 10^6 karyoplasts added to T75 flasks resulted in 5–6 colonies after 2 weeks of incubation. Thus the plating efficiency of whole cells under these conditions is about ~50% and 10^6 karyoplasts purified using Ta and a Ficoll gradient contain at most 1–2 entities capable of producing a colony of cells. When karyoplasts that had been purified using the Ta–Ficoll gradient technique were observed using the cell sorter we found considerable enrichment for the smallest karyoplasts (Fig. 6). Approximately one-third of the karyoplasts purified by differential adhesion were larger than the karyoplasts purified using the Ta technique. In addition, the population of karyoplasts obtained by differential adhesion contained fewer of the smallest karyoplasts, probably because the smaller karyoplasts degenerate more quickly. In several experiments in which L929 karyoplasts, purified by the differential adhesion procedure, were incubated for 24 hr we observed between 1170 and 1470 viable cells per 10^6 purified karyoplasts initially plated and by 2 weeks the T75 flasks were confluent with whole cells.

3.2. Results from Cell Sorter-Purified Karyoplasts

Karyoplasts purified using the cell sorter were prepared using the enucleation procedure as described in Section 2. The enucleation pellet containing karyoplasts was then collected in Puck's A salt solution and placed on ice. The karyoplasts were then sorted on the basis of light scatter. When 10^6 of the resulting smallest karyoplasts were plated out and incubated for 2 weeks, no colonies were in four flasks, one colony was in one flask, and three colonies were in one flask. Whole cells that were examined using the cell

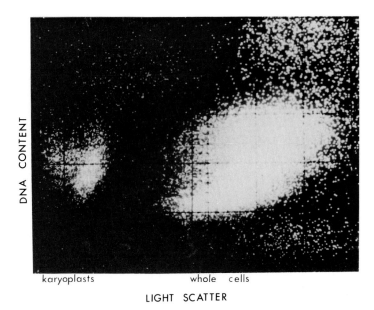

Figure 7. A dot plot taken using the cell sorter set to correlate size vs. DNA content. In this experiment 1×10^6 of the smallest karyoplasts and 5×10^6 whole cells were mixed and then stained with Mithramycin, a compound which binds quantitatively to DNA. Note that even these smallest nonregenerating karyoplasts contain varying amounts of DNA.

sorter were found to have about the same plating efficiency as cells not placed in the cell sorter. When a mixed population of karyoplasts and whole cells were stained using Mithramycin and the amount of DNA per entity vs. light scatter was analyzed we found that even the smallest karyoplasts contained varying amounts of DNA (Fig. 7) indicating that cell cycle does not influence karyoplast size.

3.3. Results from Transmission Electron Microscopy

Micrographs of whole cells and large and small karyoplasts purified using the cell sorter revealed that only whole cells contained microtubules and microfilaments (Figs. 8 and 9). The endoplasmic reticulum found in small karyoplasts (Figs. 12 and 13) was denuded of ribosomes compared to the large karyoplasts (Figs. 10 and 11) and whole cells (Figs. 8 and 9).

3.4. Use of Purified Karyoplasts for Cell Reconstructions

The ability of purified karyoplasts to participate in successful Sendai virus-induced fusions with enucleate cytoplasts was compared to that of unpurified karyoplasts. Table I gives the results of one such comparison and indicates that, while purified karyoplasts can be used in cell reconstruction,

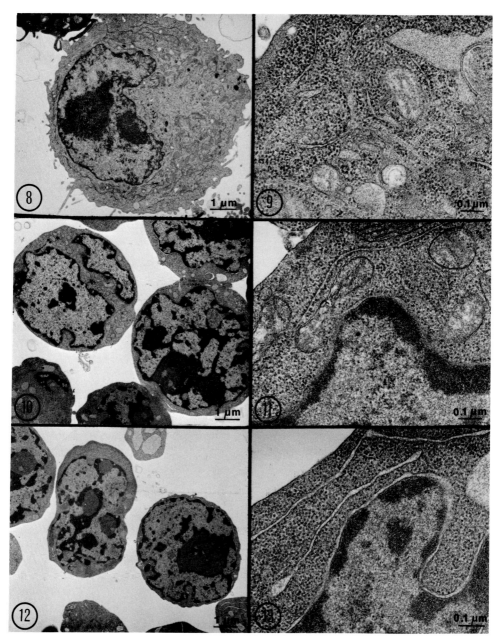

Figures 8–13. Micrographs taken using the transmission electron microscope. Whole cells (Figs. 8 and 9) contain a larger amount of cytoplasm, microtubules, and rough endoplasmic reticulum. Large karyoplasts (Figs. 10 and 11) contain less cytoplasm than whole cells, and also had rough endoplasmic reticulum; however, we did not see any microtubules. Small karyoplasts (Figs. 12 and 13) contain only 2–4% of the whole-cell cytoplasm; in addition, we did not see nearly as many ribosomes associated with the endoplasmic reticulum.

Table I. The Yield of Reconstructed Cells Using Different Preparations as Nuclear Donors[a]

Nuclear donors	Net reconstruction
Unpurified karyoplasts (3×10^6)	16%
Purified karyoplasts (3×10^6)	11%
Whole cells (1.5×10^5)	6%

[a]The indicated number of nuclear donors (purified and unpurified karyoplasts and whole cells were added to 3×10^6 cytoplasts in each instance. The "net reconstruction" is the percentage of cytoplasts containing a nucleus *minus* the number of nucleated cells found in parallel flasks of enucleate cytoplasts to which *no* nuclear donors were added.

they may be somewhat less efficient participants than unpurified karyoplasts. It is important to repeat that the unpurified karyoplast preparations contain about 5% whole cells, and experience indicates that whole cells are more effective than cell parts in cell fusion. To test this, 5% whole cells were added to 3×10^6 purified karyoplasts to determine how many viable cells could be produced. The result shown in Table I reveals that enough cells have been "reconstructed" to account for the difference between the yields of the two kinds of karyoplast preparations. (The net yield of "viable cells" in the last experiment was 6% of $3 \times 10^4 = 1.8 \times 10^3$, or far less than the 1.5×10^5 whole cells added. Note that with only about 1/20 of the number of donors, the yield was half as many reconstructed cells, which makes the efficiency of fusion about ten times as great for whole cells.) Thus, using *unpurified* karyoplasts results in ~1/3 of the presumed reconstructed cells being derived from the fusion of whole cells to cytoplasts. We conclude that purified karyoplasts, although unable to regenerate cytoplasms, are effective nuclear donors in cell reconstructions.

4. Discussion

The most important results from these investigations are: (1) karyoplasts purified by either of these methods are viable by two criteria, they exclude trypan blue and they are effective participants in cell reconstruction; (2) the purified karyoplasts are the smallest karyoplasts as determined by light scatter. When we used the previously described methods (Lucas et al., 1976) for quantitating the amount of cytoplasm we found that these smallest karyoplasts contain 2–4% of the cytoplasm found in whole cells; and (3) that in 10^6 purified karyoplasts only ~1 entity is able to form a viable colony of whole cells. This latter result is in contrast with a previous report (Lucas, Szekely, and Kates, 1976) that a small but significant proportion of karyoplasts are able to regenerate their lost cytoplasm and go on to form a viable colony of cells. A likely explanation of this discrepancy may lie in the variable amount of cytoplasm that surrounds the nucleus of the karyoplasts. We have shown that almost all of the smallest karyoplasts are unable to regenerate and it may be that there is a critical amount of cytoplasm that a karyoplast must have in order to regenerate.

We have further shown that when karyoplasts are incubated at 37°C the viability is greatly reduced after 4 hr. For example, after 4 hr of incubation at 37°C, only 45% of the karyoplasts continued to exclude the dye, whereas if the karyoplasts are maintained at room temperature or colder for 4 hr ~75% of karyoplasts will continue to exclude trypan blue. Thus any attempt to purify karyoplasts by incubation at 37°C for long periods of time will result in reduced viability.

To overcome this problem we have developed two methods that are effective in purifying karyoplasts at room temperature or colder. Each of these methods has distinct advantages. Advantages of using the Ta purification method include: (1) only inexpensive equipment is needed; and (2) only 1 hr is necessary for the purification; this is in contrast to the differential adhesion purification technique, which requires 2–4 hr at 37°C. The advantages of using the cell sorter to purify karyoplasts include: (1) cells do not need to be incubated with Ta before enucleation, and (2) one can select for any size of karyoplast desired.

Acknowledgments

This research was supported in part by grants from NSF PCM-8023070 and NIH GM 29261.

References

Clark, M. A., Crenshaw, A. H., and Shay, J. W., 1977, Fusion of mammalian somatic cells with polyethylene glycol 400 M.W., *Tissue Culture Assoc. Manual* **4**(2):801–804.
Crissman, H. A., and Tobey, R. A., 1974, Cell cycle analysis in 20 minutes, *Science* **184**:1297–1298.
Hayflick, L., 1965, Tissue cultures and mycoplasmas, *Texas Rep. Biol. Med.* **23**:285–303.
Lucas, J. J., Szakely, E., and Kates, J. R., 1976, The regeneration and division of mouse L-cell karyoplasts, *Cell* **7**:115–130.
Prescott, D. M., Myerson, D., and Wallace, J., 1972, Enucleation of mammalian cells with cytochalasin B, *Exp. Cell Res.* **71**:480–485.
Schneider, E. L., Stanbridge, E. J., and Epstein, C. J., 1974, Incorporation of ^{3}H-uridine and ^{3}H-uracil into RNA, *Exp. Cell Res.* **84**:311–318.
Shall, S., 1975, Selection synchronization by velocity sedimentation separation of mouse fibroblasts, in: *Methods in Cell Biology*, Volume VII (D. M. Prescott, ed.), Academic Press, New York, pp. 269–285.
Shay, J. W., 1977, Selection of reconstituted cells from karyoplasts fused to chloramphenicol resistant cytoplasts, *Proc. Natl. Acad. Sci. USA* **74**:2461–2464.
Shay, J. W., and Clark, M. A., 1977, Morphological studies on the enucleation of colchicine treated L-929 cells, *J. Ultrastruct. Res.* **58**:155–161.
Shay, J. W., and Clark, M. A., 1979, Nuclear control of tumorigenicity in cells reconstructed by PEG induced fusion of cell fragments, *J. Supermol. Struct.* **11**:33–49.
Tai, Y. H., and Quinn, P. A., 1977, Rapid detection of mycoplasma contamination in tissue cultures by SEM, *Scanning Electron Microscopy* **2**:291–299.
Veomett, G., Prescott, D. M., Shay, J., and Porter, K. R., 1974, Reconstruction of mammalian cells from nucleic and cytoplasmic components separated by treatment with cytochalasin B, *Proc. Natl. Acad. Sci. USA* **71**:1999–2002.
Wright, W. E., and Hayflick, L., 1972, Formation of anucleate cells in normal and SV40 transformed WI-38 by cytochalasin B, *Exp. Cell Res.* **74**:187–194.

Chapter 21

Techniques for Isolating Nuclear Hybrids

LAMONT G. WEIDE, MIKE A. CLARK, and JERRY W. SHAY

1. Introduction

Nuclear–cytoplasmic interactions have been studied in mammalian cells through the formation of cybrids (cytoplasts fused to whole cells) using a mutation for cytoplasmically inherited chloramphenicol resistance (CAP^r) (Bunn et al., 1974; Wallace et al., 1975; Wallace and Eisenstadt, 1979). McBurney and Strutt (1979) fused karyoplasts to whole cells, a product they termed karyobrids. However, their isolation technique did not distinguish nuclear hybrids from whole-cell hybrids. Muggleton-Harris and Palumbo (1979) produced nuclear hybrids, identified on the basis of nuclear morphology. Because both of these procedures lacked genetic verification, we have developed genetic and physical–genetic techniques for isolating nuclear hybrids. These techniques permit the isolation and verification of nuclear hybrids with the cytoplasmic and mitochondrial content of predominantely one cell.

2. Methods and Results

2.1. Cell Lines and Media

The MRR (TK^-/CAP^r) TL-1 (TL-1) mouse cell line (Dr. G. Getz, Chicago, Illinois) was used as the nuclear donor in all experiments. The whole-cell recipients were $HGPRT^-$ and CAP^s and consisted of the PCC_4-Azal mouse cell line (PCC_4) (Dr. W. Wright, Dallas, Texas) and the HT10806TG human cell line (Dr. C. Croce, Philadelphia, Pennsylvania). All cells were grown in antibiotic-free Dulbecco's modified Eagle's medium supplemented with 10% fetal calf serum in a humidified atmosphere containing 95% air and 5% CO_2 at 37°C. All cells were tested and found to be free of mycoplasma by uridine to uracil ratio (Schneider et al., 1974) and Hoechst staining (Chen, 1977).

LAMONT G. WEIDE, MIKE A. CLARK, and JERRY W. SHAY • Department of Cell Biology, University of Texas Health Science Center, Dallas, Texas 75235. Present address for Dr. Clark: Wistar Institute of Anatomy and Biology, Philadelphia, Pennsylvania 19104.

2.2. Cell Enucleation and Fusion

The TL-1 cell line was enucleated by a procedure modified from Shay and Clark (1978) (see Veomett, this volume, Chapter 6). The cells were centrifuged at 13,000g for 60 min in the presence of 3 μg/ml of cytochalasin B at 37°C. All fusions were done using polyethylene glycol 400 MW (Clark et al., 1978) (see Crenshaw and Murrell, this volume, Chapter 22).

2.3. Isolation of Intraspecific Nuclear Hybrids; Genetic Selection

The genetic selection of nuclear hybrids utilized the resistance to chloramphenicol (CAP^r) to identify contaminating whole-cell hybrids. The karyoplasts were obtained from CAP^r cells and fused to intact CAP^s cells. Because some karyoplasts contained few mitochondria, the locus of the CAP^r mutation, it was possible to isolate nuclear hybrids that were CAP^s. Karyoplasts were fused to whole cells and the fusion products were initially selected in HAT medium (Marin and Littlefield, 1968). All cells that survived were either whole-cell hybrids or nuclear hybrids (Fig. 1). The clones were then tested for sensitivity to CAP by dividing each clone in half, placing one-half in 75 μg/ml of CAP plus HAT, and the other half of the clone in HAT medium (see Fig. 1). Three viable fusion products were produced in the above fusions: (1) TL-1 whole cells fused to PCC_4, resulting in products which were CAP^r; (2) TL-1 karyoplasts containing appreciable numbers of mitochondria fused to PCC_4, resulting in products which were CAP^r; and (3) TL-1 karyoplasts, containing too few mitochondria to transfer CAP^r, fused to PCC_4, resulting in products which were CAP^s. The clones that died in CAP were true nuclear hybrids and the surviving half of the clone that was not tested in CAP was then isolated from the HAT medium and further analyzed.

Although this genetic isolation technique eliminated all of the whole-cell hybrids from consideration, its criteria were extremely stringent, and many nuclear hybrids were discarded as well. Table I shows that approximately 14% of the fusion products obtained were CAP^s.

2.4. Isolation of Nuclear Hybrids; Physical–Genetic Selection

The second method of obtaining nuclear hybrids involved physical selection of small karyoplasts using the Becton-Dickinson fluorescence-activated cell sorter (FACS III) (see Clark and Shay, this volume, Chapter 20). A vital dye (rhodamine 123) stains mitochondria in living cells (Johnson et al., 1980, 1981) and can be used in combination with the FACS III to isolate karyoplasts containing few, if any, mitochondria.

Figure 2 illustrates mitochondria in TL-1 cells stained with 10 μg/ml of rhodamine 123 (R_{123}) (Eastman Kodak, Rochester, New York) for 15 min. The cell sorter was set to the following specifications: laser wavelength 514 nm, PMT voltage 700 V, PMT filter 620 nm cut on. Cells to be enucleated were

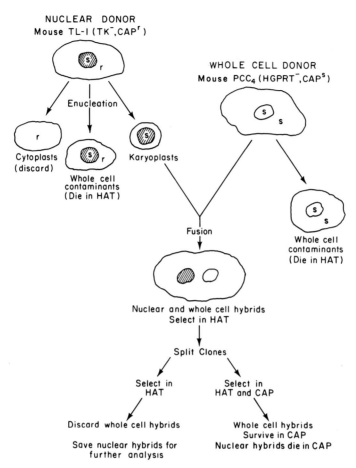

Figure 1. Diagrammatic representation of the genetic selection technique for intraspecific nuclear hybrids. Karyoplasts from HAT^s, CAP^r nuclear donors were fused to HAT^s, CAP^s whole-cell recipients and HAT^r, CAP^s products were isolated and identified as described in the text.

Table I. The Number and Percent of CAP^s Colonies in Four Independent Experiments[a]

Experiment	CAP^s colonies	Total colonies	Percent CAP^s
1	2	14	14.3
2	3	26	11.5
3	2	17	11.8
4	15	96	15.6

[a] Approximately 3×10^5 cells were plated out per 100-mm dish prior to HAT selection, resulting in 3–8 viable colonies/dish.

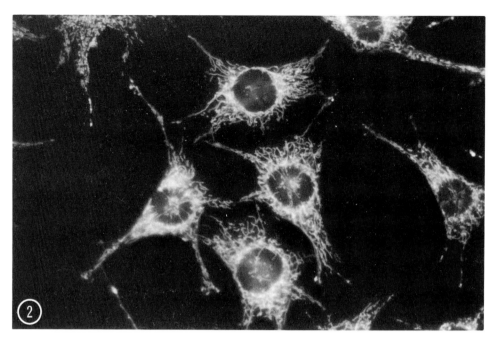

Figure 2. The TL-1 cell mitochondria stained for 15 min with 10 μg/ml rhodamine 123.

stained with R_{123} immediately prior to enucleation as described. The karyoplast fraction was placed on ice in Puck's saline A, pH 6.5, and sorted on the FACS III within 1 hr to avoid degeneration and subsequent clumping. Based on an average of 15 separate experiments, greater than 96% of the cells were stained by R_{123}.

Figure 3 illustrates the size and fluorescence profiles of unstained TL-1

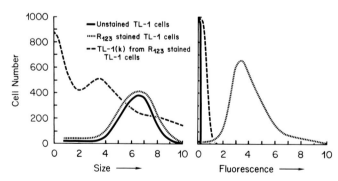

Figure 3. Summary of the cell sorter data for the unstained TL-1 cells, R_{123}-stained TL-1 cells, and TL-1$_{(k)}$ from R_{123}-stained TL-1 cells. Stained and unstained TL-1 cells exhibited identical size distribution, while the TL-1$_{(k)}$ exhibit smaller particles, as expected. Unstained TL-1 cells exhibited no fluorescence, while the R_{123}-stained TL-1 cells exhibited a distinct fluorescence peak. The TL-1$_{(k)}$ fraction contained detached contaminating whole cells with fluorescence similar to R_{123} stained TL-1 cells (data not shown) in addition to the karyoplasts, which exhibited minimal fluorescence. The FACS was set to show only the fluorescence of the smallest particles. Karyoplasts were purified by selecting particles of small size and minimal fluorescence.

cells, R_{123}-stained TL-1 cells, and TL-$1_{(k)}$ from R_{123}-stained cells. Unstained TL-1 cells had a normal size distribution and no fluorescence, whereas the R_{123}-stained TL-1 cells were the same size and brightly fluorescent due to staining of the mitochondria. The TL-$1_{(k)}$ fraction had a biphasic size distribution. The smallest peak contained karyoplasts, while the larger peak contained the whole-cell contaminants that detached from the flask during enucleation. These whole-cell contaminants exhibited fluorescence similar to R_{123}-stained cells; however, as illustrated in Fig. 3, the FACS was set to analyze only the fluorescence of the smallest particles. Most of the karyoplasts, in addition to small size, also exhibited minimal fluorescence. When TL-$1_{(k)}$ derived from R_{123}-stained cells were analyzed by the FACS, approximately 30% of the karyoplasts contained mitochondria (as determined by R_{123} fluorescence). The mitochondria-free karyoplasts were isolated and used for the production of nuclear hybrids as diagrammed in Fig. 4. The TL-$1_{(k)} \times PCC_4$ intraspecific products obtained by this procedure were found

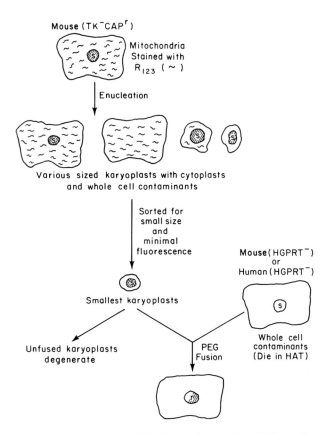

Figure 4. Diagrammatic representation of the physical–genetic isolation technique. Karyoplasts from R_{123}-stained TL-1 cells were purified by the FACS by selecting small particles exhibiting minimal fluorescence. These karyoplasts were then fused to whole-cell recipients and selected in HAT. The "s" within the nucleus represents sensitivity to HAT, while the "r" represents resistance to HAT selection.

to be CAP^s nuclear hybrids. This technique also allowed the isolation of interspecific nuclear hybrids (TL-1$_{(k)}$ × HT10806TG).

2.5. Interspecific Fusion Product Survival Is Dependent on Cytoplasmic Effects

Table II shows the 4-week survival rate of interspecific nuclear hybrids. Less than 20% of the whole-cell hybrids exhibited long-term survival, while approximately 50% of the interspecific nuclear hybrids from these experiments exhibited long-term survival. Table II also demonstrates that CAP^r mouse × CAP^s human hybrids (TL-1 × HT10806TG) may be isolated. However, none of these hybrids were observed to survive greater than 8 weeks post-fusion. In these experiments all nuclear hybrids eliminated human chromosomes.

2.6. Restriction Enzyme Analysis of Mitochondrial DNA from Nuclear Hybrids

Nuclear hybrids initially had a reduced mitochondrial complement from the karyoplast donor. Several interspecific nuclear hybrids (TL-1$_{(k)}$ × HT10806TG) were tested for their resistance to 0–200 µg/ml CAP and it was observed that the majority of the interspecific nuclear hybrids tested were CAP^r.

Restriction enzyme digestion (EcoRI) of human mtDNA results in three sizes of fragments, approximately 8, 7, and 1 kb being produced (Brown and Vinograd, 1974; Robberson et al., 1977). Digestion of mouse mtDNA also produces three distinct size fragments (13, 1.8 and 0.4 kb) after EcoRI restriction enzyme digestion (Brown and Vinograd, 1974; Robberson et al., 1977). These differences allowed identification of the species of mtDNA within the nuclear hybrids.

Table II. Long-Term Survival of Interspecies Fusion products

Fusions	Experiment	Initial number of clones	Number of clones after 4 weeks	Percent 4-week survival
TL-1 × HT10806TG	1	25	4	16
	2	3	1	33
	3	1	0	0
	4	2	0	0
TL-1$_{(k)}$ × HT10806TG	1	30	18	60
	2	21	10	48
	3	26	12	46
	4	36	20	55
HT10806TG$_{(k)}$ × TL-1	1	12	7	58
	2	22	7	32
	3	22	12	54

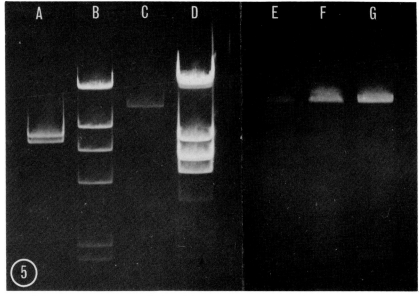

Figure 5. Restriction enzyme analysis of human (column A) and mouse (column C) mtDNA. Column B shows HindIII restricted λ DNA, while column D shows EcoRI restricted λ DNA, which were used as molecular weight standards. The restriction enzyme analysis of nuclear hybrid mtDNA by EcoRI is also shown (NH-2, column E; NH-8, column F; and NH-13, column G). The sizes of the standards are given in the text.

The mtDNA of three of the nuclear hybrid clones (NH-2, NH-8, NH-13) was analyzed by restriction enzyme analysis with EcoRI. Figure 5 shows the EcoRI restriction enzyme patterns obtained in these experiments for human and mouse mtDNA. Lambda (λ) DNA was cleaved by EcoRI (Fig. 5, column D) and HindIII (Fig. 5, column B) restriction enzymes and used as molecular weight standards. The EcoRI restriction enzyme-cleaved λ DNA produced five bands (column D) (21, 7.2, a doublet 5.7 and 5.3, 4.6, and 3.2 kb). The HindIII restriction enzyme-cleaved λ DNA produced six bands (column B) (23.5, 9.7, 6.6, 4.3, 2.2, 2.1 kb). Column A of Fig. 5 shows human EcoRI restricted mtDNA and column C contains mouse EcoRI restricted mtDNA. Figure 5 shows the EcoRI restriction enzyme patterns for nuclear hybrids (NH-2, column E; NH-8, column F; NH-13, column G). Only mouse mtDNA was detected in these nuclear hybrids. Thus, these nuclear hybrids, in addition to eliminating human chromosomes, also eliminated human mitochondria.

3. Discussion

The techniques described in this chapter for isolating nuclear hybrids ensure that the hybrids isolated are the fusion products of whole cells and the smallest karyoplasts containing minimal numbers of mitochondria. These karyoplasts do not regenerate their missing cytoplasm if isolated and replated in fresh or conditioned cultured medium (see Clark and Shay, this volume, Chapter 20).

In these experiments the nuclear hybrids were different from the corresponding whole-cell hybrids. The most dramatic difference was that by decreasing the mitochondrial complement of one of the parents, the long-term survival of the nuclear hybrid was increased as compared to the long-term survival of hybrid cells. This is in agreement with the results of Ziegler and Davidson (1981; also Ziegler, this volume, Chapter 15), who chemically inactivated one of the parental cells mitochondrial complement with rhodamine 6G (R6G) prior to fusion to mouse cells and observed better long-term survival. The hybrids made from R6G-treated cells exhibited 100% long-term survival, while the untreated whole cell hybrids exhibited less than 20% long-term survival. Thus, it appears that the reduction of the mitochondrial complement of one species enhances hybrid cell survival.

These techniques may also allow further characterization of nuclear–nuclear interactions in the presence of predominantly one cytoplasm. These techniques should also permit the manipulation of mitochondrial populations and enhance the study of mitochondrial genetics in mammalian cells.

Acknowledgment

This work was supported by a grant from NSF PCM-8023070 to Jerry W. Shay.

References

Brown, W. M., and Vinograd, J., 1974, Restriction endonuclease cleavage maps of animal mitochondrial DNAs, *Proc. Natl. Acad. Sci. USA* **71**:4617–4621.

Bunn, C. L., Wallace, D. C., and Eisenstadt, J. W., 1974, Cytoplasmic inheritance of chloramphenicol resistance in mouse tissue culture cells, *Proc. Natl. Acad. Sci. USA* **71**:1681–1685.

Chen, T., 1977, In situ detection of mycoplasma contamination in cell cultures by fluorescent Hoechst 33258 stain, *Exp. Cell Res.* **104**:255–262.

Clark, M. A., Crenshaw, A. H., and Shay, J. W., 1978, Fusion of mammalian somatic cells with polyethylene glycol 400 mw, *Tissue Culture Association Manual* **4**:801–804.

Clark, M. A., Shay, J. W., and Goldstein, L., 1980, Techniques for purifying L-cell karyoplasts with minimal amounts of cytoplasm, *Somat. Cell Genet.* **6**:455–464.

Johnson, L., Walsh, M., and Chen, L., 1980, Localization of mitochondria in living cells with rhodamine 123, *Proc. Natl. Acad. Sci. USA* **77**:990–994.

Johnson, L., Walsh, M., Bockus, B., and Chen, L., 1981, Monitoring of relative mitochondrial membrane in living cells by fluorescence microscopy, *J. Cell Biol.* **88**:526–535.

Marin, G., and Littlefield, J., 1968, Selection of morphologically normal cell lines from polyoma-transformed BHK21/13 Hamster fibroblasts, *J. Virol.* **2**:69–77.

McBurney, M. W., and Strutt, B., 1979, Fusion of embryonal carcinoma cells to fibroblast cells, cytoplasts, and karyoplasts, *Exp. Cell Res.* **124**:171–180.

Muggleton-Harris, A. L., and Palumbo, M., 1979, Replicative potentials of various fusion products between WI-38 and SV40 transformed WI-38 cells and their components, *Somat. Cell Genet.* **6**:689–698.

Robberson, D., Wilkins, C., Clayton, D., and Doda, J., 1977, Microheterogeneity detected in circular dimer mitochondrial DNA, *Nucleic Acids Res.* **4:**1315–1338.

Schneider, E., Stanbridge, E., and Epstein, C., 1974, Incorporation of ^{3}H-uridine and ^{3}H-uracil into RNA, *Exp. Cell Res.* **84:**311–318.

Shay, J. W., and Clark, M. A., 1978, Nuclear control of tumorigenicity in cells reconstructed by PEG-induced fusion of cell fragments, *J. Supramol. Struct.* **11:**33–49.

Wallace, D., and Eisenstadt, J., 1979, Expression of cytoplasmically inherited genes for chloramphenicol resistance in interspecific somatic cell hybrids and cybrids, *Somat. Cell Genet.* **5:**373–396.

Wallace, D., Bunn, C., and Eisenstadt, J., 1975, Cytoplasmic transfer of chloramphenicol resistance in human tissue culture cells, *J. Cell Biol.* **67:**174–188.

Ziegler, M. L., and Davidson, R. L., 1981, Elimination of mitochondrial elements and improved viability in hybrid cells, *Somat. Cell Genet.* **7:**73–88.

Chapter 22

Monolayer Enucleation of Colcemid-Treated Human Cells and Polyethylene Glycol 400-Mediated Fusion of Microkaryoplasts (Microcells) with Whole Cells

ANDREW H. CRENSHAW, JR. and LEONARD R. MURRELL

1. Introduction

Modern somatic cell genetics began with the observation that hybridization can be achieved by cell to cell fusion in culture (Barski et al., 1961). In recent years, this new era in the study of gene expression has produced many advances in hybridization technology. Hybridization techniques can be grouped into two general classes: (1) whole genome transfer, involving cell to cell fusion, and (2) partial genome transfer, involving the introduction of genome fragments into whole cells. Mouse–human whole–cell hybridization has been used extensively for human gene mapping (Ruddle and Creagan, 1975; McKusick and Ruddle, 1977) and has relied on the fact that human chromosomes are spontaneously eliminated from these hybrids. Lengthy cultivation, recloning, and analysis are necessary before accurate chromosome assignment is possible (Weiss and Green, 1967).

Techniques for introducing isolated metaphase chromosomes (McBride and Ozer, 1973) and microcells (microkaryoplasts) (Fournier and Ruddle, 1977) into recipient cells have increased the efficiency of hybridization studies. The introduction of single isolated metaphase chromosomes into recipient cells by phagocytosis is relatively simple, but generally less than 1% of the transferred genetic material is maintained (Willecke and Ruddle, 1975). Only microkaryoplast-mediated gene transfer provides a practical method for transferring a single intact chromosome into a recipient cell. Micro-

---

ANDREW H. CRENSHAW, JR. and LEONARD R. MURRELL • Department of Anatomy, University of Tennessee Center for the Health Sciences, Memphis, Tennessee 38163.

karyoplast-mediated gene transfer employs micronuclei produced by the treatment of somatic cells with the plant alkaloid colchicine or its analog Colcemid (N-desacetyl-N-methyl-colchicine) (Stubblefield, 1964). Colchicine prevents the assembly of microtubules which form the mitotic spindle and renders cells blocked in metaphase. In rodent cells under prolonged treatment (36–48 hr), scattered individual chromosomes and small groups of chromosomes serve as sites for the reassembly of nuclear membranes forming micronuclei [for review see Margulis (1973)]. Under conditions of maximum micronucleation, each individual chromosome appears to form one micronucleus (Stubblefield, 1964; Phillips and Phillips, 1969). Genetically competent microkaryoplasts containing one micronucleus surrounded by a thin rim of cytoplasm and a limiting plasma membrane can be isolated by centrifugation of micronucleate rodent cultures in the presence of cytochalasin B (Ege et al., 1974; Ege and Ringertz, 1974; Shay and Clark, 1977). These microkaryoplasts can be fused to whole recipient cells with inactivated Sendai virus or polyethylene glycol. Murine chromosome transfer to several recipient cell lines has been accomplished using microkaryoplasts (micro cells) and inactivated Sendai virus (Fournier and Ruddle, 1977) (see also Fournier, Chapter 23, this volume).

Micronucleation of human cells with colchicine is more difficult. Even prolonged treatment under conditions that are effective for rodent cells does not produce significant micronucleation. For this reason, a technique for producing "minisegregants" from transformed human cells (HeLa) was developed (Johnson, et al., 1975, 1978; Schor et al., 1975; Mullinger and Johnson, 1976). In this system, small membrane-bound packets of chromatin were produced by arresting cells in mitosis with hyperbaric (8 atm) nitrous oxide and then subjecting the cells to a prolonged cold shock. When the cells are returned to 37°C, a severely altered cytokinesis ensues, resulting in the formation of clusters of small daughter (minisegregant) cells. The depolymerization of microtubules and alterations in the structural and contractile properties of actin filaments appear necessary for the production of human minisegregants (Mullinger and Johnson, 1976). HeLa minisegregants have been successfully transferred into murine cells with the retention of functioning human genes (Sunder et al., 1977; Tourian et al., 1978 (see Johnson and Mullinger, Chapter 24, this volume).

The formation of membrane-bound packets of chromatin similar to minisegregants and 20–30% micronucleate cells from human diploid fibroblasts was achieved by Cremer et al. (1976) using low concentrations of colchicine plus incubation in colchicine-free medium. Based on this observation, we produced human micronucleated cells with 50–70% efficiency from both normal diploid fibroblasts (Detroit 550) and transformed aneuploid cells ($D98/AH_2$) by treatment with 0.1 μg/ml Colcemid plus incubation in Colcemid-free medium (Crenshaw and Murrell, 1980; Crenshaw et al., 1981a,b). Comparable results have been achieved by treatment of human foreskin fibroblasts with high-dose (20 μg/ml) Colcemid, and viable human chromosomes have been transferred to recipient murine cells (McNeill and Brown, 1980).

In this chapter, we review the techniques of Colcemid-induced micronucleation in human cells and present two specific micronucleation techniques with details of large-scale microkaryoplast isolation and fusion to whole cells with polyethylene glycol 400 MW.

2. Micronucleation of Human Somatic Cells with Colcemid

2.1. Microkaryoplasts Containing Condensed Chromosomes

Rodent cell lines routinely form attached micronucleate cells with 60–90% efficiency in the presence of colchicine (Fournier and Ruddle, 1977; Shay and Clark, 1977). Moreover, murine fibroblasts containing 12–18 micronuclei can be prepared routinely by treatment with 2.0 μg/ml colchicine (Shay and Clark, 1977) or 0.1 μg/ml Colcemid (unpublished observation) for 36–48 hr. However, the treatment of human cells under similar conditions and drug concentrations yields either nonviable cells or poor micronucleation, depending on the human cell line used.

Cremer et al. (1976) demonstrated that viable cell fragments can be produced from normal human fibroblasts by treatment with 0.02–0.04 μg/ml colchicine for 24 hr followed by incubation in colchicine-free medium. They showed by time-lapse cinematography that after cells became arrested in metaphase, protrusions of different sizes formed at various sites on the cell surface. These protrusions either regressed into the cell surface, remained extruded and attached to each other by cytoplasmic bridges, or became completely separated. The majority of the cell fragments produced contained single or clusters of condensed metaphase chormosomes. Attached micronucleate cells and small cell fragments containing single decondensed karyomeres were also produced at a frequency of 20–30%.

More recent experiments (Crenshaw and Murrell, 1980; Crenshaw et al., 1981a) demonstrated that human cell fragments can be produced from normal diploid fibroblasts (Detroit 550) and aneuploid cells (D98/AH$_2$) with 0.1 μg/ml Colcemid for 20–48 hr. Time lapse videotape recording of Colcemid-treated D98/AH$_2$ cells revealed that the fragments result from extrusion development from the cell surface (unpublished observation) similar to the process described by Cremer et al. (1976) in normal human fibroblasts. During Colcemid treatment, almost all cells became round and detached from the substratum at entrance into mitosis. The cells remained blocked in mitosis throughout the incubation period and did not reattach to or spread on the substrate as do Colcemid-treated rodent cultures. Twenty to 30% of the arrested cells routinely formed readily dispersible clusters of small cell fragments by protrusion from the cell surface (Crenshaw and Murrell, 1980; Crenshaw et al., 1981a). Electron microscopic observation of cell fragments from D98/AH$_2$ cells (Fig. 1) revealed small membrane-bound cytoplasmic packets containing one or more condensed chromosomes or chromosome fragments.

Cell fragmentation following Colcemid treatment has not been described

294   Chapter 22

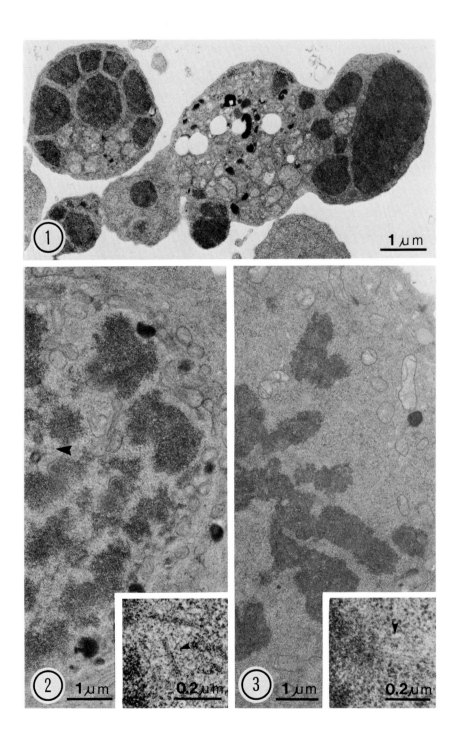

in rodent cell lines. Stubblefield (1964) used time-lapse cinematography to demonstrate multiple cleavage furrows which formed and then regressed in Colcemid-treated Chinese hamster cells. When murine L-929 cells are treated with Colcemid, similar furrows and protrusions form and regress as observed by time-lapse videotape recording (unpublished observation).

Thin sections of round, detached human $D98/AH_2$ cells arrested with 0.1 µg/ml Colcemid for as long as 24 hr reveal centrally placed condensed chromosomes surrounded by concentric rows of mitochondria and endoplasmic reticulum, numerous granules (Fig. 2), and bundles of 10-nm filaments. Paired centrioles (Fig. 2, arrow) can be observed in some cells among the condensed chromosomes. Untreated mitotic cells harvested by mitotic selection (Peterson et al., 1968) do not show a comparable high degree of cytoplasmic organization. Bundles of 10-nm filaments are rarely observable (Fig. 3), but numerous chromosomal microtubules are evident (Fig. 3, insert). In Colcemid-treated cells, chromosomes aggregate progressively, and numerous Colcemid-resistant microtubules coursing from condensed chromosomes are readily apparent (Fig. 2, inset).

The induction of 10-nm filaments with colchicine has been previously reported (Croop and Holtzer, 1975; Blose and Chacko, 1976; Cheung et al., 1978). The increase in perinuclear 10-nm filaments in $D98/AH_2$ cells in the presence of Colcemid supports the hypothesis that these filaments may inhibit the random movement of condensed chromosomes. Colcemid-blocked murine L-929 cells also show an increase in 10-nm filament bundles, but a lesser degree of organization is evident, and many of the filament bundles are among condensed chromosomes in the cell center (Fig. 4).

A Colcemid-resistant component of the mitotic spindle may be a factor in the inhibition of movement of condensed chromosomes in human cells in the presence of Colcemid concentrations that produce micronuclei in rodent cells. The mitotic spindle is not completely intact in $D98/AH_2$ cells in the presence of 0.1 µg/ml Colcemid, since there are numerous cells with paired centrioles among condensed chromosomes (Fig. 2), but no anaphase cells in such cultures. The presence of a Colcemid-resistant portion of the spindle may be dose-related.

Microtubules are required for mitotic spindle function, and colchicine acts by binding to tubulin subunits of microtubules (Borisy and Taylor, 1967a,b). Microfilaments form the contractile ring necessary for cytokinesis,

---

Figure 1. Membrane-bound cell fragments and fragment clusters derived from 20-hr Colcemid-blocked (0.1 µg/mg) mitotic human $D98/AH_2$ cells. Reprinted with permission from Crenshaw et al. (1981a).

Figure 2. Colcemid-blocked mitotic human $D98/AH_2$ cell treated with Colcemid (0.1 µg/ml) for 20 hr. Arrow indicates paired centrioles among condensed metaphase chromosomes. Insert: Arrow indicates longitudinal section of Colcemid-resistant microtubules. Reprinted with permission from Crenshaw et al. (1981a).

Figure 3. Untreated mitotic human $D98/AH_2$ cell obtained by mitotic selection. Insert: Arrow indicates chromosomal microtubules. Reprinted with permission from Crenshaw et al. (1981a).

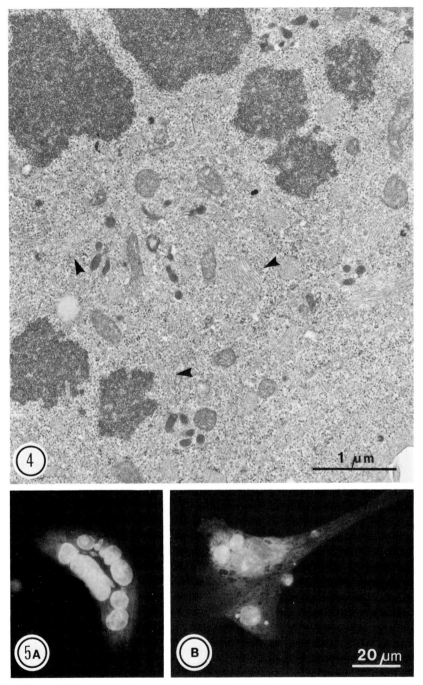

Figure 4. Colcemid-blocked mitotic murine L-929 cell treated with Colcemid for 20 hr. Arrows indicate bundles of 10-nm filaments among condensed metaphase chromosomes.

Figure 5. Micronucleate human cells. Cells were stained with the nuclear fluorescent stain Hoechst 33258 (20 μg/ml) in 0.83% NaCl. Cells were treated with 0.1 μg/ml Colcemid for 20 hr, then washed and incubated in Colcemid-free medium for 12 hr. A. Micronucleate D98/AH$_2$ cells. B. Micronucleate normal human (Detroit 550) fibroblasts.

Table I. Colcemid Induction of Micronucleation of Human Cells[a]

Cell type[b]	Post-Colcemid incubation	Cell status			
		Attached	Spread	Mononucleate,[c] %	Micronucleate,[d] %
D98/AH$_2$	With Colcemid	None	None	ND	ND
D98/AH$_2$	No Colcemid	Yes	Yes	31.6 ± 3.1	68.4 ± 3.1
RPMI-2650	With Colcemid	Yes	Yes	67.3 ± 2.8	32.7 ± 2.8
RPMI-2650	No Colcemid	Yes	Yes	77.0 ± 5.5	23.0 ± 5.5
Detroit-550	With Colcemid	None	None	ND	ND
Detroit-550	No Colcemid	Yes	Yes	51.2 ± 5.2	48.8 ± 5.2

[a] Reprinted with permission from Crenshaw et al. (1981a).
[b] All cells were treated with Colcemid (0.1 µg/ml, 20 hr), then the rounded detached cells were further incubated with or without Colcemid for 24–28 hr (column 2) before status (columns 3–6) was determined.
[c] The percentage of multinucleate cells in exponentially growing cultures was: D98/AH$_2$, 1.0 ± 0.2; RPMI-2650, 4.8 ± 0.8; Detroit 550, less than 0.5. Data are corrected to exclude multinucleate cells. ND, not determinable.
[d] Percentage (±SEM) based on mean of two or more experiments with at least 250 cells counted per experiment.

and the orientation of the mitotic spindle determines the site at which the contractile ring forms (Wessels et al., 1971). It appears likely that a Colcemid-resistant component of the mitotic spindle prevents the complete disruption of the spindle and random movement of chromosomes throughout the cytoplasm. The partially disrupted spindle may then serve as an orientation site for multiple contractile microfilament rings. Cytokinesis might then occur in several planes, resulting in cell fragments.

2.2. Micronucleate Cells Containing Decondensed Chromatin

Colchicine inhibits attachment and spreading of peritoneal macrophages, and the degree of the response is proportional to the colchicine concentration (Cheung et al., 1978). We have shown (Crenshaw et al., 1981a) that human Detroit 550 fibroblasts and D98/AH$_2$ aneuploid cells do not attach to and spread on the substrate in the presence of 0.1 µg/ml Colcemid and that if Colcemid-inhibited mitotic cells are plated in drug-free medium, micronucleate cells can be produced in significant number. (For details regarding method, see the Appendix.) Table 1 shows the ability of several human cell lines to micronucleate during and after Colcemid treatment, and that not all human cell lines behave in a similar fashion. The micronuclei produced by incubating Colcemid-treated cells in drug-free medium are morphologically similar to those in rodent cells treated with Colcemid alone (Fig. 5).

The formation of micronuclei, at least in D98/AH$_2$ cells, may be the result of the rapid assembly of chromosomal microtubules following the removal of Colcemid from blocked mitotic cells. Large numbers of microtubules form within 4 hr after the removal of Colcemid (Fig. 6). Possibly, divergent anaphasic movement of chromosomes that are not on an organized mitotic spindle is responsible for the formation of micronuclei in these cells (Crenshaw et al., 1981a).

Results previously reported (Crenshaw t al., 1981a) suggest that the

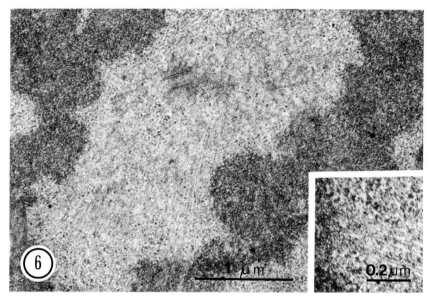

Figure 6. Recovery of Colcemid-blocked mitotic human D98/AH$_2$ cell in Colcemid-free medium. Cells were treated with 0.1 μg/ml Colcemid for 20 hr, then washed and incubated in Colcemid-free medium for 4 hr. Reprinted with permission from Crenshaw et al. (1981a).

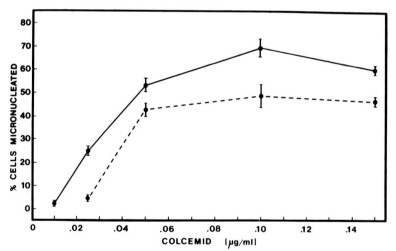

Figure 7. The response of human D98/AH$_2$ (—) and Detroit 550 (- - -) cells to increasing concentrations of Colcemid. At Colcemid concentrations of 0.05–0.15 μg/ml round, blocked mitotic cells were produced which did not attach and spread unless incubated in Colcemid-free medium. At initial Colcemid concentrations of 0.01 and 0.025 μg/ml, attachment and spreading occurred without removal of Colcemid, and numerous normal mitotic figures were observed. Cells were incubated for 20 hr in Colcemid (0.05–0.15 μg/ml) and floating round cells were incubated in Colcemid-free medium for an additional 24 hr. For cells treated with 0.01 and 0.025 μg/ml Colcemid, incubation in Colcemid was extended to 48 hr before counting. Vertical bars represent the standard error of the mean in two or more experiments with a minimum of 250 cells counted per experiment. Reprinted with permission from Crenshaw et al. (1981a).

Colcemid concentration is critical, in that drug concentrations that permit the attachment and spreading of cells following attempted mitosis yield dividing cells and a low percentage of micronucleation (Fig. 7). Unlike rodent cells, human cell lines appear to vary in their response to Colcemid. This is emphasized by the fact that human foreskin fibroblasts require treatment with 20 µg/ml Colcemid to achieve acceptable micronucleation frequency (40–60%) (McNeill and Brown, 1980). At the other end of the spectrum are RPMI-2650 quasidiploid tumor cells, which will micronucleate at a frequency of only 30%, will attach and spread in the presence of Colcemid, and are killed at Colcemid concentrations greater than 0.8 µg/ml (Crenshaw et al., 1981a).

Thus, the failure of human cells to undergo conventional micronucleation under the conditions and drug concentrations effective in rodent cells appears related to (1) differences in the effects of Colcemid on the human cell cytoskeleton in comparison with that of rodent cells and (2) the inability of certain human cells to attach and spread in the presence of Colcemid (Crenshaw et al., 1981a).

3. Isolation and Purification of Human Microkaryoplasts

3.1. Microkaryoplasts Produced by Abnormal Cytokinesis

Human microkaryoplasts formed by abnormal cytokinesis in the presence of colchicine or Colcemid contain condensed chromosomes and a small amount of cytoplasm, surrounded by limiting plasma membrane (Fig. 1). The viability of these microkaryoplasts has been shown by the exclusion of trypan blue, the incorporation of [^{3}H]leucine (Cremer et al., 1976), and their ability to rescue 8-azaguanine-resistant RPMI-2650 quasidiploid tumor cells in the hypoxanthine–aminopterin–thymidine (HAT) selection system of Littlefield (1966) (unpublished observation).

These microkaryoplasts occur singly or in small clusters connected by small cytoplasmic bridges which are readily dispersible with gentle pipetting (Cremer et al., 1976; Crenshaw et al., 1981a). Purification to remove Colcemid-blocked whole mitotic cells is easily accomplished by gravity filtration through 3-µm membrane filters (Ege et al., 1977).

3.2. Microkaryoplasts Produced by Cytochalasin B Enucleation of Micronucleatic Cells

Cytochalasin B, a mold metabolite from Helminthosporium dematoideum, has long been known to cause extrusion of a small percentage of nuclei from mammalian cells (Carter, 1967). Prescott et al. (1972) demonstrated that the percentage of cells that are enucleated can be increased to 99+% by centrifugation of cells in the presence of cytochalasin B. The centrifugation of whole cells in the presence of cytochalasin B yields

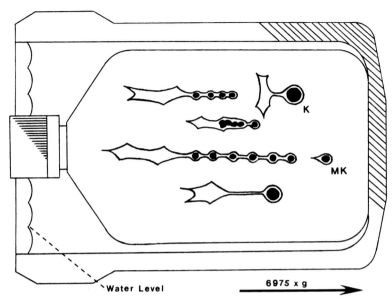

Figure 8. Cross-section of Sorvall HG-4 swinging bucket with enucleation flask in position. Only 75-cm² Falcon (#3024) flasks are acceptable. The flask is filled with cytochalasin B medium up to the middle of the neck, and the cap is sealed tightly. The water level should at least touch the flask cap. The arrow indicates the direction of centrifugal force. The maximum recommended rotor speed is 5000 rpm. K, forming karyoplast on cytoplasmic stalk; MK, microkaryoplast.

membrane-bound nucleated fragments (karyoplasts) and attached membrane-bound cytoplasmic remnants (cytoplsts) (Shay et al., 1973).

When colchicine or Colcemid is added to induce micronucleation, centrifugation of micronucleate cell monolayers in the presence of cytochalasin B yields a microkaryoplast fraction instead of karyoplasts (Ege and Ringertz, 1974). Shay and Clark (1977) demonstrated that during centrifugation the karyomeres are pulled out like "beads on a string" by the centrifugal force (Fig. 8).

Numerous techniques have been developed for cytochalasin B-induced enucleation of cell monolayers (Prescott et al., 1972; Wright and Hayflick 1972; Poste, 1973; Prescott and Kirkpatrick, 1973; Wright, 1973; Lucas et al., 1976; Veomett et al., 1976; Crenshaw et al., 1981b). Efficiencies of 80–99+% enucleation are easily achieved, depending on the technique and cell line used.

In the preparation of microkaryoplast × whole-cell hybrids, large numbers of microkaryoplasts are necessary. Only three techniques have been described for large-scale, high-efficiency enucleation of cell monolayers (>100 cm² total) with a minimum of cell loss due to growth surface breakage during centrifugation. One method involves the use of specially prepared 40-cm² glass cylinder growth surfaces with polystyrene cushions and ultracentrifugation (Wright, 1973). The method of Veomett et al. (1976) utilizes specially constructed acrylic centrifuge bottles which support 25-cm² culture flasks and permit the enucleation of 150 cm² of growth surface (six 25-cm² culture flasks) with high-speed centrifugation. Recently, we described a method for the enucleation of 300 cm² of cell monlayers (four 75-cm² culture flasks) without the need of special growth surfaces or culture flask holders (Crenshaw et al., 1980).

Briefly, cells plated in 75-cm² culture flasks (Falcon Plastics, Oxnard, California) were centrifuged at 6975g for 60 min in a Sorvall Model HG-4 swinging bucket rotor (Dupont Instruments/Sorvall, Newton, Connecticut) at 30–37°C in medium containing 10 μg/ml cytochalasin B (Fig. 8). (For details of enucleation procedure, see the Appendix.) Enucleation frequencies of 90–95% can be achieved routinely with good karyoplast and microkaryoplast viability and <2% contamination of the nucleated fraction with detached whole cells (Crenshaw et al., 1980). Once isolated, the microkaryoplast pellet is resuspended, and the contaminating karyoplasts and detached whole cells can be removed easily by gravity filtration through a 3-μm membrane filter (Ege et al., 1977).

Cytochalasin B-isolated human microkaryoplasts have been used to rescue human LMTK⁻ cells (McNeill and Brown, 1980) and 8-azaguanine-resistant human quasidiploid tumor cells (RPMI-2650) (unpublished observation).

4. Fusion of Microkaryoplasts to Whole Cells with Polyethylene Glycol

There are two basic techniques for microkaryoplast × whole-cell fusions. The first involves the use of β-propiolactone or ultraviolet light-inactivated Sendai virus as the fusing agent (Harris and Watkins, 1965). The second and currently more widely used technique employs various molecular weight polyethylene glycols (PEG) to induce cell fusion [Pontecorvo (1975); reviewed by Davidson and Gerald (1977)]. The PEG technique avoids any possible complication or uncertainty from viral DNA or other viral components.

Of the PEGs available, 1000 MW and 6000 MW are the most widely used. Clark et al. (1978) demonstrated that PEG 400 MW is just as efficient and no more cytotoxic than PEG 1000 MW and PEG 6000 MW. In addition, PEG 400 MW is easier to use, in that it is a liquid at room temperature and can be mixed directly with complete growth medium and filter-sterilized. PEG 1000 MW and PEG 6000 MW are solids at room temperature and must be weighed, autoclaved, and then mixed with serum-free growth medium while still warm.

Ten microkaryoplasts per whole cell are generally used in hybridization experiments. The fusion (described in detail in the Appendix) is carried out in suspension.

5. Conclusion

Two methods for the production of human microkaryoplasts with Colcemid have been presented. The first technique involves the use of Colcemid for the generation of cell fragments containing condensed chromosomes. The second method relies on the fact that some Colcemid-blocked human cell lines form attached micronucleated cells when plated in Colcemid-free growth medium. These micronuclei contain interphase-like decondensed chromatin and visible nucleoli and may be better suited for fusion with interphase recipient cells.

There is no single Colcemid dose which is uniformly effective for all human cell lines. Systematic titration of Colcemid concentration and length of treatment may be necessary to achieve adequate micronucleation in certain cell lines.

Microkaryoplast-mediated chromosome transfer should prove to be an effective technique in answering questions concerning human gene expression and regulation. The chromosome loss inherent in interspecific whole cell hybrids used for gene mapping purposes may be eliminated by the use of human microkaryoplasts as donor genetic material. The study of human malignant tranformation, cell differentiation, and aging represent only a few of the broad areas which may be addressed using human microkaryoplasts.

Appendix

A1. Titration of Colcemid Dose (Crenshaw et al., 1981b)

1. Grow cells in 75-cm^2 culture flasks in complete growth medium to a density of approximately 2×10^4 cells/cm^2.
2. Add Colcemid (from 10 μg/ml stock solution prepared in medium without serum, or in balanced salt solution) to final desired concentrations, initially in the range 0.02–0.2 μg/ml. Higher concentrations may be necessary for some cell lines.
3. Incubate cells at 37°C for 24–36 hr, depending upon cell doubling time.
4. Screen culture flasks for floating blocked mitotic cells with an inverted microscope. Save flasks containing floating blocked cells. Discard flasks containing normal mitotic figures.
5. For subsequent experiments with the same cell line, use a Colcemid concentration at which cells do not attach and spread following attempted mitosis.

A2. Isolation of Microkaryoplasts Containing Condensed Chromosomes

1. Once cells have been incubated in Colcemid for 24–36 hr, up to one-third of the cells will form small clusters of cell fragments with attempted mitosis. Disperse cell fragments with a Pasteur pipette and decant detached cell fragments and whole cells into sterile 15-ml conical centrifuge tubes.
2. Centrifuge the mixture (whole cells plus microkaryoplasts) at 800g for 3 min at room temperature. By centrifugation, wash cell pellet twice with complete growth medium without Colcemid.
3. Purify microkaryoplasts by filtration as described in Section A4.

A3. Isolation of Microkaryoplasts Containing Decondensed Chromosomes (Crenshaw et al., 1981b)

A3.1. Production of Attached Micronucleate Cells

a. Harvest floating blocked mitotic cells by decanting from flasks selected in Section A1 and centrifuge at 800g for 2 min at room temperature in sterile 15-ml conical centrifuge tubes.

b. By centrifugation, wash cell pellet twice in complete growth medium without Colcemid.

c. Resuspend cell pellet in complete growth medium and plate out in Falcon 75-cm² flasks at a density of 1.0–1.5 × 10⁴ "cells"/cm².

d. Incubate flasks at 37°C for 12–24 hr or until cells attach to and spread on the substrate.

A3.2. Determination of Percentage Micronucleate Cells

a. Plate out cells isolated in Section A3.1 on sterile 18 × 18 mm coverslips at a density of 1.0–1.5 × 10⁴ "cells"/cm².

b. Incubate coverslips at 37°C for 12–24 hr or until cells attach to and spread on the substrate.

c. Rinse coverslips in phosphate-buffered saline (pH 7.4).

d. Fix cells in 3% glutaraldehyde in PBS (pH 7.4) or other suitable fixative and stain for 5 min with 20 µg/ml Hoechst 33258 in normal saline.

e. Wet-mount coverslips with water and seal with nail polish.

f. Determine the percentage of the population consisting of micronucleate cells (cells with ≥2 nuclei) by counting at least 250 cells from randomly selected fields with a fluorescence microscope.

g. The formula for calculating percentage micronucleation is as follows:

$$\text{percent micronucleate} = \frac{\text{no. micronucleate}}{\text{total counted}} \times 100$$

$$- \% \text{ multinucleate cells in untreated controls}$$

A3.3. Mass Enucleation of Micronucleate Cells with Cytochalasin B and Centrifugation (Crenshaw et al., 1980)

a. *Preparation of cytochalasin B enucleation medium:*
   (i) Dissolve cytochalasin B in absolute ethanol (5 mg/ml).
   (ii) Add cytochalasin B–ethanol solution to 1.5 liters sterile growth medium (pH 7.4) without serum to a final concentration of 10 µg/ml.
   (iii) Warm cytochalasin B medium to 37°C before use.

b. *Preparation of Sorvall centrifuge with HG-4 rotor:*
   (i) Prewarm rotor to 30–37°C by running centrifuge at 4000 rpm. Maximum recommended operating temperature varies with centrifuge model.
   (ii) If centrifuge has been operated at low temperature, remove rotor once

temperature has reached room temperature and remove accumulated moisture from the rotor and centrifuge tank. If water is allowed to remain in the centrifuge, leakage onto drive motor brushes is possible. Replace rotor and continue spinning until desired temperature is reached.

 c. *Preparation of micronucleate cells for enucleation:*

  (i) Use micronucleate cells plated out in Falcon 75-cm² flasks prepared as in Section A3.1.

  (ii) Decant growth medium and replace with cytochalasin B medium (220 ml). Fill entire flask up to the middle of the neck. Seal caps tightly.

 d. *Enucleation and harvest of microkaryoplasts:*

  (i) Remove swinging buckets from HG-4 rotor and fill with approximately 440 ml water prewarmed to 37°C.

  (ii) Insert flasks into buckets with cap upward and balance opposite bucket by adding or removing water. Water level should touch the flask cap (see Fig. 8).

  (iii) Centrifuge flasks at 5000 rpm (6975g) for 50–60 min. (Time necessary for optimum enucleation may vary with cell line used.) Higher speeds will decrease necessary centrifugation time, but do not exceed the maximum speed recommended by the manufacturer (5000 rpm for Sorvall HG-4 rotor).

  (iv) Harvest microkaryoplasts and detached whole-cell contaminants by gently decanting cytochalasin B medium so as not to disturb pellet at end wall of the flask. Leave approximately 10 ml medium in the flask. Resuspend pellet with a Pasteur pipette in remaining medium.

  (v) Used cytochalasin B medium can be filter-sterilized and used approximately 15 times provided the same cell line is enucleated.

A4. Purification of Microkaryoplasts (Based on Ege et al., 1977)

 1. Microkaryoplasts isolated by either method (Section A2 or A3) can be purified by membrane filtration. Sterilize 3-μm pore size Nucleopore filters (47 mm diameter) in filter holders at 15 psi steam pressure for 15 min.

 2. Fit a 30-cm³ sterile syringe barrel to filter holder and remove whole-cell contaminants from microkaryoplast fraction by gravity filtration. Several fresh filters may be required to purify an entire preparation.

 3. Centrifuge purified microkaryoplasts at 800g at room temperature for 2 min. By centrifugation, wash microkaryoplasts twice in complete growth medium.

A5. Fusion of Microkaryoplasts to Whole Cells with Polyethylene Glycol 400 MW (Clark et al., 1978)

A5.1. Preparation of Fusion Solution

 a. Add polytheylene glycol 400 MW liquid to complete growth medium to concentration of 50% (v/v).

b. Filter-sterilize solution by passage through a 0.22-μm membrane filter under positive syringe pressure.

A5.2. Suspension Fusion Procedure

   a. Mix purified microkaryoplasts with freshly trypsin-dispersed whole recipient cells at a ratio of approximately 10:1 microkaryoplasts to whole cells in complete growth medium in a sterile 15-ml conical centrifuge tube.
   b. Centrifuge cell mixture at 800g at room temperature for 2 min.
   c. Remove as much medium as possible with a Pasteur pipette and carefully (so as not to disturb the cell pellet) add 1 ml of fusion solution.
   d. Lightly stir the cell pellet with a Pasteur pipette for 1 min at room temperature.
   e. Quickly add 10 ml of complete growth medium, disturbing the cell pellet as little as possible and centrifuge at 800g for 2 min.
   f. Decant the supernatant, gently add 2 ml of complete growth medium, and repeat centrifugation.
   g. Repeat wash procedure and incubate cell pellet in fresh growth medium at 37°C for 30–45 min.
   h. Gently resuspend fusion product using a Pasteur pipette and plate out in culture flasks at low density.

References

Barski, G., Sorieul, S., and Cornefert, F., 1961, Hybrid type cells in combined cultures of two different mammalian cell strains, *J. Natl. Cancer Inst.* **26:**1269–1291.
Blose, S. H., and Chacko, S., 1976, Rings of intermediate (100 A) filament bundles in the perinuclear region of vascular endothelial cells, *J. Cell Biol.* **70;**459–466.
Borisy, G. G., and Taylor, E. W., 1967a, The mechanism of action of colchicine: I. Binding of colchicine-^{3}H to cellular protein, *J. Cell Biol.* **34:**525–533.
Borisy, G. G., and Taylor, E. W., 1967b, The mechanism of action of colchicine: II. Colchicine binding to sea urchin eggs and the mitotic apparatus, *J. Cell Biol.* **34:**535–548.
Carter, S. B., 1967, Effects of cytochalasins on mammalian cells, *Nature* **213:**261–264.
Cheung, H. T., Cantarow, W. D., and Sundhasadas, G., 1978, Colchicine and cytochalasin B effects on random movement, spreading and adhesion of mouse macrophages, *Exp. Cell Res.* **111:**95–103.
Clark, M. A., Crenshaw, A. H., and Shay, J. W., 1978, Fusion of mammalian somatic cells with polyethylene glycol 400 MW, *Tissue Culture Assoc. Manual* **4:**801–804.
Cremer, T., Zorn, C., and Zimmer, J., 1976, Formation of viable cell fragments by treatment with colchicine, *Exp. Cell Res.* **100:**345–355.
Crenshaw, A. H., and Murrell, L. R., 1980, Micronucleation of human fibroblasts with Colcemid, *In Vitro* **16:**257–258.
Crenshaw, A. H., Shay, J. W., and Murrell, L. R., 1980, Mass enucleation of tissue culture cell monolayers, *J. Tissue Culture Methods* **6:**127–130.
Crenshaw, A. H., Shay, J. W., and Murrell, L. R., 1981a, Colcemid-induced micronucleation in cultured human cells, *J. Ultrastruct. Res.* **75:**179–186.
Crenshaw, A. H., Shay, J. W., and Murrell, L. R., 1981b, Micronucleation of human somatic cells with Colcemid, *J. Tissue Culture Methods* (in press).
Croop, J., and Holtzer, H., 1975, Response of myogenic and fibrogenic cells to cytochalasin B and Colcemid: I. Light microscopic observations, *J. Cell Biol.* **65:**271–285.

Davidson, R. L., and Gerald, P. S., 1977, Induction of mammalian somatic cell hybridization by polyethylene glycol, in: *Methods in Cell Biology*, Volume XV (D. M. Prescott, ed.), Academic Press, New York, pp. 325–338.

Ege, T., and Ringertz, N. R., 1974. Preparation of microcells by enucleation of micronucleate cells, *Exp. Cell Res.* **87**:378–382.

Ege, T., Krondahl, U., and Ringertz, N. R., 1974, Introduction of nuclei and micronuclei into cells and enucleated cytoplasms by Sendai virus induced fusion, *Exp. Cell Res.* **88**:428–432.

Ege, T., Ringertz, N. R., Hamberg, H., and Sidebottom, E., 1977, Preparation of microcells, in: *Methods in Cell Biology*, Volume XV (D. M. Prescott, ed.), Academic Press, New York, pp. 339–357.

Fournier, R. E. K., and Ruddle, F. H., 1977, Microcell-mediated transfer of murine chromosomes into mouse, Chinese hamster, and human somatic cells, *Proc. Natl. Acad. Sci. USA* **74**:319–323.

Harris, H., and Watkins, J. F., 1965, Hybrid cells derived from mouse and man: Artificial heterokaryons of mammalian cells from different species, *Nature* **205**:640.

Johnson, R. T., Mulliger, A. M., and Skaer, R. J., 1975, Perturbation of mammalian cell division: I. Human mini segregants derived from mitotic cells, *Proc. R. Soc. Lond. B* **189**:591–602.

Johnson, R. T., Mullinger, A. M., and Downes, C. S., 1978, Human minisegregant cells, in: *Methods in Cell Biology*, Volume XX (D. M. Prescott, ed.), Academic Press, New York, pp. 255–315.

Littlefield, J. W., 1966, The use of drug-resistant markers to study the hybridization of mouse fibroblasts, *Exp. Cell Res.* **41**:190–196.

Lucas, J. J., Szekely, E., and Kates, J. R., 1976, The construction of viable nuclear–cytoplasmic hybrid cells by nuclear transplantation, *Cell* **7**:397–405.

Margulis, L., 1973, Colchicine-sensitive microtubules, *Int. Rev. Cytol.* **34**:333–361.

McBride, O. W., and Ozer, H. L., 1973, Transfer of genetic information by purified metaphase chromosomes, *Proc. Natl. Acad. Sci. USA* **70**:1258–1262.

McKusick, V. A., and Ruddle, F. H., 1977, The status of the gene map of the human chromosomes, *Science* **196**:309–405.

McNeill, C. A., and Brown, R. L., 1980, Genetic manipulation by means of microcell-mediated transfer of normal human chromosomes into recipient mouse cells, *Proc. Natl. Acad. Sci. USA* **77**:5394–5398.

Mullinger, A. M., and Johnson, R. T., 1976, Perturbation of mammalian cell division: III. The topology and kinetics of extrusion subdivision, *J. Cell Sci.* **22**:243–285.

Peterson, D. F., Anderson, E. C., and Tobey, R. A., 1968, Mitotic cells as a source of synchronized cultures, in: *Methods in Cell Physiology*, Volume III (D. M. Prescott, ed.), Academic Press, New York, pp. 347–370.

Phillips, S. G., and Phillips, D. M., 1969, Sites of nucleolus production in cultured Chinese hamster cells, *J. Cell Biol.* **40**:248–268.

Pontecorvo, G., 1975, Production of independently multiplying mammalian somatic cell hybrids by polyethylene glycol (PEG) treatment, *Somat. Cell Genet.* **1**:397–400.

Poste, G., 1973, Anucleate mammalian cells: Applications in cell biology and virology, in: *Methods in Cell Biology*, Volume VII (D. M. Prescott, ed.), Academic Press, New York, pp. 211–249.

Prescott, D. M., and Kirkpatrick, J. B., 1973, Mass enucleation of cultured animal cells, in: *Methods in Cell Biology*, Volume VII (D. M. Prescott, ed.), Academic Press, New York, pp. 189–202.

Prescott, D. M., Myerson, D., and Wallace, J., 1972. Enucleation of mammalian cells with cytochalasin B, *Exp. Cell Res.* **71**:480–485.

Ruddle, F. H., and Creagan, R. P., 1975, Parasexual approaches to the genetics of man, *Ann. Rev. Genet.* **9**:407–486.

Schor, S. L., Johnson, R. T., and Mullinger, M. A., 1975, Perturbation of mammalian cell division: II. Studies on the isolation and characterization of human mini segregant cells, *J. Cell Sci.* **19**:281–303.

Shay, J. W., and Clark, M. A., 1977, Morphological studies on the enucleation of colchicine treated L-929 cells, *J. Ultrastruct. Res.* **58**:155–159.

Shay, J. W., Porter, K. R., and Prescott, D. M., 1973, The surface morphology and fine structure of CHO (Chinese hamster ovary) cells following enucleation, *Proc. Natl. Acad. Sci. USA* **71:**3059–3063.

Subblefield, E., 1964, DNA synthesis and chromosome morphology of Chinese hamster cells cultured in media containing N-desacyl-N-methyl-colchicine (Colcemid), in: *Cytogenetics of Cells in Culture*, Volume 3 (R. J. C. Harris, ed.), Academic Press, New York, pp. 223–298.

Sundar Raj, C. V., Church, R. L., Klobutcher, L. A., and Ruddle, F. H., 1977, Genetics of the connective tissue proteins: Assignment of the gene of human type I procollagen to chromosome 17 by analysis of cell hybrids and microcell hybrids, *Proc. Natl. Acad. Sci. USA* **74:**4444–4448.

Tourian, A., Johnson, R. T., Burg, K., Nicholson, S. W., and Sperling, K., 1978, Transfer of human chromosomes via human mini segregant cells into mouse cells and the quantitation of the expression of hypoxanthine phosphoribosyl-transferase in the hybrids, *J. Cell Sci.* **30:**193–209.

Veomett, G., Shay, J. W., Hough, P. V. C., and Prescott, D. M., 1976, Large scale enucleation of mammalian cells, in: *Methods in Cell Biology*, Volume XIII (D. M. Prescott, ed.), Academic Press, New York, pp. 1–3.

Weiss, M. C., and Green, H., 1967, Human–mouse hybrid cell lines containing partial complements of human chromosomes and functioning human genes, *Proc. Natl. Acad. Sci. USA* **58:**1104–1111.

Wessells, N. K., Spooner, B. S., Ash, J. F., Bradley, M. O., Luduena, M. A., Taylor, E. L., Wrenn, J. T., and Yamada, K. M., 1971, Microfilaments in cellular and developmental processes: Contractile microfilament machinery of many cell types is reversibly inhibited by cytochalasin B, *Science* **171:**135–143.

Willecke, K., and Ruddle, F. H., 1975, Transfer of the human gene for hypoxanthine-guanine phosphoribosyltransferase via isolated metaphase chromosomes into mouse L-cells, *Proc. Natl. Acad. Sci. USA* **72:**1792–1796.

Wright, W. E., 1973, The production of mass populations of anucleate cytoplasms, in: *Methods in Cell Biology*, Volume VII (D. M. Prescott, ed.), Academic Press, New York, pp. 203–210.

Wright, W. E., and Hayflick, L., 1972, Formation of anucleate and multinucleate cells in normal and SV40 transformed WI-38 by cytochalasin B, *Exp. Cell Res.* **74:**187–194.

Chapter 23
Microcell-Mediated Chromosome Transfer

R. E. K. FOURNIER

1. Introduction

Somatic cell genetics is a field which encompasses a number of parasexual techniques that can be used to transfer genetic information from one mammalian cell to another. These techniques include somatic cell hybridization (Ruddle and Creagan, 1975), microcell-mediated chromosome transfer (Fournier and Ruddle, 1977a), chromosome-mediated gene transfer (McBride and Ozer, 1973), and DNA-mediated gene transfer (Wigler et al., 1978). The distinguishing features of each approach are (a) the form in which donor genetic material is presented to recipient cells, and (b) the fraction of the donor genome ultimately retained by the cells. Microcell-mediated chromosome transfer occupies a unique position in this hierarchy: although only a fraction of the donor genome is transferred to the recipients, the introduced genetic material is retained as intact chromosomes. Thus, karyo-typically simple clones can be constructed which contain only one or a few introduced chromosomes. Since the transferred genetic material can be identified using cytogenetic tests, microcell hybrids constitute powerful gene mapping tools.

The steps involved in microcell-mediated chromosome transfer are illustrated diagrammatically in Fig. 1. Populations of donor cells are incubated under conditions which induce an aberrant mitosis. In particular, nuclear division is perturbed such that many small micronuclei are produced when the cells enter interphase. This micronucleation step serves to partition the donor chromosome complement into discrete subnuclear packets which can be physically isolated from the cells. Enucleation is accomplished using standard procedures, i.e., the micronucleate cells are centrifuged in the presence of cytochalasin B. During centrifugation, the micronuclei are drawn from the cell on long stalks, which subsequently break to yield free microcells (see Chapter 22, this volume). A microcell, then, consists of a single

---

R. E. K. FOURNIER • Department of Microbiology and the Comprehensive Cancer Center, University of Southern California School of Medicine, Los Angeles, California 90033.

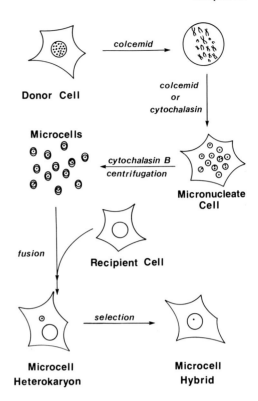

Figure 1. Diagrammatic representation of the steps involved in the preparation of microcell hybrids.

micronucleus surrounded by a thin rim of cytoplasm and an intact plasma membrane. The isolated microcells are fused with intact recipients using inactivated Sendai virus or polyethylene glycol (PEG). Thus, individual micronuclei can be delivered to a new cell type. Under appropriate selective conditions, a fraction of the microcell heterokaryons will proliferate to yield microcell hybrid clones. Such clones typically retain 1–5 introduced donor chromosomes in addition to the recipient cell chromosome complement (Fournier and Ruddle, 1977a,b,c).

The construction of microcell hybrid clones is technically straightforward. In the following sections, the steps involved in microcell-mediated chromosome transfer are discussed in detail. In general, a number of alternate techniques can be used at each step of the procedure. Therefore, a single experimental protocol which is optimal for all cell types cannot be described. Nonetheless, the methodological details presented below should facilitate the definition of optimal conditions for microcell fusions involving a wide variety of donor and recipient cells.

2. Micronucleation of the Donor Cells

Most established cell lines contain micronucleate cells, but they usually constitute only a few percent of the population. In order to generate populations composed predominantly of micronucleate cells, two alternate procedures can be employed. The first technique involves prolonged mitotic arrest (Levan, 1954; Stubblefield, 1964; Phillips and Phillips, 1969). Micronucleate populations can also be obtained by plating mitotic cells in low concentrations of cytochalasin B (Fournier, 1981).

2.1. Prolonged Arrest Micronucleation

The majority of microcell fusions reported to date have used micronucleate donor cells generated by prolonged mitotic arrest. This can be accomplished by incubating growing cultures in the presence of colchicine, Colcemid, or other inhibitors of microtubule polymerization. Initially, cells arrested in metaphase will accumulate since functional mitotic spindles cannot form in the presence of the drug. Within 8–20 hr, however, many of the mitotic cells will reenter interphase, a large fraction of them becoming micronucleate in the process (Stubblefield, 1964; Ege et al., 1977). Further incubation will generate populations composed primarily of micronucleate cells.

Although a number of mitotic arrest agents (colchicine, Colcemid, vinblastine sulfate, nitrous oxide at high pressure) can be used to induce micronucleation in mammalian cells, Colcemid has been the agent of choice owing to its low cytotoxicity. Two variables should be defined in order to optimize micronucleation conditions for a given cell line. First, the appropriate Colcemid (or other mitotic arrest agent) dose should be determined. In general, the effective concentration range is quite narrow. For example, mouse B82 cells can be micronucleated by exposure to 0.10 μg/ml Colcemid. Treating the cells with 0.05 μg/ml Colcemid has no effect, while 0.50 μg/ml results in excessive cytotoxicity. Simple calibration experiments can be performed to determine the appropriate Colcemid dose; phase contrast examination of the living cultures is usually sufficient to define that concentration.

The effective Colcemid concentration for micronucleation of all rodent cells tested has been in the range of 0.01–0.10 μg/ml. Cells that can be micronucleated under these conditions include a variety of established lines as well as diploid fibroblast cultures explanted directly from mice, hamsters, or rats. In contrast, human cells are more difficult to micronucleate by prolonged mitotic arrest. As first shown by McNeill and Brown (1980), however, diploid human fibroblasts can be rendered micronucleate by incubation in the presence of high concentrations (10–20 μg/ml) of Colcemid. Micronucleate human cells can also be prepared by treating the cultures sequentially with Colcemid and cytochalasin B (Fournier, 1981), as discussed below. (Also see Crenshaw and Murrell, Chapter 22, this volume.)

The second important variable is the duration of Colcemid arrest. We previously demonstrated that this factor critically influenced the ability of micronucleate populations to serve as donors in microcell fusions (Fournier and Ruddle, 1977a). Specifically, microcells derived from donor L cells treated with Colcemid for 48 hr were more than ten times as efficient in terms of hybrid yield than microcells isolated from donors arrested for 72 or 96 hr. Therefore, the donor cells should be incubated in Colcemid for the least possible period consistent with good micronucleation. Within the constraints imposed by the necessity of minimizing the duration of Colcemid arrest, conditions should be defined under which (a) the percent of the population that is micronucleate is maximized, and (b) the micronucleate cells are well attached to the substratum for enucleation. These conditions, which usually correspond to incubation in the presence of Colcemid for 1–2 cell generations, can be defined from calibration experiments similar to those used to determine the effective Colcemid dose (above).

The kinetics of induction of micronucleate mouse L cells in the presence of 0.02 µg/ml Colcemid is shown in Fig. 2. Mitotic cells begin to accumulate in the population upon addition of the drug, with a corresponding decrease in the frequency of mononucleate cells in the culture. Micronucleate cells appear 8–12 hr later as mitotic cells escape the metaphase block. At 28 hr of incubation, 60% of the cells have become micronucleate; this corresponds to about 80% of the attached population at this time (the mitotic cells are free-floating or tenuously attached and would be removed prior to enucleation). For the next 16 hr, the frequency of micronucleate cells decreases as some of these cells enter a second mitosis, usually generating cells with even more micronuclei (Stubblefield, 1964; Ege et al., 1977). By 48 hr, greater than 95% of the attached cells (approximately 55% of the population) have become micronucleate. Thus, a similar fraction of the *total* population was micronucleate at 28 and 48 hr. However, only attached cells are processed for

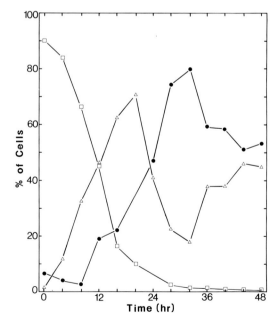

Figure 2. Kinetics of induction of micronucleate cells by prolonged mitotic arrest. At $t = 0$, Colcemid was added to a series of parallel cultures of exponentially growing mouse L cells (A9). The populations were harvested at intervals; both adherent and nonadherent cells were collected. The cell suspensions were stained with aceto-orcein, and 200 cells scored for each point. Mononucleate cells (□), mitotic cells (△), micronucleate cells (●).

enucleation, and micronucleate cells constitute a larger percentage of this population at 48 than at 28 hr. Furthermore, relatively few cells are firmly attached at 28 hr, which results in inefficient enucleation. Therefore, L-cell cultures are generally micronucleated by exposure to Colcemid for 40–48 hr. In contrast, 60–80% of the cells in vigorously growing diploid fibroblast cultures prepared from mouse embryos can be micronucleated by incubation in 0.02–0.05 µg/ml Colcemid for 24 hr. Since these cells are firmly attached to the substratum at this time, they are enucleated after only 24 hr of mitotic arrest.

A general protocol for the production of micronucleate populations suitable for use as donors in microcell fusions is outlined below. In order to ensure that the cultures are in exponential growth during Colcemid treatment, the cells are plated 1 day prior to addition of the mitotic arrest agent. The cells are usually seeded directly onto the surfaces from which they will be enucleated; alternately, they can be micronucleated in normal culture vessels and replated for enucleation (see following section). In either case, the cultures should be 20–40% confluent at the time the mitotic arrest agent is added, and should be growing vigorously. At 16–24 hr after plating, Colcemid is added to the appropriate final concentration (Colcemid stocks are stored as frozen aliquots of 10 µg/ml saline). The cultures are incubated 24–48 hr and processed for enucleation. A total of 70–90% of the attached cells in the cultures should be micronucleate at this time.

2.2. Sequential Treatment Micronucleation

Microcell fusions are commonly less efficient than whole-cell hybridizations performed under similar conditions. In a systematic study, we recently observed that micronucleate cells generated by prolonged mitotic arrest were only one-tenth as efficient in hybridization experiments as the mononucleate cells from which they were derived (Fournier, 1981). This finding prompted us to develop a micronucleation protocol which did not depend upon prolonged mitotic arrest. The strategy employed was to generate pure populations of mitotic cells by limited Colcemid arrest and to replate the cells under conditions known to perturb normal progress through mitosis. Plating mitotic cells in the presence of low concentrations of cytochalasin B was found to induce micronucleation most effectively.

Populations of mitotic cells displayed a tendency to become progressively more micronucleate when replated in normal medium as the duration of Colcemid arrest was increased. For example, 7% of mitotic LMTK⁻ cells collected after 2-hr arrest with 0.02 µg/ml Colcemid became micronucleate when the drug was removed. In contrast, 50% of the cells became micronucleate upon replating after exposure to Colcemid for 16 hr. The tendency of mitotic populations to undergo micronucleation could be augmented by plating the cells in the presence of 1 µg/ml cytochalasin B. Thus, micronucleation could be induced in 70 and 90% of the mitotic cells collected after

Colcemid arrest for 2 and 16 hr, respectively. Although some mitotic cells collected after limited Colcemid exposure (e.g., 2 hr) became binucleate rather than micronucleate when replated in cytochalasin B-containing medium (Dickerman and Goldman, 1973; Fournier and Pardee, 1975), this could be minimized by using mitotic cells collected after Colcemid arrest for 8–16 hr.

Micronucleate populations generated by sequential exposure to Colcemid and cytochalasin B were ten times as efficient in hybridization experiments as micronucleate cells produced by prolonged mitotic arrest, i.e., they were as efficient as the mononucleate cells from which they were derived (Fournier, 1981). Furthermore, micronucleation can be induced in a variety of cell lines using this technique. This includes human cells (HT1080 and D98), which are difficult to micronucleate by prolonged mitotic arrest.

The kinetics of induction of micronucleate cells in the presence of cytochalasin B is shown in Fig. 3. Mitotic A9 cells were collected after arrest with 0.02 µg/ml colcemid for 16 hr and replated in medium containing 1 µg/ml cytochalasin B. A relatively synchronous release from metaphase with concomitant micronucleation was observed upon replating. Within 4 hr, 80% of the cells had become micronucleate. In contrast to the asynchronous nature of micronucleate populations generated by prolonged mitotic arrest, these cultures were quasisynchronous (G1) populations of micronucleate cells.

A general procedure for micronucleation of mammalian cells by sequential treatment with Colcemid and cytochalasin B is outlined below. Exponentially growing cultures are incubated in the presence of the appropriate concentration of Colcemid (see preceding section) for 8–16 hr. Cells arrested

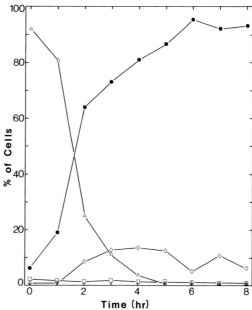

Figure 3. Kinetics of induction of micronucleate cells by sequential Colcemid–cytochalasin B treatment. Mitotic mouse L cells (A9) were collected after arrest with 0.02 µg/ml Colcemid for 16 hr and replated in medium contain 1 µg/ml cytochalasin B ($t = 0$). The populations were harvested and scored as described in the legend to Fig. 2. Mitotic cells (△), micronucleate cells (●), mononucleate cells (□), binucleate cells (◇).

in metaphase are collected by selective mitotic detachment. Ideally, >90% of the collected cells should be in mitosis. The population is pelleted by centrifugation at 600g for 5 min at room temperature, and immediately resuspended in complete medium containing 1 μg/ml cytochalasin B. The cells are replated and incubated at 37°C; in general, they will be plated onto the surfaces from which they will be enucleated at this time. After 4–6 hr, most of the cells will have completed mitosis, become micronucleate, and flattened on the substratum. The cultures can then be processed for enucleation.

2.3. Micronucleation—General Observations

Either of the procedures outlined above can be used to generate micronucleate cells suitable for use as donors in microcell-mediated chromosome transfer experiments. In the sequential protocol, metaphase-arrested cells are plated in the presence of cytochalasin B. Therefore, the utility of this approach is limited to instances where it is difficult to obtain relatively pure populations of mitotic cells, e.g., for slow-growing cultures or for cells in which selective mitotic detachment is difficult. In these cases prolonged mitotic arrest may be the method of choice. We routinely use this technique for donor cells that can be micronucleated by exposure to Colcemid for one cell generation (e.g., mouse embryo fibroblasts). However, some human cell lines are difficult to micronucleate by prolonged mitotic arrest; in these cases the sequential protocol can be employed. Although microcells derived from sequential treatment donors are more efficient in microcell fusions than those isolated from prolonged arrest donors, it is usually possible to generate microcell hybrid clones at an acceptable frequency by either route.

3. Enucleation of Micronucleate Populations

Cytochalasin B is a fungal metabolite that affects mammalian cells in a number of interesting ways (Carter, 1967). When incubated in the presence of relatively high concentrations of the drug (10 μg/ml), some cells undergo nuclear extrusion. This phenomenon forms the basis of techniques developed for mass enucleation of mammalian cells (Prescott et al., 1972; Poste and Reeve, 1972; Wright and Hayflick, 1972). The basic approach is to centrifuge monolayers attached to plastic or glass surfaces in the presence of cytochalasin B. Under these conditions, the nuclei are drawn from the cells on long cytoplasmic stalks which are broken by centrifugal force. The anucleate cytoplasms (cytoplasts) remain attached to the substratum. The nuclei, surrounded by a thin rim of cytoplasm and an intact plasma membrane (karyoplasts), are recovered from the bottom of the tube. Subsequently, a suspension enucleation procedure was described which employs cytochalasin B-containing density gradients (Wigler and Weinstein, 1975). By applying these enucleation techniques to micronucleate populations of cells, isolated microcell preparations may be obtained (Ege and Ringertz, 1974).

3.1. Monolayer Enucleation Techniques

The most generally useful enucleation techniques are those in which cell monolayers are centrifuged in the presence of cytochalasin B. Plastic or glass surfaces to which the cells are attached are spun with the monolayer facing away from the axis of rotation. The nature of the surfaces on which the cells are grown and the geometry of the cultures during enucleation constitute the main differences among the monolayer enucleation techniques that have been described. Procedures used to obtain isolated microcells from micronucleate populations are outlined below.

3.1.1. Enucleation from Plastic Disks

In early microcell fusions (Fournier and Ruddle, 1977a,b,c), micronucleate populations attached to circular plastic disks were employed for enucleation. The disks are cut from tissue culture plates using a 1-inch-diameter cork borer that has been heated by flaming. The disks are immediately removed from the cork borer, and the outside surface (not treated for cell attachment) marked with a scratch. The edges of irregular disks are filed smooth, and the disks stored in 95% EtOH.

For plating, eight disks are transferred, appropriate side up, to a 100-mm tissue culture dish and dried in a 60°C oven. In general, two plates (16 disks) are processed per fusion. Then 15 ml complete medium is added to each plate, and the disks pressed to the bottom of the dish. A suspension of donor cells (2–5 ml) is added, taking care to distribute the suspension evenly over the entire surface. For prolonged arrest micronucleation, the cells are seeded at 20–40% confluence and Colcemid added after 16–24 hr. Alternately, mitotic cells can be seeded onto the disks in medium containing 1 μg/ml cytochalasin B (above).

After micronucleation, the cultures growing on disks are processed as follows. For some cell lines it is advantageous to spin the populations first in the absence of cytochalasin B to remove loosely adherent cells which would otherwise contaminate the microcell preparation. Each disk is transferred to a 50 ml polycarbonate centrifuge tube containing 6 ml serum-free medium. The disks are placed in the tube, cell side down; it is not necessary to fix the disks in place, as they will assume an orientation perpendicular to the centrifugal force during centrifugation. The tubes are spun in an SS-34 rotor of a Sorvall RC-5B centrifuge at 34°C (this temperature can be obtained by setting the temperature at 28–33°C and spinning the empty rotor at 15,000 rpm for 15–20 min). For many cell lines (e.g., L cells), appropriate pre-enucleation centrifugation conditions are 12,000g (10,000 rpm) for 10 min.

For enucleation, the disks are transferred to a second set of tubes, each containing 6 ml serum-free medium + 10 μg/ml cytochalasin B (stock solutions of cytochalasin B in dimethyl sulfoxide, 2 mg/ml, are stable for long periods at 4°C). The tubes are spun at 39,000g (18,000 rpm) for 20 min. After centrifugation, a small pellet should be visible in each tube. The disks are removed with sterile forceps, the medium decanted, and each microcell pellet suspended in 0.5–1.0 ml serum-free medium and pooled.

For cells that attach firmly to plastic (e.g., diploid fibroblasts), centrifuging the monolayers in the absence of cytochalasin B prior to enucleation makes no detectable difference in the composition of the isolated microcell preparation, and need not be employed. In other cases the pre-spin may actually loosen the monolayer and thereby increase contamination of the microcell preparation with intact donor cells. Therefore, appropriate centrifugation conditions must be determined empirically.

Enucleation of cell monolayers on plastic disks is a relatively convenient technique. The main disadvantage of this approach is that only a small surface area (approximately 39 cm^2) can be processed per spin, thus limiting the number of cells enucleated and the number of microcells obtained.

3.1.2. Enucleation from Plastic "Bullets"

This technique is very similar to that described above, but the surfaces that can be processed are considerably larger. In this procedure, bullet-shaped pieces of plastic are cut from tissue culture plates. The bullets are approximately 24 × 86 mm and have one rounded end; they fit round end down into 50-ml centrifuge tubes (see Hightower and Lucas, Chapter 19, this volume).

The bullets are cut from tissue culture plates using a hot-wire cutter. This consists of a nichrome wire connected to a variable-voltage transformer. The wire should be hot enough to cut through the plastic without the use of excessive force, but cool enough not to warp the edges of the bullets. After cutting, the edges of the bullets are filed smooth.

The bullets are stored and sterilized as described above for plastic disks. For plating, four bullets are placed in a 150-mm tissue culture plate, and cells seeded as described above. A total of 40–50 ml of medium will be needed per plate. For enucleation, each bullet is transferred to a 50-ml centrifuge tube containing 38 ml serum-free medium + 10 µg/ml cytochalasin B, and 8–16 bullets are processed per fusion. The bullets will not withstand g forces comparable to those used for enucleating cultures attached to disks; therefore, longer spins at slower speeds are employed. For example, mouse L cells and diploid fibroblasts are enucleated by centrifugation at 27,000g (15,000 rpm) for 30 min at 34°C in an SS-34 rotor of a Sorvall RC-5B. After enucleation the bullets are removed, the medium decanted, and the microcell pellets resuspended in serum-free medium and pooled.

Plastic bullets (and disks) can be reused many times, provided surfaces containing cytoplasmic remnants are prevented from drying. After enucleation, the bullets are removed from the tubes and placed in 0.5% SDS. The bullets are wiped with a soft cloth and stored in 95% EtOH. AFter 20–30 centrifugations, the plastic will fatigue and new bullets should be prepared.

The same considerations with respect to utility of a prespin and effective enucleation conditions apply to bullets and to disks, i.e., appropriate centrifugation conditions should be determined empirically. The main advantage of enucleation using bullets is the greater surface area which can be processed per spin, 162 cm^2 vs. 39 cm^2 for disks. It is also possible to place two bullets back to back in a single tube, doubling the amount of material obtained per spin. Bullet enucleation thus combines the ease of manipulation of disks (above) with the greater capacity of enucleation from T-flasks (below), and we consider it the technique of choice for enucleating micronucleate cell populations.

3.1.3. Enucleation from T-Flasks

Cell monolayers in 25-cm^2 T-flasks can be enucleated by centrifuging the flasks themselves. This can be done by fabricating special adaptors for the T-flasks which fit into a centrifuge rotor or by taking precautions to ensure the integrity of the flasks during centrifugation. Only the latter appoach will be discussed briefly here (see also Veomett, Chapter 6, this volume).

The 25-cm^2 flasks fit into the wells of large, fixed-angle rotors, e.g., the GSA rotor of a Sorvall RC-5B centrifuge. Two steps are necessary to minimize the possibility of breakage of the flasks during centrifugation. First, the flasks are wrapped with tape such that 4–5 layers of tape cushion the (upper) corners of the flask which will be pressed against the rotor well during centrifugation. Second, the wells are filled with water to the level of the neck of each flask. Under these conditions, breakage will be minimized.

A typical protocol would be as follows. Flasks containing micronucleate cells are filled completely with serum-free medium containing 10 μg/ml cytochalasin B. The corners of the flask are wrapped with tape, and each flask placed in a well of a Sorvall GSA rotor with the upper surface of the flask facing away from the axis of rotation. Water is added to each well to the neck of the flask. For mouse L cells and diploid fibroblasts, the flasks are spun at 9900g (8000 rpm) for 60 min at 34°C (this approaches the upper limit of g forces the flasks will withstand). After centrifugation, the microcell "pellet" is smeared over the entire upper surface of the flask; it may be necessary to use trypsin to remove this material from the flask.

Enucleation from T-flasks is rather cumbersome, and prespins without cytochalasin are largely precluded. The main advantage of this technique is that 150 cm^2 of surface can be processed per spin. However, 162 cm^2 (324 cm^2 if two bullets per tube are used) can be processed per spin using bullets. In addition, enucleations using T-flasks takes twice as long, since they cannot be centrifuged at comparable g forces. Thus, bullet enucleation is recommended over the use of T-flasks.

3.1.4. Enucleation Using Concanavalin A-Coated Surfaces

The techniques outlined above require that donor cells be firmly attached to the substratum for efficient enucleation. However, it may be desirable to prepare microcells from cultures that attach only poorly to plastic or that grow in suspension. In these cases, the cells may be plated onto plastic surfaces to which concanavalin A (Con A) has been covalently linked (Gopalakrishnan and Thompson, 1975).

The procedure described below concerns the preparation and use of Con A-coated bullets; with suitable modifications it could be employed for any of the surfaces discussed above. Four sterile, dry bullets in each of 2–4 150-mm plates are processed. Each bullet is overlaid with 0.6 ml of a freshly prepared solution of the cross-linking agent 1-cyclohexyl-3-(2-morpholinoethyl)-carbodiimide metho-p-toluenesulfonate (Aldrich) (75 mg/ml saline). A 15 mg/ml solution of Con A in saline is prepared, filtered through Whatman #1 paper, and filter-sterilized. Then 0.6 ml of this solution is added to each bullet and the solution distributed over the entire surface. After 1–2 hr at room temperature, the bullets are rinsed three times with phosphate-buffered saline (PBS). Bullets prepared in this manner can be used immediately or stored in PBS.

Micronucleate donor cells are pelleted by centrifugation and resuspended in sterile PBS at a cell concentration of about 10^6/ml. A 1.2-ml cell suspension is pipetted onto the surface of each bullet and allowed to settle for 10–15 min at room temperature. Most of the cells adhere to the Con A-coated surface in this period. Complete medium is added to each plate and the cultures are incubated at 37°C. Within 30–60 min, the cells will be well spread and ready for enucleation.

This is a valuable technique, which should be explored as an alternative to suspension enucleation (below) for cells that attach poorly to plastic. For example, H4IIEC3 rat hepatoma cells cannot be enucleated as described in the preceding sections, since the g forces required for enucleation are greater than those that strip intact cells from substratum. By plating these cells onto Con A-coated surfaces, effective enucleation can be achieved.

3.2. Suspension Enucleation

Mammalian cells can be enucleated in discontinuous Ficoll gradients containing cytochalasin B (Wigler and Weinstein, 1975). Cells centrifuged in such gradients experience a shear force owing to the different buoyant densities of nucleoplasm vs. cytoplasm. In the presence of cytochalasin B, this force can be sufficient to cause enucleation, and the cytoplasmic and nuclear particles can be recovered from different parts of the gradient. The advantages of this technique are (a) large numbers of cells can be processed, and (b) attachment to a substratum is not required (see Veomett, Chapter 6, this volume).

Not all cells can be enucleated efficiently using this technique. For example, cells with scant cytoplasm (e.g., lymphocytes) tend to band in the region of the gradient normally occupied by karyoplasts. When micronucleate cells are centrifuged in these gradients, only partial enucleation generally results, i.e., some micronuclei are removed from the cells and the partially enucleated cells, still containing micronuclei, band in the vicinity of totally enucleated cytoplasts. Thus, the yield of free microcells is considerably less than expected, and seldom exceeds that which can be obtained by enucleating cultures attached to plastic bullets (above). Therefore, the general utility of this approach for enucleating micronucleate populations is limited. However, this procedure can be employed in cases where monolayer enucleation is not feasible.

3.3. Enucleation—General Observations

The crude microcell preparations obtained by enucleating micronucleate cells include a number of different kinds of particles. In addition to microcells, small cytoplasmic vesicles devoid of nuclear material will be present. The level of contamination with intact donor cells (mononucleate and micronucleate) generally varies from line to line. Isolated karyoplasts, derived from enucleation of mononucleate donor cells, may also be obtained. Generally, 80–90% of the particles in the preparation will exclude trypan blue.

The particle composition of the isolated preparation can be determined by examining stained material. One convenient stain is aceto-orcein (0.5% orcein in 50% acetic acid). Wet mounts can be prepared by placing one drop of the particle suspension and one drop of aceto-orcein on a microscope slide and quickly covering the material with a coverslip. After 2–5 min, nuclear material will stain red and cytoplasmic material pink. This is also a useful procedure for resolving mitotic, micronucleate, and mononucleate cells of the donor cell preparation. Alternately, an aliquot of the suspension can be fixed in 3:1 methanol:acetic acid and dropped onto slides. The dried material can be stained with acridine orange, Hoechst 33258, or other fluorochromes. The total particle yield can be determined using a hemocytometer, and the number of microcells can be calculated. Quantitations of this type are useful for defining optimal centrifugation conditions for a given cell line (see Crenshaw and Murrell, Chapter 22, this volume).

4. Purification of Isolated Microcell Preparations

Isolated microcell preparations are heterogeneous collections of particles, which include microcells, cytoplasmic vesicles, karyoplasts, and intact cells. The karyoplasts are derived from enucleation of mononucleate cells present in the population of micronucleate donors. These particles can fuse with recipient cells; if they constitute a significant fraction of the population, whole-cell hybrids containing many donor chromosomes will be generated along with microcell hybrid clones. Such hybrids can also be produced by fusion of intact donors with recipient cells. Furthermore, if half-selective conditions that kill recipients but not donors are used, colonies will result from intact donors present as contaminants in the microcell preparation. Therefore, it may be desirable to purify the crude microcell preparation prior to fusion. Three purification procedures are outlined below.

4.1. Purification of Nonadherent Particles

Microcells are essentially nonadherent particles which do not attach to culture surfaces. In contrast, intact donor cells and some of the karyoplasts may form firm attachments to a substratum. Therefore, by collecting nonadherent particles after replating, an enriched microcell preparation can be obtained.

The time required for intact donor cells to attach to culture vessels varies among different cell lines, and the appropriate interval between replating and collection of nonadherent particles should be determined empirically. This can be accomplished by suspending the crude microcell preparation in complete medium and plating aliquots into a series of flasks. Nonadherent particles are collected at intervals, replated in complete medium, and colonies of surviving intact donor cells counted after 10–20 days. The appropriate intervals for replacing a variety of lines fall in the range of 1–4 hr, and intact donor colonies can generally be reduced 50- to 200-fold.

This simple technique can be very effective for eliminating intact donor cells from crude microcell preparations, but karyoplasts are not removed in this procedure.

4.2. Purification by Membrane Filtration

Microcells can be purified from karyoplasts and intact cells by separating the particles according to size and collecting the smallest particles for fusion. One way to accomplish this is by membrane filtration (McNeill and Brown, 1980; Fournier, 1981). Polycarbonate membrane filters are employed; Uni-Pore filters (Bio-Rad) with pore sizes of 8 or 5 μm have proved satisfactory.

The crude microcell preparation should be fairly dilute for filtration. Typically, the material is suspended in 10–20 ml serum-free medium [1–5 $\times$ 10^6 particles/ml] and 5-ml aliquots filtered separately. For filtration, polycarbonate filters are mounted in Swinnex (25-mm) adaptors (Millipore), autoclaved, and attached to sterile syringes. The suspended preparation is gently pushed through the filters, and the filtrate recovered.

The appropriate pore size for microcell preparations isolated from a given donor cell line should be determined empirically. For preparations in which large particles (karyoplasts and intact cells) are abundant (10–20% of particles), sequential filtration (e.g., 8-μm filters followed by 5-μm filters) may be required.

Membrane filtration is a simple and fast purification technique, but it may not be possible to remove all intact cells from heavily contaminated preparations without significant loss of microcells. In many instances, however, this approach is entirely satisfactory, and constitutes the purification scheme of choice.

4.3. Purification by Unit Gravity Sedimentation

Microcell preparations can be size-fractionated by sedimentation through unit gravity density gradients (Fournier and Ruddle, 1977a). Linear gradients of 1–3% bovine serum albumin (BSA) are employed.

Tubing from a gradient mixing device (LKB, 8121) is run through a peristaltic pump (e.g., Cole Parmer 7545 with 7013 head) to a three-way valve (B-D). One outlet of the valve is connected to a 5-ml syringe which serves as a bubble trap. The other outlet is connected by a short length of tubing to a second three-way valve. The upper outlet of this valve is attached to the bottom of a 50-ml syringe which constitutes the gradient chamber. Three small glass beads are positioned over the inlet (bottom) of the chamber. A short piece of tubing connected to the other outlet is the sampling tube. The entire apparatus can be autoclaved before use.

Sterile PBS is added to the gradient mixing device and pumped to fill the tubing just to the bottom of the gradient chamber. Excess PBS is pumped into the bubble trap. Then 25 ml 3% BSA in PBS is added to one side of the mixing device, and 25 ml 1% BSA in PBS (containing phenol red) to the other. The connection between the chambers is opened and the stirrer motor started. The crude microcell preparation suspended in 2 ml 0.5% BSA in PBS is added to the gradient chamber from the top. The pump is started, and PBS in the lines pumped into the bubble trap. When the BSA solution (red) reaches the trap, the valve is turned so that the solution enters the gradient chamber. The BSA gradient is pumped into the chamber from under the sample layer at a rate of 2–3 ml/min. Total volume of the gradient is 50 ml.

The microcell preparation is allowed to sediment through the gradient for 3–3.5 hr, and fractions collected by dripping through the sampling tube. It is important to avoid collecting the original sample layer. This fractionation scheme efficiently separates the initially heterogeneous mixture into various size classes. Virtually all whole cells and karyoplasts are confined to the bottom 15–20 ml of the gradient. The purified microcell preparation can be recovered from the top 20–25 ml.

Unit gravity sedimentation is a very efficient technique for microcell purification, and the purified material that can be obtained is virtually free of karyoplasts and intact cells. Thus, even in fusions in which colonies are isolated under half-selective conditions, only true microcell hybrid clones are generally obtained. The main disadvantage of this approach is that is is time-consuming. Nonetheless, for those fusions in which stringent purification of microcells is required, this approach should be explored.

4.4. Microcell Purification—General Observations

Microcell preparations isolated from donors that attach firmly to plastic (e.g., embryo fibroblasts) can be used for fusion without prior purification. In other cases, intact donor cell colonies and/or whole-cell hybrids can be eliminated by direct selective pressure. However, the ability to generate relatively pure populations of microcells can be useful when a single selectable marker (in either donor or recipient cells) is available for the genetic cross of interest.

5. Fusion of Microcells with Intact Recipients

The most efficient microcell hybridization protocols are those in which donor microcells in suspension are fused with monolayers of recipient cells. Two such suspension/monolayer techniques are described below; one uses polyethylene glycol (PEG) and one uses inactivated Sendai virus as the fusogen.

5.1. Suspension/Monolayer Fusion with Phytohemagglutinin-P and Polyethylene Glycol

Phytohemagglutinin-P (PHA-P) enhances polyethylene glycol (PEG) mediated fusion of mammalian cells (Mercer and Schlegel, 1979). In the following protocol, PHA-P is used to agglutinate isolated microcells to a recipient monolayer, and fusion is subsequently induced by exposing the culture to PEG (Fournier, 1981).

An isolated microcell preparation [typically $0.5-2 \times 10^7$ microcells] is pelleted by centrifugation and resuspended in 2 ml serum-free medium containing 100 µg/ml PHA-P (Difco; only 50% of the weight of the lyophilized powder is PHA-P). The microcell suspension is added to a 70-80% confluent monolayer of recipient cells in a 25-cm^2 flask that has been rinsed with serum-free medium. The culture is incubated at 37°C to allow the microcells to agglutinate to the recipient monolayer, usually 10-15 min. The PHA-P-containing medium is completely removed, 1 ml PEG (PEG 1540, Baker) solution in serum-free medium is added, and the PEG solution allowed to cover the monolayer. After 1.0 min, the PEG solution is aspirated off, and the monolayer rinsed quickly with three 5-ml washes of serum-free medium. Complete medium is added, and the culture incubated 16-24 hr at 37°C. The cells are harvested by trypsinization and distributed into 10-20 25-cm^2 flasks in selective medium. Microcell hybrid clones should be visible in 2-4 weeks.

A critical variable in this technique is the PEG concentration. In general, the appropriate concentration will be less than that employed in the absence of PHA-P. Effective concentrations of PEG 1540 fall in the range 35-50% (wt/wt), with many fusions requiring 42-44% PEG. The appropriate PEG concentration is determined empirically for each recipient cell line.

5.2. Suspension/Monolayer Fusion with Inactivated Sendai Virus

Microcells can be fused to recipient monolayers with the use of inactivated Sendai virus as follows (Fournier and Ruddle, 1977 a,b). Microcells are pelleted by centrifugation and suspended in 0.5 ml serum-free medium. The suspension is added to a confluent monolayer of recipient cells in a 25-cm² flask that has been rinsed with serum-free medium. Then 0.5 ml inactivated Sendai virus suspension is added to the flask. Since different batches of virus differ in their fusogenic activity, the appropriate titer of virus must be determined empirically. This will usually be similar to the titer used to fuse intact donors and recipients in a particular cross, e.g., 500–5000 hemagglutination units per ml. The culture is incubated on ice 15–30 min, transferred to 37°C, and incubated 30–90 min. The virus-containing medium is removed, complete nonselective medium added, and the flask incubated overnight. Subsequent processing is as described in the preceding section.

5.3. Microcell Fusion—General Observations

Monolayer fusion protocols are generally more efficient than suspension hybridizations involving the same parental cells (O'Malley and Davidson, 1977). Thus, if the recipient cells of interest grow as attached cultures, the fusion protocols outlined above can be employed. PEG has largely replaced inactivated Sendai virus as the fusogen of choice, as it is very effective and readily available. For recipients that are particularly fusogenic (e.g., Chinese hamster ovary cells), efficient fusion can also be obtained using Sendai virus.

Although fusion protocols in which both donor microcells and intact recipients are hybridized in suspension are less efficient than the procedures outlined above, these techniques may be used if the recipients grow only in suspension or if they are very fusogenic. Suspension fusion procedures developed for cell–cell hybridizations can be employed (e.g., Davidson and Gerald, 1976). Microcells have been successfully fused in suspension with PEG in both the presence (McNeill and Brown, 1980) and absence (Worton et al., 1981) of PHA-P.

6. Concluding Remarks

The microcell hybrid clones that can be constructed using the techniques outlined in this chapter are useful genetic tools. One advantage of this approach is that every clone is karyotypically simple, and contains only one or a few introduced donor chromosomes. Thus, the extensive cultivation and reanalysis often required before interspecific whole-cell hybrids can be used for genetic studies is largely circumvented. Furthermore, the direction of chromosome segregation in microcell hybrids is predetermined by experimental design. Therefore, clones can be generated with chromosome sets not generally obtained with traditional hybrid cells. One example of such a situation is microcell hybrids that retain one or a few mouse chromosomes in a human background. We previously used clones of this type for mapping murine genes (Kozak et al., 1979; J. R. Landolph and R. E. K. Fournier, submitted for publication). Microcell hybridization also provides an approach for gene mapping and genetic analysis using intraspecific hybrid cells (Worton et al., 1981).

In addition to providing an effective means for mapping cellular genes, microcell hybrids have been extensively used to define the chromosomal integration sites of foreign DNA (Fourier and Ruddle, 1977c; Fournier et al., 1979; Smiley et al., 1978). In these studies, heteroploid cell lines containing a foreign gene encoding a selectable trait were used as donors in microcell fusions. The identities of the donor chromosomes containing the integrated sequences were established in a straightforward manner by analyzing the microcell hybrid clones. Since these chromosomes were derived from heteroploid donors, they were usually distinct from those of the normal diploid complement. Karyotypically simple microcell hybrids permitted their easy identification; this would have been exceedingly difficult using traditional whole-cell hybrids. Furthermore, microcell hybrid clones containing a specific donor chromosome into which a foreign selectable marker had integrated have been used for mapping cellular genes (Leinwand et al., 1978).

Recent studies have demonstrated that microcell hybrids can be useful for the regional localization of mammalian genes (R. E. K. Fournier and R. G. Moran, submitted for publication). Specifically, diploid donor chromosomes can be arranged *de novo* when introduced into particular recipient cell lines by microcell-mediated chromosome transfer. Many such rearrangements are simple deletions and interspecific translocations. Thus, this forms the basis of a deletion mapping strategy for the regional localization of mammalian genes.

Microcell-mediated chromosome transfer can be used to generate simple hybrid clones with precisely defined karyotypes that should prove useful in a host of genetic studies. The most advantageous materials are those clones that retain a single donor chromosome which is fixed in the cells by direct selective pressure. The karyotypes of such clones are not only simple, they are also both stable and homogeneous. One strategy for fixing different donor chromosomes in a series of microcell hybrid clones is to use wild-type donors in microcell fusions with a series of mutant recipients harboring recessive lesions. Microcell hybrids in which complementation has occurred can be directly selected (R. E. K. Fournier and R. G. Moran, submitted for publication). Alternately, cells containing defined translocations between chromosomes carrying a selectable marker and other autosomes can be employed as microcell donors (Fournier and Frelinger, 1982). Either of these approaches will allow us to construct novel monochromosomal hybrid panels (MHP) in which each clone retains a single, specific donor chromosome. Such panels should prove useful tools with wide applicability in mammalian somatic cell genetics.

Acknowledgments

Work performed in the author's laboratory was supported by grants GM26449 from the National Institutes of Health and IN-21-T from the

American Cancer Society. The author is the recipient of an American Cancer Society Junior Faculty Research Award (JFRA-25).

References

Carter, S. B., 1967, Effects of cytochalasins on mammalian cells, *Nature* **213**:261-264.
Davidson, R. L., and Gerald, P. S., 1976, Improved techniques for the induction of mammalian cell hybridizations by polyethylene glycol, *Somat. Cell Genet.* **2**:165-176.
Dickerman, L. H., and Goldman, R. D., 1973, A rapid method for production of binucleate cells, *Exp. Cell Res.* **83**:433-436.
Ege, T., and Ringertz, N. R., 1974, Preparation of microcells by enucleation of micronucleate cells, *Exp. Cell Res.* **87**:378-382.
Ege, T., Ringertz, N. R, Hamberg, H., and Sidebottom, E., 1977, Preparation of microcells, in: *Methods in Cell Biology*, Vol. XV (D. M. Prescott, ed.), Academic Press, New York, pp. 339-358.
Fournier, R. E. K., 1981, A general high efficiency procedure for production of microcell hybrids, *Proc. Natl. Acad. Sci. USA* **78**:6349-6353.
Fournier, R. E. K., and Frelinger, J. A., 1982, Mol. Cell. Biol., in press.
Fournier, R. E. K., and Pardee, A. B., 1975, Cell cycle studies of mononucleate and cytochalasin B-induced binucleate fibroblasts, *Proc. Natl. Acad. Sci. USA* **72**:869-873.
Fournier, R. E. K., and Ruddle, F. H., 1977a, Microcell-mediated transfer of murine chromosomes into mouse, Chinese hamster, and human somatic cells, *Proc. Natl. Acad. Sci. USA* **74**:319-323.
Fournier, R. E. K., and Ruddle, F. H., 1977b, Microcell-mediated chromosome transfer, in: *Molecular Human Cytogenetics* (R. S. Sparkes, D. E. Comings, and C. F. Fox, eds.), Academic Press, New York, pp. 189-199.
Fournier, R. E. K., and Ruddle, F. H., 1977c, Stable association of the human transgenome and host murine chromosomes demonstrated with trispecific microcell hybrids, *Proc. Natl. Acad. Sci. USA* **74**:3937-3941.
Fournier, R. E. K., Juricek, D. K., and Ruddle, F. H., 1979, Somatic cell genetic analysis of transgenome integration, *Somat. Cell Genet.* **5**:1061-1077.
Gopalakrishnan, T. V., and Thompson, E. B., 1975, A method for enucleating cultured mammalian cells, *Exp. Cell Res.* **96**:435-439.
Kozak, C. A., Fournier, R. E. K., Leinwand, L. A., and Ruddle, F. H., 1979, Assignment of the gene governing cellular ouabain resistance to *Mus musculus* chromosome 3 using human/mouse microcell hybrids, *Biochem. Genet.* **17**:23-34.
Leinwand, L., Fournier, R. E. K., Nichols, E. A., and Ruddle, F. H., 1978, Assignment of the gene for adenosine kinase to chromosome 14 in *Mus musculus* by somatic cell hybridization, *Cytogenet. Cell Genet.* **21**:77-85.
Levan, A., 1954, Colchicine-induced C-mitosis in two mouse ascites tumors, *Hereditas* **40**:1-64.
McBride, O. W., and Ozer, H. L., 1973, Transfer of genetic information by purified metaphase chromosomes, *Proc. Natl. Acad. Sci. USA* **70**:1258-1262.
McNeill, C. A., and Brown, R. L., 1980, Genetic manipulation by means of microcell-mediated transfer of normal human chromosomes into recipient mouse cells, *Proc. Natl. Acad. Sci. USA* **77**:5394-5398.
Mercer, W. E., and Schlegel, R. A., 1979, Phytohemagglutinin enhancement of cell fusion reduces polyethylene glycol cyotoxicity, *Exp. Cell Res.* **120**:417-421.
O'Malley, K. A., and Davidson, R. L., 1977, A new dimension in suspension fusion techniques with polyethylene glycol, *Somat. Cell Genet.* **3**:441-448.
Phillips, S. G., and Phillips, D. M., 1969, Sites of nucleolus production in cultured Chinese hamster cells, *J. Cell Biol.* **40**:248-268.
Poste, G., and Reeve, P., 1972, Enucleation of mammalian cells by cytochalasin B. II. Formation of hybrid cells and heterokaryons by fusion of anucleate and nucleated cells, *Exp. Cell Res.* **73**:287-294.

Prescott, D. M., Myerson, D., and Wallace, J., 1972, Enucleation of mammalian cells with cytochalasin B, *Exp. Cell Res.* **71**:480–485.
Ruddle, F. H., and Creagan, R. P., 1975, Parasexual approaches to the genetics of man, *Ann. Rev. Genet.* **9**:407–486.
Smiley, J. R. Steege, D. A., Juricek, K. D., Summers, W., and Ruddle, F. H., 1978, A Herpes Simplex Virus 1 integration site in the mouse genome defined by somatic cell genetic analysis, *Cell* **15**:455–468.
Stubblefield, E., 1964, *Cytogenetics of Cells in Culture*, Volume 3 (R. J. C. Harris, ed.), Academic Press, New York, pp. 223–298.
Wigler, M. H., and Weinstein, I. B., 1975, A preparative method for obtaining enucleated mammalian cells, *Bioch. Biophys. Res. Commun.* **63**:669–674.
Wigler, M., Pellicer, A., Silverstein, S., and Axel, R., 1978, Transfer of single copy eukaryotic genes using total cellular DNA as donor, *Cell* **14**:725–731.
Worton, R., Duff, C., and Flintoff, W., 1981, Microcell-mediated cotransfer of genes specifying methotrexate resistance, emetine sensitivity, and chromate sensitivity with Chinese hamster chromosome 2, *Mol. Cell. Biol.* **1**:330–335.
Wright, W. E., and Hayflick, L., 1972, Formation of anucleate and multinucleate cells in normal and SV-40 transformed WI-38 by cytochalasin B, *Exp. Cell Res.* **74**:187–194.

Chapter 24

Techniques for Isolating Chromosome-Containing Minisegregant Cells

R. T. JOHNSON and A. M. MULLINGER

1. Introduction

In solid tumors *in vivo* cells frequently undergo spontaneous disintegration. This often takes the form of fragmentation of the cell body into a number of membrane-bound pieces which are subsequently removed by phagocytosis. The phenomenon has been given the name apoptosis and it is common enough to influence the kinetics of tumor growth (Kerr et al., 1972; Bird et al., 1976). Though apoptosis is particularly associated with malignant cells, similar behavior has been observed in embryogenesis and during changes in tissue growth rate (Kerr et al., 1972).

The mechanism of apoptotic destruction *in vivo* remains uncertain, but it may have a homolog in the extraordinary behavior of human tumor cells *in vitro* that we have called extrusion subdivision (Johnson et al., 1975). Here, mitotic cells can be treated so that the normal pattern of cleavage is grossly perturbed and, instead, a cluster of tiny cells is produced. These fragments, many of which contain DNA, have been called minisegregants. Though mitotic cells are especially prone to this behavior, it can also be induced in the interphase state (Johnson et al., 1978).

In this chapter we will describe methods for the production, separation, and analysis of HeLa minisegregants, and show how they have been used to produce somatic cell hybrids. For many somatic cell geneticists the construction of hybrids remains a major goal. Minisegregant cells offer one way of directing the initial composition of hybrid cells, perhaps thereby simplifying subsequent analysis.

R. T. JOHNSON and A. M. MULLINGER • Department of Zoology, University of Cambridge, Cambridge CB2 3EJ, England.

2. Standard Method for Production of Minisegregants from HeLa Cells

The following procedure has been used routinely for producing HeLa minisegregant cells (Schor et al., 1975).

2.1. Production of Mitotic Cells

Minisegregants are produced from cells in mitosis. Mitotic HeLa cells are obtained by a variety of methods, including: (1) collection of mitotic cells shaken off from monolayer cultures previously exposed for up to 8 hr to the mitotic blocking agent Colcemid (0.05 μg/ml); and (2) partial synchronization by accumulation of cells either in S phase or at the G_1–S boundary, by growth in the presence of excess thymidine, followed by release of this population and its subsequent accumulation as mitotic cells in the presence of Colcemid or in an atmosphere of nitrous oxide (Rao, 1968). The second method (our method of choice) is particularly convenient for the production of large numbers of mitotic HeLa cells from suspension cultures grown in Eagle's minimal essential medium (MEM) (Eagle, 1959) supplemented with non-essential amino acids, sodium pyruvate, antibiotics, and 5% fetal calf serum (MEMFC). A total of 2.5 mM thymidine is added for 20 hr to a culture at an initial density of 2×10^5 cells/ml. Cells are released from the thymidine arrest by centrifugation and resuspension in growth medium; 4 hr later they are transferred to plastic dishes at a density of 6×10^6 cells per 150-mm dish (Falcon Plastics). They are then placed in a 5.4 atm (5.066×10^5 N/m^2) nitrous oxide atmosphere for 9 hr (for details of pressure chamber see Johnson et al., 1978). During the 9-hr period more than 90% of the cells round up and usually detach from the surface. Cytologic examination of the floating cells, using monolayers prepared by means of a Cytocentrifuge (Shandon Instruments), shows that more than 95% are usually mitotic.

2.2. Induction of Extrusion Division in Mitotic HeLa Cells

A total of $4-5 \times 10^7$ cells in MEMFC released from mitotic block are rapidly cooled to 4°C, where they remain for 9–12 hr. After gentle agitation and plating them into 150-mm plastic petri dishes at a concentration of 5×10^6 cells per dish, the cells are incubated at 37°C in a humidified carbon dioxide incubator. During the next few hours the behavior of different cells in the population is highly variable and complex (Mullinger and Johnson, 1976). Cells divide asynchronously, many displaying abnormal patterns of division. The most common type of behavior, which we have called extrusion division, results in the formation of clusters of small cells (minisegregants) resembling bunches of grapes (BOGs), as shown in Figs. 1 and 2. A few cells divide by

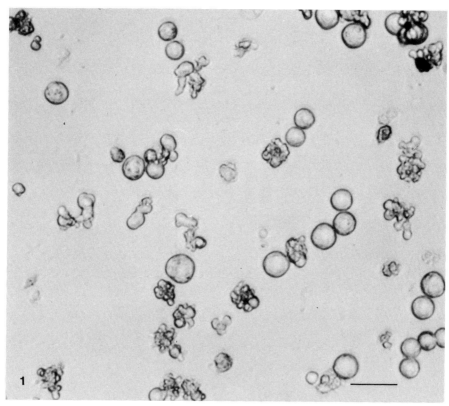

Figure 1. Photomicrograph of a population of mitotic cells stored for 6 hr at 4°C and then plated out in Hanks basal salt solution (BSSH) for 3 hr at 37°C. While some cells remain normal, others have extruded. Bar represents 25 μm. [Reproduced from Johnson et al. (1975) by courtesy of the Royal Society of London.]

normal cleavage into two daughter cells, while others produce two, three, or more daughters. Under these conditions extrusion activity can be detected within 2 hr and usually by 4 hr up to 60% of the cells form BOGs. There is a good deal of variation in the size and number of minisegregants produced from one cell and also in the way the DNA is redistributed during extrusion division (Schor et al., 1975; Johnson et al., 1978). This is revealed by Feulgen staining of cold-stored mitotic cells plated out onto polylysine-coated coverslips (PL-coverslips; clean coverslips are dipped in 0.01% Sigma grade 1B poly-L-lysine hydrobromide and then washed thoroughly; Mazia et al., 1975). Cells are fixed after various times of incubation in neutral, buffered formaldehyde for 24 hr (Lillie, 1965) and stained according to standard procedures (Pearse, 1968). Examples are shown in Fig. 3. Not all minisegregants contain Feulgen-positive material and those that do may have either "dense nuclei," which stain uniformly with an intense magenta color, or "normal nuclei," which stain less uniformly with a paler color, or occasionally

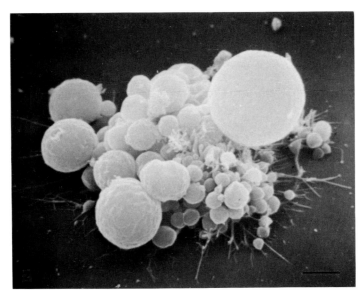

Figure 2. Scanning electron micrograph of a cluster of HeLa minisegregants, or BOG. Nitrous oxide-arrested mitotic HeLa cells stored at 4°C for 11 hr, plated out on PL-coverslips, incubated at 37°C in BSSH, pH 7.2, and fixed 3 hr after the start of incubation. Bar represents 2 μm. [Reproduced from Mullinger and Johnson (1976) by courtesy of the Company of Biologists.]

both. Examples of those two types of nuclei are shown in transmission electron micrographs of BOGs (Figs. 4 and 5); dense nuclei do not appear to be enclosed by nuclear envelopes. The genetic capacity of the different types of nuclei in these cells is uncertain. Clearly, perturbation of division by cold storage or Colcemid treatment results in abnormal deposition of nuclear lamina proteins around individual chromosomes (Jost and Johnson, 1981); this activity may determine which fragments of the genome are subsequently enclosed in nuclear envelope.

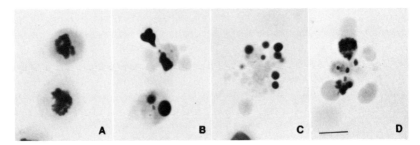

Figure 3. Photomicrographs of material stained with the Feulgen DNA technique and counterstained with fast green. Nitrous oxide-arrested HeLa cells were stored at 4°C for 9 hr, plated out on PL-slides, incubated in MEMFC at 37°C, and fixed 1–1.5 hr later. Several different stages are shown. (A) Chromosomes in a mitotic cell start to aggregate. (B) Extruding cell with "dense nuclei" which are possibly fragmenting. (C, D) BOGs with "dense nuclei" in some minisegregants. Bar represents 10 μm. [Reproduced from Johnson et al. (1978) by courtesy of Academic Press.]

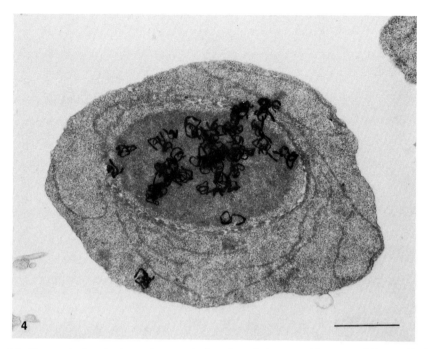

Figure 4. Transmission electron micrograph of section of nitrous oxide-arrested HeLa cell which had been stored at 4°C for 9 hr and incubated in MEMFC for 1.5–2.5 hr at 37°C. After gentle centrifugation cells were fixed in HEPES (N-2-hydroxyethylpiperazine-N'-2-ethanesulfonic acid) or phosphate-buffered glutaraldehyde, post-fixed in osmium tetroxide, and stained with lead citrate and uranyl acetate by methods similar to those described by Johnson et al. (1975). Shown is an autoradiograph of minisegregant obtained by extrusion subdivision of a [^{3}H]thymidine-prelabeled mitotic cell. Silver grains are concentrated over the "dense nuclei." Bar represents 1 μm. Ilford L4 emulsion, developed in Microdol-X. [Reproduced from Johnson et al. (1978) by courtesy of Academic Press.]

2.3. Separation of Minisegregant Cells According to Size

Populations of minisegregant cells contaminated by whole cells can be used at this stage for fusion studies and other purposes. However, the heterogeneous mixture of minisegregants can be separated according to cell diameter by sedimentation under gravity through a 1–2% Ficoll-buffered step density gradient (Schor et al., 1975), using the apparatus shown in Fig. 6 [modified from Miller and Phillips (1969) and Denman and Pelton (1973)]. All operations are carried out at room temperature. The population of extruding cells, produced as described above after 4 hr of incubation at 37°C, is *gently* pipetted to break up aggregates and to liberate as many minisegregants as possible, the process being monitored by phase contrast microscopy. Disaggregated cells are concentrated by low-speed centrifugation, gently resuspended in approximately 30 ml of MEMFC, and added to the gradient apparatus. The Ficoll gradient is introduced below the cells, as described in the legend to Fig. 6, and the cells allowed to sediment through the gradient for

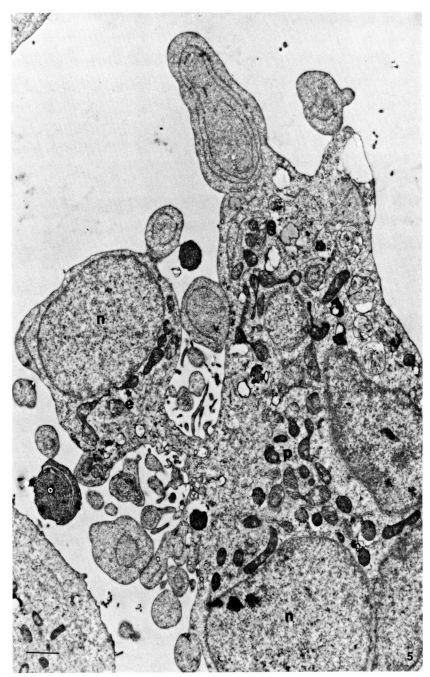

Figure 5. Extruding cell with membrane-bound "normal nuclei" (n) in both parent cell (p) and extrusion (e). See Fig. 4 for details. Bar represents 1 μm. [Reproduced from Johnson et al. (1978) by courtesy of Academic Press.]

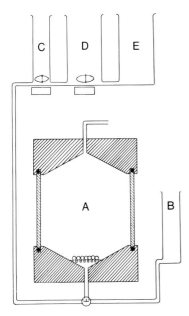

Figure 6. Apparatus for separating minisegregant cells. The sedimentation chamber (A in diagram: 14.5 cm diameter, 17 cm height, approximately 1800 ml capacity) consists of a cylindrical plastic (Rohm and Hass Plexiglass) midpiece fitted into two Plexiglass blocks which form the lower and upper portions of the chamber. Rubber O-rings ensure water-tight connections, and the entire chamber is bolted together by four metal rods. Liquid enters the chamber from below through a perforated plastic disk covered with glass beads which reduce the turbulence and increase the resolution of the gradient. Reservoirs B and C are connected to the chamber by means of Teflon tubing via a three-way valve. Screw clamps are located between all the compartments to control liquid flow. The contents of reservoirs C and D are mixed by means of magnetic stirrers. The tube connecting A with both B and C is first primed with medium (Eagle's MEM) and 50 ml is also introduced into A. Reservoirs C, D, and E are filled with, respectively, 0.33% (150 ml), 1% (800 ml), and 2% (800 ml) Ficoll in MEMFC. [Gradients can also be made in saline G (Kao and Puck, 1971) supplemented with 5% FCS.] The cell suspension containing the minisegregants is then added to B and allowed to flow slowly into A. Fresh medium is used to clear the connecting tube of cell suspension. The linear step gradient is formed by allowing the contents of reservoirs C, D, and E to flow into the sedimentation chamber. The flow rate is adjusted by a screw clamp between A and C to approximately 4 ml/min for the first 150 ml until the meniscus reaches the vertical walls of the chamber and a narrow band containing the cells lies on top of the Ficoll. The flow rate is then increased to 15 ml/min until the meniscus reaches the top of the cylindrical section. The cells are allowed to sediment for the desired time and then fractions are collected by replacing the contents of reservoirs C, D, and E with 1 M sucrose, which is allowed to flow into the chamber. The contents of the gradient are thus forced through the top exit tube at a rate of approximately 10 ml/min and are collected in a series of 50-ml tubes. [Reproduced from Schor et al. (1975) by courtesy of the Company of Biologists.]

up to 5 hr. Since it has been shown (Miller and Phillips, 1969) that sedimentation velocity depends primarily on the cell radius, it is critical to start with a population containing the maximum number of spherical cells and the minimum amount of irregular debris; for this reason extruding mitotic cells must be handled gently during the break-up of the minisegregant aggregates.

After the period of sedimentation, 1 M sucrose is introduced from below in the manner described in the legend to Fig. 6, and the contents of the gradient forced out through the top exit tube and collected as a series of about 30 fractions of 50 ml each. For subsquent use and analysis of the minisegregant cells, fractions are centrifuged at low speed, and the supernatant is removed and replaced with complete MEMFC. All steps in the fractionation process can be carried out in a laminar air flow cabinet to ensure sterile conditions. Autoclavable tubing and a combination of alcohol and UV irradiation ensure that the gradient equipment is sterile.

The cells and minisegregants are stable in MEMFC at 4°C for at least 12 hr. Moreover, 95–100% of the cells routinely exclude trypan blue and can thus

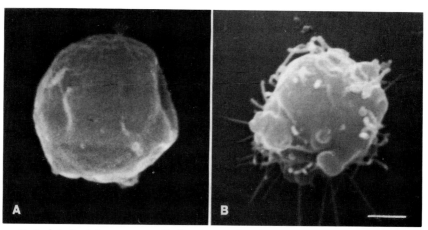

Figure 7. Isolated minisegregant cells fixed immediately after collection from a gradient, as described in the text, and prepared for scanning electron microscopy as described in Mullinger and Johnson (1976). Bar represents 1 μm. [Reproduced from Johnson et al. (1978) by courtesy of Academic Press.]

be judged "viable." Scanning electron micrographs of isolated minisegregant cells are shown in Fig. 7.

2.4. Properties of Gradient Fractions

The extent to which the initial heterogeneous mixture of cells can be separated by means of the gradient into a number of fractions, each enriched in cells of a particular size range, is shown in Figs. 8–10. Cells in the initial mixture vary in size between about 1 and 30 μm, whereas fractions are enriched in particular size ranges.

Feulgen staining of fixed material enables both the distribution of DNA and the cytologic appearance of the DNA-containing bodies (i.e., whether "normal" or "dense" nuclei) to be assessed in the various gradient fractions. The results of a representative separation are shown in Table I. Of the cells in the inital mixture, 58% contain one or more Feulgen-positive areas and in any given cell these are usually either all "normal" or all "dense," although occasionally cells contain both types. The percentage of cells containing DNA is lower in fractions from the top of the gradient than those from the bottom. A single dense nucleus is particularly common in Feulgen-positive cells from the top fractions.

Figure 8. Photomicrographs of the initial mixture layered on the gradient (A), and of isolated gradient fractions 1–2 (B), fractions 9–10 (C), fractions 19–20 (D), and HeLa cells arrested in mitosis (E). Bar represents 20 μm. [Reproduced from Schor et al. (1975) by courtesy of the Company of Biologists.]

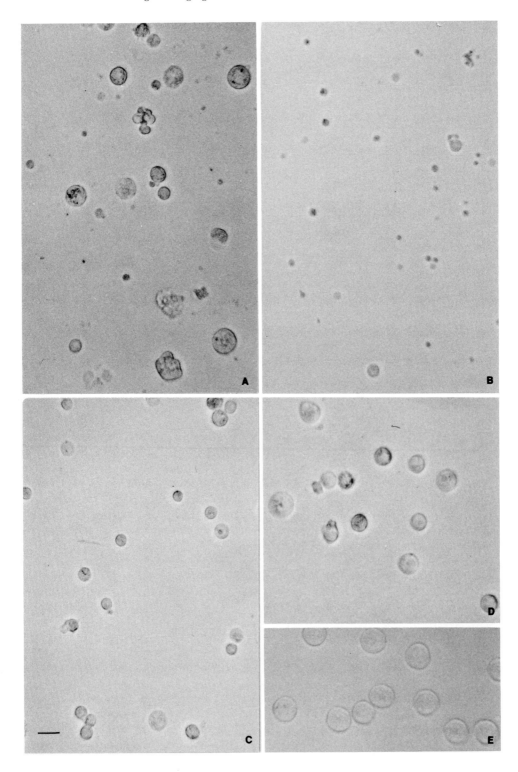

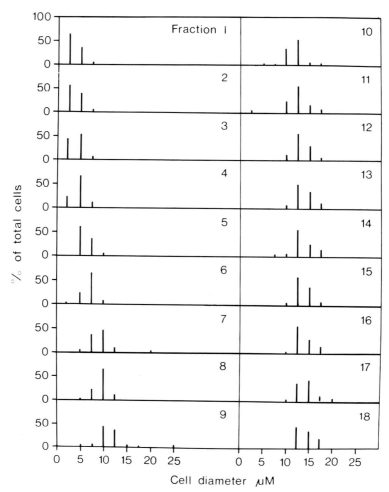

Figure 9. Histograms showing the distribution of cells with different diameters in the first 18 fractions of a gradient separation. Cell diameter was determined on a minimum of 100 cells with a calibrated eyepiece micrometer. [Reproduced from Johnson et al. (1978) by courtesy of Academic Press.]

The quantity of DNA in a minisegregant has been estimated by microdensitometry of the Feulgen-stained material. As shown in Fig. 11, the Feulgen values for the initial mixture vary between 0.01C and 4C, compared with a value of 0.03C for an average HeLa cell chromosome [data recalculated from du Praw, (1970)]. However, it is important to note that not only is the method insensitive to values below about 0.01C, but also it suffers from the "Feulgen artefact," particularly for the "dense nuclei" (Garcia, 1970); precise DNA values cannot thus be assessed. A more accurate and speedy method of analyzing size and DNA and RNA content of the various fractions is now available in flow cytofluorimetry.

The chromosome constitution of minisegregants and the nature of the chromosomes cannot be established by conventional techniques since these

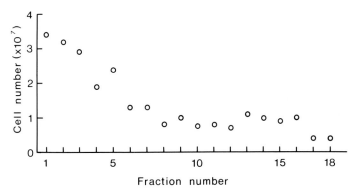

Figure 10. The number of minisegregants in each of the first 18 fractions of a gradient separation. Cell counts were made using a hemocytometer. [Reproduced from Johnson et al. (1978) by courtesy of Academic Press.]

cells do not divide. However, by fusing minisegregants with whole mitotic cells, prematurely condensed chromosomes (PCCs) are induced (Johnson and Rao, 1970; Rao and Johnson, 1972), and can be revealed cytologically in the manner shown in Fig. 12 (Schor et al., 1975). Counts of minisegregant chromosome numbers in such preparations are shown in Fig. 13 for several pooled fractions from one gradient; in this experiment minisegregant populations were fused with a fourfold excess of mitotic cells in order to reduce the probability of multiple minisegregant–single mitotic cell fusions. Figure 13 shows that smaller cells generally contain fewer chromosomes than do the larger cells. Few spreads with only a single minisegregant chromosome are observed, despite the fact that Feulgen measurements suggest many of the smallest cells contain only one chromosome.

Further examination of minisegregant PCCs reveals great variety in structure; they are either mono- or bivalent, i.e., derived from separated or nonseparated mitotic chromosomes, and they range in length from short to extended (Fig. 12). In addition, many of the PCCs are abnormal, showing damage in the form of localized breaks, despiralization, or fragmentation (Fig. 12). The smaller cells tend to produce a greater proportion of bivalent PCCs than do the larger ones, and damaged PCCs are also more common in the smaller cells (Table II).

3. Modification of Basic Technique for Producing Minisegregants

We have routinely used the above procedure for producing minisegregants from HeLa cells. There are, however, a number of modifications which change the speed of extrusion and/or the proportion of cells that extrude (Johnson et al., 1975; 1978; Mullinger and Johnson, 1976). Little is known about the way these modifications change the nature of the DNA packaging in minisegregants. Some treatments produce fragile cells, a property that

Table I. Classification of Different Cell Types in the Initial Mixture and in Isolated Minisegregant Fractions[a]

| | Percent of cells with Feulgen DNA stain | Percent of total population of Feulgen-staining cells which belong to each category ||||||||||||
| | | Cells with "normal" nuclei only ||||| Cells with "dense nuclei" only ||||| Cells with both "normal" and "dense nuclei" | Cells with mitotic chromosomes only |
Sample		1	2	3	4	>4	1	2	3	4	>4		
Initial mixture	58	39.6	15.9	8.4	3.0	3.0	10.8	2.7	0.0	2.7	5.5	2.0	6.4
Fractions 1–3 pooled	15	25.9	3.3	0.3	0.0	0.0	53.0	9.0	3.5	1.6	2.2	0.9	0.3
Fractions 4–5 pooled	29	33.0	4.3	0.7	0.2	0.0	37.0	10.7	4.5	3.8	2.4	2.4	1.0
Fractions 11–12 pooled	84	51.4	15.0	1.3	0.0	0.0	14.4	6.1	2.6	1.9	5.1	1.9	0.3
Fraction 14	78	39.9	12.7	3.7	0.6	0.0	20.8	7.2	2.9	2.3	8.4	0.9	0.6
Fraction 20	81	31.4	15.9	6.5	1.0	0.0	21.1	5.7	3.9	1.0	12.2	0.8	0.5

[a] Counts of 300 cells were made for each sample. No distinction has been made between cells with nuclei of different sizes. [Reproduced from Schor et al. (1975) by courtesy of the Company of Biologists.]

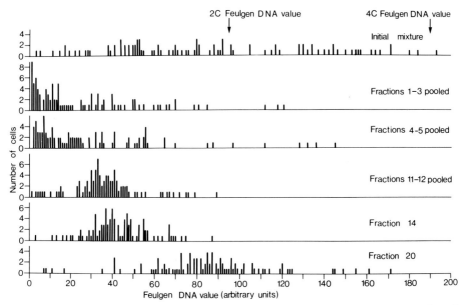

Figure 11. Feulgen DNA values of cells in both the initial mixture, which was layered on the gradients and in isolated fractions. Each sample included both mononucleate and multinucleate cells and also nuclei of both the "normal" and "dense" types (see text). One hundred cells with Feulgen-positive staining were scored for each sample. The mean DNA Feulgen values with standard deviations are, in arbitrary units: initial mixture, 84.7 ± 48.9; fractions 1–3 pooled, 25.8 ± 27.2; fractions 4–5 pooled, 31.6 ± 32.1; fractions 11–12 pooled, 35.0 ± 16.1; fraction 14, 42.6 ± 14.4; and fraction 20, 85.7 ± 30.0. Arrows indicate the 2C and 4C Feulgen values for HeLa cells. [Reproduced from Schor et al. (1975) by courtesy of the Company of Biologists.]

counterbalances their usefulness. Nevertheless, some of these modifications may prove useful for certain purposes, in particular when working with other cell types that may be more refractory to extrusion than HeLa. It is worth pointing out that care should be taken in assessing the effects of such modifications, since (1) there is a wide variety in both the time of onset and type of division exhibited by different cells in a given population, and (2) one agent often interacts with and potentiates the effect of another.

The factors we know to influence extrusion are:

(i) *Length of time in mitostatic agent.* The proportion of cells that extrude is increased by the time of exposure to nitrous oxide or Colcemid (Johnson et al., 1978). If the time in nitrous oxide is very prolonged, some cells extrude without a period of cold storage. The combined effects of nitrous oxide and Colcemid are not additive, but resemble the effects of Colcemid alone.

(ii) *Length of time in cold storage.* The proportion of cells that extrude is greatly increased if the time in cold storage is increased. For example, 28% of mitotic HeLa cells extrude at 37°C if they have been stored at 4°C for 6 hr, whereas the percentage increases to 73% after 10 hr in the cold. [Figures apply to nitrous oxide-arrested HeLa cells incubated in MEMFC, pH 7.2; from Mullinger and Johnson (1976).]

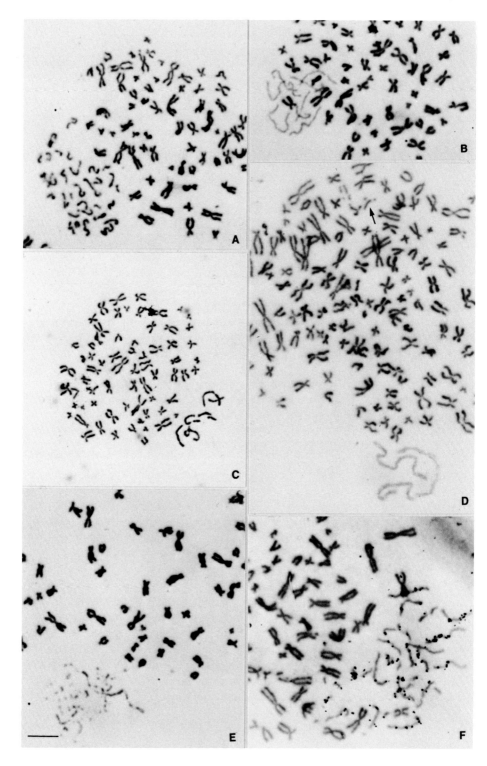

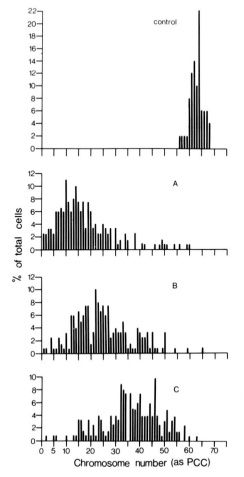

Figure 13. Histograms showing the distribution of minisegregant cells containing different numbers of PCCs in various pooled fractions: (A) From gradient fractions 1–8. (B) From gradient fractions 9–16. (C) From gradient fractions 17–24. Each pooled fraction was fused with mitotic HeLa cells to induce PCC. Chromosome preparations were made and the number of prematurely condensed chromosomes in at least 200 PCC clusters was counted for each fusion sample. The mean chromosome numbers with standard deviations in the pooled fractions are (A) 17.3 ± 11.7, (B) 26.0 ± 12.9, and (C) 35.7 ± 12.2. Few spreads containing PCC were observed in control mitotic × mitotic fusions. These were invariably of standard G_2 morphology and displayed a tight distribution around 64, the modal chromosome number for this strain of HeLa cell. [Reproduced from Schor et al. (1975) by courtesy of the Company of Biologists.]

(iii) *pH of incubation medium.* The number of cells dividing into BOGs is increased by incubation in medium of high rather than low pH, though the lag period before extrusion is initiated is longer at higher pH, as is the case for normal cleavage (Johnson et al., 1975).

(iv) *Presence of thiol reagents.* Thiol reducing agents, such as dithiothreitol (DTT), when used at 2 mM, potentiate extrusion by a factor of three to six at pH 7.4. The effect is greater after cold storage and in media of lower pH (Johnson et al., 1975). Even without a period of cold storage 5% of mitotic HeLa cells extrude within 4 hr of incubation in 2 mM DTT in MEMFC.

Figure 12. Chromosome spreads containing PCCs induced from minisegregant cells after fusion with mitotic HeLa cells. (A) Minisegregant PCCs with about 30 condensed G_1 elements. (B) PCCs with about 10 attenuated G_1 elements. (C) Seven condensed PCCs, probably bivalent. (D) Three extended and damaged bivalent PCCs and a separate group of damaged, condensed G_1 PCCs (arrow). (E) Highly fragmented PCCs. (F) Autoradiograph of G_1 PCCs derived from a mitotic cell which had been prelabeled with [³H]thymidine. Bar represents 5 μm. [Reproduced from Schor et al. (1975) by courtesy of the Company of Biologists.]

Table II. Classification of PCCs Derived from Isolated Minisegregant Fractions[a]

Pooled fractions	Percentage of PCC in each morphologic category						Ambiguous PCC
	G_1 PCC			G_2 PCC			
	Total = normal + damaged			Total = normal + damaged			
A (gradient fractions 1–8)	47	(25)	(22)	39	(12)	(27)	14
B (gradient fractions 9–14)	38	(29.5)	(8.5)	46	(21)	(25)	16
C (gradient fractions 17–24)	77.5	(64)	(13.5)	19.5	(5)	(14.5)	3

[a] At least 200 chromosome spreads containing PCCs were scored for each sample. Examples of the various morphologic categories are shown in Fig. 12. [Reproduced from Schor et al. (1975) by courtesy of the Company of Biologists.]

In order to induce extrusion in mitotic human fibroblasts we use a combination of 9 hr in nitrous oxide, 12 hr at 4°C, followed by incubation at 37°C in 2 mM DTT in Hanks basal salt solution (Hanks and Wallace, 1949), pH 8.0.

(v) *Trypsin*. Treatment of mitotic HeLa cells with 0.25% trypsin for periods of up to 30 min increases the proportion of cells that form BOGs (Johnson et al., 1978).

4. The Use of Minisegregants in Somatic Cell Genetics

Minisegregants are readily fused with whole cells by means of Sendai virus (Schor et al., 1975). In this laboratory HeLa minisegregants have been used to produce two series of mouse–human hybrid cells. Tourian et al. (1978) and Melton (1980, 1981) fused populations of small minisegregants with HGPRT$^-$ mouse A_9 and C-1300 neuroblastoma cells, respectively.

For A_9-minisegregant fusions, approximately 10^7 minisegregants from fraction 4 of the standard Ficoll gradient preparation described earlier were added to 2.5×10^6 A_9 cells in the presence of 250 hemagglutinating units of UV-inactivated Sendai virus. The minisegregants used ranged in size up to 12 μm, but 80% had a diameter of 7.5 μm or less. Feulgen-positive material was present in 20 and 26% of the minisegregant cells used in two separate fusions, respectively; earlier, quantitative Feulgen microdensitometry of *pooled* fractions 4 and 5 (i.e., a population of larger average diameter than was used for these fusions) showed that 97% of Feulgen-positive cells contained less than the HeLa 2C value and that 80% had much less than 1C. The two fusions described by Tourian et al. (1978) yielded, respectively, two and about 50 hybrids expressing the human HGPRT phenotype and containing up to 11 human chromosomes.

For mouse neuroblastoma (NB) fusions, approximately 3 and 1.5×10^6 minisegregants with mean cell diameters of 10.2 and 12.5 μm, respectively, were incubated with 1.8×10^6 NB cells plus Sendai virus. Six independent

Table III. Human Chromosomal Constitution of HeLa Minisegregant × A₉ Hybrids[a]

Hybrid	Chromosomal group								
	X	A	M^X	B	C	D	E	F	G
A_1	1	—	1	—	2	1	1	—	1
α_a	1	—	—	1	2	2	1	1	1
δ_f	1	—	1	—	1	1	1	—	1
δ_i	1	—	—	1	4	1	1	1	1
θ_e	1	1	1	—	2	2	1	—	1
θ_g	1	1	—	1	—	1	1	—	1
θ_h	1	—	1	1	2	1	1	—	1
ω_c	1	—	—	1	3	2	2	—	1
ω_e	1	—	1	—	3	2	1	—	1
$R\delta_f$	—	—	1	—	1	2	1	1	1
$R\theta_{g2}$	—	—	1	—	1	1	—	1	2
$R\theta_{g3}$	—	—	1	—	2	1	1	—	1
$R\theta_{iB}$	—	1	1	1	3	2	1	—	2

[a] Results refer to two separate fusions between HeLa minisegregants and HGPRT-deficient mouse A₉ cells. HAT-resistant hybrids were selected: A_1 arose from the first fusion, and the remaining hybrids from the second. Hybrids α, δ, θ, and ω were isolated from different plates and a number of independent colonies was picked for each of the δ, θ, and ω plates ($\delta_{f,i}$; $\theta_{e,g,h,i}$; $\omega_{c,e}$) and grown up for analysis. Hybrids that had been maintained in HAT selective medium were transferred into nonselective medium and several independently derived 6-thioguanine-resistant subclones (prefixed with an R) were derived. The data represent the average human chromosome constitution of each hybrid analyzed from a minimum of five metaphase spreads. M^X is the large submetacentric HeLa marker chromosome (Heneen, 1976). [Reproduced from Tourian et al. (1978) by courtesy of the Company of Biologists.]

colonies were selected in HAT (hypoxanthine, aminopterin, thymidine) (Littlefield, 1964), but only one of these proved to be a true hybrid since it not only expressed HeLa HGPRT but also contained human chromosomes. The remaining HAT-selected colonies proved to be unusual HGPRT⁺ mouse neuroblastoma revertants (Melton, 1980, 1981).

The use of minisegregants as fusion partners should result in an early numerical bias in favor of the karyotype of the whole cell. In the A₉–HeLa minisegregant hybrids each hybrid contained a single human X chromosome plus various numbers of chromosomes from all other human groups (Table III). The human chromosome component of these hybrids may reflect the composition of the particular fraction of minisegregants used in the fusions, though post-fusion segregation must have occurred because the number of human chromosomes varies between the cells of any particular cloned hybrid. However, initial distribution of chromosomes in the minisegregants used for the fusions may help explain why, for example, the E-group chromosome 17 was usually present in the hybrids but not chromosome 9, though both are present in multiple copies in HeLa cells. In addition, the large submetacentric HeLa marker chromosome, which is partially derived from chromosome 5 (Heneen, 1976) and is present as a singly copy in HeLa, was retained in most of the hybrids; if it was absent, a normal B-group chromosome (no. 5) was found.

At present no formal proof exists that the hybrid cells described by Tourian et al. (1978) and Melton (1980, 1981) were not derived from fusion between whole HeLa, or HeLa minisegregants with a complete genome, and mouse cells. However, this possibility is considered unlikely for a number of reasons. First, the minisegregants were taken from a position in the gradient which is only rarely contaminated by large cells; DNA densitometry of 200 cells from this region of the gradient indicated a uniformly reduced DNA content compared with parental mitotic cells (Schor et al., 1975). Second, analysis of 200 prematurely condensed chromosome sets from minisegregants with a diameter up to 12.5 μm showed that less than 1% contained as many chromosomes as whole cells (Schor et al., 1975). Third, incubation of up to 10^7 minisegregants from fractions used in the A_9 fusions did not yield surviving cells (Tourian et al., 1978).

The involvement, if any, of minisegregants in promoting reversion at the HGPRT locus of mouse neuroblastoma described by Melton (1980, 1981) remains enigmatic. Five out of six of the HGPRT-positive colonies arising out of mini-NB fusions proved to be mouse revertants. For whole HeLa cell–NB fusions the comparative figure was two out of 12. In these cells HGPRT$^-$ to HGPRT$^+$ reversion was always associated with a characteristic translocation involving the X chromosome, and not only is it likely that these events were causally related, but also that they occurred at a single time (Melton, 1980, 1981). Now further work is needed to clarify whether association or fusion with minisegregants predisposes host cells to karyotypic instability.

Acknowledgments

We are grateful to the Cancer Research Campaign, of which RTJ is a Research Fellow, and to the Medical Research Council for their support of this work.

References

Bird, C. C., Wyllie, A. M., and Currie, A. R., 1976, Ageing in tumours, in: *Scientific Foundations of Oncology* (T. Symington, ed.), Heinemann, London, pp. 52–62.

Denman, A. M., and Pelton, B. K., 1973, Cell separation by size, in: *Methodological Developments in Biochemistry*, Volume 2 (E. Reid, ed.), Longman, London, pp. 185–199.

du Praw, E. J., 1970, *DNA and Chromosomes*, Holt, Rinehart, and Winston, New York.

Eagle, H., 1959, Amino acid metabolism in mammalian cell cultures, *Science* **130**:432–437.

Garcia, A. M., (1970), in: *Introduction to Quantitative Cytochemistry* (G. L. Wied and G. F. Bahr, eds.), Volume 2, Academic Press, New York, pp. 153–170.

Hanks, J. H., and Wallace, R. E., 1949, Relation of oxygen and temperature in the preservation of tissues by refrigeration, *Proc. Soc. Exp. Biol. Med.* **71**:196–200.

Heneen, W. K., 1976, HeLa cells and their possible contamination of other cell lines: Karyotype studies, *Hereditas* **82**:217–248.

Johnson, R. T., and Rao, P. N., 1970, Mammalian cell fusion: Induction of premature chromosome condensation in interphase nuclei, *Nature* **226**:717–722.

Johnson, R. T., Mullinger, A. M., and Skaer, R. J., 1975, Perturbation of mammalian cell division: Human minisegregants derived from mitotic cells, *Proc. R. Soc. Lond. B* **189**:591–602.

Johnson, R. T., Mullinger, A. M., and Downes, C. S., 1978, Human minisegregant cells, in: *Methods in Cell Biology* (D. M. Prescott ed.), Volume XX, Academic Press, New York, pp. 255–314.

Jost, E. and Johnson, R. T., 1981, Nuclear lamina assembly, synthesis and disaggregation during the cell cycle in synchronized HeLa cells, *J. Cell Sci.* **47**:25–53.

Kao, F.-T., and Puck, T. T., 1974, Induction and isolation of auxotrophic mutants in mammalian cells, in: *Methods in Cell Biology* (D. M. Prescott, ed.), Volume VIII, Academic Press, New York, pp. 23–39.

Kerr, J. F. R., Wyllie, A. H., and Currie, A. R., 1972, Apoptosis: A basic biological phenomenon with wide-ranging implications in tissue kinetics, *Br. J. Cancer* **26**:239–257.

Lillie, R. D., 1965, *Histopathologic Technic and Practical Histochemistry*, McGraw-Hill, New York.

Littlefield, J. W., 1964, Selection of hybrids from matings of fibroblasts in vitro and their presumed recombinants, *Science* **145**:709–710.

Mazia, D., Schatten, G., and Sale, W., 1975, Adhesion of cells to surfaces coated with poly-lysine. Applications to electron microscopy, *J. Cell Biol.* **66**:198–200.

Melton, D. W., 1980, Reversion at the HPRT Locus in Mouse Neuroblastoma, Ph.D. Thesis, University of Cambridge, England.

Melton, D. W., 1981, Cell fusion-induced mouse neuroblastoma HPRT revertants with variant enzyme and elevated HPRT protein levels, *Somat. Cell Genet.* **7**:331–344.

Miller, R. J., and Phillips, R. A., 1969, Separation of cells by velocity sedimentation, *J. Cell. Physiol.* **73**:191–201.

Mullinger, A. M., and Johnson, R. T., 1976, Perturbation of mammalian cell division III. The topography and kinetics of extrusion subdivision, *J. Cell Sci.* **22**:243–285.

Pearse, A. G. E., 1968, *Histochemistry Theoretical and Applied*, vol. 1, Churchill, London.

Rao, P. N., 1968, Mitotic synchrony in mammalian cells treated with nitrous oxide at high pressure, *Science* **160**:774–776.

Rao, P. N., and Johnson, R. T., 1972, Cell fusion and its application to studies on the regulation of the cell cycle, in: *Methods in Cell Physiology* (D. M. Prescott, ed.), Volume V, Academic Press, New York, pp. 75–126.

Schor, S. L., Johnson, R. T., and Mullinger, A. M., 1975, Perturbation of mammalian cell division II. Studies on the isolation and characterization of human minisegregant cells, *J. Cell Sci.* **19**:281–303.

Tourian, A., Johnson, R. T., Burg, K., Nicolson, S., and Sperling, K., 1978, The transfer of human chromosomes via human minisegregant cells into mouse cells and the quantitation of the expression of hypoxanthine phosphoribosyltransferase in the hybrids, *J. Cell Sci.* **30**:193–209.

Chapter 25
Techniques to Isolate Specific Human Metaphase Chromosomes

WAYNE WRAY and ELTON STUBBLEFIELD

1. Introduction

The physical and chemical nature of chromosomes and the adherence of contaminating cellular materials under isolation conditions are important factors to be considered in selection of a method for mass isolation of chromosomes. The application for which the chromosomes are to be used may ultimately determine the procedure chosen.

Many publications describe the isolation of metaphase chromosomes (Prescott and Bender, 1961; Somers et al., 1963; Chorazy et al., 1963; Cantor and Hearst, 1966; Huberman and Attardi, 1967; Franceschini and Giacomoni, 1967; Mendelsohn et al., 1968; Burkholder and Mukherjee, 1970; W. Wray and Stubblefield, 1970; W. Wray et al., 1972; Stubblefield et al., 1978; Blumenthal et al., 1979; Adolph, 1980; Goyanes et al., 1980; Langlois et al., 1980). Variables in the buffers used in these methods include pH, ion concentrations, buffering agents, and the presence or absence of hexylene glycol, detergents, and polyamines. Most of the early methods (Prescott and Bender, 1961; Somers et al., 1963; Chorazy et al., 1963; Cantor and Hearst, 1966; Huberman and Attardi, 1967; Franceschini and Giacomoni, 1967; Mendelsohn et al., 1968; Burkholder and Mukherjee, 1970) utilize low pH, generally between 3.0 and 3.7. In low-pH isolation media, the chromosomes are more compact than chromosomes in neutral pH media. This could be the result of denaturation and precipitation of some chromosomal proteins and this effect in turn could contribute to increased resistance of the isolated chromosomes to mechanical damage in these media. In acidic media there is some extraction of the lysine-rich histones, which may be undesirable for certain types of studies (Huberman and Attardi, 1966). At a pH below 6.0 aggregates of materials of high RNA content form which are difficult to remove from the final preparation (Hamilton and Peterman, 1959). These aggregates, rich in

---

WAYNE WRAY • Department of Cell Biology, Baylor College of Medicine, Houston, Texas 77030. ELTON STUBBLEFIELD • University of Texas System Cancer Center, Department of Cell Biology, M.D. Anderson Hospital and Tumor Institute, Houston, Texas 77030.

ribosomes, sediment simultaneously with chromosomes at low centrifugal forces but can be removed by sedimentation of the chromosomes through sucrose gradients.

Chromosome isolation at neutral pH has a major advantage over techniques using acidic extraction conditions (Hearst and Botchan, 1970), because of the mild conditions for cell lysis and chromosome purification. A chromosome stabilizing agent is generally present to aid morphology and structure preservation. Hexylene glycol is the stabilizing agent in a simple solution of hexylene glycol–PIPES buffer–$CaCl_2$ widely used for chromosome isolation since the early 1970s (W. Wray and Stubblefield, 1970). The use of PIPES buffer allows a precise control of divalent ion concentration.

Metaphase chromosomes isolated by the Wray and Stubblefield method have a protein:DNA ratio of 2.2:1.0. The RNA:DNA ratio is very low, less than 0.1:1.0 (W. Wray and Stubblefield, 1970). These values are the same as those of isolated chromatin (V. P. Wray et al., 1980). There is no adventitiously adsorbed RNA or protein after the chromosomes have been sedimented through a sucrose gradient (W. Wray, 1976). Essentially no difference is noted in histone electrophoretic patterns of CHO and HeLa interphase chromatin or metaphase chromosomes (V. P. Wray et al., 1980). The molecular complexity of the metaphase nonhistone chromosomal proteins (NHCP) is as great as that of interphase nuclear NHCP (V. P. Wray et al., 1980; Peterson and McConkey, 1976). NHCP from metaphase chromosomes prepared by different methods have been examined and compared (V. P. Wray and Wray, 1982). There are several landmark proteins which appear to be present in all preparations, so there is a general similarity in the band patterns observed. However, the electrophoretic profiles indicate that there is definite quantitative variation in band profiles after different isolation methods (V. P. Wray and Wray, 1982).

Lipids are a constant component of isolated chromosomes (Nikolayenko and Pinaev, 1981). Triglycerides and cholesterol esters prevail, while diglycerides, which are a characteristic component of cellular debris, are absent from isolates except at pH 2.

2. Methods

2.1. Tissue Culture and Cell Lines

Mammalian cells are cultured in our laboratories as monolayers in McCoy's medium 5a supplemented with 10% fetal calf serum. The medium is buffered by a bicarbonate system requiring a 10% CO_2 atmosphere. All mammalian cell lines are grown at 37°C. Chicken cells are grown at 41°C.

2.2. Cell Synchronization

In order to increase the number of cells in mitosis at the time of harvest, mitotic cells were accumulated with Colcemid (0.06 µg/ml), and, if possible, they were selectively removed from the remaining interphase cells. The cell populations obtained by this method routinely contained 94–97% metaphase cells.

2.3. Isolation Buffers

The chromosome isolation buffer (W. Wray and Stubblefield, 1970) contains 1.0 M hexylene glycol (2-methyl-2,4-pentanediol, Eastman Organic); 5×10^{-4} M $CaCl_2$; and 10^{-4} M piperazine-N,N'-bis(2-ethanesulfonic acid) monosodium monohydrate (PIPES, Calbiochem) at a pH of 6.8. The buffer is most conveniently made by a 1:10 dilution of a 10× concentrate, which has been adjusted to a pH of 6.8 with 1.0 N sodium hydroxide. Hexylene glycol is then added after dilution. (It is important that the pH be adjusted before the addition of hexylene glycol!! Natural fiber porous pins in some types of electrodes adsorb the hexylene glycol, causing sluggish electrode response.)

For experiments where DNA of molecular weight greater than 200×10^6 daltons is desired, a buffer system which contains 1.0 M hexylene glycol, 2×10^{-3} M $CaCl_2$, and 1×10^{-3} M cyclohexylaminopropane sulfonic acid (CAPS Schwarz/Mann) at a pH of 10.5, is recommended (W. Wray et al., 1972). The buffer can be made as a 10× concentrate as above, and the pH should be adjusted to 10.5 before addition of the hexylene glycol.

2.4. Chromosome Isolation

A flow chart which outlines all steps in these isolation procedures and indicates pertinent explanations for some of these steps is shown in Fig. 1. The best starting material for chromosome isolation is a relatively pure mitotic cell population. Cooling the cells to 4°C in fresh medium inactivates trypsin, dilutes any remaining Colcemid, and causes any microtubules remaining after Colcemid treatment to disassemble spontaneously (Inoue, 1964), thereby helping to minimize aggregation of chromosomes and contamination. After the cells are pelleted by centrifugation and the medium removed with a Pasteur pipette, the cells are washed rapidly in 4°C chromosome isolation buffer. After centrifugation, the pellet is gently resuspended in about 50 volumes of the cold buffer and incubated in a 37°C water bath for 10–15 min. Following incubation, the cells may be broken by nitrogen cavitation after a 3-min exposure to 250 psi. While nitrogen cavitation is recommended as the gentlest way to break the cells, satisfactory results may also be obtained by homogenization. For small volumes of cells, gentle shearing through a 22-gauge needle on a 5-ml syringe breaks the cell membrane and frees the chromosomes into the buffer. The number of times the solution is passed through the needle and the amount of force vary with cell line, cell incubation time, and thoroughness of the washing. This and all steps of these procedures should be monitored using phase contrast microscopy. It is better to shear too gently than too vigorously, since the structure of damaged chromosomes cannot be repaired. The temperature should not be permitted to drop from 37°C until the desired breakage is accomplished. After the cells are broken

A. pH 6.8 Isolation Buffer
   1.0 M hexylene glycol
   0.5 mM $CaCl_2$
   0.1 mM PIPES
B. pH 10.5 Isolation Buffer
   1.0 M hexylene glycol
   2.0 mM $CaCl_2$
   1.0 mM CAPS
C. Procedure for Chromosome Isolation
   1. Exponentially growing culture
   2. Block cells with Colcemid (0.06 µg/ml: 3–6 hr)
   3. Differentially trypsinize or shake off mitotic cells
   4. Centrifuge 1000 rpm; 2 min
   5. Suspend in media to inactivate trypsin
   6. Cool 20 min or longer at 4°C to dissolve mitotic apparatus
   7. Centrifuge 1000 rpm; 2 min
   8. Wash in 4°C isolation buffer, either pH 6.8 or 10.5
   9. Centrifuge 2000 rpm; 3 min
   10. Suspend gently in cold chromosome buffer and incubate in a water bath at 37°C, 10 min
   11. Lyse cells by nitrogen cavitation (250 psi; 3 min) or Dounce homogenizer or syringe-shear gently through 22-gauge needle
   12. Centrifuge 3000 rpm; 5–10 min

Figure 1. Procedure for isolation of metaphase chromosomes.

and the chromosomes liberated, further operations can be done in the cold. All steps in the procedure should be accomplished as rapidly as possible.

Convenient equipment for these isolation procedures with a small cell pellet (0.1 ml volume) includes 15-ml Pyrex or plastic conical centrifuge tubes, 5-ml plastic B-D syringes with 1.5-inch 22-gauge needles, and a clinical centrifuge. For larger volumes of cells, comparable results may be obtained with a glass Dounce homogenizer from Kontes. To facilitate removal of nuclei and unbroken cells, Nucleopore membrane filters (8 and 5 µm pore sizes) may be used. The entire procedure is conveniently done within 1 hr. The appearance of the cells and chromosomes before, during, and after the isolation procedure is shown in the phase contrast micrographs of Fig. 2. Differential centrifugation yields a chromosome preparation like that shown in Fig. 2D. Figure 3 demonstrates by transmission electron microscopy the ultrastructure of typical Chinese hamster chromosomes which are unstained and critical point-dried on a Formvar film. Figure 4 shows the ultrastructure of an isolated Chinese hamster chromosome in the scanning electron microscope. The chromatids are seen to consist of numerous compact microconvules averaging about 520 Å.

2.5. Chromosome Identification

Isolated metaphase chromosomes provide a powerful resource for the study of chromosome morphology and biochemistry, but progress has been

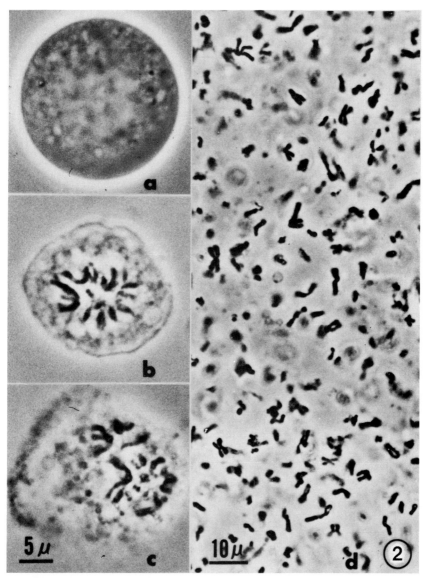

Figure 2. (a) A Chinese hamster metaphase cell in chromosome isolation buffer before equilibration. The chromosomes are not visible until the cell membrane ruptures. (b) The appearance of the chromosomes after a 5 min incubation at 37°C. (c) Cell disintegration upon further incubation, releasing chromosomes, which can then be freed of contaminating cytoplasm by gentle shearing forces. (d) A chromosome suspension prepared according to the methodology described here and in the text. Phase contrast. Reference bar: a, b, and c, 5 μm; d, 10 μm. [Stubblefield and Wray (1971).]

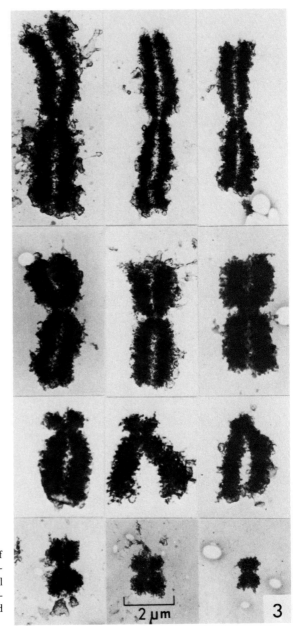

Figure 3. Electron micrographs of typical Chinese hamster chromosomes. Reference bar, 2.0 μm. Critical point-dried, by the method of Anderson unstained. [Stubblefield and Wray (1971).]

hampered by the inability to identify individual isolated chromosomes in a mixture of thousands of similar size.

Even in whole chromosome spreads, the identification of the chromosomes was subjective prior to the development of chromosomal banding techniques. Now, numerous methods for longitudinal differentiation of chromosomes are available for characterizing and karyotyping whole-cell chromosome spreads

Isolation of Human Metaphase Chromosomes 355

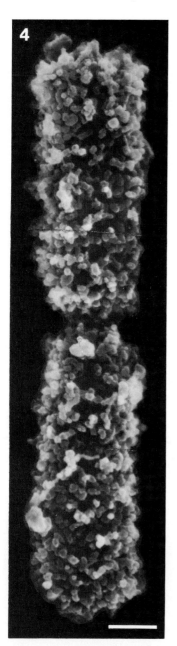

Figure 4. Scanning electron micrograph of an isolated Chinese hamster (Don cell) metaphase chromosome. The chromosome was isolated using nitrogen cavitation, stabilized in 2% uranyl acetate, dehydrated using a graded series of acetone solutions, critical point-dried, and sputter-coated with gold–palladium. The chromatids are characterized by highly coiled topical microconvules and axial coils and appear to be connected by interchromatidal fibers. The scale bar represents 0.5 μm. [This micrograph was taken by Dr. Susanne Gollin.]

or squashes; these have been reviewed by Hsu (1973). W. Wray and Stefos (1975) and Krajca and Wray (1977) have adapted these banding techniques to isolated metaphase chromosomes. Karyotypes of cells stained using the different banding techniques are already established, and a simple visual inspection can match an isolated chromosome with its karyotypic identity. While G-banding with trypsin treatment is relatively rapid and produces

1. Rinse slides of cytocentrifuged preparations in cold 0.025 M KH_2PO_4 buffer
2. Fix in Carnoy's (three parts methanol; one part acetic acid) for 15 min at room temperature
3. Air dry the slides quickly and heat in a 70–75°C oven for 2.5 min
4. Treat chromosomes with 0.0025% trypsin in Hanks basic salt solution with 1.3 mM Ca^{++} and 0.9 mM Mg^{++} at pH 6.5 for 20 sec at room temperature
5. Rinse slides in 70% and 95% ethanol; air dry
6. Stain with 4% Giemsa in 0.025 M KH_2PO_4 buffer for 5.5 min
7. Rinse in distilled water; air dry

   Slides for G-banded controls are completed here; the protocol continues for G/Q-bands

8. Stain with 0.05% aqueous quinacrine dihydrochloride for 5 min at room temperature
9. Rinse well in glass-distilled water; air dry
10. Mount under a coverslip in chromosome isolation buffer containing 10^{-4} M quinacrine dihydrochloride and 2.0 M sucrose

   G/Q-banding of fixed whole cell chromosome preparations follows the same procedure, but begins at step 3

Figure 5. Procedure for G/Q chromosome banding.

easily characterized results, it is a difficult procedure to perform with consistency on isolated chromosomes. Although Q-banding has the advantage of both rapidity and reliability, the fluorescent bands of the isolated chromosomes fade rapidly under UV exposure, which makes prolonged observation and photography difficult. Since dark G-bands and fluorescent Q-bands generally correspond to each other, we developed a sequential technique of quinacrine staining over Giemsa banding to produce slow-fading fluorescent G/Q-bands (Krajca and Wray, 1977). In these preparations, the

Figure 6. Examples of G/Q-banded isolated Chinese hamster Don-3 chromosomes. Procedure as outlined in Fig. 5. The bar represents 5 μm. [Krajca and Wray (1977)].

Isolation of Human Metaphase Chromosomes

background fluorescence quickly fades under continued UV exposure, while the chromosomes remain brightly banded. A single field under the microscope can be observed and photographed for at least 5 min under continuous UV exposure, which allows ample time for careful selection and study of the chromosomes as well as for photography. A flow chart which outlines all steps in the G/Q chromosome banding procedure is shown in Fig. 5. Typical fields of G/Q-banded isolated chromosomes are shown in Fig. 6. Specific examples compared to Q-banding are shown in Fig. 7. A discussion of the variables tested and their results is given in Krajca and Wray (1977).

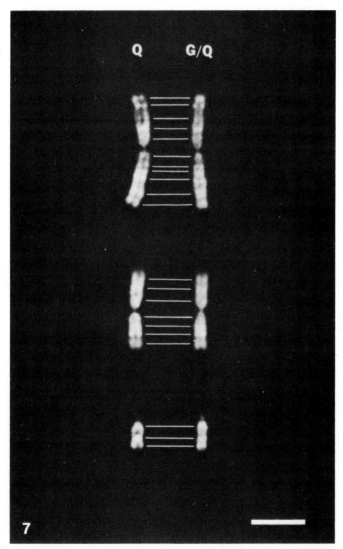

Figure 7. Specific comparisons of identically banded Chinese hamster Don-3 fixed, whole-cell chromosomes by Q- and G/Q-banding techniques. Lines designate common bands between the chromosomes. The bar represents 5 μm. [Krajca and Wray (1977).]

The G/Q-banding method has several advantages for metaphase chromosome identification: (1) it is simple to perform, (2) it is a rapid technique (requiring less than 2 hr), (3) the quality of G-bands in the G/Q preparations does not affect the quality of the superimposed Q-bands (i.e., poor G-bands will still produce good G/Q-bands), (4) destaining the Giemsa before quinacrine application is not necessary, (5) it works equally well on both isolated metaphase chromosomes and whole-cell chromosome preparations, and (6) the chromosome fluorescence is long-lived as compared to previous chromosome Q-band procedures, which allows more time for observation and photography of the chromosomes. (Also see Dev and Tantravahi, Chapter 34, this volume.)

2.6. Chromosome Fractionation

Isolated metaphase chromosomes can be separated into groups according to their size. Sedimentation velocity sucrose gradients of Chinese hamster chromosome suspensions give patterns like that seen in Fig. 8. The contaminating whole cells and nuclei sediment to the 85% sucrose-buffer cushion. The mitotic apparatus and aggregates of chromosomes also sediment far down the gradient. The broad absorbance peak which occurs from the middle of the gradient to near the top is a rough distribution of chromosomes according to size. For Chinese hamster cells the large metacentric chromosomes (A group), which include the 1's and 2's, are sedimented the greatest distance into the gradient. The medium-sized metacentric chromosomes (B group) are the X, Y, 4's and 5's. The majority of these are located above the A group and below the acrocentric chromosomes (6's, 7's, and 8's) which comprise the C group. The D-group chromosomes, which are the small metacentric chromosomes (9's, 10's, and 11's), are sedimented the least distance into the gradient and run only slightly ahead of a large peak of ultraviolet-absorbing material composed of small cellular debris. If the gradient is monitored closely by phase contrast microscopy, four fractions corresponding to the four groups may be obtained

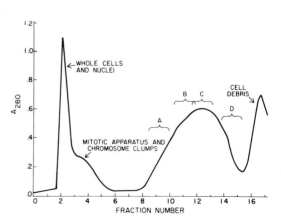

Figure 8. Sucrose velocity gradient of isolated chromosomes. Three milliliters of an isolated chromosome suspension were layered on a 24-ml linear 10–40% (w/v) buffered sucrose gradient with a bottom cushion of 85% sucrose–buffer. This cushion prevented a pellet of whole cells and nuclei from forming and occluding the puncture needle. Coating with Siliclad minimized the adherence of chromosomes to the nitrocellulose centrifuge tubes and glassware. The gradient was centrifuged 90 min at 1000 rpm in an HB-4 swinging bucket rotor in a Sorvall RC-2B centrifuge, after which the tube was punctured and absorbance of the fractions monitored [W. Wray (1973).]

which include the majority of the chromosomes in the appropriate group, but contaminated with those from adjoining groups. Analysis of the centrifugation pattern has shown that the chromosomes that are separated by more than one group are only minor contaminants of each other (i.e., group B and group D). Further purification is accomplished by repeating the velocity sedimentation centrifugation on each of the obtained fractions.

These same principles of sucrose velocity sedimentation gradients hold for cells with complex karyotypes, such as human. However, sedimentation techniques do not provide clean preparations of a single chromosome type, as a single fraction of a gradient will contain several different chromosomes, unless the karyotype is very simple, as in the case of Indian muntjac. Visual identification by phase contrast microscopy of the chromosomes in a fraction is difficult, but flow cytometry (Gray et al., 1975; Stubblefield et al., 1975), coupled with computerized data analysis (Stubblefield et al., 1978), has made rapid advances in this field possible.

2.7. Flow Cytometry

Flow cytometry provides a rapid and quantitative technique for chromosome analysis (Gray et al., 1975; Stubblefield et al., 1975). In this procedure, the chromosomes are stained with ethidium bromide (25 μg/ml) and then passed single file in a flow stream through a laser beam tuned to 514 nm, a wavelength that will cause the dye to fluoresce. As the chromosomes pass through the laser beam, a pulse of fluorescence occurs that is approximately proportional to the DNA content of the chromosome. The fluorescence is detected by a photomultiplier tube and amplified for final storage in a multichannel analyzer. A frequency distribution of the number of particles versus their DNA content is accumulated in this way, and in cases where the DNA content of a chromosome is unique, a single peak representing that chromosome may be observed in the DNA distribution profile for that karyotype.

Chromosome sorting by electrostatic deflection is possible at speeds approaching 1000 particles/sec. After passing the laser beam, the flow stream emerges from the flow chamber and is broken into small droplets by a sonication device. As the droplets break free, a charge is applied to the droplet if it contains a particle of interest, as determined by evaluation of the fluorescent pulse at a fixed time interval earlier. The droplets then fall between two charged deflector plates, which cause the charged droplets to move laterally from the major flow path of uncharged droplets, allowing collection of the desired chromosomes in one or two vessels separated from the unsorted stream. [For a more complete discussion of flow cytometry principles and applications see Horan and Wheeless (1977).]

The full potential of flow cytometry in chromosome sorting has not yet been realized. The upper limit on yield is imposed by the sorting speed and the stability of the flow cytometry equipment. For isolation of a population of a single chromosome species, sedimentation velocity centrifugation fraction preparation followed by chromosome sorting using flow cytometry technology is a better approach than either single technique. Prefractionation of chromosomes increases the flow cytometer sorting efficiency so that sorting of single chromosomes becomes feasible. Runs of 1 hr can be done with the present stability of instrumentation. Since analytical rates approaching 1000 particles/sec are achieved in practice, in 1 hr about 3.6×10^6 chromosomes would be processed. This would correspond to about 0.4 μg of DNA from the sorted chromosome as starting material for biologic experiments.

2.8. Chromosome Sorting of the Human Karyotype

The usefulness of sorting chromosomes as they emerge from the cytometry flow chamber has been demonstrated (Horan and Wheeless, 1977; Carrano et al., 1976). We have applied this technology to sort human chromosomes taken from sucrose gradient fractions. It is possible to sort a chromosome from the total karyotype, but the efficiency is improved if the sort is made from a gradient fraction where the desired chromosome is a large fraction of the total.

In the human karyotype, shown in Fig. 9, many chromosomes have nearly identical DNA contents. The standard flow cytometry equipment commonly used is capable of only a 3–5% coefficient of variation (CV) with isolated

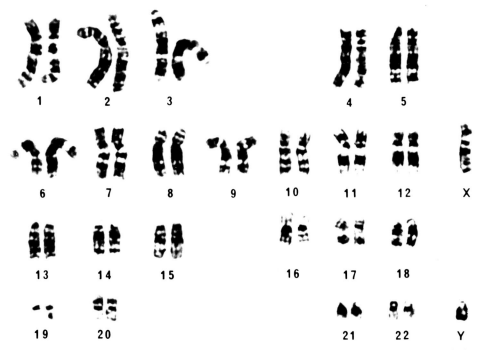

Figure 9. G-banded human karyotype. Note that the banding pattern allows each of the chromosomes to be distinguished.

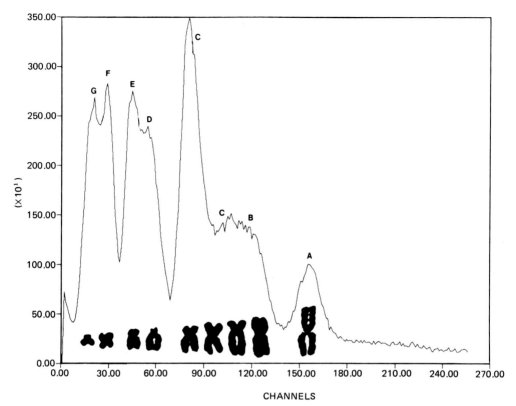

Figure 10. Flow cytometry analysis of isolated metaphase chromosomes from the diploid human lymphocyte cell line CRG. The abscissa is the channel number reflecting the ethidium bromide fluorescence (proportional to DNA content), and the ordinate is the number of chromosomes per channel. The regions of the curve corresponding to specific human chromosome groups are indicated by the letters A–G, and examples of the size of the chromosomes contained in each group are shown below the curve. Usually chromosome number 3 is included in group A; however, by DNA content it is smaller than chromosomes 1 and 2, and is accordingly found as the largest chromosome in the B region of the profile. There is no marked separation between groups B and C, and chromosomes 9–12 have nearly the same DNA content, resulting in the sharp peak between channels 80 and 90. Group D, E, F, and G each produce distinct peaks. [Stubblefield and Wray (1978).]

chromosomes, so only the major chromosome groups are discernible, as in Fig. 10. However, Van Dilla et al. (1975) have shown that a reduction of CV to about 1% would resolve single peaks for about 18 of the 24 different human chromosomes.

Flow cytometric analysis of the different fractions taken from a sucrose gradient quickly reveals the distribution of chromosomes throughout the gradient (Fig. 11). Such data are readily assembled by computer manipulation of the numbers accumulated in the multichannel analyzer (Stubblefield et al., 1978) and do not require long periods of time for data processing.

The human karyotype is readily divided into the major chromosome groups: A, B, C, D, E, F, and G. The groups B and C are not clearly delineated, and chromosome number 3 is in the B group rather than in group A with

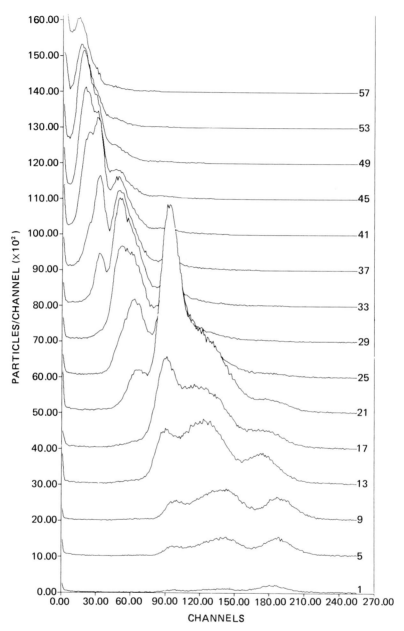

Figure 11. Flow cytometry profiles of CRG chromosomes fractionated according to size. Isolated CRG human chromosomes were sorted roughly into size classes by zonal centrifugation on a sucrose gradient. An SZ-14 Sorvall zonal rotor is loaded at rest with a linear 1-liter gradient of 20–40% sucrose made up in a 1.0 M hexylene glycol, 5×10^{-4} M $CaCl_2$, and 1×10^{-4} M PIPES buffer at pH 6.5. The isolated chromosomes are diluted with an equal volume of 20% sucrose, immediately loaded onto the sucrose gradient spinning at 3000 rpm, and centrifuged for 60 min. Subsequently, the rotor is slowly brought to rest over a 10-min interval to allow reorientation of the gradient. It is then pumped out dense end first and collected in about 50 fractions of 20 ml

chromosomes 1 and 2. Otherwise, the traditional classification holds for the human flow cytometric karyotype (Fig. 10).

2.9. Isolation of Human Chromosome Group F

The chromosomes found in fraction 36 of the gradient shown in Fig. 11 provided a good source for isolation of human F-group chromosomes. Flow cytometric analysis of this fraction (Fig. 12A) revealed two major peaks, chromosome groups E and F, with minor quantities of groups C, D, and G chromosomes. This was visually confirmed (Fig. 13A). Peak F was sorted out, setting the upper and lower discriminator gates as shown in Fig. 12A. Since sorted chromosomes are diluted by several volumes (~1:60) of sheath fluid, it is necessary to add stain in buffer to restore the ethidium bromide concentration before reevaluating the sorted sample. Figure 12B shows the flow cytometric profile of the sorted chromosomes. Note that a single major peak with minor contaminants of larger and smaller chromosomes was seen. A visual analysis (Fig. 13B) of the contents of this fraction (N_{total} = 300 chromosomes) revealed that it consisted of 86% F chromosomes, 9.7% G chromosomes, 3.3% E chromosomes, and 1% D chromosomes. Because the F group contains only two chromosomes (19 and 20) and represents about 4.5% of the total genome, a 20-fold enrichment of a small portion of the genome has been accomplished. Similar fractions can be obtained from the other major flow cytometry peaks of the human karyotype.

The resolution of individual human chromosomes by flow cytometric techniques is difficult, considering their relative homogeneity in size. The newest two-parameter technology, however, allows resolution far superior to single-parameter analysis and sorting on the basis of any given parameter. Using Hoechst 33258 to preferentially react with A–T base pairs and Chromomycin A_3 to preferentially react with G–C base pairs, it is possible to resolve practically every chromosome in the human karyotype. Figure 14 shows the results obtained by Dr. Richard Langlois at the Lawrence Livermore Laboratory plotted on three-dimensional isocontour graphs. Essentially all of the chromosomes may be resolved using this system of data analysis and display. Chromosome sorting using this technology in combination with velocity sedimentation and/or other chromosome separation techniques will allow complete resolution and isolation of each of the chromosomes of the human karyotype in sufficient quantity for biochemical analysis.

---

each. Every fourth fraction was analyzed, and the baseline for each succeeding plot is elevated by 1000 on the ordinate. The data were normalized by counting a constant volume of each sample. It is apparent that any one chromosome group is distributed through about 20 fractions of the gradient, but considerable enrichment can be found in certain fractions. For example, chromosomes 9–12 make up about half of fraction 21, whereas they comprise only 17% of the starting mixture. [Stubblefield and Wray (1978).]

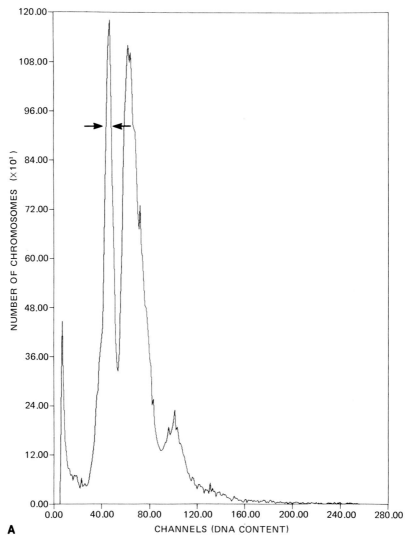

Figure 12. (A) Flow cytometry profile of fraction 36 from the gradient shown in Fig. 11. The peak positions do not correspond with those in Fig. 10 and 11 because of changes in the signal amplification between the two runs. The peak near the origin is electronic noise. The tallest peak contains mostly F chromosomes (numbers 19 and 20), and the next peak represents chromosome group E with some group D chromosomes as a shoulder on the right. The next peak out is contaminating group C chromosomes with minor quantities of larger chromosomes. A photomicrograph of a sample of these chromosomes is shown in Fig. 13A. The arrows at channels 45 and 55 represent the discriminator settings for sorting peak F. [Stubblefield and Wray (1978).] (B) Flow cytometry profile of sorted chromosomes from fraction 36 with the sorter set to select the chromosomes in peak F. Most of the C, D, and E group chromosomes were eliminated. The peak at channel 20 is electronic noise. [Stubblefield and Wray (1978).]

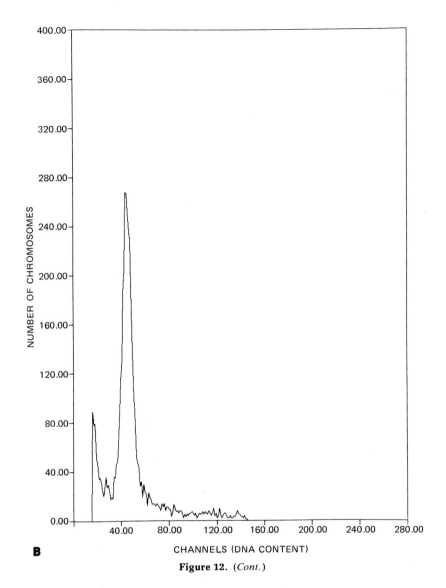

Figure 12. (*Cont.*)

3. Applications for Isolated Chromosomes

The development of recombinant DNA technology provides the tools for detailed examination of specific chromosomal gene segments of an organism. In this approach it is possible to correlate the new data with those of classical genetics, since the chromosome(s) of origin will be known for each DNA sample. However, the enormous complexity of mammalian genomes has made such work a formidable task. The technology for fractionation of isolated metaphase chromosomes into single species makes feasible the isolation of populations of individual chromosomes in yields exceeding one million copies. This amount of DNA from specific isolated chromosomes is sufficient

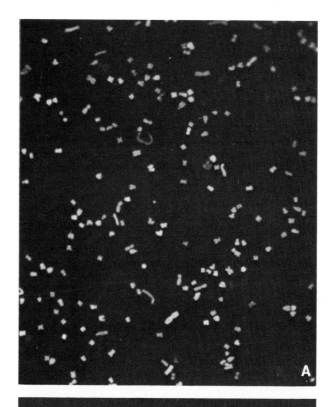

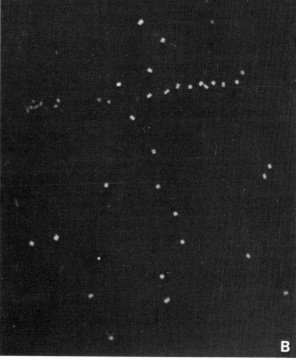

Figure 13. (A) Chromosomes found in fraction 36. Most of the chromosomes are from groups D, E, and F with a few contaminating chromosomes from groups B, C, and G. The chromosomes were sedimented onto a circular coverslip and stained with ethidium bromide for photography in a fluorescence microscope. [Stubblefield and Wray (1978).] (B) Chromosomes found in the sorted material shown in Fig. 12B. By actual count of 300 chromosomes, 86% were group F chromosomes. [Stubblefield and Wray (1978).]

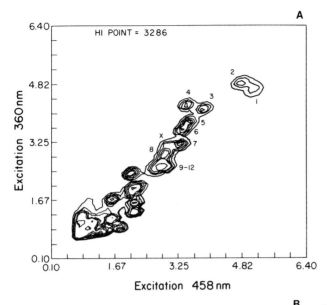

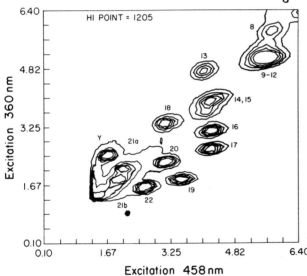

Figure 14. (A) Contour plot of data from double-stained chromosomes derived from human lymphocytes analyzed on a dual laser flow cytometer. (B) An expanded view of the same data set showing the smaller chromosome types in greater detail. Chromosomes were stained in suspension with 5 μM Hoechst 33258 (excitation 360 nm) and 70 μM chromomycin A_3 (excitation 458 nm). All chromosome types except 9–12, 14, and 15 are separately resolved, and individual homologs of chromosome 21 are also separately resolved. [Unpublished data kindly provided by R. Langlois, L. C. Yu, J. Gray, A. Carrano, Lawrence Livermore National Laboratory.]

for plasmid amplification, and reduces the complexity of mammalian genomes by more than an order of magnitude. We can now directly map the chromosomal site(s) of any gene for which a nucleic acid probe can be made. The isolated chromosomes are separated roughly according to size on a sucrose gradient in a zonal rotor. DNA isolated from each of the gradient fractions is digested using a restriction endonuclease and the DNA fragments separated according to size by agarose gel electrophoresis. Over a dozen fractions are processed on each gel and transferred by the Southern blotting method (Southern, 1975) to nitrocellulose paper. The gene fragments derived from a total genome can then be tested for complementarity to a radioactive molecular probe, perhaps a cDNA copy of purified mRNA from a specific gene. Molecular hybridization, followed by autoradiography, is used to reveal the fractions containing the chromosome on which the gene corresponding to the probe is located. The pattern of restriction fragments verifies the gene identity. In the case where only a single gene exists, this information is very interesting, but for multigene families, the restriction pattern can be essential in identifying the various members of the family.

When pure chromosome fractions are obtained using a multiparameter flow cytometer, gene mapping can be done with more precision. DNA isolated from a pure chromosome fraction can be readily tested for its homology with a molecular probe. If each of the chromosomes of a karyotype can be purified, then mapping a gene to a chromosome is accomplished by simply testing the DNA from each chromosome for complementarity with the probe. This has not yet been accomplished, because no karyotype has been completely dissected using these techniques.

3.1. Chicken Gene Mapping

To date, the best examples of gene localization using isolated metaphase chromosomes come from the avian cell line MSB-1. Although the MSB-1 cell line is derived from a lymphoma caused by Marek's disease virus, the karyotype is quite typical of a normal chicken (Fig. 15), and consists of a mixture of large chromosomes (macrochromosomes) grading into tiny, almost invisible, chromosomes (microchromosomes). MSB-1 chromosomes are prepared and fractionated in a zonal rotor on a sucrose gradient as described earlier. The fractions when analyzed by flow cytometry give profiles as shown in Fig. 16. Stubblefield and Oro (1982) sorted quantities of the ten largest chicken macrochromosomes sufficient for analysis using DNA probes. Cleveland et al. (1981) then took the DNA from the purified chromosomes and tested for complementarity with a molecular probe made from tubulin mRNA. Using DNA from chromosomes separated on a gradient, they were able to determine that the tubulin genes were scattered throughout the chicken genome. At least six different chromosomes contain tubulin genes; chromosomes 1, 2, and 8 contain four of the tubulin genes, according to these studies. Using a similar approach, W. Wray et al. (1981) have localized α-actin to a chromosome fraction highly enriched in chromosome 6 and shown that the α-actin gene was not linked to other members of the actin multigene family. Other chicken genes that have been studied using isolated chromosomes include globin, ovalbumin, ovotransferrin, ovomucoid, ribosomal RNA, and several integrated viral genes (Hughes et al., 1979; Padget et al., 1977).

3.2. Human Gene Mapping

An experiment has been done using isolated human chromosomes to map the α-globin gene. Human chromosomes were isolated from a cultured

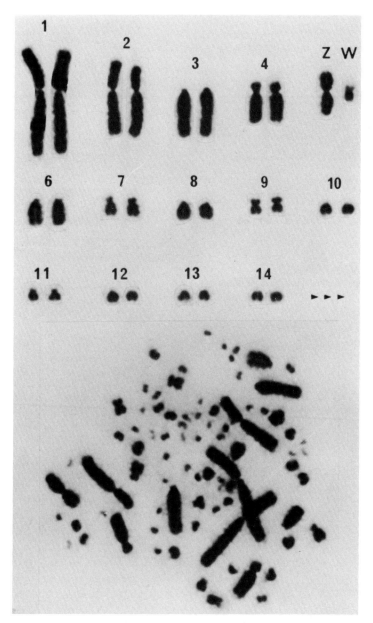

Figure 15. The chromosomes of cell line MSB-1. Only the largest dozen or so pairs can be accurately distinguished. Although this cell line is derived from a Marek's disease lymphoma, the chromosomes are indistinguishable from a normal chicken karyotype. [This photograph was provided by Dr. Steven Bloom.]

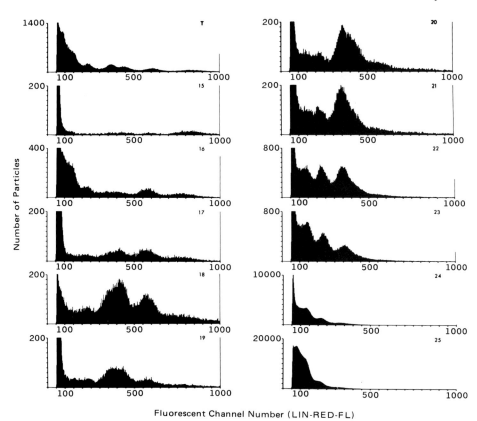

Figure 16. Flow cytometry analysis of isolated metaphase chromosomes from MSB-1 chicken cells fractionated according to size. Centrifugation conditions are essentially the same as in the legend for Fig. 11. The absissa is the channel number reflecting the ethidium bromide fluorescence (proportional to DNA content), and the ordinate is the number of chromosomes per channel. The top left panel marked T is the profile from the unfractionated total chromosome population. The chromosomes were distributed in fractions 15–25 of the linear 1-liter sucrose gradient.

lymphocyte cell line, and were then separated on a sucrose gradient in a zonal rotor as previously described. DNA from the bottom 12 fractions was digested with EcoRI and the restriction fragments were separated by agarose gel electrophoresis followed by Southern blotting (Southern, 1975). A molecular probe (cDNA), prepared from the mRNA of α-globin and labeled with ^{32}P, was then hybridized to the DNA in the blot. A single EcoRI fragment located in the fractions containing E-group chromosomes, as determined by flow cytometry, reacts as seen in the autoradiographs of Fig. 17. Obviously a probe for any other gene on the same chromosome would have this same distribution through the gradient fractions, while a gene on a smaller or larger chromosome would be distributed differently. The gene for α-globin has already been mapped to human chromosome 16 (Deisseroth et al., 1977), so this experiment is consistent with those results. Once we have pure fractions of each chromosome sorted by flow cytometry, precise mapping to a single chromosome becomes possible.

Isolation of Human Metaphase Chromosomes

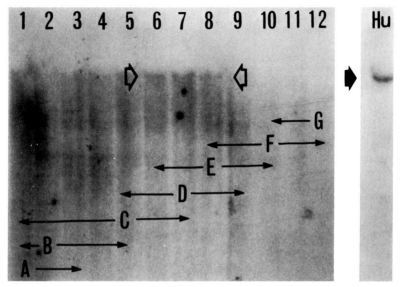

Figure 17. Mapping of α-globin gene using gradient-fractionated chromosomes. The DNA was isolated from the bottom 12 fractions of a 20–40% sucrose gradient, run for 40 min in a SZ-14 Sorvall zonal rotor at 3000 rpm. The chromosome distribution through the gradient was analyzed by flow cytometry, and the distribution of the major chromosome groups is indicated along the bottom of the figure. After restriction digestion with EcoRI and agarose gel electrophoresis, a Southern blot was made and the DNA fragments were then annealed to a ^{32}P probe for human α-globin. Lane 13 shows the probe reacted to total human DNA. A single fragment, about 23,000 base pairs in size, is found following EcoRI digestion. The corresponding band is also seen in fractions 6–9, which would indicate that this gene is either a D- or E-group chromosome. It has previously been mapped to chromosome E-16. [This experiment was done in collaboration with Dr. Tien Kuo.]

Since the chromosome is a precise subunit of the genome, recombinant DNA libraries derived from single isolated chromosomes will certainly be of value, both in gene mapping and in gene isolation. As the technology of molecular genetics is refined and expanded, isolated chromosomes will clearly play a role, especially in the exploration of the human genome.

Acknowledgments

This work was supported by grants CA-18455 and GM-26415 from NIH and grant PCM 79-05428 from NSF. One of the authors (WW) is the recipient of Research Career Development Award CA-00532 from the National Cancer Institute.

The authors are grateful for the assistance of Dr. Steven Bloom in analyzing the MSB-1 chicken karyotype, Dr. Tien Kuo for assistance with molecular hybridization experiments, Dr. Susanne Gollin for providing the scanning electron microscopy and careful reading of the manuscript, Dr. Richard Langlois for providing unpublished two-dimensional flow cytometry data, Dr. Virginia Wray for helpful and critical comments on the organization and development of the manuscript, Scott Smith for photographic assistance, and Claudia Payne for excellent typing and editorial assistance.

References

Adolph, K. W., 1980, Isolation and structural organization of human mitotic chromosomes, Chromosoma **76**:23–33.

Blumenthal, A. B., Dieden, J. D., Kapp, L. N., and Sedat, J. W., 1979, Rapid isolation of metaphase chromosomes containing high molecular weight DNA, J. Cell Biol. **81**:255–259.

Burkholder, G. D., and Mukherjee, B. R., 1970, Uptake of isolated metaphase chromosomes by mammalian cells in vitro, Exp. Cell Res. **61**:413–422.

Cantor, K. P., and Hearst, J. E., 1966, Isolation and partial characterization of metaphase chromosomes of a mouse ascites tumor, Proc. Natl. Acad. Sci. USA **55**:642–649.

Carrano, A. V., Gray, J. W., Moore II, D. H., Minkler, J. L., Mayall, B. H., Van Dilla, M. A., and Mendelsohn, M. L., 1976, Purification of the chromosomes of the Indian muntjak by flow sorting, J. Histochem. Cytochem. **24**:348–354.

Chorazy, M., Bendich, A., Borenfreund, E., and Hutchison, D. J., 1963, Studies on the isolation of metaphase chromosomes, J. Cell Biol. **19**:59–69.

Cleveland, D. W., Hughes, H. E., Stubblefield, E., Kirchner, M. W., and Varmus, H. E., 1981, Multiple and tubulin genes represent unlinked and dispersed gene families, J. Biol. Chem. **256**:3130–3134.

Deisseroth, A., Nienhuis, A., Turner, P., Velez, R., Anderson, W. F., Ruddle, F., Lawrence, J., Creagan, R., and Kucherlapati, R., 1977, Localization of the human α globin structural gene to chromosome 16 in somatic cell hybrids by molecular hybridization assay, Cell **12**:205–218.

Franceschini, P., and Giacomoni, D., 1967, Isolation and fractionation of metaphase chromosomes from HeLa cells, Atti Assoc. Genet. Ital. **12**:248–258.

Frenster, J. H., 1963, Constraints on isolation of mammalian chromosomes, Exp. Cell Res. Suppl. **9**:235–238.

Goyanes, V. J., Matsui, S.-I., and Sandberg, A. A., 1980, The basis of chromatin fiber assembly within chromosomes studied by histone–DNA crosslinking followed by trypsin digestion, Chromosoma **78**:123–135.

Gray, J. W., Carrano, A. V., Steinmetz, L. L., Van Dilla, M. A., Moore, D. H., Mayall, B. H., and Mendelsohn, M. L., 1975, Chromosome measurement and sorting by flow systems, Proc. Natl. Acad. Sci. USA **72**:1231–1234.

Hamilton, M. G., and Peterman, M. L., 1959, Ultracentrifugal studies on ribonucleoprotein from rat liver microsomes, J. Biol. Chem. **234**:1441–1446.

Hearst, J. E., and Botchan, M., 1970, The eukaryotic chromosome, in: Annual Review of Biochemistry, pp. 151–182, Annual Review Inc., Palo Alto, California.

Horan, P. K., and Wheeless, L. L., 1977, Quantitative single cell analysis and sorting, Science **198**:149–157.

Hsu, T. C., 1973, Longitudinal differentiation of chromosomes, Ann. Rev. Genet. **7**:153–156.

Huberman, J. A., and Attardi, G. J., 1966, Isolation of metaphase chromosomes from HeLa cells, J. Cell Biol. **31**:95–105.

Huberman, J. A., and Attardi, G. J., 1967, Studies of fractionated HeLa cell metaphase chromosomes. I. The chromosomal distribution of DNA complementary to 28S and 18S ribosomal RNA and to cytoplasmic messenger RNA, J. Mol. Biol. **29**:487–505.

Hughes, S. H., Stubblefield, E., Payvar, F., Engel, J. D., Dodgson, J. B., Spector, D., Cordell, B., Schimke, R. T., and Varmus, H. E., 1979, Gene localization by chromosome fractionation: Globin genes are on at least two chromosomes and three estrogen-inducible genes are on three chromosomes, Proc. Natl. Acad. Sci. USA **76**:348–1352.

Inoue, S., 1964, Organization and function of the mitotic spindle, in: Primitive Motile Systems in Cell Biology (R. D. Allen and N. Kamiya, eds.), Academic Press, New York, pp. 549–598.

Kane, R. E., 1965, The mitotic apparatus. Physical-chemical factors controlling stability, *J. Cell Biol.* **25**(Suppl.):136–144.

Krajca, J. B., and Wray, W., 1977, Banding isolated metaphase chromosomes by a sequential fluorescent G/Q technique, *Histochemistry* **51**:103–111.

Langlois, R. G., Carrano, A. V., Gray, J. W., and Van Dilla, M. A., 1980, Cytochemical studies of metaphase chromosomes by flow cytometry, *Chromosoma* **77**:229–251.

Maio, J. J., and Schildkraut, C. L., 1967, Isolated mammalian metaphase chromosomes. I. General characteristics of nucleic acids and proteins, *J. Mol. Biol.* **24**:29–39.

Mendelsohn, J., Moore, D. E., and Salzman, N. P., 1968, Separation of isolated Chinese hamster metaphase chromosomes into three size groups, *J. Mol. Biol.* **32**:101–112.

Nikolayenko, N. S., and Pinaev, G. P., 1981, Lipid composition of metaphase chromosomes dependent on the method of their isolation, *Tsitologiya* **23**:185–192.

Padget, T. G., Stubblefield, E., and Varmus, H. E., 1977, Chicken macrochromosomes contain an endogenous provirus and microchromosomes contain sequences related to the transforming gene of ASV, *Cell* **10**:649–657.

Peterson, J. L., and McConkey, E. H., 1976, Non-histone chromosomal proteins from HeLa cells. A survey by high resolution, two dimensional electrophoresis, *J. Biol. Chem.* **251**:548–554.

Prescott, D. M., and Bender, M. A., 1961, Preparation of mammalian metaphase chromosomes free of cytoplasm, *Exp. Cell Res.* **25**:222–223.

Somers, C. E., Cole, A., and Hsu, T. C., 1963, Isolation of chromosomes, *Exp. Cell Res.* (Suppl.) **9**:220–234.

Southern, E., 1975, Detection of specific sequences among DNA fragments separated by gel electrophoresis, *J. Mol. Biol.* **98**:503–517.

Stubblefield, E., and Oro, J., 1982, The isolation of specific chicken macrochromosomes by zonal centrifugation and flow sorting, *Cytometry* **2**:273–281.

Stubblefield, E., and Wray, W., 1971, Architecture of the Chinese hamster metaphase chromosome, *Chromosoma* **32**:262–294.

Stubblefield, E., and Wray, W., 1973, Biochemical and morphological studies of partially purified Chinese hamster chromosomes, *Cold Spring Harbor Symp. Quant. Biol.* **38**:835–843.

Stubblefield, E., Cram, S., and Deaven, L., 1975, Flow microfluorometric analysis of isolated Chinese hamster chromosomes, *Exp. Cell Res.* **94**:464–468.

Stubblefield, E., Linde, S., Franolich, F. K., and Lee, L. Y., 1978, Analytical techniques for isolated metaphase chromosome fractions, in: *Methods in Cell Biology*, Volume XVII (D. M. Prescott, ed.) Academic Press, New York, pp. 101–113.

Van Dilla, M. A., Carrano, A. V., and Gray, J. W., 1975, Flow karyotyping: Current status and potential development, in: *Automation of Cytogenetics* (M. L. Mendelsohn, ed.), U.S. ERDA Publication, pp. 145–164.

Wray, V. P., and Wray, W., 1982, Proteins of the metaphase chromosome, in: *Biochemistry and Biology of the Cell Nucleus II. Nonhistone Proteins.* (L. S. Hnilica, ed.), CRC Press.

Wray, V. P., Elgin, S. C. R., and Wray, W., 1980, Proteins of metaphase chromosomes and interphase chromatin, *Nucleic Acids Res.* **8**:4155–4163.

Wray, W., 1973, Isolation of metaphase chromosomes, mitotic apparatus, and nuclei, in: *Methods in Cell Biology*, Volume VI (D. M. Prescott, ed.), Academic Press, New York, pp. 283–306.

Wray, W., 1976, Isopycnic centrifugation of mammalian metaphase chromosomes in Metrizamide, *FEBS Lett.* **62**:202–207.

Wray, W., and Stefos, K., 1976, Quinacrine bands in isolated chromosomes, *Cytologia* **41**:729–732.

Wray, W., and Stubblefield, E., 1970, A new method for the rapid isolation of chromosomes, mitotic apparatus, or nuclei from mammalian fibroblasts at near neutral pH, *Exp. Cell Res.* **59**:469–478.

Wray, W., Stubblefield, E., and Humphrey, R., 1972, Mammalian metaphase chromosomes with high molecular weight DNA isolated at pH 10.5, *Nature New Biol.* **238**:237–238.

Wray, W., Zimmer, W., Berglund, E., and Schwartz, R., 1981, Chromosomal localization of the skeletal muscle α actin gene in chicken, *J. Cell Biol.* **91**:68a.

Chapter 26
Techniques of Chromosome-Mediated Gene Transfer

O. WESLEY McBRIDE

1. Introduction

Functional genes can be transferred to eukaryotic cells by incubation with isolated metaphase chromosomes. Uptake occurs by phagocytosis and most of the ingested chromosomal DNA is degraded to small inactive fragments. The frequency of transfer of any selected marker ranges from about 10^{-5} to 10^{-7}. Hence, the isolation of transformants requires a sensitive selection system. A free functional chromosome fragment can be retained in the progeny of a transformant for many generations after uptake if selection is maintained. In the absence of selection, the transferred genes are lost from progeny at a rate of about 2–10% per generation (Klobutcher and Ruddle, 1979). Stabilization of transferred genes occurs at a much lower frequency and results from integration of the fragment into nonhomologous sites on recipient chromosomes (Fournier and Ruddle, 1977). The size of the transferred fragment is usually considerably decreased during the process of integration (Klobutcher and Ruddle, 1979; Olsen et al., 1981).

Closely linked genes are cotransferred by this technique. Although intact chromosomes have not been observed in transformants, the transferred fragment can be detected cytologically in a small fraction (10–20%) of the transformants (Miller and Ruddle, 1978; Olsen et al., 1981; Klobutcher and Ruddle, 1979). The size of the transferred fragment (transgenome) has been evaluated by genetic analysis (Miller and Ruddle, 1978; Klobutcher and Ruddle, 1979; Willecke et al., 1976; McBride et al., 1978) and cytogenetic (Miller and Ruddle, 1978; Klobutcher and Ruddle, 1979) and nucleic acid hybridization techniques (Olsen et al., 1981). These results indicate that the fragment can range from less than 0.1% to about 1% of the haploid genome, representing less than 3000 to about 30,000 kbp DNA. Multiple transferred fragments are observed in some transformants (Klobutcher and Ruddle, 1979;

O. WESLEY McBRIDE • Laboratory of Biochemistry, National Cancer Institute, National Institutes of Health, Bethesda, Maryland 20205.

Olsen et al., 1981) and these sometimes originate from multiple donor chromosomes (Klobutcher and Ruddle, 1979).

The specific requirements for application of this technique will be discussed. These include procedures for the bulk isolation of metaphase chromosomes, incubation conditions for chromosome transfer, selective conditions for isolation of transformants, and methods for analysis of the transformants.

2. Methods for Metaphase Chromosome Isolation and Purification

Metaphase chromosomes can be isolated from a population of cells with a mitotic index less than 10%. However, the ease of isolation and purification of chromosomes is markedly enhanced when a greater proportion of mitotic cells are present. Nearly pure preparations of mitotic cells can be obtained by their selective detachment from monolayers of rapidly proliferating fibroblasts after mitotic arrest with Colcemid. Alternatively, a major fraction of metaphase cells can be obtained by mitotic arrest of asynchronous, or parasynchronous, populations of cells in suspension. A large quantity of cells can be isolated from suspension cultures with minimum effort.

The cells are swollen in a hypotonic medium to disperse the chromosomes and facilitate subsequent disruption of the cell membrane. Many methods have been described which are suitable for isolating metaphase chromosomes in high yield. To disrupt cells and maintain the structural integrity of released chromosomes and interphase nuclei, all methods require the use of some mechanical force and an isolation medium containing acidic buffers, high concentrations of divalent cations, solvents containing hexylene glycol, or various combinations. Unfortunately, acidic buffers and divalent cations also promote the aggregation of cytoplasmic debris which is difficult to remove from the chromosomes. Adequate stabilization of the chromosomes and nuclei is essential to prevent the formation of nucleoprotein gels during subsequent steps in the chromosome isolation. These sticky gels trap and remove all chromosomes, nuclei, and debris.

Chromosomes are isolated under sterile conditions for use in gene transfer experiments. Two methods of chromosome isolation which have been used extensively are described below. (Also see Wray and Stubblefield, Chapter 25, this volume).

2.1. Chromosome Isolation at pH 3

A method for chromosome isolation at pH 3 (Mendelsohn et al., 1968) has been slightly modified (Fig. 1). Either asynchronous populations of cells are labeled with [^{3}H]thymidine (TdR) for several generations and arrested at mitosis with Colcemid, or synchronized cells (Fig. 1) are used. The incorporation of a small amount of [^{3}H]TdR (0.05–0.1 cpm/cell) into the chromosomal DNA is useful when estimating the final chromosome concentration and following the course of purification.

1. Arrest cells in S phase and at G_1-S boundary with HAG (10^{-6} M amethopterin, 10^{-4} M hypoxanthine, 10^{-4} M glycine) for an interval equal to the generation time minus the duration of S phase
2. Reverse block with TdR (5×10^{-6} M TdR + 0.2 µCi/ml [^{3}H] TdR) and arrest cells at mitosis with Colcemid (0.2 µg/ml)
3. Harvest cells (after an interval equal to the duration of S phase and G_2) by sedimentation at 500 g and wash cells with cold PBS; determine cell concentration, mitotic index, and cpm/cell
4. Swell cells [$2-4 \times 10^6$/ml] in 0.075 M KCl (room temperature for 10-20 min; monitor by phase microscopy; subsequent steps performed at 5°C
5. Centrifuge 5 min at 220 g
6. Transfer the sticky pellets (10^7 cells/ml) to blender in pH 3 buffer [0.05 M NaCl–0.05 M acetic acid (pH 3.0) containing 0.1 M sucrose, 1 mM $CaCl_2$, 1 mM $MgCl_2$]
7. Adjust Powerstat and homogenize for 15-sec intervals until cells are disrupted; monitor by phase microscopy
8. Transfer lysate (plus buffer rinses) to polypropylene tubes; centrifuge 30 min at 1000 g
9. Disperse the sediment in 15 ml pH 3 buffer (2×10^7 cell equivalents/ml) in a tight Dounce homogenizer; add 75 ml (98.6 g) 85% sucrose–pH 3 buffer and mix; transfer suspension to three $1 \times 3\frac{1}{2}$ inch polyallomer tubes; rinse homogenizer with 10 ml pH 3 buffer and use it to overlay suspensions and to balance the tubes
10. Centrifuge for 2 hr at 100,000 g in a swinging bucket rotor to remove debris
11. Resuspend pelleted chromosomes and nuclei [0.02 M Tris-HCl (pH 7.0)–3 mM $CaCl_2$] and centrifuge 30 min at 1000 g to remove sucrose
12. Resuspend pellet (2×10^6 cell equivalents/ml) in 50 ml 0.02 M Tris-HCl (pH 7.0)–3 mM $CaCl_2$ in a tight Dounce homogenizer
13. Separate chromosomes from nuclei by unit gravity sedimentation [sucrose density gradient containing 0.02 M Tris-HCl (pH 7)–2 mM $CaCl_2$–0.05% Triton X-100] for 4-6 hr
14. Displace suspension, collect fractions, and assess by scintillation counting
15. Centrifuge suspensions 30 min at 1000 g; pool chromosome fractions and wash; estimate chromosome concentration from counts per minute

Figure 1. Flow chart for chromosome isolation (pH 3) and purification. (See text for details.)

Swollen cells are disrupted in pH 3 buffer in a blender. The course of homogenization is evaluated by phase microscopy. Intact nuclei, large amorphous debris, and well-dispersed, highly refractile, compact chromosomes are observed after homogenization. It is possible to disrupt nearly all cells and release the free metaphase chromosomes without disrupting either nuclei or metaphase chromosomes.

Chromosomes and nuclei are sedimented at 1000g. Most of the debris is also recovered in the sediment, but it can be separated from the chromosomes and nuclei by isopycnic centrifugation in 70% sucrose containing pH 3 buffer.

The solution of 85% sucrose–pH 3 buffer (Fig. 1) can be sterilized by 0.45-µm Millipore filtration at 50°C using air pressure. This solution is extremely viscous at 5°C, and it is added by weight to the dispersed suspension of chromosomes, nuclei, and debris. The particulate material is dispersed in 70% sucrose, rather than overlaying it, to prevent trapping of chromosomes and nuclei in a dense layer of large debris.

After centrifugation of the suspension at 100,000g, a thick layer of debris and intact cells is observed at the solvent interface near the top of the tubes and a pellet of chromosomes and nuclei can be seen on the bottom. Phase contrast microscopic examination of the resuspended pellet of chromosomes and nuclei indicates that nearly all debris is removed by this procedure.

The chromosomes are separated from nuclei by unit gravity sedimentation (McBride and Ozer, 1973) in Lucite chambers of various sizes. The chamber consists of a cylindrical separation section (typically 6 cm high and 20 cm diameter) fused to conical upper and lower sections. Liquid enters the bottom, which contains a baffle, and it exits through the top. The gradient is produced by connecting the chamber through a mixing bottle containing 250 ml 0.5% sucrose to a two-chamber linear gradient device containing 700 ml 3.5% and 700 ml 10% sucrose. This gradient is followed by 500-ml aliquots of 10% and 12% sucrose in the linear device. All solutions contain 0.02 M Tris HCl (pH 7.0), 2 mM $CaCl_2$, and 0.05% Triton X-100. This system produces a shelf and a short exponential density gradient followed by two quasilinear gradients.

The suspension of chromosomes and nuclei is applied to the bottom of the chamber and inflow of the density gradient is started immediately. Flow is interrupted after the suspension is lifted to the top of the cylindrical separation section. The course of sedimentation can be observed visually by light scattering. Within about 1 hr, the broadening band of particles can be observed to split into two bands. The lower band contains sedimenting nuclei and the upper band contains chromosomes. Overloading of the density gradient must be avoided to prevent the appearance of small streaming elements of suspension and resultant diminished resolution between bands of chromosomes and nuclei. Recent advances in the design of sedimentation chambers and chromosomes fractionation have been described (Tulp et al., 1980).

The band of chromosomes is adequately separated from nuclei after less than 4–6 hr of sedimentation. Inflow of solvent (13% sucrose) is resumed to displace the suspension from the outlet in the top conical section. Fractions (50 ml) are collected and an aliquot (0.5–1 ml) of each fraction is used to determine the distribution of [^{3}H]-labeled chromosomes and nuclei by scintillation counting. Chromosomes are pooled after recovery from appropriate fractions by sedimentation. The material is again centrifuged after resuspension in 0.02 M Tris-HCl (pH 7.0)–3 mM $CaCl_2$. About 30% of the chromosomes are recovered as a purified fraction from the mitotic cells. These chromosomes retain a highly refractile, compact shape under phase microscopy and they have a normal appearance after fixation and Giemsa staining. This isolation method provides excellent stabilization of liberated chromosomes and nuclei. Hence, there is little or no free nucleoprotein released, and the preparations are not "sticky."

2.2. Chromosome Isolation at pH 7

A method used for chromosome isolation at neutral pH (Fig. 2) is basically that described by Maio and Schildkraut (1969). Cells are labeled

> 1. Sediment cells after Colcemid arrest; wash with cold PBS; determine mitotic index, cell concentration, and cpm/cell
> 2. Resuspend cells $1-2 \times 10^7$; ml in 0.02 M Tris-HCl (pH 7.0)–3 mM $CaCl_2$ and swell cells for 20–30 min (0–5°C); add 1% Triton X-100 and disrupt cells in Dounce homogenizer with tight pestle; evaluate by phase microscopy
> 3. Centrifuge lysate (plus rinse solution) for 30 min at 1000 g
> 4. Separate chromosomes from nuclei by unit gravity sedimentation; pool chromosomes and determine recovery; wash in isolation buffer without Triton X-100

Figure 2. Procedure for chromosome isolation at neutral pH. (See text for details.)

with [^{3}H]TdR to permit estimation of chromosome recovery. Mitotic and interphase cells are washed, swollen in hypotonic buffer, and disrupted after adding nonionic detergent. Disruption of cells and preservation of released chromosomes and nuclei are best accomplished by Dounce homogenization. The process is evaluated by phase microscopic analysis.

Chromosomes and nuclei released during homogenization are stabilized moderately well by the presence of divalent cations. Separation of debris from chromosomes liberated at pH 7 is a difficult problem. Some debris can be removed by isopycnic centrifugation in 70% sucrose–0.02 M Tris-HCl, pH 7.0 (containing 0 or 2 mM $CaCl_2$), and the chromosomes and nuclei are quantitatively recovered in the resultant pellet. However, a large fraction of the cytoplasmic debris is also found in the material pelleted from 70% sucrose, and isopycnic centrifugation is usually omitted.

Chromosomes are separated from interphase nuclei and cells by unit gravity sedimentation as described (Section 2.1 and Fig. 1). Chromosomes isolated at pH 7 are rather "sticky" and tend to aggregate with other chromosomes and debris during sedimentation. Thus, the overall recovery of chromosomes does not usually exceed 50% and considerable cytoplasmic debris persists in the final preparations.

No morphologic changes are apparent after storage of chromosomes, isolated at either pH 3 or 7, for several weeks at 5°C. In contrast, there is a progressive decrease in the chromosomal DNA size during storage, as indicated by sedimentation analysis in alkaline sucrose density gradients. This effect probably arises from endonuclease contamination. It is advisable to use chromosomes for gene transfer immediately after isolation.

3. Metaphase Chromosome Uptake

3.1. Uptake in Suspension

Purified chromosomes are incubated with recipient cells in suspension at high concentrations; the details are described in Fig. 3. The ratio of recipient cells to cell equivalents of donor chromosomes is usually one. The transfer frequency of any selected functional gene marker is about $10^{-6}-10^{-7}$.

1. Centrifuge recipient cells in sterile, siliconized, screw-cap culture tube for 5 min at 220 g and discard supernatant medium
2. Suspend purified chromosomes in complete MEM spinner medium (10% fetal bovine serum), centrifuge 30 min at 1000 g, and discard supernatant; suspend chromosome pellet in complete medium (50–75% of final volume) and disperse chromosomes in small Dounce homogenizer with tight pestle
3. Transfer chromosome suspension to tube containing recipient cells
4. Rinse homogenizer with small quantity of medium and transfer to incubation tube; add 0.1 volume of 110 µg/ml poly-L-ornithine (mol. wt. 70,000) to the tube, disperse cells by pipetting, equilibrate with 5% CO_2, and warm to 37°C
5. Incubation mixture contains 6×10^6 recipient cells/ml and 6×10^6 cell equivalents/ml of chromosomes
6. Incubate for 90–120 min at 37°C while rolling the tube gently to maintain cells in suspension
7. Dilute with complete medium and dispense aliquots of cell suspension to plastic culture dishes; apply selective media after growth for three cell generations

Figure 3. Flow diagram for chromosome transfer to cells in suspension.

3.2. Calcium Phosphate-Precipitated Chromosome Uptake

Miller and Ruddle (1978) applied the calcium phosphate coprecipitation technique of Graham and van der Eb (1973) to chromosome uptake; their method is described in Fig. 4. They observed that the frequency of functional gene transfer was increased tenfold when chromosomes were applied to cell monolayers by this procedure. Combination of this technique and postincubation of the monolayers with dimethyl sulfoxide (DMSO) was reported to enhance functional transfer by 100-fold.

1. Inoculate 2×10^6 recipient cells into (10-cm-diameter) plastic petri dishes 18 hr before uptake; culture cells and remove medium just before application of chromosomes
2. Suspend chromosomes in HEPES-buffered saline (2 mM HEPES pH 7.1, 137 mM NaCl, 5 mM KCl, 0.7 mM Na_2HPO_4, 28 mM glucose) at a concentration of 2.5×10^6 cell equivalents/ml; Add 1/16 volume 2 M $CaCl_2$ while mixing to form a coprecipitate of chromosomes and calcium phosphate
3. After 10 min incubation, add 2 ml chromosome suspension to each monolayer (about 5×10^6 cells) and incubate 30 min (room temperature)
4. Add 20 ml complete growth medium (MEM containing 10% fetal bovine serum) to each dish and incubate 4 hr (37°C) in incubator with CO_2 control
5. Add 5 ml 50% DMSO (in medium) or 10 ml 30% DMSO to each dish and incubate 30 min (room temperature)
6. Aspirate medium, add 10 ml fresh culture medium, and return dishes to incubator for 18 hr
7. Aspirate medium, remove cells by trypsin treatment, and transfer cell suspension from each dish to ten petri dishes in selective medium

Figure 4. Procedure for uptake of calcium phosphate-precipitated chromosomes by monolayers of cells.

There are two potential problems in using this procedure. If the chromosomes are exposed to high concentrations of salts during the precipitation procedure, chromosomal proteins and DNA dissociate and a nucleoprotein gel forms. Direct exposure of recipient cells to high concentrations of dimethyl sulfoxide also results in nonviability. Thus, the DMSO must be diluted several-fold before contact with the cells.

4. Isolation and Analysis of Transformants

A sensitive selection system is required to demonstrate chromosome-mediated gene transfer and to isolate transformants. Recipient cells usually carry a conditional enzyme deficiency which can be complemented by a gene from the donor line, and the HAT (hypoxanthine, aminopterin, thymidine, and glycine) system if often used (Fig. 5). In this system, only cells that have ingested a chromosome carrying the required gene will survive. Other selection systems have been described and summarized (McBride and Peterson, 1980), including those involving ouabain or methotrexate resistance or virus rescue. Transformants expressing new surface antigens potentially could be isolated from unselected populations by fluorescence-activated cell sorting.

The reversion frequency of mutant recipient cell lines which are employed must be low. Usually, an interspecies transfer system is used to permit identification of a transferred gene or gene product as the chromosomal donor species. Either nucleic acid hybridization (Scangos et al., 1979) or isozyme analytical techniques can be used for this purpose. Demonstration of instability of the putatively transferred phenotype or high frequencies of

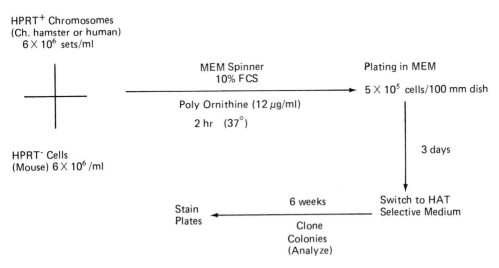

Figure 5. Typical method for chromosome-mediated gene transfer (CMGT) in suspension employing the HAT selection system and HPRT as the selected marker.

presumed transformants are less satisfactory criteria for gene transfer (McBride and Peterson, 1980). The size of transferred chromosomal fragments can be evaluated by cytogenetic (Klobutcher and Ruddle, 1979) or nucleic acid hybridization (Olsen et al., 1981) techniques. Integration of the transferred genes into the chromosomal DNA of a recipient cell can be detected by analysis of phenotypic stability of the transferred marker (Klobutcher and Ruddle, 1979).

Transformants can be isolated most satisfactorily from dishes that remain subconfluent during selection. Transformed colonies are cloned in glass or stainless steel cylinders and expanded in selective media. Alterations in phenotypic stability and size of a transferred fragment are frequently observed at intervals after the transformants are isolated. These changes accompany integration of the transferred fragment. Thus, viable frozen stocks of isolated transformants should be prepared early.

5. Applications and Discussion

Moderate-sized segments of chromosomal DNA are introduced into eukaryotic cells by chromosome-mediated gene transfer (Olsen et al., 1981; McBride and Peterson, 1980). This technique provides a method for regional mapping of closely linked genes. It could be most useful for an intermediate range of fine structural mapping with greater resolution than can be attained by somatic cell hybridization methods, but extending greater distances than are currently accessible by recombinant DNA technology. Intergenic distances should be inversely related to the frequency of cotransfer of genes by this procedure. Deletion mapping following stabilization of visible chromosome fragments has been proposed (Klobutcher and Ruddle, 1979) as an alternative method to provide quantitative intergenic mapping data after cotransfer of genes. Three loci were mapped to a short region on human chromosome 17 by this method. Mapping results obtained by either method should still be interpreted cautiously. Transferred fragments can be quite large and cotransfer can involve multiple fragments which may originate from more than one chromosome (Klobutcher and Ruddle, 1979).

Chromosome-mediated gene transfer could be used for isolation of limited portions of eukaryotic genomes. Isolation of the nonintegrated chromosomal fragments from unstable transformants by physical separation methods would be advantageous for mapping large chromosomal segments by recombinant DNA techniques. Recombinant DNA could also be prepared from transformant DNA and cloned. Most recombinants containing transferred chromosomal DNA sequence could be identified by hybridization with a repetitive DNA probe prepared from donor DNA (Gusella et al., 1980).

The current major limitation to chromosome-mediated gene transfer relates to the paucity of selection systems. In contrast to somatic cell hybridization, a selectable marker is needed for each chromosome. There are presently good selection systems for isolation of transformants containing human chromosomes 16 (adenine phosphoribosyl transferase), 17 (thymidine kinase), and the X chromosome (hypoxanthine phosphoribosyl transferase).

Recent advances in fractionating metaphase chromosomes suggest that specific chromosomes could be used for transfer to establish syntenic relationships directly. This application also requires development of additional selection systems or improvements in the frequency of functional gene transfer.

References

Fournier, R. E. K., and Ruddle, F. H., 1977, Stable association of the human transgenome and host murine chromosomes demonstrated with trispecific microcell hybrids, *Proc. Natl. Acad. Sci. USA* **74**:3937–3941.

Graham, F. L., and van der Eb, A. J., 1973, A new technique for the assay of infectivity of human adenovirus 5 DNA, *Virology* **52**:456–467.

Gusella, J. F., Keys, C., Varsanyi-Breiner, A., Kao, F., Jones, C., Puck, T. T., and Housman, D., 1980, Isolation and localization of DNA segments from specific human chromosomes, *Proc. Natl. Acad. Sci. USA* **77**:2829–2833.

Klobutcher, L. A., and Ruddle, F. H., 1979, Phenotype stabilization and integration of transferred material in chromosome-mediated gene transfer, *Nature* **280**:657–660.

Maio, J. J., and Schildkraut, C. L., 1969, Isolated mammalian metaphase chromosomes: II, Fractionated chromosomes of mouse and Chinese hamster cells, *J. Mol. Biol.* **40**:203–216.

McBride, O. W., and Ozer, H. L., 1973, Gene transfer with purified mammalian chromosomes, in: *Possible Episomes in Eukaryotes*; Le petit Colloquia on Biology and Medicine, Volume 4 (L. G. Silvestri, ed.), North-Holland, Amsterdam, pp. 255–267.

McBride, O. W., and Peterson, J. L., 1980, Chromosome-mediated gene transfer in mammalian cells, *Ann. Rev. Genet.* **14**:321–345.

McBride, O. W., Burch, J. W., and Ruddle, F. H., 1978, Cotransfer of thymidine kinase and galactokinase genes by chromosome-mediated gene transfer, *Proc. Natl. Acad. Sci. USA* **75**:914–918.

Mendelsohn, J., Moore, D. E., and Salzman, N. P., 1968, Separation of isolated Chinese hamster metaphase chromosomes into three size-groups, *J. Mol. Biol.* **32**:101–112.

Miller, C. L., and Ruddle, F. H., 1978, Co-transfer of human X-linked markers into murine somatic cells via isolated metaphase chromosomes, *Proc. Natl. Acad. Sci. USA* **75**:3346–3350.

Olsen, A. S., McBride, O. W., and Moore, D. E., 1981, Number and size of human X chromosome fragments transferred to mouse cells by chromosome-mediated gene transfer, *Mol. Cell. Biol.* **1**:439–448.

Scangos, G. A., Huttner, K. M., Silverstein, S., and Ruddle, F. H., 1979, Molecular analysis of chromosome-mediated gene transfer, *Proc. Natl. Acad. Sci. USA* **76**:3987–3990.

Tulp, A., Collard, J. G., Hart, A. A. M., and Aten, J. A., 1980, A new unit gravity sedimentation chamber, *Anal. Biochem.* **105**:246–256.

Willecke, K., Lange, R., Kruger, A., and Reber, T., 1976, Cotransfer of two linked human genes into cultured mouse cells, *Proc. Natl. Acad. Sci. USA* **73**:1274–1278.

Chapter 27

Transfer of Macromolecules Using Erythrocyte Ghosts

MARTIN C. RECHSTEINER

1. Introduction. Comparison of Injection Using Microneedles, Liposomes, and Red Blood Cell (RBC) Ghosts

Microinjection using RBC ghosts is accomplished by first introducing molecules into RBCs during hypotonic hemolysis and then fusing the RBCs to cultured cells with Sendai virus or polyethylene glycol. In this manner molecules can be injected into large numbers of cultured cells within a short time. The method was developed several years ago in three laboratories (Furusawa et al., 1974; Schlegel and Rechsteiner, 1975; Loyter et al., 1975), and it has been used to study a variety of cellular, genetic, and biochemical problems. Since anyone reading this chapter probably has an application in mind, the text emphasizes methodology. Several articles are available which review previous applications and present a broader discussion of the technique (Kulka and Loyter, 1979; Furusawa, 1980). Those who wish to use RBC-mediated injection should consider microneedles or liposomes as alternate injection procedures. Each technique offers certain advantages and disadvantages, and these are briefly reviewed.

Microneedle injection is clearly the most versatile of the three methods. There is no limitation on the size of the macromolecule that can be injected, and molecules can be selectively introduced into nucleus, cytoplasm, or even into individual mitochondria. Microneedles also permit introduction of large amounts of material. Disadvantages of microneedle injection include requirements for specialized equipment and skill on the part of the experimenter. The major disadvantage, however, is the relatively small number of cells that can be injected—approximately 1000 per hr. In some cases this disadvantage can be circumvented by prefusing recipient cells before microneedle injection (see Graessman and Graessmann, this volume, Chapter 32, for further discussion of microneedle injection).

MARTIN C. RECHSTEINER • Department of Biology, University of Utah, Salt Lake City, Utah 84112.

Liposomes offer the dual advantages that there are no constraints on the size of macromolecules encapsulated and large numbers of cells can be treated. Major disadvantages are the small volume and poor fusability of unilamellar liposomes. To date only modest quantities of macromolecules have been transferred using this technique. Moreover, macromolecules may be transferred to cultured cells by endocytosis of liposomes as well as by fusion of liposomes with the plasma membrane. Nevertheless, liposome technology is continually improving and holds promise (see Straubinger and Papahadjopoulos, this volume, Chapter 28. for further information).

The principal advantage of RBC-mediated injection is that an almost unlimited number of cells can be injected with substantial amounts of certain macromolecules. Injection is efficient, with as many as 90% of the recipient cells fusing with RBCs. The procedure suffers in comparison to microneedle injection in that macromolecules can only be injected into the cytoplasm. Cotransfer of RBC contents and membrane presents another possible problem. The major drawback, however, is that RBC-mediated injection is only suitable for transfer of proteins, small RNAs or DNA fragments. Nevertheless, if one must inject large numbers of cells and the molecule to be injected is sufficiently small, RBC-mediated injection could be the method of choice.

The operations required for red cell-mediated injection can be completed without pause in a morning or afternoon. About 3 hr is required for loading macromolecules into RBCs and 1 hr is required for fusion of loaded RBCs with cultured cells. This chapter considers loading and fusion separately. It also discusses the removal of unadsorbed RBCs, and the identification of and enrichment for injected culture cells.

2. Loading Macromolecules into RBCs

2.1. Mechanism

Mammalian RBCs are non-nucleated biconcave disks with a volume of 90 μm^3 (Fischer et al., 1981). Because most cultured mammalian cells have volumes between 1000 and 3000 μm^3, RBCs are of ideal size for use as injection vehicles. They are large enough to accommodate significant amounts of macromolecules, yet small enough that the injected cell generally increases in volume by less than 10%.

When RBCs are placed in sufficiently hypotonic saline, they rupture and release soluble macromolecules, metabolites, and ions. Whereas detergent-induced hemolysis is irreversible, RBC plasma membranes regain impermeability after hypotonic hemolysis. Seeman (1967) used electron microscopy to study morphologic changes in RBCs during this process. Examination of RBCs fixed while hemolyzing in the presence of ferritin showed a cluster of about 40 holes in the plasma membrane of hemolyzing RBCs. The holes, which have diameters between 200 and 500 Å, were present from 15 to 25 sec after the onset of hemolysis. It seems clear that these transient holes in the RBC membrane permit molecules to enter RBCs, and it is likely that the diffusion rate of a molecule determines how much of it enters during hypotonic hemolysis.

2.2. Quantitative Aspects

Ihler et al. (1973) first demonstrated the progressive sieving of proteins by loading a crude extract of ^{35}S-labeled *Escherichia coli* proteins. When the molecular weight distribution of the starting proteins was compared to that of the loaded proteins by gel filtration, they found preferential uptake of smaller proteins. The average molecular weight of the input proteins was 200,000 daltons (D) and that of the loaded proteins was 90,000 D. Over the past 6 years we have loaded a number of ^{125}I-labeled proteins enough times to be confident of their partitioning. These data, which are presented in Table I, confirm progressive exclusion of larger proteins.

Table I also presents data on the partitioning of RNA and DNA molecules. Small RNAs, such as tRNAs and small nuclear (sn) RNAs, have been loaded and successfully injected into culture cells (Kaltoft et al., 1976; Capecchi et al., 1977; Schlegel et al., 1978). With ribosomal RNAs we observed extensive degradation during loading (unpublished observation), and we found apparent sieving among the snRNAs. Thus, while it may be possible to inject small mRNAs by ghost-fusion (Anderson et al., 1978), larger mRNAs will probably be difficult to load in any quantity.

Table I. Partitioning of Macromolecules during Hypotonic Hemolysis

Protein	Molecular weight	Fraction internalized	Molecules per RBC^a
Chromosomal protein HMG1	26,000	0.35	8×10^6
Bovine serum albumin	68,000	0.25	2×10^6
Lactate dehydrogenase	140,000	0.20	9×10^5
Immunoglobulin G	150,000	0.20	8×10^5
Pyruvate kinase	180,000	0.15	5×10^5
Ferritin	500,000	0.08	10^5
RNAs	Bases	Fraction internalized	Molecules per RBC^a
tRNA	80	0.2	5×10^6
snRNAs	100–200	0.15	$>10^6$
DNA	Base pairs	Fraction internalized	Molecules per RBC^a
Micrococcal nuclease limit	140	0.04	3×10^5
Sonicated	400	0.02	5×10^4

^{a} These values are based on loading from a solution containing 10 mg/ml of the macromolecule.

Under certain conditions, DNA molecules assume highly condensed states, termed Ψ forms, where the DNA concentration can exceed 300 mg/ml (Maniatis et al., 1974; Laemmli, 1975; Eickbush and Moudrianakis, 1978). Although Ψ forms assume dimensions that should enter a 500-Å hole in an RBC membrane, we have been unable to load them (M. Rechsteiner and A. O. A. Miller, in preparation). DNA fragments generated by micrococcal nuclease digestion of chromatin can be loaded with no evidence of degradation. Approximately 4% of 140-bp fragments and 2% of 400-bp fragments remain associated with RBC after hemolysis. Only trace amounts of DNA molecules larger than 800 bp were observed in DNA extracted from loaded RBC (M. Rechsteiner and A. O. A. Miller, in preparation).

These results indicate that RBC-mediated injection, as presently practiced, is not suitable for transfer of large amounts of high-molecular-weight DNA to culture cells. Straus and Raskas (1980) reported that RBC-mediated injection could be used to transfer adenovirus DNA into KB cells. It is possible that a small fraction of the hemolytic events permit entry of high-molecular-weight DNA. Alternatively, DNA adsorbed to RBC membrane may be transferred during fusion.

The number of molecules that can be loaded per RBC is of obvious interest to those who might use this method of injection. In two studies where protein concentration was varied, the partitioning of IgG or BSA was not affected (Rechsteiner, 1975; Wasserman et al., 1976). Therefore, the number of molecules loaded per RBC appears to depend upon the partitioning behavior of the molecule and its concentration outside the RBCs during hypotonic hemolysis. The last column in Table I contains calculations based on loading from solutions of 10 mg/ml, a concentration readily obtained for most macromolecules. It can be seen that values range from approximately 10^7 for small proteins and tRNAs to several hundred for 800-bp DNA fragments.

2.3. Preparation of RBCs

Furusawa (1980) has reported that human RBCs fuse better than erythrocytes from dog, guinea pig, or cow. This and the availability of humans in most laboratories make human erythrocytes the cells of choice. We use RBCs freshly drawn from the forearms of volunteers. Blood type does not appear to have any influence on subsequent loading. Coagulation can be prevented with either heparin or EDTA (100 μl of Lipo-Hepin at 10,000 USP units/ml or 100 μl of 250 mM Na_2 EDTA pH 7.2 are more than sufficient to prevent clotting of 10 ml of blood). At least 5 ml of blood should be drawn to facilitate removal of white cells during washes. Blood cells can be washed by suspension in phosphate-buffered saline (PBS) and centrifugation in 15-ml conical plastic centrifuge tubes in a tabletop clinical centrifuge. The white cells form a layer atop the RBC pellet, and they should be removed by aspiration at each centrifugation. Washed RBCs should be loaded as soon as possible since loading is significantly reduced following overnight storage of washed RBCs in either Hanks solution or PBS. If one cannot easily obtain freshly drawn blood, outdated human RBC can be used. In this case, however, it may be necessary to add BSA or cytochrome C to the loading mixture to stabilize the RBC membrane during fusion (Loyter et al., 1975).

2.4. Preparation of Molecules for Loading

The choice of loading procedure determines how the molecules should be prepared. For preswell loading (Rechsteiner, 1975) the molecules must be in a suitably dilute buffer. We use 10 mM Tris pH 7.4 whenever possible. For dialysis loading (Furusawa et al., 1974; Loyter et al., 1975), the molecule can be dissolved in physiologic saline.

2.5. Choice of Loading Procedure

Three methods have been described for loading macromolecules into RBCs. They can be termed rapid-lysis, dialysis, and preswell-loading. They differ with respect to the volume of hypotonic saline used and the rate at which RBCs swell before lysis. There is no evidence, however, that the ultimate hemolytic events are any different among the three procedures.

In rapid-lysis loading, RBCs are diluted into large volumes of hypotonic saline containing macromolecules (Ihler et al., 1973). Hemolysis begins within seconds and molecules become trapped within resealed RBCs. Although the procedure is simple enough, it is not recommended. The use of large volumes of hypotonic saline is wasteful of macromolecules. Moreover, RBCs loaded by rapid lysis are fragile and fuse poorly with Sendai virus (Peretz et al., 1974). Consequently, this procedure is not suited for microinjection.

Dialysis-loading is accomplished by placing RBCs and macromolecules in dialysis tubing in physiologic saline and then dialyzing against dilute saline. After hemolysis, concentrated salt solution can be added to restore isotonicity. With the exception of molecules that might leak from the dialysis bag, this procedure is quite suitable for loading molecules prior to microinjection. However, because we have not used dialysis-loading, we will omit details of the method. Anyone wishing to use this method should consult Furusawa (1980) or Loyter et al. (1975).

In preswell loading, RBCs are suspended in physiologic saline diluted with 0.7–0.9 volumes of water. Under these conditions red cells swell but do not lyse. The swollen RBCs are packed into a small pellet by centrifugation and the overlying dilute saline is removed. The macromolecule to be loaded is added in dilute buffer, the tube contents are mixed by vortexing, and concentrated salt is added 2 min later. The resealed RBCs are incubated at 37°C for 60 min to stabilize the RBC membrane. Unincorporated macromolecules and molecules released upon lysis are then removed by washing in PBS. This procedure, which is used in our laboratory, is now described in detail.

2.6. Preswell-Loading Procedure

One milliliter of washed, packed human RBCs consists of approximately 10^{10} cells. Efficient microinjection requires 50- to 100-fold excess RBCs during fusion, so one should start with 100 µl of packed RBCs for every 10^7 culture cells to be injected. typically we suspend packed RBCs in an equal volume of PBS and add 0.3 ml of this 50% suspension to a 15-ml conical centrifuge tube containing 6 ml of PBS and 5 ml of water. The RBCs are centrifuged at greater than 1000g for 10 min, and the overlying saline is removed. Maximal loading is achieved when the external volume is minimized during hemolysis; therefore we use a Pasteur pipette drawn to a fine tip to remove as completely as possible traces of overlying saline. We then add 150 µl of macromolecule solution to the swollen RBC pellet (the 1.5×10^9 cells have now swollen to almost 0.3 ml). The contents of the centrifuge tube are vortexed immediately upon addition of macromolecule and placed on ice. (The solution should become deep transparent red after addition of macromolecule. If the solution scatters light, lysis has not been complete. One may find it necessary to vortex several times to obtain complete lysis. If this never occurs, one should check for excess salt in the macromolecule solution.) After 2 min, 22 µl of tenfold-concentrated saline is added with vortex mixing, and the solution should again scatter light like a typical RBC suspension.

Tenfold-concentrated PBS and 1.5 M KCl have been used to restore isotonicity after hemolysis. We use a tenfold-concentrated Hanks solution, which is described in Table II. It should be noted that during hypotonic hemolysis the RBC membrane spontaneously reseals even without addition of the concentrated salt solution. High salt serves to shrink RBCs rather than reseal them.

Following addition of the tenfold-concentrated salt solution, RBCs are placed in a 37°C water bath for 30–60 min before washing with PBS. The extent of loading is determined by sampling the suspension obtained upon addition of 5 ml of PBS to the resealed RBCs. The RBCs are washed three times with 5 ml of PBS and once with 5 ml of Tris-saline (150 mM NaCl, 20 mM Tris, pH 7.4). A second sample is taken after suspension of the washed RBCs in

Table II. Preparation of the Concentrated Hanks Solution Used to Reverse Hypotonic Hemolysis

Tenfold-concentrated Hanks solution is prepared by mixing four solutions in the following order and proportions:
  A. 1 ml Hanks V
  B. 0.3 ml 20 mg/ml $NaHCO_3$
  C. 0.1 ml 20 mg/ml glucose
Right before use:
  D. 1 ml Hanks VI

The compositions of Hanks V and Hanks VI are:
  Hanks V: NaCl 105.5 g, KCl 6.27g, Na_2HPO_4 2.86 g, KH_2PO_4 1.75 g, to 1 liter with H_2O
  Hanks VI: $MgSO_4 \cdot 7H_2O$ 1.22 g, $MgCl_2 \cdot 6H_2O$ 2.16 g, $CaCl_2 \cdot 2H_2O$ 3.5 g, to 1 liter with H_2O

Tris-saline, and the amount of macromolecule remaining with the RBCs is compared to that found in the original suspension. Although molecules may adsorb to RBC membranes, we generally find that adsorption accounts for less than 10% of the molecules retained by the RBCs. If one suspects adsorption, one can relyse RBCs or treat loaded RBCs with nucleases, proteases, or other appropriate enzymes to test for degradation of surface-bound molecules.

2.7. Properties of Loaded RBCs

Examination of loaded RBCs under phase microscopy reveals that they are dark gray, in contrast to the higly refractile normal RBCs and the virtually transparent empty ghosts. Loaded erythrocytes can be found in a variety of shapes, including spheres, footballs, cup-shapes, and the biconcave disks typical of nonhemolyzed RBC. These different shaped ghosts vary in volume from 90 μm^3 for biconcave disks to nearly 150 μm^3 for spheres (Fischer et al., 1981). Loaded RBCs are reasonably stable. They can be stored overnight on ice in Tris-saline containing 2 mM $MnCl_2$ with only a trace of hemolysis. In fact, we have obtained successful injections with loaded RBCs stored on ice for 8 days in Tris-saline containing 20 mM $MnCl_2$. There is also little lysis of loaded RBCs incubated for 24 hr in culture medium at 37°C. Longer incubation, however, results in lysis.

It may be necessary to remove RBC contents if they interfere with subsequent measurements or degrade the loaded molecules. This can be accomplished by cycling the RBCs through several hemolyses in 10 mM Tris before adding the macromolecule. In this case, loaded RBCs become increasingly fragile with each susequent hemolysis. Presumably, normal RBC contents stabilize the RBC membrane.

2.8. Stability of Molecules within RBCs

In general, proteins are stable within loaded RBCs, particularly if the cells are kept on ice. Upon incubation of loaded RBCs in tissue culture medium at 37°C, however, one can observe degradation rates up to several percent per day for some iodinated proteins. Native tRNAs are stable in loaded RBCs, as shown by sedimentation in DMSO–sucrose gradients, but glyoxalated tRNAs are significantly degraded after loading (Schlegel et al., 1978). Moreover, M. Capecchi found that it was necessary to remove most of the contents of RBCs by several rounds of hypotonic hemolysis before loaded yeast or E. coli suppressor tRNAs retained their activity. Small nuclear RNAs are stable in RBCs (T. Gurney and M. Rechsteiner, unpublished). The instability of ribosomal RNA has been mentioned, and it is not known whether the observed ribonuclease activity which degrades rRNA originates with RBCs or whether it is released from contaminating leukocytes. If the latter is true, one should be able to overcome the problem by meticulous removal of leukocytes prior to loading. There does not seem to be significant DNAse activity to RBC lysates or within loaded RBCs.

To summarize, most proteins and small RNAs can be loaded efficiently, and they are stable within RBCs. DNA fragments longer than 800 bp are not loaded to any significant extent. Ribosomal RNAs are degraded during loading, and the fate of messenger RNAs has not been studied in detail.

3. Fusion of Loaded RBCs and Cultured Mammalian Cells

3.1. Choice of Fusogen and Fusion Protocol

Two decisions must be made before fusing RBCs and culture cells—whether to use Sendai virus or polyethylene glycol (PEG) as the fusogen and whether to fuse with culture cells in suspension or in monolayer. Certain studies may require that fusions be done in monolayer. For example, studies involving quiescent cells require monolayer fusion, since quiescent fibroblasts are induced to proceed through the cell cycle upon trypsinization. However, we routinely use suspension fusion because it is more efficient than several monolayer protocols we have tried.

PEG is a very attractive fusogen, since it comes in reagent bottles, whereas one must spend time growing, titering, and inactivating Sendai virus. On the other hand, Sendai receptors bring RBCs and culture cells into close apposition, thereby promoting microinjection. With PEG as fusogen one may have to resort to centrifugation (Kaltoft and Celis, 1978) or the use of nonviral agglutinins, such as phytohemagglutinin (Mercer et al., 1979), to achieve efficient microinjection. For this reason, we continue to use Sendai virus as the fusing agent, and our present protocol is described below.

3.2. Sendai-Mediated Fusion with Cells in Suspension

Description of the preparation of Sendai virus is beyond the scope of this review, and one should consult Watkins (1971) for a thorough discussion of this subject. Two points can be made, however. First, the preparation of Sendai virus is not very difficult. Second, some strains of Sendai virus, which fuse culture cells efficiently, produce poor fusion of RBCs. Therefore, one should test the stock of Sendai virus before attempting microinjection. A suitable preparation of Sendai virus at 1000 hemagglutinating units in 0.5 ml of Tris-saline pH 7.4 containing 2 mM $MnCl_2$ will fuse more than 80% of 10^9 RBCs within 20 min at 37°C. Fusion can be detected by phase microscopy.

Sendai virus is prepared by the method of Harris and Watkins (1965), inactivated for 3 min by ultraviolet irradiation at 30 cm from a General Electric G8T5 bulb, and stored at −70°C in 1 ml aliquots of 4000 HAU/ml in Tris-saline pH 7.4. Culture cells are harvested in 0.1% trypsin in Ca^{++}-, Mg^{++}-free saline, and washed twice in culture medium and once in PBS to remove residual trypsin. The cells are then washed in 10 ml of Tris-saline pH 7.4 to remove phosphate ($MnPO_4$ is very insoluble, so components of the fusion mixture are washed with Tris-saline before and after fusion to prevent its formation). For addition to the fusion mixture, culture cells are suspended in Tris-saline at 10^8 cells/ml.

Fusion mixtures are assembled on ice in 15-ml conical centrifuge tubes as follows: (a) each pellet of loaded RBCs (1.5×10^9 cells is suspended in 0.2 ml of tris-saline pH 7.4 and divided equally among two centrifuge tubes; (b) 10^7 culture cells in 0.1 ml of Tris-saline is added to each tube and mixed with the RBC by vortex; (c) 1 ml of Sendai virus (4000 HAU/ml) is diluted with 0.1 ml of 20 mM $MnCl_2$ in Tris-saline, and depending upon cell type, 0.2–0.4 ml is added to each tube. The tube is mixed with a flip of the wrist and placed on ice. One minute or so after virus addition one should see agglutination of RBCs and culture cells in the centrifuge tube. After 10 min on ice, tubes are transferred to a reciprocating water bath at 37°C and incubated for 20 min. Fusion is stopped by addition of 5 ml of cold Tris-saline, and the cells are collected by centrifugation. The supernate should be slightly red, due to lysis of some loaded RBC during fusion. Procedures for removing RBCs are described in Section 3.4.

3.3. PEG Fusion in Monolayer

Polyethylene glycol is a very efficient fusogen (See Mercer and Baserga, this volume, Chapter 3) and there are several published procedures which use it to fuse loaded RBCs to cells in monolayer (Kriegler and Livingston, 1977; Mercer et al., 1979). I will briefly describe the method of Mercer et al. (1979).

Monolayer cultures are prepared by inoculating 35-mm petri dishes containing a 22-mm-square coverslip with 10^5 cells in 2 ml of media. RBCs are loaded by the preswell method as described above and suspended in Hanks balanced salt solution (HBSS) without glucose to a final concentration of 0.03% (v/v). Phytohemagglutinin-P (Difco) is dissolved in HBSS and added to the RBC suspension to a final concentration of 100 µg/ml. Immediately after mixing, 2 ml of the PHA-treated RBCs is added to cultures from which the media has been drained. The cultures are incubated at 37°C for 1 hr and rinsed with HBSS at room temperature, and fusion is initiated by adding 1 ml of a 44% solution of PEG 6000 (Koch-Light Laboratories) in 0.15 M NaCl, 20 mM Tris pH 7.4, and 5 mM $MnCl_2$. After 60 sec, 3 ml of media without serum is added and cultures are incubated at room temperature for 30 min. Cultures are then washed four times in medium containing serum and cultured in medium at 37°C.

3.4. Removal of Unfused RBCs

Efficient microinjection requires a large excess of loaded RBCs during fusion, and removal of these RBCs is usually necessary. If one plans a functional assay following injection or one examines individual cells by microscopy, then residual RBC may not interfere, because of their low metabolic activity and microscopically distinct shape. However, if one plans to follow the fate of a radiolabeled molecule, residual RBC present a significant problem. Unfortunately, there is no efficient solution to this problem.

While NH_4Cl lysis (Kaltoft et al., 1976) and Nitex–fetal calf serum step gradients (Wasserman et al., 1976) have been used to remove residual RBCs after fusion, in most cases differential sedimentation and the lack of RBC adherence to tissue culture plastic provide an adequate method for removing RBCs. Culture cells are significantly larger than RBCs and under appropriate conditions sediment faster. We find that centrifugation of fusion mixtures suspended in 10 ml of culture medium for 3 min at setting three in an IEC clinical tabletop centrifuge produces a tight pellet of culture cells and an overlying fluffy coat of RBCs, many of which can be decanted. Several rounds of this procedure reduce the RBC to culture cell ratio to less than two with no more than 10% loss of culture cells. Following centrifugation, cells are plated into a minimal volume of medium containing 10% serum. This promotes rapid attachment, and injected cells can generally be rinsed free of RBCs within 2 hr of plating. One should rinse away RBCs as soon as possible. Upon prolonged incubation RBCs settle onto and adhere to culture cells, presumably because of residual Sendai virus particles adsorbed to both cell types. If RBC contamination is still significant, one may remove them by trypsin treatments brief enough so that culture cells are not detached.

3.5. Expected Results

Using the procedures outlined above, one should be able to microinject over half of the cells from most established cell lines. At a ratio of 100 RBCs per culture cell, one transfers roughly 1% of the macromolecules present in the starting RBCs. Culture cells that have fused with more than three RBCs are not observed in the plated cell population, and they are presumed to be nonviable. Calculations show that each RBC adds 20% new plasma membrane area to a typical culture cell (volume 2000 μm^3), so the absence of multiple fusion products may result from a loss of membrane integrity. In practical terms, one cannot inject cells with more than three times the number of macromolecules presented in Table I without increasing the concentration of the macromolecule.

4. Identification and Enrichment for Microinjected Cells

Phase microscopic examination of the post-fusion cell population provides a crude, but rapid method for judging the success of RBC-mediated injection. One should observe polyghosts and culture cells with phase-dark "blebs" following successful fusion (see Fig. 1 for examples). This is not

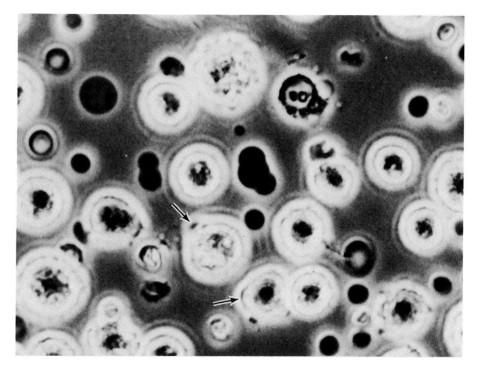

Figure 1. Phase contrast micrograph of the post-fusion mixture. HeLa cells were fused with loaded RBCs under standard conditions and washed extensively, and photomicrographs were taken just prior to plating. A phase-dark polyghost is shown near the center. Arrows point to "blebs" on the surface of HeLa cells. These "blebs" are presumed to arise from fused RBCs. Clearer examples can be found in Schlegel and Rechsteiner (1975).

usually sufficient, since experiments require a reliable quantitative method for determining the proportion of injected cells. Fortunately, a number of options are available. Benzidine staining can be used to detect transferred hemoglobin. While the technique lacks sensitivity, it is relatively easy. Fluoresceinated BSA is often used as a marker, but we prefer horseradish peroxidase (HRP), which presents a number of advantages. HRP is very soluble (solutions of 50 mg/ml are easily obtained), it is readily loaded into RBCs, and the histochemical detection of HRP is probably tenfold more sensitive than detection of fluoresceinated BSA. Moreoever, one does not need a fluorescence microscope (see Fig. 2 for examples of HRP-positive cells). Table III presents protocols for benzidine staining of hemoglobin and histochemical detection of HRP. The former is adapted from LoBue et al. (1963); the latter is taken from Steinman et al. (1974).

Since RBC membrane antigens are incorporated into the plasma membrane of injected culture cells, one should be able to separate injected cells by fluorescense-activated cell sorting after staining with fluorescent antibodies. As these machines are not readily available, another approach is considered here. Yamaizumi et al. (1978) showed that injection of antidiphtheria toxin

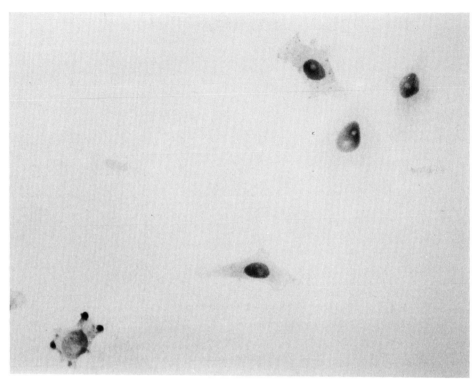

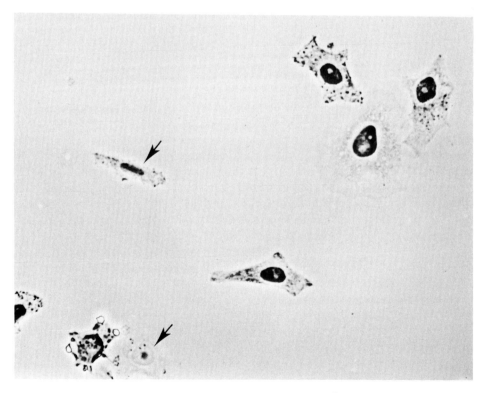

Table III. Histochemical Detection of Injected Cells

A. Benzidine staining for hemoglobin
Injected cells are allowed to attach to coverslips, rinsed in PBS, and stained for 1–5 min in 1% Benzidine in methanol; the coverslips are then immersed in 10% H_2O_2 in 70% ethanol; injected cells turn a deep gold, while uninjected cells are unstained.

B. Histochemical detection of horseradish peroxidase (HRP)
RBCs are loaded with HRP at 4–40 mg/ml and fused with culture cells; cells are allowed to attach to coverslips and are then fixed in 2.5% glutaraldehyde in PBS for 10 min; cells are rinsed thoroughly with PBS and stained in 50 mM Tris pH 7.6 containing 0.01% H_2O_2 and 0.5 mg/ml diaminobenzidine; injected cells turn a deep gold within 20 min, while uninjected cells remain unstained.

C. All reagents can be purchased from Sigma Chemical Co.; horseradish peroxidase type II should be used.

---

antibodies into human KB cells spared the cells from the lethal effects of the toxin. We have confirmed this result and found that the same is true for anti-ricin antibody injection and subsequent ricin challenge (Rogers and Rechsteiner, unpublished observation). Unlike diphtheria toxin, moderate levels of ricin appear to kill cells from most, if not all, established culture lines. Ricin killing is also fast; at 200 ng of ricin/ml virtually all uninjected HeLa cells are killed within 24 hr. Thus, in principle, inclusion of anti-ricin IgG in any loading followed by subsequent ricin challenge should remove uninjected cells from the population.

5. Summary

The methods described above allow one to inject large numbers of cultured mammalian cells with macromolecules. Calculations in Table I show that a single loaded RBC may contain as many as 10^7 small proteins or RNA molecules. Most cultured mammalian cells contain 5×10^6 ribosomes, many chromosomal proteins are present at 10^3–10^5 copies per cell, and there are about 2×10^5 copies of HGPRT per mouse L929 cell. Therefore, RBC-mediated injection should allow one to transfer physiologically significant amounts of most macromolecules.

References

Anderson, W. F., Deisseroth, A., Nienhuis, A. W., Gopalakrishnan, T. V., Huang, A., and Krueger, L., 1978, Cellular and molecular studies on globin gene expression, *Natl. Cancer Inst. Monogr.* **48**:65.

Figure 2. Identification of cells injected with horseradish peroxidase. RBCs were loaded in the presence of 4 mg/ml HRP and fused with HeLa cells. The HeLa cells were plated onto coverslips and stained for HRP 24 hr later. The upper micrograph, which was taken under bright field illumination, shows HRP-positive cells. The bottom micrograph shows the same field under phase contrast. Arrows identify uninjected HeLa cells.

Capecchi, M. R., Vonder Haar, R. A., Capecchi, N. E., and Sveda, M. M., 1977, The isolation of a suppressible nonsense mutant in mammalian cells, Cell **12**:371.

Eickbush, T. H., and Moudrianakis, E. N., 1978, The compaction of DNA helices into either continuous supercoils or folded fiber rods and torroids, Cell **13**:295.

Fischer, T. M., Hoest, C. W. M., Stohr-Liesen, M., Schmid-Schonbein, H., and Skalak, R., 1981, The stress-free shape of the red blood cell membrane, Biophys. J. **34**:409.

Furusawa, M., 1980, Cellular microinjection by cell fusion: Technique and application in biology and medicine, Int. Rev. Cytol. **62**:29.

Furusawa, M., Nishimura, T., Yamaizumi, M., and Okada, Y., 1974, Injection of foreign substances into single cells by cell fusion, Nature **249**:449.

Harris, H., and Watkins, J. F., 1965, Hybrid cells derived from mouse and man: Artificial heterokaryons of mammalian cells from different species, Nature **205**:640–646.

Ihler, G., Glew, R., and Schnure, F., 1973, Enzyme loading of erythrocytes, Proc. Natl. Acad. Sci. USA **70**:2663.

Kaltoft, K., and Celis, J. E., 1978, Ghost-mediated transfer of human hypoxanthine-quanine phosphoribosyl transferase into deficient Chinese hamster ovary cells by means of polyethylene glycol-induced fusion, Exp. Cell Res. **115**:423.

Kaltoft, K., Zeuthen, J., Engback, F, Piper, P. W., and Celis, J. E., 1976, Transfer of tRNAs to somatic cells mediated by Sendai virus induced fusion, Proc. Natl. Acad. Sci. USA **73**:2793.

Kriegler, M. P., and Livingston, D. M., 1977, Chemically facilitated microinjection of proteins into intact monolayers of tissue culture cells, Somat. Cell Genet. **3**:603.

Kulka, R. G., and Loyter, A., 1979, The use of fusion methods for the microinjection of animal cells, in: Current Topics in Membranes and Transport, Volume 12 (F. Bronner and A. Kleinzeller, eds.), Academic Press, New York, p. 365.

Laemmli, U. K., 1975, Characterization of DNA condensates induced by poly(ethylene oxide) and polylysine, Proc. Natl. Acad. Sci. USA **72**:4288.

LoBue, J., Dornfest, B. S., Gordon, A. S., Hurst, J., and Quastler, H., 1963, Marrow distribution in rat femurs determined in cell enumeration and Fe^{59} labeling, Proc. Soc. Exp. Biol. Med. **112**:1058.

Loyter, A., Zakai, N., and Kulka, R., 1975, "Ultramicroinjection" of macromolecules or small particles into animal cells, J. Cell Biol. **66**:292.

Maniatis, T., Venable, Jr., J. H., and Lerman, L. S., 1974, The structure of Ψ DNA, J. Mol. Biol. **84**:37.

Mercer, W. E., Terefinko, D. J., and Schlegel, R. A., 1979, Red cell-mediated microinjection of macromolecules into monolayer cultures of mammalian cells, Cell Biol. Internatl. Rep. **3**:265.

Peretz, H., Toister, Z., Laster, Y., and Loyter, A., 1974, Fusion of intact human erythrocytes and erythrocyte ghosts, J. Cell Biol. **63**:1.

Rechsteiner, M., 1975, Uptake of proteins by red blood cells, Exp. Cell Res. **93**:487.

Schlegel, R., and Rechsteiner, M., 1975, Microinjection of thymidine kinase and bovine serum albumin into mammalian cells by fusion with red blood cells. Cell **3**:371.

Schlegel, R. A., Iverson, P., and Rechsteiner, M. C., 1978, The turnover of tRNAs microinjected into animal cells, Nucleic Acids Res. **5**:3715.

Seeman, P., 1967, Transient holes in the erythrocyte membrane during hypotonic hemolysis and stable holes in the membrane after lysis by saponin and lysolecithin, J. Cell Biol. **32**:55.

Steinman, R. M., Silver, J. M., and Cohn, Z. A., 1974, Pinocytosis in fibroblasts: Quantitative studies in vitro, J. Cell Biol. **63**:949.

Straus, S. E., and Raskas, H. J., 1980, Transfection of KB cells by polyethylene glycol-induced fusion with erythrocyte ghosts containing adenovirus type 2 DNA, J. Gen. Virol. **48**:241.

Wasserman, M., Zakai, N., Loyter, A., and Kulka, R. G., 1976, A quantitative study of ultramicroinjection of macromolecules into animal cells, Cell **7**:551.

Watkins, J. F., 1971, Fusion of cells for virus studies and production of cell hybrids, in: Methods in Virology, Volume V (K. Maramorosch and H. Koprowski, eds.), Academic Press, New York, p. 1.

Yamaizumi, M., Uchida, T., Okada, Y., and Furusawa, M., 1978, Neutralization of diphtheria toxin in living cells by microinjection of antifragment A contained within resealed erythrocyte ghosts, Cell **13**:227.

Chapter 28
Liposome-Mediated DNA Transfer

ROBERT M. STRAUBINGER and
DEMETRIOS PAPAHADJOPOULOS

1. Introduction

The DNA-mediated transfer of genes has been used extensively as a tool for genetic study of both prokaryotic and eukaryotic cells. For those cells for which DNA transformation systems have been developed, gene transfer has yielded a wealth of detail for bacterial (Avery et al., 1944), yeast (Hinnen et al., 1978), and animal cell (McBride and Athwall, 1977; Wigler et al., 1977) molecular biology. A variety of techniques have been used to effect the introduction of relatively unpurified nucleic acids into eukaryotic cells, including cell:cell fusion (Ringertz and Savage, 1976) and uptake of isolated nuclei, microcells (Fournier and Ruddle, 1977), bacterial protoplasts (Schaffner, 1980), viruses (Hamer et al., 1979; Mulligan et al., 1979), or whole chromosomes (McBride and Athwall, 1977) by cells. Purified nucleic acids can be introduced into cells as calcium phosphate coprecipitates (Graham and Van der Eb, 1973) or polycation complexes (McCutchen and Pagano, 1968), which may be endocytosed. In addition, cryoprotectants and polyalcohols (Hinnen et al., 1978; Stow and Wilkie, 1976) also facilitate nucleic acid uptake. Direct introduction of genetic material into cells can be accomplished by microinjection with small capillary needles (Capecchi, 1980; Anderson et al., 1980) or by prepackaging the nucleic acid in carriers such as liposomes (phospholipid vesicles*) or red blood cell ghosts (Straus and Raskas, 1980). While specific successes have been attained by all the methods mentioned, this chapter will deal with the methods, advantages, and potential of liposomes as carriers of nucleic acids.

*The terms "liposome" and "phospholipid vesicle" are used synonymously in this chapter. The names and acronyms for various types of liposomes follow the convention suggested at the New York Academy of Sciences meeting on Liposomes in Biology and Medicine (Papahadjopoulos, 1978).

ROBERT M. STRAUBINGER and DEMETRIOS PAPAHADJOPOULOS • Cancer Research Institute and Department of Pharmacology, University of California–San Francisco, San Francisco, California 94143.

Both RNA and DNA molecules can be encapsulated within the aqueous interior of phospholipid vesicles [for review, see Fraley and Papahadjopoulos (1982)]. Intracellular delivery of the encapsulated macromolecules occurs by a complex mechanism not fully understood (Tyrrell et al., 1976; Pagano and Weinstein, 1978; Poste, 1980), though endocytosis and fusion of liposomes with cellular membranes are two possible and nonexclusive pathways. Bacterial, mammalian, and plant cells have all been shown to incorporate and express liposome-encapsulated nucleic acids, and in some cases the efficiency of liposome-mediated delivery is higher than that reported for other methods of DNA or RNA transfer (Mukherjee et al., 1978; Wilson et al., 1979; Dellaporta et al., 1981; Fraley et al., 1982). Because cellular uptake of liposome contents seems independent of the usual uptake mechanisms for macromolecules, such as receptor-mediated endocytosis or specific transport systems, and independent of surface restrictions imposed by the cell membrane, liposomes can be used to introduce materials that normally cannot enter cells. As a result, cells that are outside the host range of a particular virus can be infected by the liposome-encapsulated virus (Wilson et al., 1977) and cells that are resistant to particular toxins can be killed by liposome-entrapped toxins (Nicolson and Poste, 1978).

Other advantages of liposomes as a carriers for introducing RNA and DNA molecules into cells are simplicity of preparation and long-term stability, low toxicity, and ability to protect encapsulated nucleic acids from degradation. Since some methods of liposome preparation allow solutes to be entrapped to a degree independent of molecular weight, purified genes and conceivably chromosomes can be encapsulated and introduced into cells. In addition, such parameters as liposome size, charge, fluidity and stability may be varied to suit particular applications. The liposome surface may be modified to carry glycolipids (Jonah et al., 1978; Mauk et al., 1980), lectins (Szoka et al., 1981), or covalently bound antibodies (Heath et al., 1980a,b, 1981; Huang et al., 1980; Martin et al., 1981; Leserman et al., 1980), both to confer selectivity and to improve overall magnitude of liposome binding to specific target cell types. Obvious applications of liposome-mediated delivery of nucleic acids will be to maximize expression in existing transformation and transfection systems and to extend the ability for genetic manipulation to cell types that do not repond to existing methods for introducing nucleic acids into cells. Further, liposomes may be useful as carriers for delivering nucleic acids to cells *in vivo* if such a course of therapy can be envisaged for the treatment of inborn errors of metabolism.

In the present chapter we will review current methods for preparation of liposomes and their separation from unencapsulated material, and for interaction of liposomes with cells and the conditions which favor uptake and expression of entrapped nucleic acids (Fraley et al., 1980, 1981).

2. Liposome Physical Properties and Preparation

As liposomes were developed originally as model membranes for biochemical and biophysical studies, there exists a vast literature which deals with the variety of component amphipaths from which liposomes may be made, different methods of liposome preparation, and the physicochemical properties of liposomes [for reviews see Szoka and Papahadjopoulos (1980) and Deamer and Uster (1980)]. However, the encapsulation of nucleic acids for the purpose of functional delivery to cells imposes certain restrictions on method and composition, so that a brief review of the relevant principles is sufficient.

2.1. Lipids

Though lipids of varying degrees of purity will allow liposome formation, such common contaminants as lyso compounds, fatty acids, and metal ions can alter vesicle charge and permeability, and can be toxic to cells. Polyunsaturated fatty acids peroxidize readily, yielding products which may damage DNA (Pietronigro et al., 1977). Lipids obtained from a variety of natural sources contain polyunsaturated fatty acids and must be stored under an inert atmosphere (argon or nitrogen) at low temperature ($-70°C$) to prevent peroxidation. Products of cholesterol oxidation also have been shown to be toxic to cells (Chen et al., 1974), so the same storage conditions are recommended for that lipid.

High-purity lipids are obtained from a number of commercial suppliers, but purity should be checked by thin-layer chromatography in several solvent systems. Kates (1972) is a succinct source of information on thin-layer chromatography of lipids.

One common source of difficulty encountered in preparing liposomes from acidic lipids such as phosphatidylglycerol (PG) and phosphatidylserine (PS) results from the fact that lipid obtained from some suppliers contains a significant amount of divalent cations. Since divalent cations and low pH alter drastically the physical properties of these lipids (Papahadjopoulos et al., 1976), it is important to convert acidic phospholipids to the sodium salt form by washing with EDTA (ethylenediaminetetraacetate) and NaCl (Papahadjopoulos and Miller, 1967; Papahadjopoulos and Kimelberg, 1973).

The concentration of lipids is important in some procedures, and the reader is directed to Bartlett (1959) for a convenient method for lipid phosphorus determination. Colorimetric or enzymatic methods can be used for cholesterol quantitation, and a variety of other techniques for lipid determination are reviewed by Kates (1972).

2.2. Liposome Preparation

The methods for encapsulating DNA in liposomes are nearly as diverse as the applications to which the technique has been applied [for review see Fraley and Papahadjopoulos (1982)]; however, as some of those methods of liposome preparation were developed for purposes quite different from that of entrapping large, often precious molecules and delivering them to cells, they suffer certain shortcomings and can be summarized briefly.

2.2.1. Multilamellar Vesicles (MLV)

Hydration of a thin lipid film with aqueous solutions of DNA is perhaps the simplest method of liposome preparation. The heterogeneous population of particles produced (1.0–5.0 μm diameter) are composed of multiple, concentric lamellae separated by aqueous layers (Bangham et al., 1964). Though some lipids, notably those having charged head groups, will allow some DNA to be entrapped (unpublished observation), the efficiency of entrapping gene-sized macromolecules is expected to be low. Furthermore, the fact that MLV formation may damage nucleic acids, and that nucleases may digest some MLV-associated material (Hoffman et al., 1978; Lurquin, 1979), suggests that incomplete encapsulation and significant lipid/nucleic acid interaction occur during formation of this type of vesicle.

2.2.2. Small Unilamellar Vesicles (SUV)

Sonication of multilamellar vesicles produces a reasonably homogeneous population of liposomes which are small (0.025 μm diameter) and unilamellar (Papahadjopoulos and Miller, 1967; Papahadjopoulos and Watkins, 1967). During sonication, disruption of the large vesicles allows external solute to be entrapped in the small vesicles which form, and some investigators have exploited this phenomenon to load SUV with DNA fragments (Wong et al., 1980). However, the rigors of extended sonication necessary to produce small unilamellar vesicles make it unlikely that nucleic acids are encapsulated intact, and the small internal aqueous volume of SUV makes them unsuitable for efficient encapsulation of high-molecular-weight nucleic acids.

2.2.3. Large Unilamellar Vesicles (LUV)

A number of methods have been developed to exploit the carrier potential of liposomes, methods which produce large, unilamellar vesicles. Most commonly used for nucleic acid encapsulation are Ca^{++}-EDTA chelation (Papahadjopoulos et al., 1975), ether injection (Deamer and Bangham, 1976), detergent dialysis (Enoch and Strittmatter, 1979), and reverse-phase evaporation (REV) (Szoka and Papahadjopoulos, 1978). Of these methods, the REV method is well suited for capturing nucleic acids. A small volume of a highly concentrated DNA or RNA solution can be captured at high efficiency (up to 50% of the original aqueous volume) in a wide variety of lipid mixtures and without appreciable degradation of the solute (Fraley et al., 1980). Efficiency of encapsulation is independent of solute molecular weight, and DNA of approximately 10^8 daltons has been encapsulated (Dellaporta et al., 1981). The liposomes produced are relatively homogeneous in size (0.45 μm diameter) (Olson et al., 1979) and subsequent extrusion of the vesicles through polycarbonate filters of defined pore diameter can be used to ensure sterility (Fraley et al., 1980) or to limit the maximum liposome diameter (Szoka et al., 1980).

2.3. DNA Encapsulation

The REV method was originally described (Szoka and Papahadjopoulos, 1978) for 1.0–2.0 ml aqueous phase and 50–100 µmole phospholipid; for DNA, we routinely use 0.3 ml aqueous phase with 10 µmole lipid, though the proportions can be scaled down to as little as 0.033 ml and 1.0 µmole lipid without adversely affecting vesicle formation. In the brief summary of the method given below, volumes and concentrations of various reagents are included as an example. The procedure can be scaled to fit any application simply by maintaining the given aqueous/lipid/solvent ratio.

Five micromoles of the phospholipid mixture to be used (most commonly phosphatidylserine and cholesterol in a 1:1 mole ratio) is transferred to a 10×100 mm screw-cap glass tube, and the lipid is dried from the original solvent on a rotary evaporator. In order to prevent loss of material in the case of small-scale preparations, the procedure is carried out in a small tube which can be inserted into the much larger evaporation tube commonly used on rotary evaporators. A few milliliters of glycerol or H_2O are added to the larger tube to ensure thermal contact between the sample tube and the evaporator bath. Once the lipid has dried to a thin film on the walls of the sample tube, 0.6 ml of diethyl or isopropyl ether is added to resuspend the lipid. It is essential to use peroxide-free ether, especially if isopropyl ether is used. Redistillation of the ether over a small amount of sodium bisulfite or extraction with a volume of bisulfite solution just before use are two ways to ensure that peroxide is eliminated. The choice of ether is arbitrary in most cases, save those in which lipids with a high transition temperature are used; as the lipid must be in the liquid-crystalline state for vesicle formation, the higher boiling point of isopropyl ether makes it the solvent of choice.

DNA dissolved in an appropriate aqueous buffer is added in a volume of 0.175 ml; only the solubility of high-molecular-weight DNA (approximately 4.0–10.0 mg/ml) limits the concentration at which the material can be captured. If negatively charged phospholipids are used in the preparation, the buffer must not contain millimolar concentrations of divalent cations. The buffer must also be isotonic with respect to the final buffer in which the incubation with cells will take place, as unilamellar vesicles are osmotically sensitive. Finally, it is possible to determine the efficiency of encapsulation by including in the DNA buffer a small amount of a radioactive tracer such as sucrose, inulin, or polyadenylic acid. The process of encapsulation proceeds independently of the Stokes radius of the solute, so such small markers are convenient and accurate markers of aqueous volume trapped.

A stable emulsion is generated by brief sonication of the two-phase system. A bath-type sonicator is recommended to allow the sonication to be done in a closed tube purged with inert gas. Thus sterility, biohazard, and lipid peroxidation concerns may be satisfied. A stable emulsion can result from as little as 5–7 sec of sonication, though poorly tuned sonicators with low power may necessitate much longer irradiation and thus may contribute to nicking or shearing of large DNA molecules. The sample tube is returned to the rotary evaporator and the bulk ether is evaporated under low vacuum (300–400 mm Hg). When a viscous gel has formed, the sample is vortexed briefly and then returned to a higher vacuum (500–600 mm Hg) until the gel breaks and a slightly turbid vesicle suspension is formed. Residual ether can be removed in several minutes at high vacuum (700–750 mm Hg), though care must be taken to ensure that small-volume preparations do not evaporate. An additional volume of buffer may be added and is recommended to prevent dehydration. At this point, encapsulation of the material is complete, and the preparation may be stored under an inert atmosphere at 4°C until used.

2.4. Separation of Liposomes from Free Material

While it may not be important in many applications to separate entrapped material from that remaining free, a number of methods have been devised to attain such an end. Separation is required to determine efficiency of encapsulation, conserve nucleic acid, or optimize liposome-mediated delivery in a particular system. Nuclease digestion followed by gel chromatography has been a common method, though destruction of unencapsulated material, sample dilution, difficulty in maintaining sterility, and lipid adsorption to polyacrylamide (F. Martin, unpublished observation) or agarose gels (Huang, 1969) are undesirable features of the method. Differential centrifugation with reclamation of material in the supernatant is also possible (Wilson et al., 1979), though good separation of free from entrapped material requires several washes and increases the time spent and material lost (up to 40%) in preparation. Flotation of the vesicles on discontinuous polymer gradients has also been described (Fraley et al., 1980; Heath et al., 1981), offering the advantages of clean separation without destruction of free material, rapidity and adaptability to multiple samples, ease of maintaining sterility, and low sample dilution.

For experiments involving substantial amounts of DNA, we use a combination of the two latter methods: centrifugation once (20 min at 120,000g, SW50.1 Beckman rotor) to reclaim most of the unencapsulated DNA, followed by flotation on a metrizamide, Ficoll, or dextran gradient. Though Ficoll and dextran are suitable for negatively charged liposomes (Fraley et al., 1980; Heath et al., 1981), neutral liposomes flocculate in gradients of those polymers. Thus the metrizamide method is summarized here, applicable to all lipids and differing from the other methods only in the concentration of polymer needed for a given solution density.

The pellet from centrifugation is resuspended in 1.5 ml of 7% metrizamide, isotonic with respect to the liposome aqueous contents. If the pelleting step is not used, then the original liposome solution is diluted to a volume of 0.5 ml and mixed with 1.0 ml of a 10% metrizamide solution. As metrizamide undergoes a reversible, concentration-dependent dimerization and thus changes in osmotic activity upon dilution, it is important to verify that the polymer solution becomes isotonic upon dilution with the vesicles. Three milliliters of 5% isotonic metrizamide are overlaid, and 0.5 ml isotonic buffer forms the top gradient step. The gradient is centrifuged at 45,000 rpm for 30 min, 20°C, with no brake and minimal acceleration. The liposomes are removed from the buffer–7% metrizamide interface as a tight, easily pipetted band. While metrizamide is known to be nontoxic, it may be removed by dialysis or desalting on Sephadex G25.

3. Liposome–Cell Interaction

The work of Wilson et al. (1977) was the first to show that large macromolecular assemblies could be encapsulated in LUV and delivered to cells. In that work, purified polio virions encapsulated by the Ca^{++}-EDTA chelation method were shown to be infectious either in the presence of high titers of neutralizing antiserum or in cells of nonprimate origin, which lack the poliovirus-specific surface receptor. While it was clear that the contents of liposomes could be delivered to cells without contact with the external bulk medium, such an observation offers no definitive evidence for mechanism of liposome uptake. However, it is not inconsistent with the hypothesis of liposome–cell membrane fusion or endocytosis through a pathway that allows significant amounts of material taken up to escape lysosomal degradation. Subsequent work with LUV-entrapped RNA, either purified from poliovirus (Wilson et al., 1979) or globin mRNA from rabbit reticulocytes (Ostro et al., 1978; Dimitriadis, 1978), supports such an hypothesis, since the entrapped RNA was shown to be resistant to degradation by RNAse and functional when delivered to cells. In the case of globin mRNA, a globin-length peptide could be immunoprecipitated from lysates of cells exposed to entrapped, but not free, RNA. In the case of polio RNA, delivery and expression of intact RNA resulted in the production of progeny virus, which was assayed by the appearance of viral plaques on permissive cell monolayers.

Quantitation of cell–vesicle interaction in the polio RNA system (Wilson et al., 1979) showed that virtually all cells in a vesicle-exposed population could be infected, whether of primate (HeLa) or of rodent (L, CHO) origin. An analysis of infectivity in vesicle preparations loaded at varying ratios of RNA molecules per vesicle showed that infectivity increases linearly in proportion to the number of molecules per vesicle and plateaus at a ratio of one RNA per vesicle. Infectivity is a one-hit phenomenon, and a single delivery event is sufficient to cause infection. By comparing RNA infectivity, cell-associated vesicle lipid (^{3}H-labeled), and cell-associated vesicle contents (^{32}P polio RNA), the following conclusions were made (Wilson et al., 1979; Papahadjopoulos et al., 1980b): in order to ensure delivery of one functional molecule of RNA to every cell, approximately 10^4 vesicles per cell must be added; of those added, approximately 10% of the vesicle lipid becomes cell-associated, while only 1% of the RNA is found to be cell-bound. Thus, approximately 10^3 vesicles per cell are bound when a saturating concentration of vesicles is applied, and only 100 of those vesicles retain their load of RNA. From the data on infectivity, it can be deduced that the infectivity of liposome-entrapped RNA (one plaque per 100 particles) is comparable to that of a high-titer stock of whole poliovirus.

3.1. Influence of Vesicle Lipid Composition on Intracellular Delivery

With introduction of the REV technique (Szoka and Papahadjopoulos, 1978), it was possible to extend the studies of cell–vesicle interaction to include LUV of a wide variety of compositions, not just of the acidic lipids required by the Ca^{++}-EDTA chelation technique. Taking advantage of the high amplification factor inherent in assays of virus production, Fraley et al. (1980, 1981) compared a number of lipid compositions with respect to the ability to bind and to deliver SV40 DNA to African Green Monkey kidney (AGMK) cells. That particular DNA was chosen for a number of reasons, one being the growing number of recombinant DNA vectors based on the SV40 genome (Hamer et al., 1979; Mulligan et al., 1979); thus, results obtained from plaque assays with the viral DNA would give some indication of the degree of success to be expected in future experiments with the recombinants. In general, it was found that effective delivery of DNA is commensurate with cell–liposome binding and with retention of liposome contents (Table I). Negatively charged vesicles are superior to neutral or positively charged vesicles with respect to binding, though there is considerable variability in retention of contents and DNA infectivity among specific acidic lipids. The DNA-carrying liposomes of phosphatidylserine (PS) are over tenfold more infectious than those of the neutral phospholipid phosphatidylcholine (PC) and threefold more infectious than phosphatidylcholine liposomes doped with the synthetic, positively charged lipid sterylamine (SA). Phosphatidylserine is the best single lipid for intracellular delivery of contents; phosphatidylglycerol (PG), which binds

Table I

Vesicle preparation[a]	Net electrostatic charge	Cell associated lipid,[b] nmole bound	Vesicle contents,[c] % retained	Efficiency of delivery,[d] pfu/μg DNA	
				−Glycerol	+Glycerol
PC	0	0.51	28	1.3×10^2	7.2×10^2
PC-SA	+	0.57	87	0.5×10^2	1.7×10^2
PS	−	2.7	43	1.8×10^3	6.5×10^4
PG	−	2.9	8.8	1.1×10^2	6×10^3
PS-Chol	−	2.5	84.7	2.8×10^3	1.2×10^5
PG-Chol	−	2.6	49.2	1.6×10^3	6.8×10^4

[a] All preparations encapsulated between 0.3 and 0.5 μg SV40 DNA/μmole phospholipid.
[b] Amount of cell-associated lipid measured after incubation of 10 nmole vesicle lipid/plate.
[c] Determined from the amount of cell-associated vesicle contents (^{3}H-inulin) measured for each preparation at 10 nmole of vesicle lipid added/plate.
[d] The infectivity of liposome-encapsulated SV40 DNA in glycerol-treated (25% v/v) and nontreated cells was determined by plaque assays; pfu: plaque-forming units.

with equal avidity to cells, forms vesicles which rapidly leak their contents on exposure to cells. Depending on delivery conditions (to be discussed below), SV40 DNA in negatively charged vesicles can be 10^3 times more infectious than DNA in neutral vesicles.

The phenomenon of cell-induced vesicle leakage has been substantiated in a number of studies (Szoka et al., 1979; Van Renswoude and Hoekstra, 1981). Cholesterol, which reduces the effect of a variety of soluble proteins that promote vesicle leakage (Papahadjopoulos et al., 1972; Mayhew et al., 1979; Gregoriadis and Davis, 1979), likewise reduces cell-induced leakage. The magnitude of the effect is clear from Table I: addition of cholesterol (Chol), which causes a substantial decrease in the leakiness of phosphatidylglycerol vesicles, also causes a substantial increase in the infectivity of DNA in PG vesicles. Thus it is possible to attribute to retention of contents the marked variability in DNA delivery among liposome compositions of the same charge.

3.2. Incubation Conditions

The conditions under which cell–liposome interaction occurs can have a marked effect on the efficiency of liposome delivery. For a review of the importance of such parameters as length and temperature of incubation, cell–vesicle ratio, sample volume, and incubation medium, the reader is directed to Tyrrell et al. (1976), Gregoriadis (1976), Poste and Papahadjopoulos (1976), Pagano and Weinstein (1978), and Kimelberg and Mayhew (1978). The most important points to be considered are that high concentrations of liposomes (>500 μmole/ml) can be toxic to cells, though cell type and liposome composition modulate the toxicity. Charged lipids, particularly positively charged sterylamine, are the most toxic. If possible, serum is to be avoided in the initial (<30 min) period of cell–vesicle interaction, though cholesterol in the vesicles does reduce serum-induced loss of contents. If acidic liposomes are used, divalent cations must be reduced to the submillimolar range. Finally, most cell–vesicle interaction relevant to nucleic acid delivery takes place in the first 30 min of incubation (Wilson et al., 1979), with protracted incubations having perhaps an undesirable effect on cellular metabolism.

3.3. Improvement of Liposome Efficiency

A variety of perturbations have been reported to increase either liposome uptake or DNA expression in other systems [for details, see Fraley et al. (1981)]. Dimethyl sulfoxide, ethylene glycol, glycerol, and polyethylene glycol [see also Szoka et al. (1981)] are all effective in enhancing effective delivery of liposome-entrapped DNA. While polyethylene glycol enhances tenfold the infectivity of entrapped DNA, the most effective treatment found is exposure of cells to moderate concentrations of glycerol after allowing a period of cell–vesicle interaction. The increase in SV40 DNA infectivity is most dramatic for negatively charged liposomes (Table I), suggesting that not only magnitude but also site of binding is important in determining the ability of various liposome compositions to deliver their contents to cells. The glycerol effect is both dose and time dependent, effective with a variety of cell types (L, 3T6, HeLa), and promotes the intracellular delivery of liposome-encapsulated molecules other than DNA (Fraley et al., 1981; also, unpublished observations).

4. Mechanism of Delivery

As mentioned above, the mode of intracellular delivery by liposomes is as yet unclear. However, some insight relevant to improvement of DNA delivery has been gained by studies of agents that inhibit a number of cellular functions (Fraley et al., 1981; also, unpublished observations). Without glycerol treatment, infectivity of encapsulated SV40 DNA is not greatly affected by azide with 2-deoxyglucose, cytochalasin B, colchicine, or chloroquine. It is possible only to conclude that such basal infectivity is largely independent of short-term inhibition of metabolic energy, endocytosis, and phagolysosomal processing. Polyethylene glycol-mediated stimulation of infectivity generally is unperturbed by the drug treatments mentioned (unpublished observations), perhaps suggesting that polyethylene glycol promotes cell–vesicle fusion (Szoka et al., 1981).

Glycerol stimulation of DNA delivery is quite sensitive to metabolic poisons and partially to cytochalasin B. In conjunction with morphologic observations (Fraley et al., 1981), it seems likely that glycerol promotes uptake of liposomes by a mechanism similar to endocytosis. Chloroquine, ammonium chloride, and a number of lysosomotropic agents (de Duve et al., 1974) enhance the glycerol effect, suggesting that lysosomal degradation may be the fate of at least some of the material whose uptake glycerol promotes. Chloroquine may enhance SV40 DNA infectivity by blocking DNA degradation, prolonging the lifetime of intact, intracellular DNA, and thereby increasing the likelihood of nuclear access. It is not clear how the contents of endocytosed liposomes could gain access to the cytoplasm, though pH-dependent fusion of acidic liposomes with the lysosomal membrane is an attractive possibility. Thus the liposome would provide a route by which endocytosed material could escape the lysosome, similar to that suggested for Semliki Forest virus (Helenius et al., 1980) and vesicular stomatitis virus (Miller and Lenard, 1980).

5. Protocol for Liposome–Cell Incubation

As a summary, it is useful to condense the protocol we use for maximal effective delivery of SV40 DNA. The description here is for adherent cells, though suspension cells are treated in a similar fashion with respect to the liposome–cell ratio and incubation time.

Cells in their original growth medium are given a concentrated solution of chloroquine (pH 7.3) sufficient to result in a final concentration of 100 μM. As some cells are sensitive to such a concentration, it is recommended that toxicity be checked beforehand. If chloroquine treatment is used, the drug is added to all solutions until the final, post-experiment washes. After 30 min of incubation, cells are washed twice with buffer (e.g., Phosphate-buffered saline) or serum-free medium containing only enough divalent cations to keep adherent cells attached to the substrate. If medium is used and the liposomes are negatively charged, the divalent cation concentration must be kept below 1.0 mM, using EDTA if necessary.

Liposomes diluted in buffer or serum-free medium are added to the cells in a small volume. Generally, 200 μl per 5×10^6 cells is sufficient. Since cells have a finite capacity to bind liposomes, 100 nmole lipid (acidic vesicles) is usually sufficient to saturate 5×10^6 cells.

Incubation of cells with liposomes is allowed to proceed for 30 min, at which time polyethylene glycol or glycerol is added. Some prior experimentation is useful here, for cells vary in their ability to tolerate glycerol. While AGMK cells will survive 40% (v/v) glycerol for 4 min, other cells, such as L, may tolerate only 15% for the same length of time. In addition, it is possible to substitute longer periods of exposure (unpublished observations) with lower concentrations of glycerol. At the end of the exposure period, the glycerol (or polyethylene glycol) is diluted slowly, and cells are washed three times with medium or buffer before return to the normal growth medium.

6. Future Prospects

Current efforts are directed toward improving effective delivery of liposome- encapsulated DNA, increasing the specificity of liposome delivery in mixed-cell populations, and extending the technology to cell types not easily transformed by other methods of gene transfer.

Using a convenient assay which detects progeny virus in *Nicotiana* protoplasts treated with liposome-encapsulated tobacco mosaic virus RNA, Fraley et al. (1982) have found that several of the principles important for functional delivery of liposome contents to animal cells hold true for plant protoplasts: superiority of negatively charged lipids, notably phosphatidylserine, beneficial effect of cholesterol addition to vesicles, and enhancement of functional delivery by polymer treatments following liposome–cell interaction. Results on plant protoplast transformation by purified liposome-entrapped Ti DNA (Dellaporta et al., 1981) are quite exciting, suggesting that liposomes may be an important aid in genetic manipulation of plant cells. Liposome-mediated DNA transfer resulted in transformation to phytohormone independence and octopine-positive phenotype with an efficiency much greater than that of free DNA.

In the case of animal cells, liposome-entrapped HSV-TK DNA has been observed to transform L-TK⁻ cells to TK⁺ at a frequency comparable to that obtained in parallel experiments with free DNA as a calcium phosphate coprecipitate (unpublished observations). The work is being extended to S49 lymphoma cells and CHO cells, in concert with several new TK-carrying vectors which are under development.

Perhaps the most exciting development in liposome technology is the ability to couple covalently to liposomes (Heath et al., 1980a,b, 1981; Huang et al., 1980; Leserman et al., 1980; Martin et al., 1981) ligands with high affinity for specific cell-surface components. With the proper monoclonal antibody or peptide ligand, it may be possible to direct liposomes and their contents to specific cellular targets with great efficiency. It is expected that these and other such developments will provide a realistic basis for important applications of liposome-mediated transfer of macromolecules, both in basic research and in the clinic.

References

Anderson, W., Killos, L., Sanders-Haigh, L., Kretschmer, P., and Diakumakos, E., 1980, Replication and expression of thymidine kinase and human globin genes microinjected into mouse fibroblasts, *Proc. Natl. Acad. Sci. USA* **77**:5399–5403.

Avery, O. T., Macleod, C. M., and McCarty, M., 1944, Studies on the chemical nature of the substance inducing transformation of pneumococcal types. Induction of transformation by a deoxyribonucleic acid fraction isolated from pneumococcus Type III, *J. Exp. Med.* **79**:137–158.

Bangham, A., Standish, M., and Watkins, J., 1964, Diffusion of univalent ions across the lamellae of swollen phospholipids, *J. Mol. Biol.* **13**:238–52.

Bartlett, G., 1959, Phosphorous assay in column chromatography, *J. Biol. Chem.* **234**:466–68.

Capecchi, M., 1980, High efficiency transformation by direct microinjection of DNA into cultured mammalian cells, *Cell* **22**:479–88.

Chen, H. W., Kandutsch, A. A., and Waymouth, C., 1974, Inhibition of cell growth by oxygenated derivatives of cholesterol, *Nature* **251**:419–21.

Deamer, D., and Bangham, A., 1976, Large volume liposomes by an ether vaporization method, *Biochim. Biophys. Acta* **443**:629–34.

Deamer, D., and Uster, P., 1980, Liposome preparation methods and monitoring liposome fusion, in: *Introduction of Macromolecules into Viable Mammalian Cells* (R. Baserga, C. Croce, and G. Roueza, eds.), Alan R. Liss, New York, 205–20.

de Duve, C., de Barsy, T., Poole, B., Trouet, A., Tulkens, P., and van Hoof, F., 1974, Lysosomotropic agents, *Biochem. Pharmacol.* **23**:2495–531.

Dellaporta, S., Fraley, R., Giles, K., Papahadjopoulos, D., Powell, A., Thomashow, M., Nester, E., and Gordon, M., 1982, Plant protoplast transformation by liposome-encapsulated Ti plasmid of *Agrobacterium tumefaciens*, In preparation.

Dimitriadis, G., 1978, Translation of rabbit globin mRNA introduced by liposomes into mouse lymphocytes, *Nature* **274**:923–4.

Enoch, H., and Strittmatter, P., 1979, Formation and properties of 1000-A-diameter single bilayer phospholipid vesicles, *Proc. Natl. Acad. Sci. USA* **76**:145–9.

Fournier, R., and Ruddle, F., 1977, Microcell-mediated transfer of murine chromosomes into mouse, Chinese hamster, and human somatic cells, *Proc. Natl. Acad. Sci. USA* **74**:319–23.

Fraley, R., and Papahadjopoulos, D., 1981, New generation liposomes: The engineering of an efficient vehicle for intracellular delivery of nucleic acids, *Trends Biochem. Sci.* **6**:77–80.

Fraley, R., and Papahadjopoulos, D., 1982, Liposomes: The development of a new carrier system for introducing nucleic acids into plant and animal cells, in: *Current Topics in Microbiology and Immunology*: Gene Cloning in Organisms Other Than E. coli, (P. H. Hofschneider and W. Goebel, eds.), Vol. 96, pp. 171-192, Springer-Verlag, Heidelberg.

Fraley, R., Fornari, C., and Kaplan, S., 1979, Entrapment of a bacterial plasmid in phospholipid vesicles: Potential for gene transfer, *Proc. Natl. Acad. Sci. USA* **76:**3348–52.

Fraley, R., Subramani, S., Berg, P., and Papahadjopoulos, D., 1980, Introduction of liposome-encapsulated SV40 DNA into cells, *J. Biol. Chem.* **255:**10431–35.

Fraley, R., Straubinger, R. M., Rule, G., Springer, E. L, and Papahadjopoulos, D., 1981, Liposome-mediated delivery of deoxyribonucleic acid: Enhanced efficiency of delivery by changes in lipid composition and incubation conditions, *Biochemistry* **20:**6978–6987.

Fraley, R., Dellaporta, S., and Papahadjopoulos, D., 1982, Liposome-mediated delivery of TMV RNA into tobacco protoplasts: A sensitive assay for monitoring liposome–protoplast interactions, *Proc. Natl. Acad. Sci. USA* **79:**1859–63.

Graham, F., and Van der Eb, A., 1973, A new technique for the assay of infectivity of human Adenovirus 5 DNA, *Virology* **52:**456–60.

Gregoriadis, G., and Davis, C., 1979, Stability of liposomes *in vivo* and *in vitro* is promoted by their cholesterol content and the presence of blood cells, *Biochem. Biophys. Res. Commun.* **89:**1287–92.

Hamer, D. H., Smith, K. D., Boyer, S. H., and Leder, P., 1979, SV40 recombinants carrying rabbit β-globin gene coding sequence, *Cell* **17:**725–35.

Heath, T., Fraley, R., and Papahadjopoulos, D., 1980a, Antibody targeting of liposomes: Cell specificity obtained by conjugation of F(ab')$_2$ to the vesicle surface, *Science* **210:**539–41.

Heath, T. D., Robertson, D., Birback, M. S. C., and Davies, A. J. S., 1980b, Covalent attachment of horseradish peroxidase to the outer surface of liposomes, *Biochim. Biophys. Acta* **599:**42–62.

Heath, T., Macher, B. A., and Papahadjopoulos, D., 1981, Covalent attachment of immunoglobulins to liposomes via glycosphingolipids, *Biochim. Biophys. Acta.* **640:**66–82.

Helenius, A., Kartenbeck, J., Simons, K., and Fries, E., 1980, On the entry of Semliki Forest Virus into BHK-21 cells, *J. Cell. Biol.* **84:**404–20.

Hinnen, A., Hicks, J., and Fink, G., 1978, Transformation of yeast, *Proc. Natl. Acad. Sci. USA* **75:**1929–33.

Hoffman, R., Margolis, P., and Bergelson, L., 1978, Binding and entrapment of high molecular weight DNA by lecithin liposomes, *FEBS Lett.* **93:**365–8.

Huang, C., 1969, Studies on phosphatidylcholine vesicles. Formation and physical characteristics, *Biochemistry* **8:**344–52.

Huang, A. C., Huang, L., and Kennel, S. J., 1980, Monoclonal antibody covalently coupled with fatty acid. A reagent for *in vitro* liposome targeting, *J. Biol. Chem.* **255:**8015–8.

Jonah, M., Cerny, E. A., and Rahman, Y. E., 1978, Tissue distribution of EDTA encapsulated within liposomes of varying surface properties, *Biochim. Biophys. Acta* **541:**321–3.

Kates, M., 1972, Techniques in lipidology, in: *Laboratory Techniques in Biochemistry and Molecular Biology* (T. S. Work and E. Work, eds.), North-Holland, New York.

Kimelberg, H., and Mayhew, E., 1978, Properties and biological effects of liposomes and their uses in pharmacology and toxicology, *CRC Crit. Rev. Toxicol.* **6:**25–78.

Leserman, L., Barbet, J., Kourilsky, R., and Weinstein, J., 1980, Targeting to cells of fluorescent liposomes covalently coupled with monoclonal antibody or protein A, *Nature* **288:**602–4.

Lurquin, P., 1979, Entrapment of plasmid DNA by liposomes and their interactions with plant protoplasts, *Nucleic Acid Res.* **6:**3773–84.

Martin, F., Hubbel, W., and Papahadjopoulos, D., 1981, Immunospecific targeting of liposomes to cells: a novel and efficient method for covalent attachement of Fab' fragments via disulfide bonds, *Biochemistry* **20:**4429–38.

Mauk, M., Gamble, R., and Baldeschwieler, J., 1980, Vesicle-targeting: Timed release and specificity for leukocytes in mice by subcutaneous injection, *Science* **207:**309–11.

Mayhew, E., Rustum, Y., Szoka, F., and Papahadjopoulos, D., 1979, Role of cholesterol in enhancing the anti-tumor effect of 1-β-D arabinofuranosyl cytosine entrapped in liposomes, *Cancer Treatm. Rep.* **63**(11–12):1923–8.

McBride, O., and Athwall, R., 1977, Genetic analysis by chromosome-mediated gene transfer, *In Vitro* **12:**777–86.

McCutchen, J., and Pagano, J., 1968, Enhancement of the infectivity of Simian Virus 40 deoxyribonucleic acid with diethylaminoethyldextran, *J. Natl. Cancer Inst.* **41**:351–7.

Miller, D. K., and Lenard, J., 1980, Inhibition of Vesicular Stomatitis Virus infection by spike glycoprotein. Evidence for an intracellular, G protein-requiring step, *J. Cell. Biol.* **84**:430–7.

Mukherjee, A., Orloff, S., Butler, J., Triche, T., Lalley, P., and Schulman, J., 1978, Entrapment of metaphase chromosomes into phospholipid vesicles (lipochromosomes): Carrier potential in gene transfer, *Proc. Natl. Acad. Sci. USA* **75**:1361–5.

Mulligan, R. C., Howard, B. H., and Berg, P., 1979, Synthesis of rabbit β-globin in cultured monkey kidney cells following infection with a SV40 β-globin recombinant genome, *Nature* **277**:108–14.

Nicolson, G., and Poste, G., 1978, Mechanism of resistance to ricin toxin in selected mouse lymphoma lines, *J. Supramol. Struct.* **8**:235–45.

Olson, F., Hunt, C., Vail, W., and Papahadjopoulos, D., 1979, Preparation of liposomes of defined size by extrusion through polycarbonate filters, *Biochim. Biophys. Acta* **557**:9–23.

Ostro, M., Giacomoni, D., Lavelle, D., Paxton, W., and Dray, S., 1978, Evidence for translation of rabbit globin mRNA after liposome-mediated insertion into a human cell line, *Nature* **274**:921–3.

Pagano, R., and Weinstein, J., 1978, Interaction of liposomes with mammalian cells, *Ann. Rev. Biophys. Bioeng.* **7**:435–68.

Papahadjopoulos, D., 1978, *Liposomes and Their Uses in Biology and Medicine*, New York Academy of Sciences, New York.

Papahadjopoulos, D., and Kimelberg, H., 1973, Phospholipid vesicles (liposomes) as models for biological membranes, in: *Progress in Surface Science*, Volume 4 (S. G. Davidson, ed.), Pergamon Press, New York, pp. 141–232.

Papahadjopoulos, D., and Miller, W., 1967, Phospholipid model membranes. I. Structural characteristics of hydrated liquid crystals, *Biochim. Biophys. Acta* **135**:624–38.

Papahadjopoulos, D., and Watkins, J. C., 1967, Phospholipid model membranes. II. Permeability properties of hydrated liquid crystals, *Biochim. Biophys. Acta* **135**:639–52.

Papahadjopoulos, D., Nir, S., and Ohki, S., 1972, Permeability properties of phospholipid membranes: Effect of cholesterol and temperature, *Biochim. Biophys. Acta* **266**:561–71.

Papahadjopoulos, D., Vail, W., Jacobson, K., and Poste, G., 1975, Cochleate lipid cylinders: Formation by fusion of unilamellar vesicles, *Biochim. Biophys. Acta* **394**:483–91.

Papahadjopoulos, D., Vail, W., Pangborn, W., and Poste, G., 1976, Studies on membrane fusion II: Induction of membrane fusion in pure phospholipid membranes by calcium ions and other divalent metals, *Biochim. Biophys. Acta* **448**:265–83.

Papahadjopoulos, D., Wilson, T., and Taber, R., 1980a, Liposomes as macromolecular carriers for the introduction of RNA and DNA into cells, in: *Transfer of Cell Constituents into Eukaryotic Cells* (J. Celis, A. Graessman, and A. Loyter, eds.), Plenum Press, New York.

Papahadjopoulos, D., Wilson, T., and Taber, R., 1980b, Liposomes as vehicles for cellular incorporation of biologically active macromolecules, *In Vitro* **16**:49–54.

Pietronigro, D. D., Jones, W. B. G., Katy, K., Demopoulos, H. B., 1977, Interaction of DNA and liposomes as a model of membrane-mediated DNA damage, *Nature* **267**:78–9.

Poste, G., 1980, The interaction of lipid vesicles (liposomes) with cultured cells and their use as carriers for drugs and macromolecules, in: *Liposomes in Biological Systems* (G. Gregoriadis and A. Allison, eds.), Wiley, New York, pp. 101–51.

Ringertz, N., and Savage, R., 1976, *Cell Hybrids*, Academic Press, New York, pp. 1–366.

Schaffner, W., 1980, Direct transfer of cloned genes from bacteria to mammalian cells, *Proc. Natl. Acad. Sci. USA* **77**:2163–7.

Stow, N., and Wilkie, N., 1976, An improved technique for obtaining enhanced infectivity with Herpes Simplex Virus Type I DNA, *J. Gen. Virol.* **33**:447–58.

Straus, S., and Raskas, H., 1980, Transfection of KB cells by polyethylene glycol-induced fusion with erythrocyte ghosts containing Adenovirus Type II DNA, *J. Gen. Virol.* **48**:241–5.

Szoka, F., and Papahadjopoulos, D., 1978, Procedure for preparing liposomes with large internal aqueous space and high capture by reverse-phase evaporation, *Proc. Natl. Acad. Sci. USA* **75**:145–9.

Szoka, F., and Papahadjopoulos, D., 1980, Comparative properties and methods of preparation of lipid vesicles (liposomes), *Ann. Rev. Biophys. Bioeng.* **9**:467–508.

Szoka, F., Jacobson, K., and Papahadjopoulos, D., 1979, The use of aqueous space markers to determine the mechanism of interaction between phospholipid vesicles and cells, *Biochim. Biophys. Acta* **551**:295–303.

Szoka, F., Olson, F., Heath, T., Vail, W., Mayhew, E., and Papahadjopoulos, D., 1980, Preparation of unilamellar liposomes of intermediate size (0.1–0.2 µm) by a combination of reverse-phase evaporation and extrusion through polycarbonate membranes, *Biochim. Biophys. Acta* **601**:559–71.

Szoka, F., Magnusson, K. E., Wojcieszyn, J., Hou, Y., Derzko, Z., and Jacobson, K.. 1981. Use of lectins and polyethylene glycol for fusion of glycolipid-containing liposomes with eukaryotic cells, *Proc. Natl. Acad. Sci. USA* **78**:1685–9.

Tyrrell, D., Heath, T., Colley, C., and Ryman, B., 1976, New aspects of liposomes, *Biochim. Biophys. Acta* **457**:259–302.

Van Renswoude, J., and Hoekstra, D., 1981, Cell-induced leakage of liposome contents, *Biochemistry* **20**:540–6.

Wigler, M., Silverstein, S., Lee, L., Pelecer, A., Cheng, Y., and Axel, R., 1977, Transfer of purified Herpes Virus thymidine kinase gene to cultured mouse cells, *Cell* **11**:223–32.

Wilson, T., Papahadjopoulos, D., and Taber, R., 1977: Biological properties of poliovirus encapsulated in lipid vesicles: Antibody resistance and infectivity in virus resistant cells, *Proc. Natl. Acad. Sci. USA* **74**:3471–5.

Wilson, T., Papahadjopoulos, D., and Taber, R., 1979, The introduction of poliovirus RNA into cells via lipid vesicles (liposomes), *Cell* **17**:77–84.

Wong, T.-K., Nicolau, C., Hofschneider, P., 1980, Appearance of β lactamase activity in animal cells upon liposome-mediated gene transfer, *Gene* **10**:87–94.

Chapter 29

Techniques of DNA-Mediated Gene Transfer for Eukaryotic Cells

ARTHUR P. BOLLON and SAUL J. SILVERSTEIN

1. Introduction

DNA transformation in bacteria has permitted detailed genetic analysis as well as the implementation of recombinant DNA technology. The utility of manipulation and cloning of genes in bacteria is well recognized. This chapter will concern the manipulation and cloning of genes in eukaryotic systems ranging from yeast to animal systems. Gene manipulation utilizing eukaryotic cloning systems permits the analysis of regulation of eukaryotic genes in their native environment and offers certain advantages for the production of foreign gene products which may require post-transcriptional modifications which are indigenous to eukaryotic organisms.

The best developed lower eukaryotic gene cloning system involves *Saccharomyces cerevisiae*. Hinnen et al. (1978) developed a method for the transformation of yeast which permits the introduction of yeast or foreign DNA into yeast spheroplasts treated with $CaCl_2$ and polyethylene glycol (PEG). The transformed genes can reside integrated at homologous chromosomal sites if they are yeast genes, as integrated genomic units at nonhomologous sites, or as autonomous replicating units, depending on the vectors utilized (Ilgen et al., 1979). Transformed genes which exist as autonomous replicating units can exist on hybrid plasmids containing components of a bacterial plasmid such as pBR322 (Bolivar et al., 1977) and a yeast plasmid such as 2 μ DNA (Beggs, 1978), or the transformed gene can be part of a self-replicating unit containing an autonomous replicating sequence (ARS) (Stinchcomb et al., 1979). Hybrid vectors containing bacterial and yeast plasmid components can be shuttled between both organisms (Broach et al., 1979). The yeast–bacterial transformation system permits the positive selection of genes by complementation of yeast or bacterial auxotrophs, although only a 20% success rate has been achieved for the complementation

---

ARTHUR P. BOLLON • Department of Molecular Genetics, Wadley Institutes of Molecular Medicine, Dallas, Texas 75235.   SAUL J. SILVERSTEIN • Department of Microbiology, Columbia University, New York, New York 10032.

of bacterial mutations with yeast genes (Struhl et al., 1976; Ratzkin and Carbon, 1977; Broach et al., 1979).

Yeast genes which have been modified *in vitro* have been introduced back into yeast and the effect of the modifications, such as deletions of parts of putative control regions, on gene expression have been explored (Scherer and Davis, 1979). In addition to the analysis of basic mechanisms of regulation, the *Saccharomyces cerevisiae* cloning system has been utilized for the production of human leukocyte interferon (Hitzeman et al., 1981) and hepatitis B coat protein, which is glycosylated by the yeast, utilizing a vector containing the yeast alkaline phosphatase promoter (B. Hall, personal communication) or the yeast 3-phosphoglycerate promoter (R. Hitzeman, personal communication).

The initial interest in animal cell DNA transformation arose as a consequence of the desire to isolate the gene from herpes simplex virus (HSV) encoding the deoxypyrimidine kinase. The early study of Szybalska and Szybalski (1962) suggested that exogenous DNA could be introduced into an appropriate recipient cell to alter its phenotype to correspond to that of the donor DNA. These experiments were not pursued, perhaps because of the lack of a reliable technique to introduce foreign DNA into mammalian cells. McCutchan and Pagano (1968) demonstrated that purified SV40 DNA could be reproducibly introduced into mammalian cells using DEAE–dextran as a facilitator. Subsequently, Graham et al. (1974) assayed for biologic activity a DNA fragment by transfecting rat embryo fibroblasts with purified restriction endonuclease fragments of adenovirus DNA, looking for the appearance of morphologically transformed cells. Transformation was facilitated by coprecipitating DNA fragments with carrier DNA in a calcium phosphate complex. This system has been used to identify genes coding for selectable biochemical markers from a complex animal virus (TK) and from the more complex chicken (TK) (Parucho et al., 1980) and hamster genomes (adenine phosphoribosyl) transferase *aprt*) (Lowy et al., 1980). In this chapter we describe the technology developed to transfer DNA into eukaryotic cells. First we describe the basic methods for DNA transformation of yeast and then mammalian cell systems. In addition, a discussion of some of the vectors and applications of this technology is presented.

2. Transformation of Yeast

2.1. Yeast DNA Transformation Protocol

The basic method for yeast transformation involves the coprecipitation of calcium-treated yeast spheroplasts and transforming DNA by polyethylene glycol 4000 (PEG 4000) according to the methods of Hinnen et al. (1978). The transformed yeast spheroplasts are then rejuvenated in regeneration agar and grown on selective medium. An assortment of different vectors can be utilized for the cloning of DNA in yeast, as will be discussed later. The presence of vector material in transformed yeast offers the opportunity of using a probe containing vector DNA to identify the location of transformed sequences.

2.1.1. Preparation of Yeast Spheroplasts

1. Inoculation of rich medium (i.e., YEPD = 1% yeast extract, 2% bacto peptone, and 2% dextrose) with an overnight culture of yeast and growth of cells to a density of $2-3 \times 10^7$ cells/ml.
2. Wash cells once with 1 M sorbitol and resuspend in 1 M sorbitol at a concentration of $2-3 \times 10^8$ cells/ml.
3. Cells are treated with 10 µg/ml Zymolyase 60,000 (0.5 µg/ml) (Miles Laboratories, Elkhart, Indiana) or Glusulase (Endo Labs, Garden City, New York) added to a final concentration of 1%, and the mixture is incubated with gentle shaking at 30° for 1 hr.

Note. High transformation efficiency requires approximately 95–99% spheroplast formation, which can be determined by viable cell counts on YEPD agar plates not containing sorbitol. The number of spheroplasts that cannot regenerate on these plates can be compared to the number of colonies in the culture prior to enzyme treatment.

4. After treatment, the spheroplasts are washed twice in 1 M sorbitol by low-speed centrifugation at room temperature.

2.1.2. DNA Transformation of Yeast Spheroplasts

1. Take up spheroplasts in 1 M sorbitol, 10 mM $CaCl_2$, and 10 mM Tris HCl, pH 7.5; spin again and resuspend in the same solution at a concentration of $2-3 \times 10^9$ cells/ml.
2. Add DNA and incubate for 20 min at room temperature.

Note a. Transformation frequency of circular DNA is proportional to DNA concentration within a range of 1–100 µg DNA/ml (Ilgen et al., 1979).

Note b. A 5- to 20-fold increased transformation frequency can be obtained when circular plasmids are linearized (i.e., digestion of YEpl3 with Sal 1, Xho 1, or BamH1, which does not cut within the leu gene) (Ilgen et al., 1979).

Note c. A tenfold increase in circular plasmid transformation efficiency can be achieved with the addition of undigested genomic yeast DNA (one part plasmid DNA to 2.5 parts yeast genomic DNA) (Bollon and Stauver, 1980).

3. Add tenfold volume of 44% PEG with a final concentration of 40% and incubate for 10 min at room temperature.
4. Centrifuge the spheroplasts and resuspend in 1 M sorbitol.
5. Add treated spheroplasts to regeneration agar (3% agar in 1 M sorbitol and selective medium).

Note a. A useful regeneration agar consists of 1 M Sorbitol, 1–2% YEPD broth, Synthetic Medium.

Note b. Synthetic Medium consists of 0.67% yeast nitrogen base (Difco),

2% glucose, 2% Difco agar, required amino acids, purines, or pyrimidines (Whelan et al., 1979).

6. Spread 10 ml of the mixture onto selective agar plates (about 2–3 × 10^8 cells per plate.)

Note a. Viable colonies of transformed spheroplasts usually develop in 3–4 days.

Note b. The *leu 2-3 leu 2-112* allele present in the yeast strain DC5 (a, *leu 2-3*, *leu 2-112*, *his 3*, *can 1-11*) is useful for standardizing the transformation efficiency using the YEp13 plasmid since it has a reversion frequency of less than 10^{-12} (Hinnen et al., 1978).

Note c. Yeast clones containing specific gene sequences can be identified by gene complementation, detection of cross-reacting material, or colony hybridization (Ilgen et al., 1979).

2.2. Vectors for Cloning Genes in Yeast

Following the transformation of DNA into yeast, the DNA can integrate into the genome or autonomously replicate in the nucleus. Such fate is dictated to a good extent by the type of vector utilized for the yeast transformation.

As indicated in Table I, one example of an integrative vector is pYe *leu 10*, which contains the yeast *leu 2* gene integrated into the EcoRI site of the *E. coli* plasmid Col E1 (Ratzkin and Carbon, 1977). The yeast *leu B* gene can be used to complement defective *leu B* gene in *E. coli* (Ratzkin and Carbon, 1977) as well as the corresponding *leu 2* gene in *Saccharomyces cerevisiae* (Hinnen et al., 1978). As indicated in Table II, the pYe *leu 10* gives a good transformation frequency when used to transform *E. coli*, indicating that the yeast gene does not interfere with the bacterial transformation efficiency, but gives a very low transformation efficiency in yeast.

Hinnen et al. (1978) showed that leu 2⁻ yeast transformed with pYe *leu 10* and selected for *leu*⁺ contained the plasmid integrated as a tandem duplication at the homologous *leu B* site of yeast in about 60% of the cases and integrated into other chromosomes in 11% of the clones examined. Approximately 16% of the yeast *leu*⁺ clones failed to hybridize with Col E1 probes and did not segregate the *leu 2*⁻ portion of the putative tandem duplications.

Three different kinds of yeast autonomous replicating vectors can be utilized, as indicated in Table I. The vector YEp13 is a hybrid vector containing the yeast *leu 2* gene, part of the yeast 2 μ plasmid, and the *E. coli* plasmid pBR322 (Broach et al., 1979). This plasmid can be used to complement *E. coli leu B*⁻ alleles and yeast *leu 2*⁻ alleles. As indicated in Table II, YRp13 transforms with the high frequency of 10,000 transformants/μg DNA. The yeast strains transformed with 2 μ hybrid vectors are unstable in the absence of selective pressure. About 5–100 copies per cell of the 2 μ vector can exist, including one copy which may be integrated into the genome (Struhl et al., 1979).

Table I. Description of Representative Yeast Vectors

Vector type	Vector example	Composition	Reference
Integrative	pYe leu 10	Bacterial: COl E1 Yeast: leu 2 gene	Ratzkin and Carbon (1977) Hinnen et al. (1978)
Autonomous (2 μ DNA)	YEp13	Bacterial: pBR322 Yeast: leu 2 gene, 2 μ plasmid	Broach et al. (1979)
Autonomous (ARS)	YRp7	Bacterial: pBR322 Yeast: ARS (autonomous replicating sequence), trp 1 gene	Struhl et al. (1979)
Cosmid	pYcl	Bacterial: amp^R and origin of replication of pBR322 Yeast: 2 μ plasmid, his 3 gene Lambda: cos (cohesive end sites)	Hohn and Hinnen (1980)

Table II. Properties of Representative Yeast Vectors

	Plasmids			Cosmid[d]
	pYe leu 10	YEp 13	YRp7	pYcl
1. Transformation frequency[a]				
Saccharomyces cerevisiae	10	1×10^4	1×10^3	7×10^3
Escherichia coli[b]	1×10^6	18	30	—
2. Replicative form[c]	I	I, A	A	A

[a] Transformation frequency is the number of transformants per µg DNA.
[b] For the transformation of E. coli with YEp13 and YRp7, total yeast DNA isolated from yeast strains containing the plasmids was used.
[c] I, integrative; A, autonomous.
[d] Hohn and Hinnen (1980).

The second type of autonomous replicating vector is YRp7, which is a hybrid of pBR322 and the yeast *trp 1* gene which is tightly linked to the centromere of chromosome IV. This vector transforms with a frequency of about 1000 transformants/µg DNA and contains an autonomous replicating sequence (ARS). This plasmid type does not appear to have an integrated copy in the chromosome as does the 2 µ vector (Struhl et al., 1979).

The last type of autonomous replicating yeast vector is pYcl, which contains the origin of replication and the amp^R gene of pBR322, yeast 2 µ plasmid, and the cohesive end site (cos) of bacteriophage lambda. This plasmid can exist autonomously in E. coli and yeast and can be packaged *in vitro* (Hohn and Hinnen, 1980). The advantage of this vector is the ability to clone DNA fragments 30–50 kb long without much reduced transformation efficiency.

As can be seen in Table II, the autonomous replicating vectors can be recovered from yeast by transformation of E. coli with total yeast DNA obtained from a yeast strain containing one of the hybrid vectors having a selectable marker in bacteria.

3. Transformation of Mammalian Cells

3.1. The Components and Their Preparation

Transformation of mammalian cells in culture requires three components: a competent source of donor DNA, a suitably buffered medium, and an appropriately marked recipient cell and/or convenient selective media that will identify the rare transformant. Critical to achieving high rates of transformation are a source of high-molecular-weight carrier DNA, accurately buffered medium, and freshly plated recipient cells.

3.2. Isolation of Carrier DNA

Carrier DNA is routinely purified from a commercial source of salmon testes DNA or from a cell line that lacks the sequences that are to be transferred. For example, if Tk⁻ mammalian cells are to be transformed, DNA from a mouse L cell deficient in both TK and APRT enzyme activity is used. This cell line has never been reported to revert to the Tk⁺ phenotype; thus, it serves as an excellent source of carrier DNA and as a suitable recipient of TK information. The cell line does, however, revert to the Aprt⁺ phenotype at a frequency of 5×10^{-7}/generation. Therefore, these cells are maintained in the presence of 50 µg/ml diaminopurine to selectively kill any Aprt⁺ revertants.

3.2.1. Preparation of Salmon DNA Carrier

Fifty milligrams of DNA is resuspended in 100 ml of autoclaved 50 mM Tris, 250 mM NaCl, 5 mM EDTA, pH 8.3 (DNA extraction buffer). The DNA is dissolved by rocking gently overnight at 37°C on a platform shaker. When dissolved, 50 µg/ml of RNase A that was heated to 90° for 10 min is added and incubated with the dissolved DNA for 1 hr at 37°C. Then, self-digested pronase and sodium dodecyl sulfate (SDS) are added to 500 µg/ml and 0.2% (w/v), respectively, and incubated for an additional 2 hr at 37°C. At this point, an equal volume of redistilled phenol containing 0.1% hydroxyquinolinol and buffered with 10 mM Tris, pH 8.1, 1 mM EDTA is added and gently mixed until an emulsion is formed. Then, mixing is continued for 10 min. The aqueous phase containing high-molecular-weight DNA is separated from the organic phase by centrifugation and removed using a wide-bore pipette. The aqueous phase is successively reextracted with a mixture of phenol:chloroform:isoamyl alcohol 50:48:2 and then chloroform:isoamyl alcohol 24:1.

The final aqueous phase is removed and the DNA is concentrated and sterilized by the addition of two volumes of ethanol. High-molecular-weight DNA precipitates immediately upon the addition of ethanol and is removed with a sterile plastic 1-ml pipette to a sterile polypropylene tube. Autoclaved water is added and the DNA dissolved at 37°C overnight on a platform rocker. The concentration of DNA is measured and may be stored at 4°C for an indefinite period of time, providing it remains sterile.

3.2.2. Preparation of Cellular DNA

Cells in culture are harvested and washed in isotonic saline (10 mM Tris, 140 mM NaCl pH 8.0) and the cell pellet resuspended in DNA extraction buffer at a cell concentration not to exceed $1 \times /10^7$ ml. Cells are lysed by the addition of SDS to 0.2% and then incubated with 500 µg/ml of self-digested pronase as described in the previous section. After the successive phenol and chloroform extractions, the aqueous phase is dialyzed overnight against two changes of DNA extraction buffer and then digested with 50 µg/ml of heat-treated RNase A for 1 hr at 37°C. Pronase and SDS are added to 250 µg/ml and 0.2% (w/v), respectively, and the incubation continued for an additional 2 hr. At this point, the DNA is extracted as described in the previous section, ethanol-precipitated, and resuspended in sterile water. DNA isolated in this fashion is generally greater than 100 kb as judged by its mobility in 0.3% agarose gels using phage λ (50 kb) and HSV-1 (160 kb) DNAs as markers and is suitable as carrier or a source of donor DNA in transformation experiments. Ideally, the DNA concentration is adjusted to 500 µg/ml.

3.2.3. Preparation of Transformation Buffer and Calcium Phosphate DNA Precipitate

Preparation of a calcium phosphate DNA precipitate with high transforming activity requires a carefully buffered HEPES solution. We routinely prepare a twice-concentrated HEPES-buffered saline phosphate stock that is buffered to pH 7.10 ± 0.05 using a glass electrode (Sigma Cat. #E4513). It should be noted that many standard electrodes will not accurately measure the pH of a HEPES-buffered solution (HBS). As prepared 2× HBS contains 50 mM HERPES, 280 mM NaCl, 1.5 mM Na phosphate (equal amounts of mono- and dibasic). The pH is adjusted and the solution filtered through a 0.22-µm filter and stored in sterile polypropylene tubes at $-20°C$. A tube of 2× HBS is thawed prior to use, and the pH checked and adjusted as necessary; the contents are used once and the remainder is discarded.

A typical $CaPO_4$–DNA precipitate is prepared by adding 10 µl of carrier or donor eukaryotic DNA to 230 µl of sterile H_2O. The DNA is gently mixed by inverting the tube and then an equal volume (250 µl) of 250 mM ($CaCl_2$) is added and mixed again. This DNA–$CaCl_2$ solution is added dropwise to an equal volume of 2× HBS. A sterile, plugged 1-ml plastic pipette is inserted into the tube containing 2× HBS and bubbles are slowly introduced by blowing while the DNA is added dropwise. The calcium phosphate–DNA precipitate is allowed to form while standing for 30–45 min at room temperature.

3.2.4. Preparation of Recipient Cells and Transformation

Recipient cells are freshly plated from 2-day-old cultures grown in the absence of any selective pressure. For example, when L TK^- $APRT^-$ cells are to be transformed to the $APRT^+$ phenotype the cells normally maintained in 50 µg/ml diaminopurine are subcultured in drug-free media. The day prior to transformation, cells are seeded at 5×10^5/90 mm plate in 10 ml of Dulbecco's modified Eagle's medium (DME) containing 10% heat-inactivated calf serum. The following morning 1 ml of the calcium phosphate–DNA precipitate is added directly to the 10 ml of growth medium that covers the recipient cells. The plates are gently agitated and then returned to the incubator (6% CO_2, 36°) for 4–12 hr of incubation. The medium is then replaced and the cells are allowed to incubate for an additional 21 hr. At that time, selective pressure is applied and the transformants allowed to express.

3.2.5. Selective Systems

Biochemical selection is a powerful tool for identifying the rare transformant among the population of recipient cells. A variety of selective media and the appropriate source of donor DNA permit ready identification of the recipient cells that has expressed the donor gene. The following describes selective media and a source of donor DNA for gene transfer experiments; the selection schema are differentiated between recessive and dominant acting genes.

3.2.5a. Recessive Selection. Cells that are deficient in expression of an enzyme function such as APRT, HPRT, or TK are transformable to the expressor state and these transformants are identified using an appropriate selective medium. One can recognize APRT$^+$ cells by virtue of their ability to survive selection in media containing 50 μM azaserine and 100 μM adenine (Aza Ad). Azaserine inhibits *de novo* synthesis of adenine, and thus only cells with a functional transferase can survive selection.

Virtually any mammalian cell line can serve as a source of competent DNA for transformation of aprt activity. Recently, we have cloned the hamster gene coding for APRT (Lowy et al., 1980). Thus, the purified gene is readily available in large amounts. The HPRT$^-$ recipients are readily available and occur as natural mutants (cells from patients with Lesch-Nyan disease), or mouse L-cell lines are readily obtained following mutagenesis and selection in 6-thioguanine or 8-azaguanine. The HPRT$^+$ transformants are rare and are recognized by their ability to grow in HAT medium (hypoxanthine 15 μg/ml, aminopterin 1 μg/ml, thymidine 5 μg/ml). Aminopterine interferes with dihydrofolate reductase and prevents *de novo* synthesis of purines and methylation of DUMP (deoxyuracil monophosphate) to yield thymidylate. Although there are many competent sources of DNA for transfer of this locus, we and other laboratories have failed to identify a restriction endonuclease that does not destroy the donor activity of otherwise competent DNA. This may in part reflect a gene of enormous size and/or one that is liberally peppered with an abundance of restriction endonuclease sites. Based on the requirement for enormously high-molecular-weight DNA and the extremely low frequency with which HPRT$^+$ transformants are isolated, we predict that the gene when isolated will prove to be very large.

We have used a variety of TK$^-$ recipients in TK transformation studies and these various cell lines have served to demonstrate the enormous inherent differences in "competence" between TK$^-$ cell lines (Sweet et al., 1981). Some cell lines, such as mouse L TK$^-$ cells, are readily transformed. The addition of 1 ng of purified HSV *tk* DNA will result in 50 HAT-resistant colonies when 10^6 cells are plated in one 9-cm dish; other TK$^-$ cell lines, such as human 143 TK$^-$, mouse TK$^-$ teratocarcinoma cells, or TK$^-$ buffalo rat liver cells are transformed with decreasing efficiencies (10%, 1%, 0.1%, respectively, of mouse L TK$^-$ cells). When transformation efficiencies become low, because of either cell competence or gene size, it becomes necessary to use cloned DNA or transform large numbers of cells to detect the rare transformant. Almost any source of mammalian tissue will serve as an effective donor of *tk* DNA; the chicken *tk* (Perucho et al., 1980) and HSV *tk* genes are cloned. Surprisingly, DNA from vaccinia, a virus which codes for a new TK activity shortly after infecting cells in culture, is not effective at transforming TK$^-$ cells in culture.

3.2.5b. Dominant Selection. Detection of gene transfer requires the availability of a suitably marked mutant cell line deficient in the donor function. In some cases, cells resistant to metabolic inhibitors contain a dominant acting mutant gene. An interesting methotrexate (MTX) resistant Chinese hamster cell line (A29mtxR^{III}) synthesizes elevated levels of a mutant dihydrofolate reductase (DHFR) with a greatly reduced affinity for the folic acid antagonist methotrexate (Flintoff et al., 1976). We have used genomic DNA from this cell line to transfer the mutant *dhfr* gene to MTX-sensitive mouse L cells. Transformants were initially selected in 0.1 μg/ml of methotrexate. Stepwise selection of transformants in increasing levels of drug results in numerous surviving colonies that have amplified the transformed locus without concomitantly increasing the gene dosage of the endogenous locus (Wigler et al., 1980). Attempts to transform other cell lines have met with little success; most cell lines will spontaneously amplify the endogenous *dhfr* locus at about the same frequency as transformation with MTX-resistant DNA occurs. Thus, the potential usefulness of DNA-mediated transformation is readily extended to dominant acting genes.

Other dominant acting selectors result from either the reconstruction of naturally occurring prokaryotic genes or the use of other dominant acting eukaryotic genes. One such example of the latter is represented by the genes coding for morphologic transformation. Recent reports have demonstrated that the malignant phenotype can be transferred to mouse 3T3 cells using DNA from human tumors and selection in low serum. Examples of synthetic dominant acting genes are the Eco–gpt hybrid gene developed by Mulligan and Berg (1980) and a *tk–tn5* construct. The Eco–gpt vector results from fusion of an SV40 promoter and polyadenylation transferase. This vector will transform mammalian cells so xanthine can be utilized as a source of purines when the normal purine metabolic pathway is inhibited with mycophenolic acid and aminopterin. An alternative, dominant acting vector results from inserting the sequence from the bacterial transposon *tn5* coding for aminoglycoside resistance into the body of the HSV *tk* gene (Pao, Sweet, Silverstein, and Axel, unpublished observation). The prokaryotic gene makes use of virus promoter, ribosome binding, and polyadenylation sites to convert mammalian cells to G418 resistance. A dose of 400 μg/ml of G418 (an aminoglycoside analog) will inhibit the growth of mouse L cells. Such dominant vectors provide the opportunity to introduce genes into unmarked recipients. Their transfer can be readily detected using stringent selective media and verified by molecular hybridization.

3.2.5c. Cotransformation. The demonstration that cells in culture were competent to incorporate and express foreign DNA suggested that a subpopulation of cells was likely to incorporate additional unlinked DNA. Thus, experiments were performed to determine if two physically unlinked genes could be cotransformed into a competent cell. Initially, the viral tk gene was used as the selectable marker and cells cotransformed for additional DNA sequences were analyzed by molecular hybridization for the presence of the unselected sequence. The results of these early experiments demonstrated that when the nonselectable gene was present in excess (10- to 100-fold molar excess) cotransformation occurred frequently (Wigler et al., 1979). Subsequently, other cell lines were assayed for cotransformation of unlinked genes and were shown to readily acquire the nonselectable marker. These experiments provided the opportunity to introduce any genetic element into mammalian cells on the basis of its ability to be cotransformed with a selectable marker.

In principal, cotransformation is performed in a manner identical to transformation. The significant variation is that in order to identify transformants that contain the nonselectable gene of interest, it is first necessary to isolate cells transformed by a selectable marker. Accordingly, the selectable marker 1–1000 ng/plate is mixed with a 10- to 100-fold excess of nonselectable gene and a calcium phosphate precipitate is formed and applied to the recipient cells. Transformants are identified on the basis of their newly acquired phenotype and scored for the presence of the other marker by molecular hybridization; cotransformants are then screened for expression of the unselected marker via hybridization analysis to detect specific mRNA's or by immunochemical techniques to recognize new cell surface constitutents.

Using this technology and variations employing linkage of selectable to nonselectable DNA sequences, genes coding for cellular TK and APRT have been isolated. Foreign sequences such as rabbit, mouse, and human globin, human insulin and growth hormone, and genes coding for chicken ovalbumin have been introduced and expressed to different levels of mouse cells.

3.3. Identification of Transformants

There are several readily identifiable traits that serve to identify the transformant after exposure to DNA. First, transformation is a rare event, although it occurs more frequently than the spontaneous mutation rate of the recipient. Thus, when the frequency of transformation of cultures exposed to DNA containing the selective marker exceeds the natural mutation rate, one can use this as the first criterion of transformation. Additional evidence for transformation comes from the observation that the bulk of transformants are unstable when cultured in nonselective medium shortly after isolation, although Xgpt$^+$ transformants appear to be stable in the absence of selective pressure. A third criterion for distinguishing transformants from revertants depends on the identification of the foreign gene product in the putative transformant. In these instances, the investigator can utilize chemical and physical properties such as specific neutralization by defined antisera, heat lability, isoelectric point, or migration in nondenaturing gels to differentiate the protein product of a true transformant from a revertant. Finally, if a molecularly pure probe is available, Southern blot hybridization can be employed to demonstrate successful gene transfer.

4. Potential Uses of Transformation of Eukaryotic Cells

Transformation of eukaryotic genes has permitted investigators to alter the genotype and phenotype of mammalian cells in culture. Experimenters have already begun to introduce foreign genes into the genome of whole animals by transforming or microinjecting embryos and implanting the embryos back into foster mothers. In this fashion, embryos that have had foreign DNA introduced into them have been propagated into mature animals and there is evidence that in one report the foreign DNA was expressed (Wagner et al., 1981) and in another case the DNA was shown to enter the germ line and to be inherited by succeeding generations (Gordon et al., 1980). Whether transformation has a practical future in engineering the repair of genetic disorders remains to be seen. We do not as yet know enough about how and where the transformed DNA goes and what effect transforming DNA has on chromosome structure and regulation of normally expressed loci. There is ample evidence demonstrating that it enters the chromosome of the recipient (Robins et al., 1981). There also appear to be some chromosomal abnormalities associated with transformants, and whether this represents the norm or if it is a trait particular to the recipient BRL TK$^-$ cells used in this study is unknown. Finally, we do not yet understand how to correctly express foreign DNA in the bulk of transformed cells or how to guarantee that the transformed locus will remain stably associated with the recipient in the absence of selective pressure. Thus, at this point, it would seem premature to begin experiments on correcting genetic deficiencies in whole animals.

The greatest immediate value for eukaryotic cell transformation lies in its potential for dissecting complex regulatory phenomena by investigating the role specific nucleotide sequences within and about genes serve in regulating expression of these loci. In addition, transformation in concert with modern recombinant DNA technology has made it possible to identify and isolate DNA sequences coding for selectable biochemical markers, morphologic transformation, and portions of the HLA locus (Sim and Augustin, 1981). Clearly, the technology will remain useful to those interested in dissecting complex phenotypes and isolating the genes coding for cell surface markers. Finally, eukaryotic cloning systems can be used for the production of important animal proteins such as human insulin or interferon using the yeast cloning systems or using animal cloning systems.

Acknowledgments

Work by A.P.B. in the Oree Meadows Perryman laboratory was supported by grants from NIH, GM 28090, and The Meadows Foundation. Work in the laboratory of SJS was supported by a grant from NIH, CA 17477, and a Research Career Development Award from NIH, CA 00491. We thank Jerry Long for her assistance in the assemblage of this manuscript.

References

Beggs, J. D., 1978, Transformation of yeast by a replicating hybrid plasmid, *Nature* **275**:104–109.
Bolivar, F., Rodriquez, R., Green, P., Betlach, M., Heyneka, H., and Boyer, H., 1977, Construction and characterization of new cloning vehicles II. A multipurpose cloning system, *Gene* **2**:95–113.

Bollon, A. P., and Stauver, M., 1980, DNA transformation efficiency of various bacterial and yeast host-vector systems, *J. Clin. Hematol. Oncol.* **10**:39–48.

Broach, J. R., Strathern, J. N., and Hicks, J. B., 1979, Transformation in yeast: Development of a hybrid cloning vector and isolation of the CAN 1 gene, *Gene* **8**:121–133.

Flintoff, W. F., Davidson, S. V., and Siminovitch, L., 1976, Isolation and partial characterization of three methotrexate-resistant phenotypes from Chinese hamster ovary cells, *Somat. Cell Genet.* **2**:245–261.

Gordon, J. W., Scangos, G. A., Plotkin, D. J., Barbosa, J. A., and Ruddle, F. H., 1980, Genetic transformation of mouse embryos by microinjection of purified DNA, *Proc. Natl. Acad. Sci. USA* **77**:7380–7384.

Graham, F. L., Abrahams, P. J., Mulder, C., Heijneker, H. L., Warnaar, S. O., de Vries, F. A. J., Fiers, W., and van der Eb, A. J., 1974, Studies on *in vitro* transformation by DNA and DNA fragments of human adenoviruses and Simian Virus 40, *Cold Spring Harbor Symp. Quant. Biol.* **39**:637–650.

Hinnen, A., Hicks, J. B., and Fink, G. R., 1978, Transformation of yeast, *Proc. Natl. Acad. Sci. USA* **75**:1929–1933.

Hitzeman, R. A., Hagie, F. E., Levine, H. L., Goeddel, D., Ammerer, G., and Hall, B. D., 1981, Expression of human gene for Interferon in yeast, *Nature* **293**:717–722.

Hohn, B., and Hinnen, A., 1980, Cloning with cosmids in *E. coli* and yeast, in: *Genetic Engineering*, Volume 2 (J. K. Setlow and A. Hollaender, eds.), Plenum Press, New York, pp. 169–183.

Ilgen, C., Farabaugh, P. J., Hinnen, A., Walsh, J. A., and Fink, G. R., 1979, Transformation of yeast, in: *Genetic Engineering*, Volume 1 (J. K. Setlow and A. Hollaender, eds.), Plenum Press, New York, pp. 117–132.

Lowy, I., Pellicer, A., Jackson, J. F., Sim, G. K., Silverstein, S., and Axel, R., 1980, Isolation of transforming DNA: Cloning the hamster aprt gene, *Cell* **22**:817–823.

McCutchan, J. H., and Pagano, J. S., 1968, Enhancement of infection of Simian Virus 40 deoxyribonucleic acid with diethylaminoethyl-dextran, *J. Natl. Cancer Inst.* **41**:351–356.

Mulligan, R. C., and Berg, P., 1980, Expression of a bacterial gene in mammalian cells, *Science* **209**:1422–1427.

Perucho, M., Hanahan, D., Lipsick, L., and Wigler, M., 1980, Isolation of the chicken thymidine kinase gene by plasmid rescue, *Nature* **285**:207–210.

Ratzkin, B., and Carbon, J., 1977, Functional expression of cloned yeast DNA in *Escherichia coli*, *Proc. Natl. Acad. Sci. USA* **74**:487–491.

Robins, D., Ripley, S., Henderson, A., and Axel, R., 1981, Transforming DNA integrates into the host chromosome, *Cell* **23**:29–39.

Scherer, S., and Davis, R. W., 1979, Replacement of chromosomal segments with altered DNA sequences constructed *in vitro*, *Proc. Natl. Acad. Sci. USA* **76**:4951–4955.

Simm, G. K., and Augustin, A., 1981, Transfection of L-cells with cloned H-2 genes leads to expression of corresponding alloantigens, *Nature* (in press).

Stinchcomb, D. T., Struhl, K., and Davis, R. W., 1979, Isolation and characterization of a yeast chromosomal replicator, *Nature* **282**:39–43.

Struhl, K., Cameron, J. R., and Davis, R. W., 1976, Functional genetic expression of eukaryotic DNA in *Escherichia coli*, *Proc. Natl. Acad. Sci. USA* **73**:1471–1475.

Struhl, K., Stinchcomb, D. T., Scherer, S., and Davis, R. W., 1979, High frequency transformation of yeast: Autonomous replication of hybrid DNA molecules, *Proc. Natl. Acad. Sci. USA* **76**:1035–1039.

Sweet, R., Jackson, J., Lowy, I., Ostrander, M., Pellicer, A., Roberts, J., Robins, D., Sim, G.-K., Wold, B., Axel, R., and Silverstein, S., 1981, The expression, arrangement, and rearrangement of genes in DNA-transformed cells, in: *Genes Chromosomes and Neoplasia* (F. E. Arrighi, P. N. Rao and E. Stubblefield, eds.), Raven Press, New York, pp. 205–219.

Szybalska, E. H., and Szybalski, W., 1962, Genetics of human cell lines IV. DNA-mediated heritable transformation of a biochemical trait, *Proc. Natl. Acad. Sci. USA* **48**:2026–2034.

Wagner, E. F., Stewart, T. A., and Mintz, B., 1981, The human β-globin gene and a functional viral thymidine kinase gene in developing mice, *Proc. Natl. Acad. Sci. USA* **78**:5016–5020.

Whelan, W. L., Gocke, E., and Manney, T. R., 1979, The CAN 1 locus of *Sacchromyces cerevisiae*: Fine structure analysis and forward mutation rates, *Genetics* **91**:35–51.

Wigler, M., Sweet, R., Sim, G.-K., Wold, B., Pellicer, A., Lacy, E., Maniatis, T., Silverstein, S., and Axel, R., 1979, Transformation of mammalian cells with genes from prokaryotes and eukaryotes, *Cell* **16**:777–785.

Wigler, M., Perucho, M., Kurtz, D., Dana, S., Pellicer, A., Axel, R., and Silverstein, S., 1980, Transformation of mammalian cells with an amplifiable dominant acting gene, *Proc. Natl. Acad. Sci. USA* **77**:3567–3570.

Chapter 30

Viral-DNA Vectors in the Analysis of Mammalian Differentiation

ALBAN LINNENBACH and CARLO M. CROCE

1. Introduction

Advances in molecular biology and methods in mammalian cell culture have been utilized in the construction of DNA-transformed, pluripotent murine teratocarcinoma stem cells (Linnenbach et al., 1980) that have proven to be a useful *in vitro* model for the study of the regulation of gene expression during differentiation (Huebner et al., 1981; Croce et al., 1981; Linnenbach et al., 1981).

We have constructed a recombinant plasmid genome consisting of the pBR322 genome linked to a herpes simplex type 1 thymidine kinase (HSV-1 *tk*) gene and a simian virus 40 (SV40) genome, and introduced it into a thymidine kinase (TK$^-$) teratocarcinoma cell line F9 (Gmür et al., 1980), which is a homogeneous stem cell line that does not undergo significant spontaneous differentiation. Selection for HSV-1 *tk* expression resulted in the isolation of a clonal line of stem cells, designated 12-1, that stably carries one copy of plasmid DNA integrated into murine chromosomal DNA through a site on the pBR322 genome (Linnenbach et al., 1980). The 12-1 stem cell transformed has retained the phenotype of its F9 parent (Knowles et al., 1980), except that it produces HSV-1 TK.

The SV40 large T antigen is not expressed in 12-1 stem cells; however, upon differentiation of 12-1 cells into endodermal cells after exposure to retinoic acid (Strickland and Mahdavi, 1978), we obtained several daughter cell lines, one designated 12-1a, which express immunologically detectable SV40 early gene products. Because molecular mechanisms operative in the regulation of expression of viral genes in this differentiating model system may be analogous to mechanisms involved in gene regulation during embryogenesis (Martin and Evans, 1975), it is important to define the molecular basis

---

ALBAN LINNENBACH • Wistar Institute of Anatomy and Biology, Philadelphia, Pennsylvania 19104. Present address: Department of Internal Medicine, Yale University School of Medicine, New Haven, Connecticut 06510. CARLO M. CROCE • Wistar Institute of Anatomy and Biology, Philadelphia, Pennsylvania 19104.

for the suppression of viral genes in teratocarcinoma stem cells. Thus, the integrated SV40 genome in stem and differentiated cells has been characterized using techniques of DNA, transcriptional, and protein analysis.

2. Isolation and Characterization of the Recombinant DNAs

The pHSV-106 plasmid is a pBR322 (Bolivar et al., 1977)/HSV-1 tk (Wigler et al., 1977) recombinant plasmid isolated by McKnight et al. (1979). The pHSV-106 plasmid was partially digested with BamH1, and full-length linear molecules were isolated from agarose gels (Maxam and Gilbert, 1977) treated with alkaline phosphatase (Worthington) and ligated (Tonegawa et al., 1977) to BamH1-digested SV40 DNA (Bethesda Research Laboratories). The ligated DNA was transfected into *Escherichia coli* strain χ1776 (Curtiss et al., 1977) colonies containing the SV40 genome identified by colony hybridization (Grunstein and Hogness, 1975), using nick-translated (Maniatis et al., 1976) ^{32}P-labeled SV40 DNA, and then amplified. To extract plasmid DNA, cleared lysates were prepared by treatment of spheroplasts with BRIJ 58 and sodium deoxycholate followed by centrifugation to remove most of the chromosomal DNA (Clewell and Helsinki, 1960). Covalently closed circular plasmid DNA was further purified by ethidium bromide/cesium chloride density gradient centrifugation (Radloff et al., 1967). The plasmid DNA was preliminarily characterized to determine that the pBR322, HSV-1 tk, and SV40 genomes were all present. Several such recombinant DNAs were isolated and one recombinant, pC6, which was used in transformation experiments, was further characterized by digestion of pC6 with SalI, BglII, BamH1, and PvuI restriction endonucleases. Agarose (0.7% or 1.0%, containing 1 µg of ethidium bromide per ml) gel electrophoresis was carried out in 40 mM Tris/5.0 mM Na acetate/2.0 mM EDTA, pH 8.0 (Hayward and Smith, 1972).

A map of the DNA of the pBR322/HSV-1 tk/SV40 recombinant plasmid pC6 is shown in Fig. 1. The plasmid contains one copy each of the pBR322 genome, HSV-1 tk gene, and the SV40 genome and contains SV40 in the position and orientation shown. This map was prepared by the use of restriction enzymes which cut the entire plasmid once or several times (single and double digests) at known positions on the respective genomes.

3. Transformation of TK⁻ F9 Cells

3.1. Calcium Technique for DNA-Mediated Gene Transfer

Thymidine kinase-deficient (TK⁻) F9 teratocarcinoma stem cells (Gmür et al., 1980) were transfected with pC6 (pHSV-106/SV40) recombinant DNA by an adaptation (Wigler et al., 1979) of the calcium technique (Graham and Van der Eb, 1973) for DNA-mediated gene transfer, followed by selection in hypoxanthine/aminopterin/thymidine (HAT medium) (Littlefield, 1965).

A total of 10^7 recipient cells were seeded in 100-mm plastic petri dishes in RPMI at 2×10^6 cells/dish. Five hundred micrograms of pC6 DNA was

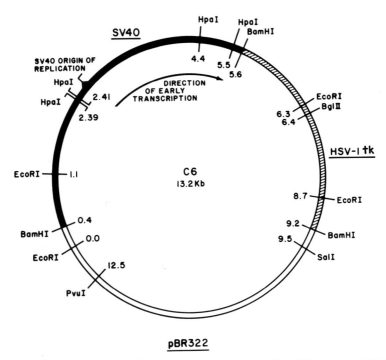

Figure 1. Restriction map of 13.2-kb plasmid C6. One copy of BamHI-digested SV40 DNA is inserted into the pBR322/HSV-1 tk vector, pHSV-106.

sterilized by precipitation in two volumes of absolute ethanol, 0.12 M ammonium acetate overnight at −20°C.

Cultures were refed with 10 ml fresh medium 4 hr prior to transfection. The DNA precipitate was centrifuged and dissolved for 3.25 hr in sterile 1 mM Tris-Cl, pH 7.9, 0.1 mM Na_2EDTA at 267 µg/ml. Sterile 2 M $CaCl_2$ was slowly added to a final 0.5 M concentration using a plastic pipette. The DNA was transferred dropwise to a plastic tube containing an equal volume of 2× HEPES-buffered saline (280.0 mM NaCl, 50.0 mM HEPES, and 1.5 mM Na_2HPO_4, pH 7.10). Simultaneously, air bubbles were introduced with another pipette. The calcium phosphate–DNA coprecipitate was placed in each dish, followed by incubation at 37°C for 4 hr. The medium was then replaced and the cultures incubated for 20 hr before HAT selective pressure was applied. Cultures were refed at 2-day intervals, and HSV-1 TK^+ transformant colonies were picked by days 10–14.

Out of 10^7 F9TK$^-$ cells transfected as described with pC6 without carrier DNA, only two colonies survived HAT selection. These two colonies, designated 12-1 and 13-1, were isolated and cultured.

3.2. HSV-1 tk Starch Gel Electrophoresis

Lyophilized pellets of exponentially growing cultures were solubilized by four cycles of freeze-thawing in 50 µl of 0.01 M Tris-Cl pH 8.0, 0.01 M

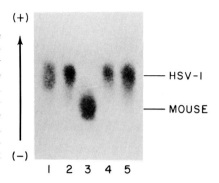

Figure 2. Starch gel electrophoresis separation of *tk* gene products. Enzyme migration is visualized by the binding of tritiated thymidine monophosphate product to DE-81 paper overlaid on electrophoretically separated TK enzymes. Enzymes in lanes 2 and 5 are from F9 stem cell transformants 13-1 and 12-1, respectively; lane 3, murine TK derived from MC57G cells (a chemically transformed mouse cell line); lane 4, HSV-1 *tk* derived from 41-1-1 cells (a mouse cell line containing the HSV-1 tk gene). Lane 1 contains HSV-1 *tk* derived from an F9 stem cell transformant not described in this report.

dithiotheitol, and 0.016 mM thymidine (Migeon et al., 1969) and centrifuged at 27,000g.

The electrode buffer consisted of 0.1 M Tris, 0.1 M maleic acid, 0.01 M EDTA, and 0.01 M $MgCl_2$ at pH 7.4 (Spencer et al., 1964); the 12% starch gel buffer was a 1:10 dilution of electrode buffer with 0.2 mM ATP added to preserve enzyme activity (Migeon et al., 1969). Electrophoresis was at 5 V/cm in the cold for not more than 5 hr.

The separation of HSV-1 *tk* and wild-type F9 TK was demonstrated by applying the reaction mix [0.1 M Tris-Cl pH 8.0/500 µCi of [methyl-³H]thymidine (55.2 Ci/mmole, New England Nuclear)/5 mM ATP/5 mM $MgCl_2$] to the cut, moistened gel by DE81 paper overlay for 1.5 hr at 37°C. The DE81 sheet was then dried, rinsed with 4 liters of deionized water, and dried again. Fluorography was carried out as described (Linnenbach et al., 1980). Areas of bound radiolabeled reaction product thymidine monophosphate, indicated the enzyme migration (Migeon et al., 1969).

As illustrated in Fig. 2, HSV-1 TK and murine TK can be distinguished by starch gel electrophoresis; 13-1 (lane 2) and 12-1 (lane 5) express only HSV-1 *tk* and are thus transformants and not revertants. Thus far, in transfection of more than 10^8 F9TK⁻ cells, we have not isolated a single revertant colony.

The 12-1 cells have now been subcultured more than 70 times and the stability of the 12-1 line has been tested by determining the plating efficiency of these cells in nonselective medium, HAT medium, and medium containing 15 µg/ml of 5-bromo-2-deoxyuridine (BrdU). The 12-1 cells, which plated at an efficiency of 4% in nonselective or HAT medium, had a reversion frequency of 10^{-4} in medium containing BrdU.

4. Viral Antigens of Transformants before and after Differentiation

4.1. Retinoic Acid Induction

Differentiation of transformed stem cell lines 12-1 and 13-1 was induced by using 0.1 µM retinoic acid (all-*trans*, Eastman Chemical). The procedure was essentially as described by Strickland and Mahdavi (1978), except that retinoic acid was dissolved in dimethyl sulfoxide. Induced 12-1 cultures were either prepared for analysis after 1–12 days of treatment, or established as continuously growing clonal lines.

4.2. Indirect Immunofluorescence

SV40 T-antigen nuclear fluorescence (Pope and Rowe, 1964) was demonstrated with either a polyclonal or a monoclonal (Martinis and Croce, 1978) anti-SV40 T-antigen antibody followed by an appropriate fluoresceinated probe.

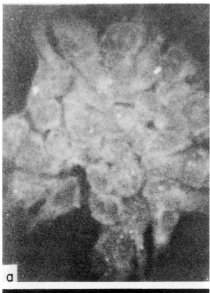

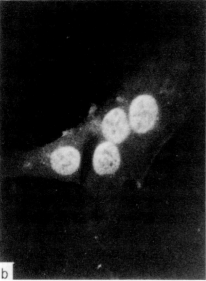

Figure 3. Indirect immunofluorescence of SV40 T antigen in transformant 12-1 cells (a) before and (b) after induction of differentiation with 0.1 μM retinoic acid.

The 12-1 stem cells were negative for expression of T antigen (Fig. 3a). When 12-1 cells were treated with retinoic acid and tested for T antigen at various times during treatment, there was a pronounced induction of the SV40 T antigen (Fig. 3b); by 3 days after addition of retinoic acid, more than 10% of the cells were positive for SV40 T antigen; by 10 days after induction, more than 50% were positive. Results were the same with polyclonal or monoclonal antiserum against SV40 T antigen. The clonal 12-1a differentiated cell is 100% positive for SV40 T antigen. The 13-1 stem and differentiated cells were negative for SV40 T antigen and were later found not to contain SV40 sequences.

4.3. Immunoprecipitation

In order to confirm the SV40 T-antigen immunofluorescence results, monoclonal antibody to SV40 T antigen was used to immunoprecipitate protein from cell lysates of 12-1, 12-1a, and on SV40-transformed monkey kidney cell line T22TK$^-$.

Subconfluent cell cultures were washed with prewarmed methionine-deficient medium supplemented with 5% dialyzed fetal bovine serum and then incubated in the same medium for 2 hr. L-[^{35}S]Methionine (50 µCi/ml, 400 Ci/mmole; New England Nuclear) was added to each culture, and the cells were incubated for another 4 hr. The cells were collected with a rubber policeman, then chilled and lysed in 0.5% Nonidet P-40/50 mM Tris, pH 8.0/8 mM EDTA/0.6 M NaCl/phenylmethylsulfonylfluoride (0.3 mg/ml) lysis buffer (Linzer and Levine, 1979) for 20 min with occasional mixing. Lysates were then sonicated four times for 5 sec each.

SV40 T antigen was immunoprecipitated from the cell lysates by the double-antibody method described by Hughes and August (1981). Briefly, lysates were clarified by centrifugation at 100,000g (1 hr; 4°C) and supernatants were stored at −70°C; aliquots containing 1.5×10^7 acid-insoluble cpm were reacted for 1 hr at 4°C with monoclonal anti-T-antigen antibody (Martinis and Croce, 1978) (1:1000 final dilution) or with nonimmune ascites fluid (1:1000) in a 100-µl reaction mix containing gelatin (1 mg/ml) and lysis buffer. Rabbit anti-mouse immunoglobulin (25 µl; Bio-Rad) was then added and the reaction was allowed to proceed overnight at 4°C. Precipitates were washed three times with 2.5 ml of 20 mM Tris, pH 7.6/1 mM EDTA/0.1 M NaCl/2.5 M KCl/0.5% Nonidet P-40 by centrifugation at 2000g for 20 min. Precipitates were then washed once in 2.5 ml of 20 mM Tris, pH 7.6/1 mM EDTA/0.1 M NaCl/0.5% Nonidet P-40/0.5% sodium deoxycholate/0.1% NaDodSO$_4$ and centrifuged as before. Pellets were dissolved in 50 µl of NaDodSO$_4$ sample buffer (Laemmli, 1970), boiled for 1.5 min, and electrophoresed on NaDodSO$_4$/12% polyacrylamide gels; molecular weight standards (Bio-Rad) were included on each gel. Gels were dried and fluorographed to locate immunoprecipitated proteins.

The 94,000-dalton T antigen was precipitated from the T22TK SV40-

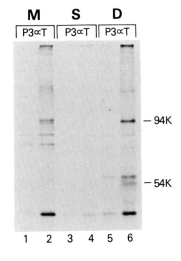

Figure 4. Immunoprecipitation of SV 40 T antigen from 12-1 and 12-1a cells. [^{35}S]Methionine-labeled extracts from SV40-transformed monkey kidney cells (M), 12-1 stem cells (S), and 12-1a differentiated cells (D) were reacted with monoclonal anti-SV40 T-antigen antibody (αT) or with nonimmune antibody (P3) and precipitated by addition of rabbit anti-mouse immunoglobulins. The 12% polyacrylamide gel was fluorographed, dried, and exposed for 7 days. Size markers are daltons $\times 10^{-3}$.

transformed monkey kidney cells and the 12-1a differentiated cells by the anti-SV40 T-antigen monoclonal antibody but not by the control P3 fluid (Fig. 4). Neither SV40 T antigen nor any other protein was precipitated from 12-1 stem cells by anti-SV40 T-antigen monoclonal antibody (Fig. 4).

5. Organization of the Plasmid Genome in Stem and Retinoic Acid-Treated Cells

5.1. Isolation of Cellular DNA

Stem cells were lysed and separated into supernatant (nonchromosomal) and pellet (chromosomal) fractions by the Hirt procedure (Hirt, 1967), which preferentially precipitates the high-molecular-weight chromosomal DNA in 1 M NaCl at 4°C. DNA was extracted from both fractions by a standard method consisting of proteinase K treatment (100 μg/ml, 37°C, 16 hr), phenol extraction, and precipitation in ethanol. Nucleic acids were dissolved in a Tris-Cl buffer, treated with RNase A (25 μg/ml, 1 hr), extracted with phenol and then with chloroform:isoamyl alcohol (24:1), and finally precipitated with 2.5 volumes of ethanol at −20°C. (Additional techniques for isolation of DNA, nick translation, DNA blotting, and filter hybridization are presented in Chapter 31, this volume).

5.2. Nick Translation

The various DNA probes were labeled according to a procedure described by Maniatis (1976). The 100-μl reactions contained 1.0 μg of DNA, 100 μCi each of deoxyadenosine 5'-(α-^{32}P) triphosphate and thymidine 5'-(α-^{32}P) triphosphate (350 Ci/mmole; Amersham), 50 mM Tris-Cl pH 7.8, 5 mM MgCl$_2$, 10 mM 2-mercaptoethanol, and 100 µg/ml bovine serum albumin. The reaction was primed with 0.02 ng of DNase I (Worthington) at room temperature for 1 min. After the addition of two units of DNA polymerase I (Boehringer-Mannheim), the mixture was incubated at 15°C for 70 min. The reaction was terminated by adding NaDodSO$_4$ to a final concentration of 1%. A Sephadex G-50-80 (Sigma) column separated the unincorporated label from the DNA. Probes prepared by this method generally had a specific activity of 1.5×10^8 cpm/µg.

5.3. Southern Blot Analysis

Ten micrograms of DNA extracted from the supernatant and pellet fractions of transformed stem cells were digested with 30 units of the restriction endonucleases *Bam*H1, *Kpn*1, and *Xba*1, for 5 hr at 37°C. Agarose (0.7%) gel electrophoresis, in 40 mM Tris, 5.0 mM Na acetate, 2.0 mM EDTA, pH 8.0 was at 55 V overnight, followed by ethidium bromide (10 µg/ml) staining for photography.

Blotting analysis (Southern, 1975) was modified (P. Botchan, personal communication) as follows. The gel was denatured in 400 ml 0.2 M NaOH, 0.6 M NaCl with rocking at room temperature for 45 min, followed by rinsing in 1200 ml of distilled water. The gel was adjusted to pH 7.5 with 400 ml of 1 M Tris-Cl, 0.6 M NaCl, pH 7.0, also with shaking at room temperature for 45 min. A plexiglass plate was wrapped with filter paper and saturated with 6× SSC (20× SSC = 3 M NaCl, 0.3 M Na citrate, pH 7.0) in a leveled pan. After transfer to the plate, the gel was covered first with a 6× SSC saturated nitrocellulose filter (type HA, pore size 0.45 µM; Millipore Corp.), second with two sheets of 6× SSC saturated Whatman 3 MM paper, then with four layers of dry Wyp-alls (Scott), and finally with a plexiglass weight. To transfer the DNAs to the filter approximately 1 liter of 6 × SSC was passed up through the gel by several changes of Wyp-alls. The filter was rinsed with a minimal amount of 2× SSC, air-dried for 30 min on Whatman 3MM paper, and baked in a vacuum oven at 80°C for 2-3 hr: filters were stored at 4°C until use. Before hybridization, the nitrocellulose filter was soaked in 200 ml of 6× SSC, 5× Denhardt's (10× Denhardt's = 0.2% Ficoll 400, 0.2% bovine serum albumin, 0.2% polyvinylpyrrolidone) at room temperature with rocking for 4 hr, followed by soaking in a 200 ml prehybridization mix of 1× Denhardt's 50% formamide and 4× SSC at 37°C with shaking for 1 hr. The ^{32}P-labeled probes were heat-denatured at 100°C for 5 min in the presence of preboiled, sonicated salmon sperm carrier DNA (Sigma) followed by rapid swirling in an ice:water (1:1) bath for 2.5 min. The 10- or 20-ml hybridization mix consisted of 50% formamide, 1× Denhardt's 4× SSC, 0.2% NaDodSO$_4$, and the denatured ^{32}P-labeled probe, plus 50 µg/ml carrier DNA. Hybridization was in a wide-mouth roller bottle (Nalgene) at 37°C for 48 hr; the roller apparatus was set at 0.5 rpm.

The filter was washed at 37°C with rocking for 20 min, first in prehybridization mix, 0.2% NaDodSO$_4$; second in 50% formamide, 3×SSC, 1× Denhardt, 0.2% NaDodSO$_4$; then in 50% formamide, 1×SSC, 1×Denhardt, 0.2% NaDodSO$_4$. The following washes were done in a 65°C shaking water bath for 30 min each: 1× SSC, 0.2% NaDodSO$_4$; 0.5× SSC, 0.2% NaDodSO$_4$; 0.3× SSC, 0.2% NaDodSO$_4$. The filter was then placed on Whatman 3MM paper, air-dried overnight, and exposed to film for 0.5–6 days at −70°C.

5.3.1. Chromosomal Integration

The first step in this analysis was to determine if SV40 DNA was present either in the supernatant (free) or pellet (chromosomal) DNA fractions of the two cell lines. Thus, 12-1 and 13-1 supernatant and pellet DNAs were digested with restriction enzyme KpnI, which cuts SV40 DNA once, and with XbaI, which does not cleave the pC6 plasmid. These digested DNAs were blotted and hybridized with ^{32}P-labeled SV40 DNA; the results are shown in Fig. 5, lanes 2, 4, 6, and 8, which contain DNA extracted from the 13-1 cell line, and do not contain sequences that hybridize with SV40 DNA; the DNA of this cell line, which does express HSV-1 tk and thus must contain part of the C6

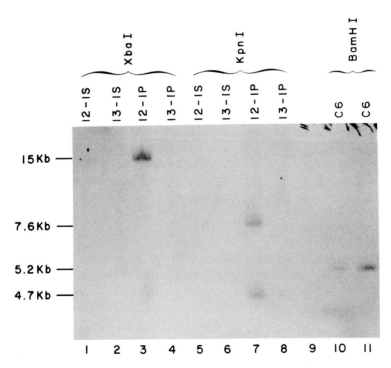

Figure 5. Hybridization of ^{32}P-labeled SV40 DNA to 13-1 and 12-1 cellular DNA after restriction endonuclease digestion and transfer from agarose gel. Hirt supernatant (S) and pellet (P) DNA (10 μg per lane) from 12-1 and 13-1 stem cells were cleaved with XbaI (lanes 1–4) and KpnI (lanes 5–8); electrophoresed in a 0.7% agarose slab gel; denatured; transferred to a nitrocellulose sheet; and hybridized to ^{32}P-labeled SV40 DNA. Here BamHI-cleaved and similarly treated pC6 DNA (lane 10, 6.25 pg; lane 11, 12.5 pg) was included as a marker.

plasmid DNA, was not further analyzed. The DNA derived from the 12-1 supernatant fraction (unintegrated DNA, Fig. 5, lanes 1 and 5) does not contain DNA that hybridizes with SV40 DNA. The 12-1 pellet DNA fraction (chromosomal DNA, lanes 3 and 7) does contain DNA that hybridizes with SV40 DNA. From these results we conclude that 12-1 cells contain integrated SV40 DNA and that there is a single copy of the pC6 plasmid per cell, since the XbaI-digested DNA (lane 3) gives only one band that hybridizes with SV40 DNA, whereas the KpnI-digested DNA (lane 7) gives two bands. These are the expected results for enzymes that either do not cut SV40 DNA (XbaI) or cut SV40 DNA once (KpnI).

5.3.2. Organization in Stem and Differentiated Cells

Since the XbaI enzyme, which does not cut the plasmid C6 DNA, gives a fragment of 15.0 kbp (which is greater than the size of the linear pC6 plasmid), it was possible that this fragment contained the entire C6 plasmid DNA. To investigate this question further, an experiment was designed to determine the location of the pBR322 DNA and the HSV-1 tk DNA in the 12-1 stem cells. Plasmid C6 DNA and 12-1 pellet DNAs were digested with XbaI and BamH1 restriction enzymes and separated by agarose gel electrophoresis. The gel was then cut into three parts, and the DNA from each of the three parts was transferred to nitrocellulose filters. Each filter was then hybridized to a different labeled probe as illustrated in Fig. 6. Filter a was hybridized to ^{32}P-labeled pBR322 DNA, filter b to ^{32}P-labeled HSV-1 tk DNA, and filter c to ^{32}P-labeled SV40 DNA.

The results shown in Fig. 6 illustrate several important points: (i) the same 15.0-kpb band, which does not comigrate with covalently closed or nicked circular forms of pC6 DNA (Fig. 6, lanes a1, b1, c1), is detected by all three probes when 12-1 DNA is digested with XbaI (lanes a2, b2, and c2), indicating that the three parts of the recombinant plasmid are present and linked to each other within a chromosomal site in the 12-1 DNA; (ii) the plasmid DNA is probably integrated into the cellular DNA through a site on the pBR322 genome, because BamH1 digestion of 12-1 DNA does not produce an intact linear pBR322 4.3-kbp genome (lane a3), but produces a 12.5-kbp fragment that hybridizes with pBR322; there should theoretically be another band in this lane (a3) containing DNA that hybridizes with pBR322 DNA. This fragment is detected in other experiments with higher specific activity probes; and (iii) the entire SV40 genome is present in these cells, since BamH1 restriction of 12-1 cellular DNA gives a 5.2-kbp fragment that hybridizes with the SV40 probe (lane c3) and comigrates with linear SV40 DNA derived by BamH1 digestion of pC6 DNA or BamH1 digestion of covalently closed circular SV40 DNA (lanes c4 and c5, respectively).

We conclude from these results that cell line 13-1 does not contain SV40 DNA, whereas the 12-1 cell line contains one copy per cell of pC6 DNA integrated through a site on the pBR322 genome, leaving the SV40 and HSV-1 tk genomes intact and possibly retaining the same relationship to each other as shown in the map of pC6 (Fig. 1).

Analysis of Mammalian Differentiation

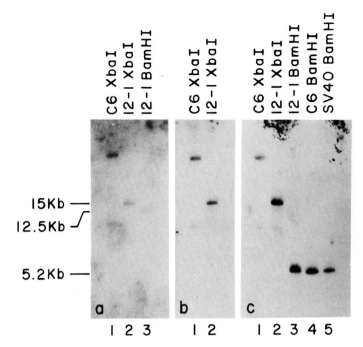

Figure 6. Hybridization of ^{32}P-labeled (a) pBR322 DNA, (b) HSV-1 tk DNA, and (c) SV40 DNA to XbaI and BamHI-cleaved 12-1 cellular DNA. Hirt pellet DNA (10 μg/lane) from 12-1 cells was cleaved with XbaI and BamHI restriction endonucleases and applied to an agarose slab gel as indicated above each lane in this figure. After electrophoresis, the gel was cut into three parts, and the DNA in each gel was denatured and transferred to nitrocellulose sheets. Each of the three nitrocellulose sheets was hybridized to a different ^{32}P-labeled probe: (a) hybridized to ^{32}P-labeled pBR322 DNA; (b) hybridized to ^{32}P-labeled HSV-1 tk DNA; and (c) hybridized to ^{32}P-labeled SV40 DNA. Here pC6 DNA, BamHI-cleaved pC6 DNA, and BamHI-cleaved SV40 DNA are included as markers.

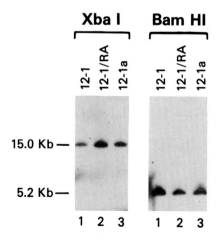

Figure 7. Arrangement of the SV 40 genome before retinoic acid induction. DNAs from uninduced 12-1 stem cells (lane 1), 12-day induced cells (lane 2), and an established differentiated cell line 12-1a (lane 3), were restricted with XbaI or BamHI and analyzed by blotting and hybridizaiton with ^{32}P-labeled Sv40 DNA.

In order to determine whether this arrangement of the integrated plasmid genome remained unchanged during the process of differentiation, we cleaved cellular DNA from 12-1 cells, 12-1 cells exposed to retinoic acid for 12 days, and 12-1a cells with XbaI and with BamHI, blotted the DNAs, and hybridized with ^{32}P-labeled SV40 DNA. DNA from the three cell types cleaved by XbaI (Fig. 7, left lanes 1–3) exhibited a single 15-kbp band containing DNA homologous to SV40. After BamHI cleavage of the DNA from the same three cell types (Fig. 7, right lanes 1–3), a single 5.2-kbp band, the same length as BamHI-linearized SV40 DNA, was detected in each of the three cell types. We concluded that no gross rearrangement in the plasmid genome occurred in these cells during the process of differentiation.

6. Transcription of the SV40 Genome in Stem and Differentiated Cells

6.1. Preparation of Cellular RNAs

Total cellular RNA was prepared either by the hot phenol technique (Scherrer, 1969), which removes DNA by partitioning into the hot phenol at low pH, or by the cesium chloride method (Glisin et al., 1974), which pellets RNA but not DNA or protein. Total cytoplasmic RNA was obtained by 0.65% Nonidet P-40 lysis in the presence of ribonucleoside–vanadyl complexes (Berger and Berkenmeir, 1979) to inhibit ribonuclease, followed by five phenol–chloroform (1:1), 8-hydroxyquinoline (0.2%) extractions. Polyadenylated mRNAs were obtained from total RNA preparations by two cycles of oligo (dT)-cellulose chromatography (Aviv and Leder, 1972) with binding in 0.1% NaDodSO$_4$, 0.5 M NaCl, 0.001 EDTA, 0.01 M Tris-Cl pH 7.5 (Bantle et al., 1976) and with elution in the same buffer without NaCl. RNAs were dissolved in double-distilled water (15 mg/ml), then stored at −20°C.

6.2. RNA Transfer to Nitrocellulose

Specific transcripts were identified by a transfer technique developed by Thomas (1980). One to ten micrograms of ethanol-precipitated RNA or marker DNA were denatured by first dissolving ethanol pellets in one-third volume 30 mM sodium phosphate, pH 7.0; then an addition of one-half volume 99.5% dimethyl sulfoxide (Fluka, puriss p.a.) and one-sixth volume of extensively deionized 40% glyoxal (Carmichael and McMaster, 1980); followed by incubation at 50°C for 1 hr. After chilling on ice for 1 min, glycerol was added to 10% and Bromophenol Blue to 0.035% from a 5× concentrated stock prepared in 10 mM PO$_4$, pH 7.0. Agarose (1.1%) gel electrophoresis was carried out in a carefully recirculated 10 mM sodium phosphate pH 7.0 buffer system at 4 V/cm for 5 hr.

Both HindIII-digested λ DNA and HaeIII-digested φX174 DNA and rRNA marker lanes were cut from the gel and stained with acridine orange (33 µg/ml). Sample RNAs were transferred to nitrocellulose (Schleicher and Schuell, BA85, 0.45 µM) in 20× SSC according to the method of Southern (1975). The filters were air-dried for 30 min on Whatman 3 MM paper, baked at 80°C for 2 hr under reduced pressure, and stored at 4°C. To reduce filter background, nick-translated probes were purified by termination of the reactions by addition of NaOH to 0.3 M and boiling for 2 min prior to chromatography on Sephadex G-50-80. The RNA filters were prehybridized at 37°C for 6 hr in 0.05 M $NaPO_4$, pH 6.5, 50% formamide (Fluka, puriss p.a.), 5× SSC, and 0.02% each of Ficoll, polyvinylpyrolidone, and bovine serum albumin and 75 µg/ml sonicated, denatured Salmon sperm DNA. Hybridizations at 37°C for 36 hr were in 16 ml of the prehybridization mix, 4 ml of 50% (w/v) dextran sulfate (Pharmacia), plus the heat-denatured ^{32}P-labeled DNA (specific activity 1.5×10^8 cpm/µg) probe. After hybridization, filters were washed four times in 2× SSC, 0.1% $NaDodSO_4$ at room temperature and twice in 0.1× SSC, 0.1% $NaDodSO_4$ at 50°C, then autoradiographed.

6.2.1. Total RNA

Total RNA was extracted from 12-1 stem and 12-1a differentiated cells with hot phenol, selected on oligo(dT)-cellulose, glyoxalated, and assayed for viral-specific RNA by transfer to nitrocellulose. Hybridization with ^{32}P-labeled SV40 DNA detected a 2900-base and a 2600-base mRNA in the poly-A^+ fractions of both stem and differentiated cells (Fig. 8, lanes 3 and 5, respectively). The observed sizes of these two SV40 transcripts are in agreement with those found early after infection of monkey cells with SV40 DNA (Alwine et al., 1979). The most important point illustrated by this gel is that the 12-1 stem cell, which does not express large T antigen, does transcribe the large T-antigen (2600-base) mRNA. The 2900-base transcript is of the size expected for small tumor (t-antigen) mRNA.

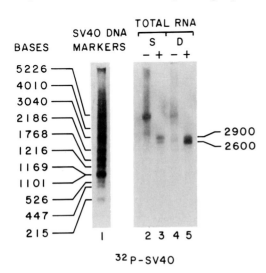

Figure 8. SV40 RNA in 12-1 and 12-1a cells. Poly-A^+ and poly-A^- RNA from 12-1 stem cells (S) and 12-1a differentiated cells (D) were denatured, separated electrophoretically, transferred to nitrocellulose filters, and hybridized to ^{32}P-labeled SV40 DNA. Lanes 1 and 3, 10 µg of poly-A^- RNA; lanes 2 and 4, 1 µg of poly-A^+ RNA.

6.2.2. Cytoplasmic RNA

When total cytoplasmic RNA from stem and differentiated cells was analyzed, the same two SV40 RNAs were detected in the 12-1 stem cell and, as expected, in the 12-1a differentiated cell (Fig. 9). Interestingly, the 2.9-kb RNA species, which is the size of small t-antigen mRNA, is present in stem cells in larger amounts than the 2.6-kb species, which is the size of T-antigen mRNA; the reverse is true for the differentiated cells.

6.3. S1 Nuclease-Resistant Duplex Analysis of RNAs

Spliced SV40 early mRNAs were assayed by the method of Berk and Sharp (1977, 1978), which was adapted to include Southern transfer and hybridization with high-specific-activity, ^{32}P-nick-translated SV40 DNA probes.

Unlabeled SV40 DNA (Bethesda Research Laboratories) was digested with EcoRI restriction endonuclease at 37°C for 6 hr and the 50-μl reaction was terminated by addition of one-tenth volume 0.15 M EDTA, 1% NaDodSO$_4$. After proteinase K addition (to 100 μg/ml) and incubation at 37°C for 15 min, the reaction was brought to 200 μl and extracted with phenol, and then with chloroform:isoamyl alcohol (24:1). Aliquots (0.2 μg) of SV40 EcoRI fragments were mixed with 100 μg of aqueous transformed cell RNA in 1.5-ml Eppendorf tubes. The mix was made 0.15 M in NaOAc (pH 6.0) and precipitated with two volumes of ethanol at −20°C. The ethanol pellets were air-dried for 15 min at room temperature; dissolved in 16 μl of 100% formamide (Matheson, Coleman, and Bell) that had been previously deionized with AG 501-X8 (D) mixed bed resin (Duesberg and Vogt, 1973); followed by addition of 4 μl of a 5× hybridization buffer (2 M NaCl, 0.2 M PIPES, pH 6.4, 0.005 M EDTA (Berk and Sharp, 1978). Denaturation was carried out by completely submerging the

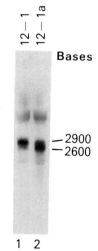

Figure 9. Cytoplasmic SV40 RNA (10 μg/lane) from 12-1 stem and 12-1a differentiated cells, electrophoresed, blotted, and hybridized to ^{32}P-labeled SV40 DNA.

tubes in an 85°C bath for 15 min (Favalora et al., 1980), followed by immediate transfer to and submersion in a 49°C bath for a 3-hr hybridization (Berk and Sharp, 1978). Keeping the tips of the tubes in the bath, the reactions were individually terminated by addition of 400 µl of 0.25 M NaCl, 0.03 M NaOAc (pH 4.50), 0.001 M $ZnSO_4$, 5% glycerol containing 100 units of S1 nuclease (Sigma, Type III) at 0°C, followed by vortexing for 1 sec and transfer to a 0°C ice/water bath. The S1 nuclease digestion was done by transfer to a 37°C bath for 30 min. The enzyme reactions were terminated by addition of one-fifth volume 0.5 M Tris-Cl, pH 9.5, 0.1 M EDTA (McKnight, 1980). The samples were divided into two parts; 20 µg of carrier tRNA (Boehringer-Mannheim) was added to each, followed by precipitation with 2.5 volumes of ethanol at −20°C. After centrifugation, one set of samples was dissolved in denaturing buffer and prepared for alkaline agarose (1.4%) gel electrophoresis in 0.03 M NaOH, 0.002 M EDTA at 2 V/cm for 16 hr; the other set of samples was dissolved in neutral buffer and electrophoresed in a 1.4% agarose gel containing 0.04 M NaAc, 0.05 M Tris, 0.002 M EDTA, pH 8.3 at 2 V/cm for 16 hr (Berk and Sharp, 1978).

The gels were prepared for transfer to nitrocellulose in the same manner

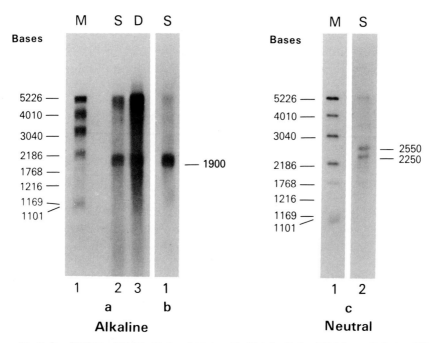

Figure 10. Spliced SV40 mRNA in 12-1 and 12-1a cells. Total cellular RNA from 12-1 stem (S) and 12-1a cells. Total cellular RNA from 12-1 stem (S) and 12-1a differentiated (D) cells was hybridized to EcoRI-cleaved SV40 DNA, treated with S1 nuclease, electrophoresed on alkaline or neutral gels, blotted, and hybridized with ^{32}P-labeled SV40 DNA. (a, b) Results after alkaline gel electrophoresis; (c) results after neutral gel electrophoresis. Marker lanes (M) contained SV40 DNA fragments prepared by digestions of SV40 DNA with BamHI, PstI, HindIII, and BamHI/HpaII; lanes labeled S contained duplexes derived from 12-1 RNA; lanes marked D contained duplexes derived from 12-1a RNA.

as described for Southern transfer, except that only the neutral gel needed to be denatured. The filters were hybridized to ^{32}P-labeled SV40 DNA, washed, and autoradiographed for 6–20 hr.

The sizes of the SV40 mRNAs determined on the RNA gels in Figs. 8 and 9 were indicative of mature transcripts; the S1 nuclease technique directly demonstrates the presence of spliced SV40 mRNAs. The alkaline gel revealed a 1900-base "body" sequence in the stem cell (Figs. 10a, lane 2, and 10b, lane 1) and, as expected, in the differentiated cell (Fig. 10a, lane 3). The 650- and 350-base "leader" sequences cannot be easily detected under the condition of this transfer. The 1900-base fragment is common to both large T- and small t-antigen mRNAs, and can only occur if spliced RNA is present (Berk and Sharp, 1978).

Splicing of the SV40 RNA in teratocarcinoma-derived stem cells was confirmed by neutral gel electrophroesis of intact S1 nuclease-digested DNA–RNA hybrids. The observed 2250- and 2550-bp duplexes (Fig. 10c, lane 2) are identical to those observed for spliced T- and t-antigen mRNAs in SV40-infected monkey kidney cells (Berk and Sharp, 1978).

7. Summary

Taking advantage of the powerful tools of molecular biology, we have exploited the F9 teratocarcinoma cell line to investigate molecular events involved in the control of gene expression during development. Transfection of the F9TK$^-$ cell line with a recombinant plasmid consisting of the SV40 genome, the HSV-1 *tk* gene, and the pBR322 genome has resulted in isolation of a stem cell clone 12-1 which has a single copy per cell of the SV40-containing plasmid integrated into chromosomal DNA. The SV40 gene products are undectable in this stem cell line, whereas, after induction with retinoic acid, the differentiated cells express SV40 early gene products as detected by immunofluorescence and immunoprecipitation (Linnenbach et al., 1980, 1981; Knowles et al., 1980).

The SV40 genome is transcribed in both stem and differentiated cells. Two SV40-specific RNAs of the sizes expected for SV40 T- and t-antigen mRNAs are produced in both cell types. In addition, SV40-spliced poly-A$^+$ mRNAs, also of the sizes expected for T and t antigens, are present in the cytoplasm of both cell types (Linnenbach et al., 1981).

The simplest interpretation of these results is that stem cells that carry integrated SV40 genomes are capable of synthesizing authentic early SV40 mRNAs and that lack of expression of SV40 early region genes is due to a post-transcriptional event. *In vitro* translation of SV40-specific mRNAs from stem cells will determine whether the block to SV40 T-antigen expression in stem cells occurs at the translational level or if the mRNAs as produced in stem cells are defective. Segal and Khoury (1979) have reported that lack of expression of SV40 T antigens in SV40-infected F9 cells was due to post-transcriptional control. In their case, however, they detected a small amount of unspliced RNA homologous to the SV40 early region and no spliced SV40 mRNA. Because most of the exogenously added SV40 DNA in infected F9 cells is probably not integrated into chromosomal DNA, it is possible that the expression of free viral DNA is regulated differently from that of integrated SV40 genomes.

Since we have shown previously, using the same methods for detection of RNA and the same 12-1 stem cells, that no RNAs homologous to H-2 and β-2 microglobulin genes are detectable in stem cells (Croce et al., 1981), it is clear that the teratocarcinoma-derived stem cells exert control over gene expression by at least two different mechanisms: one transcriptional and one post-transcriptional.

We are currently constructing various other viral-DNA vectors for transformation of F9TK⁻ cells for the purpose of describing similar or new control mechanisms of differential gene expression during the differentiation process.

References

Alwine, J. C., Kemp, D. J., Parker, B. A., Reiser, J., Renart, J., Stark, G. A., and Wahl, G. M., 1979, Detection of specific RNAs or specific fragments of DNA by fractionation in gels and transfer to diazobenzyloxymethyl paper, in: *Methods in Enzymology*, Volume 68 (R. Wu, ed.), Academic Press, New York, pp. 220–242.

Aviv, H., and Leder, P., 1972, Purification of biologically active globin messenger RNA by chromatography on oligothymidylic acid-cellulose, *Proc. Natl. Acad. Sci. USA* **69**:1408–1412.

Bantle, J. A., Maxwell, I. H., and Hahn, W. E., 1976, Specificity of oligo (dT)-cellulose chromatography in the isolation of polyadenylated RNA, *Anal. Biochem.* **72**:413–427.

Berger, S., and Birkenmeir, C., 1979, Inhibition of intractable nucleases with ribonucleoside-vanadyl complexes: Isolation of messenger ribonucleic acid from resting lymphocytes, *Biochemistry* **18**:5143–5149.

Berk, A. J., and Sharp, P. A., 1977, Sizing and mapping of early adenovirus mRNAs by gel electrophoresis of S1 endonuclease digested hybrids, *Cell* **12**:721–732.

Berk, A. J., and Sharp, P. A., 1978, Spliced early mRNAs of simian virus 40, *Proc. Natl. Acad. Sci. USA* **75**:1274–1278.

Bolivar, F., Rodriguez, R., Green, P., Betlach, M., Heyneker, H., Boyer, H., Crosa, I., and Falkow, S., 1977, Construction and characterization of new cloning vehicles. II. A multi-purpose cloning system, *Gene* **2**:95–113.

Carmichael, G., and McMaster, G., 1980, The analysis of nucleic acids in gels using glyoxal and acridine orange, in: *Methods in Enzymology*, Volume 65 (L. Grossman and K. Maldowe, eds.), Academic Press, New York, pp. 380–391.

Clewell, D. E., and Helsinki, D. R., 1960, Supercoiled DNA–protein complex in *Escherichia Coli*: Purification and induced conversion to an open circular DNA form, *Proc. Natl. Acad. Sci. USA* **62**:1159–1166.

Croce, C. M., Linnenbach, A., Huebner, K., Parnes, J., Margulies, D., Appela, E., Seidman, J., 1981, Control of expression of histocompatibility antigens (H-2) and β_2-microglobulin in F9 teratocarcinoma stem cells, *Proc. Natl. Acad. Sci. USA* **78**:5754–5758.

Curtiss, R., III, Pereira, D. A., Hsu, J. C., Hull, S. C., Clark, J. F., Maturin, L. J., Sr., Goldschmidt, R., Moody, R., Inoue, M., and Alexander, L., 1977, Biological containment: The subordination of *Escherichia coli* K-12, in: *Recombinant Molecules: Impact on Science and Society* (R. F. Beers and E. G. Bassett, eds.), Raven Press, New York, pp. 45–56.

Duesberg, P., and Vogt, P., 1973, Gel electrophoresis of avian leukosis and sarcoma viral RNA in formamide: Comparison with the other viral and cellular RNA species, *J. Virol.* **12**:594–599.

Favalora, J., Triesman, R., and Karmen, R., 1980, Transcriptional maps of polyoma virus-specific RNA: Analysis by two-dimensional nuclease S1 gel mapping, in: Methods in Enzymology, Volume 65 (L. Grossman and K. Maldowe, eds.), Academic Press, New York, pp. 718–749.

Glisin, V., Crkvenjakov, R., and Byus, C., 1974, Ribonucleic acid isolation by cesium chloride centrifugation, Biochemistry **13**:2633–2637.

Gmür, R., Solter, D., and Knowles, B. B., 1980, Independent regulation of H-2K and H-2D gene expression in murine teratocarcinoma somatic cell hybrids, J. Exp. Med. **151**:1349–1359.

Graham, F. L., and Van der Eb., A. J., 1973, A new technique for the assay of infectivity of human adenovirus 5 DNA, Virology **52**:456–567.

Grunstein, M., and Hogness, D. S., 1975, Colony hybridization: A method for the isolation of cloned DNAs that contain a specific gene, Proc. Natl. Acad. Sci. USA **72**:3961–3965.

Hirt, B., 1967, Selective extraction of polyoma DNA from infected mouse cell cultures, J. Mol. Biol. **26**:365–369.

Huebner, K., Linnenbach, A., Weidner, S., Glenn, G., and Croce, C. M., 1981, Deoxyribonuclease I sensitivity of plasmid genomes in teratocarcinoma-derived stem and differentiated cells, Proc. Natl. Acad. Sci. USA **78**:5071–5075.

Hughes, E. W., and August, T. T., 1981, Characterization of plasma membrane proteins identified by monoclonal antibodies, J. Biol. Chem. **256**:664–671.

Knowles, B. B., Pan, S. H., Solter, D., Linnenbach, A., Croce, C. M., and Huebner, K., 1980, Expression of H-2 laminin and SV40T and TASA on differentiation of transformed murine teratocarcinoma cells, Nature **288**:615–618.

Laemmli, U. K., 1970, Cleavage of structural proteins during the assembly of the head of bacteriophage T4, Nature **227**:680–685.

Linnenbach, A., Huebner, K., and Croce, C. M., 1980, DNA-transformed murine teratocarcinoma cells: Regulation of expression of simian virus 40 tumor antigen in stem versus differentiated cells, Proc. Natl. Acad. Sci. USA **77**:4875–4879.

Linnenbach, A., Huebner, K., and Croce, C. M., 1981, Transcription of the simian virus 40 genome in DNA-transformed murine teratocarcinoma stem cells, Proc. Natl. Acad. Sci. USA **78**:6386–6390.

Linzer, D. I., and Levine, A. I., 1979, Characterization of a 54K Dalton cellular SV40 tumor antigen present in SV40 transformed cells and uninfected embryonal carcinoma cells, Cell **17**:43–52.

Littlefield, J. W., 1965, The use of drug-resistant markers to study the hybridization of mouse fibroblasts, Exp. Cell. Res. **41**:190–196.

Maniatis, T., Kee, S. G., Efstratiadis, A., and Katatos, F. C., 1976, Amplification and characterization of a β-globin gene synthesized in vitro, Cell **8**:163–182.

Martin, G., and Evans, M. J., 1975, Multiple differentiation of clonal teratocarcinoma stem cells following embryonal body formation in vitro, Cell **6**:467–474.

Martinis, J., and Croce, C. M., 1978, Somatic cell hybrids producing antibodies specific for the tumor antigen of simian virus 40, Proc. Natl. Acad. Sci. USA **75**:2320–2323.

Maxam, A. M., and Gilbert, W., 1977, A new method for sequencing DNA, Proc. Natl. Acad. Sci. USA **74**:560–564.

McKnight, S., 1980, The nucleotide sequence and transcript map of the herpes simplex virus thymidine kinase gene, Nucleic Acids Res. **8**:5949–5964.

McKnight, S., Croce, C. M., and Kingsbury, R., 1979, Introduction of isolated DNA sequences into cultured eukaryotic cells, Carnegie Institute Yearbook **78**:56–61.

Migeon, B., Smith, S. W., and Leddy, C., 1969, The nature of thymidine kinase in the human–mouse hybrid cell, Biochem. Genet. **3**:583–590.

Pope, J. H., and Rowe, W. P., 1964, Detection of specific antigen in SV40 transformed cells by immunofluorescence, J. Exp. Med. **120**:121–127.

Radloff, R., Bauer, W., and Vinograd, J., 1967, A dye-buoyant-density method for the detection and isolation of closed circular duplex DNA: The closed circular DNA in Hela cells, Proc. Natl. Acad. Sci. USA **57**:1514–1522.

Scherrer, K., 1969, Isolation and sucrose gradient analysis of RNA, in: Fundamental Techniques in Virology (K. Habel and N. Salzman, eds.), Academic Press, New York, pp. 413–432.

Segal, S., and Khoury, G., 1979, Differentiation as a requirement for simian virus 40 gene expression in F9 embryonal carcinoma cells, Proc. Natl. Acad. Sci. USA **76**:5611–5616.

Southern, E., 1975, Detection of specific sequences among DNA fragments separated by gel electrophoresis, *J. Mol. Biol.* **98**:503–517.

Spencer, N., Hopkinson, D. A., and Harris, H., 1964, Phosphoglucomutase polymorphism in man, *Nature* **204**:742–745.

Strickland, S., and Mahdavi, V., 1978, The induction of differentiation in teratocarcinoma stem cells by retinoic acid, *Cell* **15**:393–403.

Thomas, P., 1980, Hybridization of denatured RNA and small DNA fragments transferred to nitrocellulose, *Proc. Natl. Acad. Sci. USA* **77**:5201–5205.

Tonegawa, S., Brack, C., Hozumi, N., and Schuller, R., 1977, Cloning of an immunoglobulin variable region gene from mouse embryo, *Proc. Natl. Acad. Sci. USA* **74**:3518–3522.

Wigler, M., Silverstein, S., Lee, L., Pellicer, A., Cheng, Y., Axel, R., 1977, Transfer of purified herpes virus thymidine kinase gene to cultured mouse cells, *Cell* **11**:223–232.

Wigler, M., Pellicer, A., Silverstein, S., Axel, R., Urlaub, G., and Chasin, L., 1979, DNA-mediated transfer of the adenine phosphoribosyltransferase locus into mammalian cells, *Proc. Natl. Acad. Sci. USA* **75**:1373–1375.

Chapter 31

Detection of Specific DNA Sequences in Somatic Cell Hybrids and DNA Transfectants

NANCY HSIUNG and RAJU KUCHERLAPATI

1. Introduction

One of the methods to study gene expression in mammalian cells is to introduce the desired gene or genes into an appropriate recipient mammalian cell. There are several methods to introduce foreign genes into mammalian cells. These include somatic cell hybridization of whole cells, fusion of whole cells with microkaryoplasts, chromosome transfer, chromatin transfer, and DNA transfer.

Somatic cell hybrids can be classified as being intra- or interspecific in nature. In both cases whole cells are fused. The intraspecific hybrids are characterized by a high degree of chromosomal stability, while the interspecific hybrids lose the chromosomes of one of the species in a gradual and random fashion. The latter hybrids have proved to be extremely useful for gene mapping purposes. The rationale of the gene mapping procedure is to correlate the expression of presence of a species-specific gene with other genes or with individual chromosomes. The availability of gene-specific nucleic acid probes makes it possible to establish the presence of a gene without regard to its expression. Hybrids generated from fusions of microkaryoplasts (microcells) can be used in a similar fashion.

DNA-mediated transfer permits introduction of defined sequences into mammalian cells. Different methods of introducing DNA into mammalian cells are described elsewhere in this volume. In this chapter we describe the methods for detecting DNA in cells containing foreign DNA sequences.

Different methods are used to detect foreign DNA sequences in mammalian cells. One of them is *in situ* hybridization. This method is most useful if several copies of the DNA to be detected are tandemly arrayed or are located

---

NANCY HSIUNG and RAJU KUCHERLAPATI • Department of Biochemical Sciences, Princeton University, Princeton, New Jersey 08544.

close to each other, although a recent report (Harper et al., 1981) indicates the feasibility of this procedure to detect single-copy genes.

The second method is liquid hybridization. The first successful use of this method for gene mapping purposes was described by Deisseroth and colleagues (1977). In their experiments labeled human α-globin cDNA was used as a probe to detect complementary sequences in mouse–human hybrids. This led to assigning the human α-globin gene to chromosome 16. The drawback of these experiments is that a relatively large amount of cellular DNA is necessary for hybridization. The development of blot-hybridization procedures (Southern, 1975) makes such analysis less complex. In this procedure the DNA from the cell line to be tested is digested with a restriction endonuclease, fractionated by gel electrophoresis, denatured, transferred to nitrocellulose, and hybridized with DNA which is labeled to a very high specific activity by nick translation. Autoradiography of the filter reveals bands characteristic of the probe utilized. Since this is the most widely used procedure for detecting foreign DNA sequences in mammalian cells, we describe the methodology employed in the various steps, including DNA isolation, fractionation, labeling, and hybridization.

2. Methods

2.1. Isolation of DNA

2.1.1. Cellular DNA [Modified after Weintraub and Groudine (1976) and Pellicer et al. (1978)]

Materials
1. 0.89% NaCl (saline).
2. Nuclei isolation buffer: 10 mM Tris-Hcl pH 7.5, 10 mM NaCl, 5 mM $MgCl_2$, 0.5% Nonidet P-40, 1 mM phenylmethylsulfonylfluoride (add fresh).
3. Nuclei lysis buffer: 10 mM Tris-HCl pH 8.2, 400 mM NaCl, 2 mM EDTA.
4. TE buffer: 10 mM Tris-HCl pH 8.0, 1 mM EDTA.
5. Stock solutions: (a) 10% SDS, (b) 1 M NaCl, (c) 10 mg/ml pronase, aliquots stored at −20°C (do not refreeze), (d) 2 mg/ml ribonuclease; boiled in water for 15 min, frozen in aliquots.
6. Distilled phenol, buffered in 0.1 M Tris pH 8.2, 10 mM EDTA, chloroform/isoamyl alcohol (24:1), ethanol, stored at −20°C.

Procedure
1. Nuclei Isolation
(a) Grow cells to confluency. Use 4–5 150-mm petri dishes (approximately 10^8 cells).
(b) Pour off media and rinse each dish with 0.89% NaCl. Add 4 ml of 0.89% saline and scrape off cells with a rubber policeman. Transfer cells to a centrifuge tube. Rinse each plate with a few milliters of saline and add to cell suspension.
(c) Centrifuge at 1000 rpm for 10 min. All remaining steps in the nuclei isolation procedure are done at 4°C.

(d) Pour off supernatant and resuspend in 5 ml of nuclei isolation buffer.

(e) Homogenize the cells with 15 strokes of the B pestle. Monitor nuclei isolation under the microscope.

(f) Transfer solution to centrifuge tube and pellet nuclei by centrifugation at 2000 rpm for 8 min.

(g) Pour off supernatant and resuspend in 4.75 ml of nuclei lysis buffer, making sure all the nuclei are resuspended.

2. DNA Isolation

(a) Add 250 μl of 10% SDS so the final concentration of the solution is 0.5% SDS. The nuclei should lyse and the release of DNA will cause the solution to become viscous.

(b) Add 100 μl of 10 mg/ml pronase to a final concentration of 200 μg/ml. Incubate at 37°C for at least 4 hr or until the solution is clear. If desired, the mixture can be incubated overnight.

(c) Dilute the solution to 7–8 ml with TE buffer. Add an equal volume of phenol and shake the tube thoroughly to mix the two layers. Centrifuge at 1500 rpm for 5 min to separate the layers. Remove the phenol layer (bottom layer) and discard.

(d) Add an equal volume of phenol to the remaining aqueous layer and repeat step (c).

(e) Add an equal volume of chloroform/isoamyl alcohol and repeat step (c).

(f) Repeat the chloroform/isoamyl alcohol step to remove all traces of phenol. After centrifugation, carefully remove aqueous (top) layer and transfer to clean tube.

(g) Add two volumes of ice-cold ethanol and gently shake the tube until the DNA forms a fluffy precipitate. Spool the DNA onto a glass rod (or sealed pasteur pipette). Dry 1–2 hr.

(h) Add 5 ml of TE buffer and resuspend the DNA by shaking at 4°C overnight or until dissolved.

(i) Add 100 μl of ribonuclease to the DNA (final ribonuclease concentration 40 μg/ml). Incubate at 37°C for at least 1 hr (up to 4 hr).

(j) Add pronase to a final concentration of 200 μg/ml and SDS to 0.5%. Incubate at 37°C for at least 2 hr or overnight.

(k) Add NaCl to a concentration of 0.3 M. Repeat steps (c)–(g).

(l) Resuspend DNA in 1–2 ml of TE buffer and measure the concentration. The DNA can be stored at 4°C.

Note. The yield ranges from 400 μg to 1 mg, depending on the number of cells at the beginning of the preparation.

2.1.2. Plasmid DNA [Modified after Clewell and Helinski (1969), Betlach et al. (1976), and Norgard et al. (1979)]

Materials
1. Growth of Bacteria (all solutions must be sterile)
(a) Media

(i) M9, 1 liter (use 2-liter flasks): Na_2PO_4, 7 g; KH_2PO_4, 3 g; NH_4Cl, 1 g; NaCl, 0.5 g; 0.01 M $FeCl_3$, 0.3 ml. Autoclave for 30 min.

(ii) To 1 liter of sterile M9, add: 1 ml of $MgSO_4$, one drop 1 M $CaCl_2$, 40 ml of 5% glucose, 40 ml 12.5% Casamino acids (filter-sterilize).

(b) Uridine 0.5 g/liter of M9.

(c) Chloramphenicol 0.2 g/liter of M9.

2. Plasmid Isolation

(a) Cell lysis

(i) TE Buffer: 10 mM Tris-HCl pH 8.0, 1 mM EDTA.

(ii) 10 mg/ml lysozyme in 0.25 M Tris-HCl pH 7.5

(iii) 250 mM EDTA pH 8.0.

(iv) Triton solution: 50 mM Tris-HCl pH 8.0, 62.5 mM EDTA, 0.1% Triton X-100.

(b) DNA isolation

(i) 10 mg/ml ethidium bromide.

(ii) Cesium chloride 26.13 g/liter of cells.

(iii) Dowex AG50W-X4 (100-200 mesh) in TE buffer.

(iv) Ethanol, stored at $-20°C$.

(v) 70% ethanol in 10 mM Tris-HCl pH 8.0, 2 mM EDTA, stored at $-20°C$.

Procedure

1. Growth of bacteria

(a) Transfer a single colony to a 50-ml culture in a 150-ml flask. Shake overnight at 37°C.

(b) Transfer the 50-ml culture to 1 liter of M9 media. Shake at 37°C until bacteria reach an $OD_{600} = 0.1$.

(c) Add 500 mg of uridine/liter suspension culture (see Note 1)

(d) Shake at 37°C until bacteria reach an $OD_{600} = 0.6$.

(e) Add 200 mg of chloramphenicol and shake the culture at 37°C for 14–17 hr

2. Plasmid Isolation

(a) Cell Lysis

(i) Add 2 ml of chloroform to stop bacterial growth. Using large centrifuge tube, spin cells at 5000 rpm for 5 min at 4°C.

(ii) Pour off supernatant and wash cells with 1/4 volume of TE buffer. Centrifuge at 5000 rpm for 5 min at 4°C.

(iii) Resuspend cells in 9 ml of TE buffer. Add 0.9 ml of 10 mg/ml lysozyme. Incubate on ice for 5 min with occasional shaking.

(iv) Add 3.6 ml of 250 mM EDTA. Incubate on ice for 5 min.

(v) Add 14.5 ml of Triton X-100 solution. Incubate on ice for 5 min with occasional shaking. The cells should lyse and the solution becomes extremely viscous.

(vi) Centrifuge at 15,000 rpm for 90 min at 4°C. Alternatively, the centrifugation can be carried out in a high-speed centrifuge at 25,000 rpm for 30 min at 4°C using an SW40, SW41, or SW27 rotor.

(vii) Decant the supernatant until you reach the very viscous material above the pellet (about 1/2 inch above pellet). If supernatant is nonviscous, decant all the liquid.

(b) DNA Isolation

(i) Adjust the volume of the supernatant to 27.5 ml with TE buffer and add 26.13 g of cesium chloride. Thoroughly dissolve. The refractive index should be between 1.390 and 1.396.

(ii) Add 2.75 ml of 10 mg/ml ethidium bromide and spin in a 60Ti rotor at 35,000–38,000 rpm at 20°C for 48–72 hr (Note 2).

(iii) View bands by illuminating the tube with long-wavelength UV. The lower band contains the supercoiled plasmid DNA, which may be collected by side puncture of the tube with a 21-gauge needle.

(iv) The DNA (usually 2–3 ml) is diluted 3× with TE buffer. To remove the ethidium bromide, pass the solution over a 2-ml column of Dowex AS50W-X4 in TE buffer (Note 3).

(v) Add 2–2½ volumes of ethanol and mix. Precipitate at −20°C for at least 5 hr. Spin at 15,000 rpm for 30 min. Pour off supernatant and wash pellet (do not resuspend) in 70% ethanol. Centrifuge at 15,000 rpm for 15 min. Decant supernatant and dry pellet 1–2 hr (Note 4).

(vi) Resuspend plasmid DNA in TE buffer. The yield should range between 800µg and 1.2 mg.

(vii) Check the DNA by agarose gel electrophoresis.

Note 1. Addition of the uridine retards maximal growth of bacteria by altering the optimal ratios of nucleotides necessary for bacterial replication. The cells at OD_{600} can range between 0.07 and 0.13 when the uridine is added.

Note 2. The longer the centrifugation, the less RNA contaminates the plasmid DNA. Bands may be visible in 24 hr. The use of vertical rotors will reduce the time of centrifugation.

Note 3. A siliconized Pasteur pipette is sufficient to hold the 2-ml Dowex column. The DNA solution is passed through the column and the liquid is collected until the column runs dry.

Note 4. If there is a lot of salt in the ethanol/water mixture, centrifuge as described and resuspend pellet in TE buffer. Dialyze the DNA against TE buffer for at least 3 hr to dilute the cesium chloride. After dialysis of the DNA is completed, ethanol-precipitate the plasmid DNA as described in step (v).

2.2. Analysis of DNA

2.2.1. Restriction Enzyme Digestion

Materials

1. Restriction endonucleases, store at −20°C; e.g., *Bam*H1.
2. 10× stock solutions of enzyme digestion buffers, store at 4°C (see Note); e.g., *Bam*H1. 10× stock solution: 300 mM Tris-HCl pH 7.5, 70 mM $MgCl_2$, 1 M NaCl, 20 mM B-mercaptoethanol.

Procedure

1. Plasmid DNA or DNA fragments: Add 1 U of enzyme for 1 µg of plasmid DNA in 25–50 µl of 1× enzyme digestion buffer; e.g., *Bam*H1 digestion: Add 2 U of *Bam* H1 restriction enzyme to 2 µg of plasmid DNA in 30 µl of 1× *Bam* H1

digestion buffer. Incubate at 37°C for 1–3 hr. Stop reaction by adding EDTA (final concentration 10 mM) or by freezing at −20°C.

2. Cellular DNA: Add 1–2 µl of enzyme for 1 µg of cellular DNA in 25–50 µl 1× enzyme digestion buffer; e.g., BamH1 digestion: Add 7.5 U of BamH1 enzyme to 5 µg of cellular DNA in 30 µl of 1× BamH1 digestion buffer. Incubate at 37°C for at least 5 hr up to 15 hr (overnight). Stop reaction by 10 mM EDTA or freezing at −20°C.

Note. It is a good policy to sterilize stock solutions and pipette tips, although daily use does not need sterile conditions.

2.2.2. Agarose Gel Electorphoresis [Modified after McDonell et al. (1977)]

Materials
1. Stock Solutions
   (a) 10× acetate buffer: 400 mM Tris-HCl pH 7.9, 200 mM sodium acetate, 20 mM EDTA.
   (b) 1% bromophenol blue.
2. Final solutions
   (a) Running buffer (1× acetate buffer); gel buffer: 40 mM Tris-HCl pH 7.9, 20 mM sodium acetate, 2 mM EDTA.
   (b) Sample buffer: 50% glycerol, 0.1% Bromphenol Blue in 1× acetate buffer.
   (c) 1 mg/ml ethidium bromide.
3. Agarose

Procedure
   (a) Autoclave (or boil in H_2O bath) 1.6 g of agarose in 180 ml of distilled water (0.8% agarose) for approximately 15 min. When agarose is dissolved, add 20 ml of 10× acetate buffer and cool to 55–60°C (Note 1).
   (b) While autoclaving agarose, set comb in horizontal gel apparatus (Note 2).
   (c) Pour agarose carefully into gel apparatus. Remove any air bubbles with a Pasteur pipette. Let agarose set for at least 30 min at room temperature.
   (d) Add running buffer until the gel is completely covered. Remove the comb gently, allowing the buffer to displace the teeth of the comb. The gel can be stored for hours if kept moist.
   (e) Prepare DNA samples by adding sample buffer to each DNA. For a 25-µl DNA sample, add 6–7 µl sample buffer (at least ¼ volume of the DNA).
   (f) Each DNA sample is loaded into a well with a clean pipette tip. Otherwise cross-contamination may result.
   (g) Run the samples into the gel at 45–55 V; this takes about 15 min. For DNA samples in acetate buffer, the direction of electrophoresis is from the cathode to anode.
   (h) When samples have migrated into the gel, reduce the voltage to approximately 2 V/cm (length of gel). For a 20-cm gel, electrophoresis at 40 V is usually completed in 14–16 hr (Note 3). The progress can be monitored by noting the migration of bromophenol blue.

(i) When electrophoresis is complete, stain the gel in 0.5 µg/ml ethidium bromide for at least 30 min. Destain in water for 10 min.

(j) Photograph gel under UV illumination.

Note 1. The percentage agarose used depends on the size of the DNA which is to be fractionated. (a) Genomic DNA, 0.5–0.8% agarose. (b) Plasmid DNA or DNA fragments (1–8 kb), 1% agarose.

Note 2. The amount of DNA loaded in each well depends on the volume of the wells. (a) Width of 3–3.5 mm, 10–15 µg of DNA/well. (b) Width of 1–1.4 mm, ⩽6 µg of DNA/well.

Note 3. An average horizontal gel is 20 cm in length. Running time for a 0.8% agarose gel varies: 30 V, 16–18 hr; 40 V, 14–16 hr; 50 V, 12–14 hr.

2.3.3. DNA Blotting [Modified after Southern (1975)]

Materials

1. Solutions
(a) 20× SSC pH 7.0: 3.0 M sodium chloride, 0.3 M sodium citrate.
(b) 2× SSC pH 7.0: 0.3 M sodium chloride, 0.03 M sodium citrate.
(c). Denaturation buffer: 1.5 M sodium chloride, 0.5 M sodium hydroxide.
(d) Neutralization buffer pH 7.9: 3.0 M sodium chloride, 0.5 M Tris–HCl.

2. Hardware
(a) Nitrocellulose paper (0.45 µm pore size).
(b) Whatman 3MM paper.
(c) Glass plates: one approximately 25 cm × 25 cm; one approximately 20 cm × 20 cm.
(d) Plastic strips (20 cm × 2 cm × 0.3 cm) or 1ml disposable pipettes (two pipettes can replace one plastic strip).
(e) Paper towels, cut to the dimension of the gel.

Procedure (Perform all operations using gloves)

1. Remove the nonessential parts of the gel so that only the section of the gel to be blotted remains. Measure this portion of the gel (Note 1).

2. Immerse the gel in 400–500 ml of denaturation buffer. Shake gently at room temperature for 45 min (up to 1½ hr).

3. Rinse gel with distilled water and transfer to 400–500 ml of neutralization buffer. Shake gently at room temperature for at least as long as the gel was denatured (45 min to 1½ hr).

4. While the gel is being processed, prepare the following:

(a) Cut a sheet of nitrocellulose the same length and 3–4 cm wider than the gel segment to be blotted. Immerse the nitrocellulose in 2× SSC by first floating it on the surface of the buffer and then submerging it when hydration is complete.

(b) Cut three Whatman 3MM sheets the same size or slightly smaller than the gel.

(c) Place the 25 cm × 25 cm glass plate on a raised support in a large dish containing 20× SSC. The 20× SSC should always remain below the level of the glass plate. Cut a sheet of Whatman 3MM approximately 25 cm wide and 40 cm long. Wet the sheet in 2× SSC and place over glass plate so that both ends hang over the edges of the plate in such a manner that wicks in contact with the 20× SSC are formed. Smooth down any air bubbles between the Whatman 3MM paper and the plate.

5. Transfer the neutralized gel to the Whatman 3MM covered glass plate and remove any air bubbles. Place the plastic strips or pipettes next to both sides of the gel.

6. Lay the moistened nitrocellulose sheet on top of the gel with the edges of the filter resting on the plastic strips or pipettes. It is essential to remove all air bubbles between the gel and the nitrocellulose paper in order to obtain complete transfer of DNA to all sections of the filter.

7. Place the Whatman 3MM sheets (previously cut to the size of the gel and moistened in 2× SSC) one by one on top of the nitrocellulose paper. Remove any air bubbles by carefully smoothing away with a gloved hand.

8. Stack about 10 cm of paper towels on top of the Whatman 3M sheets, making sure that the edges of the paper towels do not exceed the boundaries of the nitrocellulose sheet (Note 2).

9. Place plastic wrap over the towels enclosing the blotting apparatus to prevent evaporation of the buffer. Place the 20 cm × 20 cm glass plate on top and lay a weight (approximately 400–500 g) on top to promote contact throughout the layers. When the paper towels are saturated, replace with a fresh batch (usually after 6–8 hr). Add more 20× SSC to the dish if needed and allow transfer to continue overnight (Note 3).

10. After DNA transfer has been completed, cut the nitrocellulose flush with the edges of the gel, peel off the gel, and immerse the nitrocellulose in 2× SSC for 5–10 min, swirling occasionally. In order to orient the filter with respect to the gel, cut a corner to mark the direction not only of migration but also of samples, e.g., clip lower right corner.

11. Remove nitrocellulose filter and place on absorbent paper to dry. Allow to air dry for at least 4 hr (Note 4).

12. Wrap filter in aluminum foil and bake in vacuum oven for 2–4 hr at 80°C. The filter may be stored at −20°C indefinitely before use.

Note 1. A typical gel size to be blotted is 16 cm × 12 cm.

Note 2. It is important to prevent any contact of the stacked paper above the nitrocellulose sheet with the paper on the glass plate. If this happens, the towels will draw the 20× SSC directly and bypass elution of the DNA through the gel onto the nitrocellulose paper.

Note 3. For a gel (16 cm × 12 cm) containing genomic DNA, overnight transfer is preferred. Usually 200–300 ml of 20× SSC is used in the elution process. When transferring plasmid DNA instead of total cellular DNA, transfer is usually completed in 8–10 hr.

Note 4. Air drying for 4–16 hr ensures that the filter is dry and will not stick to the aluminum foil or allow reannealing of DNA on the filter during the baking process. At this point, the filter can be marked with a ball point pen; the ink will survive the oven and hybridization processes.

2.2.4. Nick Translation of DNA [Modified after Rigby et al. (1977)]

Materials

1. Reaction Solutions

   (a) 10× nick translation buffer: 0.5 M Tris-HCl pH 7.9, 0.05 M $MgCl_2$, 0.01 M dithiothreitol, 0.5 mg/ml bovine serum albumin.

   (b) Stock cold triphosphates: 100 µM dCTP, 100 µM dTTP, 100 µM dATP, 100 µM dGTP.

   (c) 0.1 µg/ml deoxyribonuclease I, aliquots stored at $-20°C$.

   (d) *E. coli* DNA polymerase (4–5 U/µl stock).

   (e) ^{32}P-triphosphates, usually 600–800 Ci/mM.

   (f) 20% trichloroacetic acid (TCA).

   (g) 250 mM EDTA pH 8.0.

   (h) 10% SDS.

   (i) 1 mg/ml salmon sperm DNA.

2. Probe Isolation

   (a) TE buffer: 10 mM Tris–HCl pH 8.0, 1 mM EDTA.

   (b) Sephadex G-75 in TE buffer.

   (c) Phenol, buffered in 0.1 M Tris-HCl pH 8.0, 10 mM EDTA.

Procedure

1. Reaction

   (a) In 20 µl: DNA to be labeled (0.5 µg); 10× nick translation buffer, 2 µl; deoxyribonuclease I, 1 µl. Incubate at 37°C for 8 min, transfer to 4°C.

   (b) In 80 µl: 10 µl dATP, 10 µl dGTP, 8 µl 10× nick translation buffer, 100 µCi ^{32}P-dCTP, 100 µCi ^{32}P-dTTP. Add this mixture to the DNA on ice (Note 2).

   (c) Add 1 µl DNA polymerase. Incubate at 15°C for 30 min up to 2 hr. Monitor the reaction by determining trichloroacetic acid precipitable counts. The average reaction time is about 1 hr (Note 3).

   (d) Stop the reaction in 50 mM EDTA and 0.1% SDS.

2. Labeled DNA Isolation

   (a) Dilute the reaction mixture to 200 µl with TE buffer. Add an equal volume of phenol and then thoroughly mix the two layers. Centrifuge at 1500 rpm for 5 min. Remove aqueous (top) layer and save.

   (b) Add 100 µl of TE buffer to phenol layer. Mix and centrifuge 1500 rpm for 5 min. Remove aqueous layer and combine with first aqueous extract.

   (c) Make a Sephadex G-75 column (in TE buffer) in a 10-ml disposable pipette (usually about 25 cm long). The radioactive probe is separated from free radioactive nucleotides by passing the aqueous mixture through the Sephadex G-75. The separation can be followed by monitoring with a Geiger counter. The first radioactive peak is the nick-translated probe.

   (d) The radioactive DNA probe should be stored at 4°C until used for hybridization.

Note 1. (a) For nick translation under these conditions, the amount of DNA can vary from 0.2 to 1 µg.

(b) For small DNA fragments (<1 kb), the deoxyribonuclease reaction can be eliminated.

Note 2. (a) Nick-translating DNA with two radioactive labels is routine, although using one label is satisfactory, especially if a high specific activity is not required.

(b) Under these nick translation conditions, the range of DNA varies from 0.2 to 1 µg. For 3–5 standard blots (6–8 lanes of samples per filter) approximately 0.5–0.6 µg of DNA is nick-translated with 200 µCi of total radioactive label.

(c) Under these reaction conditions, it is not necessary to add unlabeled dCTP and dTTP. If a higher amount of DNA is to be labeled, 2 µM of cold triphosphates should be added to supplement their equivalent radioactive triphosphates.

Note 3. The reaction is monitored by taking 1-µl aliquots at 0 time and at 20 min intervals until the reaction levels off or the desired specific activity is reached. Add each aliquot to 50 µl of carrier (i.e., 1 mg/ml bovine serum albumin or salmon sperm DNA). Spot 10 µl onto a glass fiber and count. Precipitate the remaining 40 µl with 2 ml of 20% TCA, filter, and count. Percent incorporation = (one-fourth TCA precipitable counts)/(soluble counts in 10 µl). Incorporation should range between 20 and 60%, depending on the purity and type of DNA. The range of specific activity is $5 \times 10^7 - 3 \times 10^8$ cpm/µg for DNAs varying from small fragments and plasmids to total cellular DNA.

2.2.5. Filter Hybridization and Autoradiography [Modified after Botchan et al. (1976)]

Materials
1. Filters.
2. Radioactive probe ($10^7 - 10^8$ cpm/µg).
3. Stock solutions:
(a) 20× SSC pH 7.0, 3.0 M sodium chloride, 0.3 M sodium citrate.
(b) 50× Denhardt's solution, 1% bovine serum albumin, 1% Ficoll, 1% polyvinylpyrrolidone. This solution is prone to fungal contamination. Aliquot and store at −20°C.
(c) 10% SDS.
(d) 5 mg/ml salmon sperm DNA (sonicated).
4. Final solutions:
(a) Prehybridization buffer: 10× Denhardt's solution, 6× SSC.
(b) Hybridization buffer: radioactive probe, 10× Denhardt's solution, 6× SSC, 100 µg/ml salmon sperm DNA.
(c) Wash solution: 2× SSC, 0.1% SDS.
5. Seal-a-meal (Sears) apparatus and bags or equivalent.
6. Autoradiography:
(a) Kodak XRP-5 film or Kodak XAR.
(b) Ilford or Dupont tungstate intensifying screens.
(c) Film exposure holders.
(d) Saran wrap or equivalent.

Detection of Specific DNA Sequences

Procedure

1. Hydrate filter by floating it on surface of 2× SSC and submerge when hydrated.

2. Cut a Seal-a-meal bag slightly larger than the filter on three sides and about 2 cm longer on the edge of the blot that is equivalent to the bottom of the gel. Seal to enclose the filter on three sides, leaving the extra 2 cm side open (Note 1).

3. Add approximately 5 ml of prehybridization buffer, squeeze out air bubbles, and seal, leaving room between the edge of the filter and the border of the bag. Incubate at 65–68°C for 4–24 hr.

4. Combine the nick-translated probe and sonicated salmon sperm DNA. Denature for 10 min at 100°C. Transfer to ice and add stock solutions to a final concentration of 10× Denhardt's and 6× SSC.

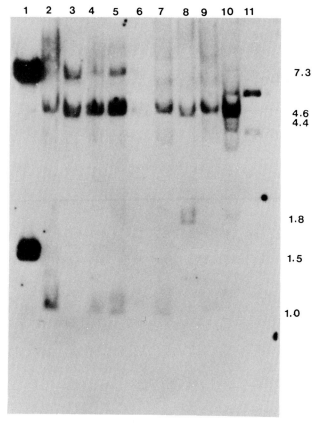

Figure 1. Detection of foreign DNA sequences in mammalian cells. Mouse LMTK⁻ cells were transfected with a plasmid containing pBR322 and a purified gene for herpes simplex virus *tk*. The cells were selected for *tk* expression. DNA from these cells is digested with EcoRI and blot-hybridized using labeled pBR322 as the probe. Lanes 2–10 contain various transfectants. Note the presence of a 4.6-kb band common to all transfectants, reflecting the presence of pBR322 sequences.

5. Cut a corner of the bag containing the filter and squeeze out the prehybridization solution. Add 3–4 ml of the hybridization solution, remove the air bubbles, and reseal. Incubate at 65°C for 30–48 hr, occasionally mixing the solution in the bag (either by rubbing the filter by hand or putting the blot on a temperature-controlled shaker) (Note 2).

6. When hybridization has been completed, remove the hybridization solution and transfer to a tube (Note 3).

7. Place the filter in a dish with 200–300 ml of wash solution. Incubate for 45 min at 65°C. Change the wash solution and repeat incubating at 65°C. The wash solution is changed at least three times. Check background radioactivity with a Geiger counter. The background can be lowered with a final 15 min wash at 65°C for 15 min in 0.1X SSC and 0.1% SDS.

8. Dry filter at room temperature for 1 hr. Cover the filter with Saran wrap and mount in film exposure holder.

9. Autoradiography is carried out by first placing a sheet of film on top of the filter and then a tungstate screen to enhance detection. Store at −70°C until ready for developing.

Note 1. Air bubbles causing hot spots on the filter can be minimized by trying to remove all bubbles by hand before sealing. A few are likely to remain, so by leaving room in the bag for air bubbles to collect, hot spots can be minimized.

Note 2. (a) Usually 3–4 ml of hybridization solution is used for a filter 16 cm $\times$ 6 cm. For genomic blots, try to keep the probe $>5 \times 10^6$ cpm/ml and with a specific activity $>5 \times 10^7$ cpm/μg. Incubation is at 65–68°C for 30–48 hr.

(b) For plasmid DNA (or DNA fragment) blots, amounts of radioactivity 10–100X less can be used. The incubation period can be reduced to 16–24 hr at 65–68°C.

Note 3. If the probe has a high specific activity, it can be reused. Denature the entire solution at 100°C for 10 min, cool on ice, and use as previously described.

A representative result obtained by the use of procedures described here is shown in Fig. 1.

References

Betlach, M. C., Hershfeld, V., Chow, L., Brown, W., Goodman, H. M., and Boyer, H. W., 1976, A restriction endonuclease analysis of the bacterial plasmid controlling the Eco RI restriction and modification of DNA, *Fed. Proc.* **35**:2037–2043.

Botchan, M., Topp, W., and Sambrook, J., 1976, The arrangement of Simian Virus 40 sequences in the DNA of transformed cells, *Cell* **9**:269–287.

Clewell, D. B., and Helinski, D. R., 1969, Supercoiled circular DNA–protein complex in *E. coli*: Purification and induced conversion to an open circular DNA form, *Proc. Natl. Acad. Sci. USA* **62**:1159–1166.

Deisseroth, A., Nienhuis, A., Turner, P., Velez, R., Anderson, W. F., Ruddle, F., Lawrence, J., Creagan, R., and Kucherlapati, R., 1977, Localization of the human α-globin structural gene to chromosome 16 in somatic cell hybrids by molecular hybridization assay, *Cell* **12**:205–218.

Harper, M. E., Ullrich, A., and Saunder, G. F., 1981, Localization of the human insulin gene to the distal end of the short arm of chromosome 11, *Proc. Natl. Acad. Sci.* **78**:4458–4460.

McDonnell, M., Simon, M. N., and Studier, W. F., 1977, Analysis of restriction fragments of T7 DNA and determination of molecular weights by electrophoresis in neutral and alkaline gels, *J. Mol. Biol.* **110**:119–146.

Norgard, M., Emigholz, K., and Monahan, J. J., 1979, Increased amplification of pBR322 plasmid deoxyribonucleic acid in *Escherichia coli* K-12 strains RRI and χ^{1776} grown in the presence of high concentration of nucleoside, *J. Bacteriol.* **138**:270–272.

Pellicer, A., Wigler, M., Axel, R., and Silverstein, S., 1978, The transfer and stable integration of the HSV thymidine kinase gene into mouse cells. *Cell* **14**:133–142.

Rigby, P. W. J., Dieckman, M., Rhodes, C., and Berg, P., 1977, Labeling DNA to high specific activity *in vitro* by nick translation with DNA polymerase I. *J. Mol. Biol.* **113**:237–251.

Southern, E. M., 1975, Detection of specific sequences among DNA fragments separated by gel electrophoresis, *J. Mol. Biol.* **98**:503–517.

Weintraub, H., and Groudine, M., 1976, Transcriptionally active and inactive confirmations of chromosomal subunits, *Science* **193**:848–856.

Chapter 32
Microinjection Turns a Tissue Culture Cell into a Test Tube

A. GRAESSMANN and M. GRAESSMANN

1. Introduction

The demand for techniques to investigate biologically important macromolecules (DNA, RNA, proteins) within the living cell has increased continually with our growing competence to purify, analyze, and modify them *in vitro*. Since these molecules are not taken up readily by a cell unless a specific recognition—internalization mechanism exists, ways to bypass cellular membranes efficiently without affecting the viability of a target cell have had to be found. For tissue culture cells, this goal has been achieved by an approach which uses glass micropipettes for physical injection of single culture cells. This microinjection technique was developed in our laboratory initially for the transplantation of cell nuclei (Graessmann 1968, 1970) and shortly thereafter adopted for the transfer of biologic macromolecules: for example, mRNA was injected into and translated in a foreign cell (Graessmann and Graessmann, 1971).

2. Procedure

2.1. General

The microinjection is performed under a phase contrast microscope (Fig. 1): a glass microcapillary, prefilled from the tip with the biologic sample, is directed into the cells to be injected with the aid of a micromanipulator, and an appropriate sample volume is transferred by gentle air pressure exerted by a syringe connected to the capillary. Recipient cells are grown on glass slides imprinted with numbered squares for convenient localization of the cells injected.

A. GRAESSMANN and M. GRAESSMANN • Institut für Molekular Biologie und Biochemie der Freien Universität Berlin, Berlin, West Germany.

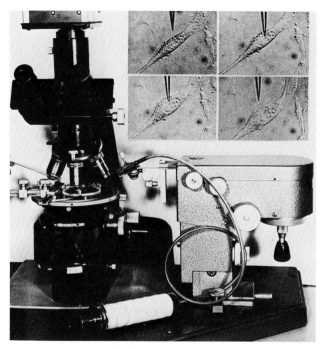

Figure 1. Assembled instruments for microinjection.

2.2. Preparation of Glass Microcapillaries

Glass tubes of 1.2 mm inner and 1.5 mm outer diameter (SGA Scientific, Bloomfield, New Jersey) are broken into pieces of about 30 cm in length and cleaned by treatment with a mixture of one-third concentrated H_2SO_4 and two-thirds concentrated HNO_3 (v/v) for 24 hr and by extensive washings with tap water, double-distilled water, and ethanol. Glass tubes pretreated in this way and air-dried at 120°C for 24 hr were kept in a closed container until use.

Before pulling of the capillary, a constriction is introduced at the middle of the glass tube. This is done by hand using the small flame of a bunsen burner: the glass is softened in the flame and the ends pulled apart outside, resulting in a constriction 8–15 mm in length and 0.3–0.5 mm in diameter (Fig. 2A). The glass tube is then clamped into a capillary puller, the middle of the constriction being surrounded by the heating wire (Fig. 2B). The shape of the capillary to be obtained is shown in Fig. 2C; the tip should be rigid and open with an outer diameter of about 0.5 μm. Conditions for the mechanical pulling must be optimized by varying the temperature of the heating wire and the pulling forces at the carriage. The puller shown in Fig. 2B was built in our workshop.

The capillary tips are further treated by connecting the capillary to a 50-ml syringe, dipping the tip into 50% HF for 1 sec, and then serially washing it by suction and pressure exerted by the syringe with double-distilled water,

Microinjection

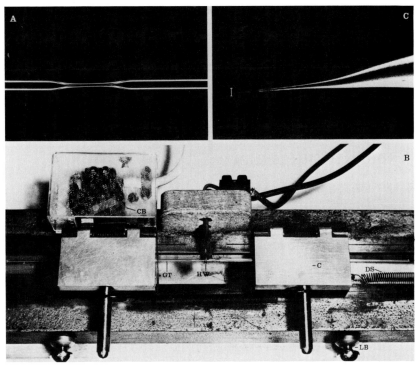

Figure 2. Pulling of glass microcapillaries. (A) Glass tube with constriction. (B) Partial view of the drawing apparatus with platinum heating wire (HW), glass tube (GT), carriages (C) connected to drawing springs with adjustable tension (DS), locking bolts (LB) to hold carriages in place before the glass tube is inserted, and circuit breaker (CB) to turn off the current when carriage moves. (C) Microcapillary tip. Bar: 10 µm.

ethanol, tetrahydrofuran (p.a.), 0.5% dichlorodimethylsilane in tetrahydrofuran (v/v), and again with tetrahydrofuran and ethanol. This treatment produces a smooth capillary tip, but for most purposes it may be omitted without affecting the microinjection procedure. Capillaries are conveniently stored in an upright position (tip end up) by inserting them into fitting holes pierced into a Perspex block. The block is placed into a petri dish and covered with a beaker. Capillaries are then air-dried at 120°C for 4 hr.

2.3. Preparation of Glass Slides

Glass slides (5 × 1 cm) are coated on both sides with a melted mixture of one-third beeswax and two-thirds stearin (w/w) using a cotton plug, resulting in a thin, supple film. This film is then scored (squares of 1 or 4 mm²) and numbered (Fig. 3) with a steel needle and the glass etched with a paste made from CaF_2 (precipitated) and 40% HF, which is spread over the slide and left there for 10–15 min. The slides are kept under running tap water overnight to split off the wax. Finally the slides are treated with a mixture of

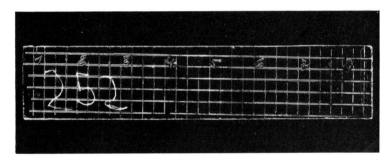

Figure 3. Scored and numbered glass slide.

one-third concentrated H_2SO_4 and two-thirds concentrated HNO_3 (v/v), washed with running tap water for 24 hr, followed by washings with double-distilled water and ethanol (p.a.). They are air-dried, wrapped in aluminum foil, sterilized in an oven at 200°C for 24 hr, and stored until use.

2.4. Cells

Cells are grown on plastic dishes or slides under standard conditions for cell culture. Cells growing in suspension culture become accessible to microinjection following binding them to a substrate by suitable linkers [e.g., concanavalin A, phytohemagglutinin P, poly L-lysine, pockweed mitogen, IgG] (Graessmann et al., 1980).

Figure 4 shows Friend leukemia cells unattached (top) and attached to the plastic surface of a Petri dish via concanavalin A (bottom). For this, a 60-mm petri dish was incubated for 2 hr with 2.5 ml of 2.5% glutaraldehyde at room temperature, washed four times with sterile water, and covered with 200 µl of concanavalin A (100 µg/ml) and kept for 1–2 hr at 37°C in an incubator. After drying, the dish was washed with PBS without Mg^{++} and Ca^{++}, and cells were plated.

2.5. Sample

Whenever feasible, the material to be injected is dissolved in injection buffer (0.048 M K_2HPO_4, 0.014 M NaH_2PO_4, 0.0045 M KH_2PO_4, pH 7.2), but other compositions, such as 0.01–0.1 M Tris-HCl, pH 7.2, will be tolerated by the cells. Concentrations up to 1 mg DNA, 5–10 mg RNA, or 10–20 mg protein per milliliter injection solution can be handled. The sample is centrifuged at 10,000–15,000g for 10 min directly before microinjection and then transferred as a small drop to a 60-mm plastic petri dish furnished with a moistened filter paper. The drop may be placed either directly on the plastic surface or on a small sheet of Parafilm, a volume of 2 µl being sufficient. The sample dish is kept on ice.

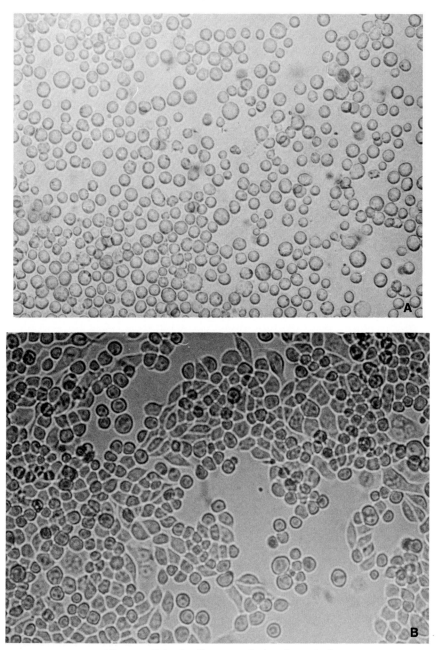

Figure 4. Friend cells unattached and attached.

2.6. Microinjection

A Leitz Ortholux microscope with phase contrast equipment (Phaco 10/0.25 objective lens, Periplan GF 16× oculars) and a Leitz micromanipulator (Fig. 1) are employed for microinjection in our laboratory. These instruments are placed in a vibration-free room reserved for the purpose which can be UV-sterilized for long-term experiments. The microcapillary is fixed to the instrument holder of the manipulator, and the tip focused under the microscope. Next the sample drop is brought into plane, and the needle is filled by capillary attraction forces supported by negative pressure exerted by the syringe. The capillary tip is slightly raised by turning the vertical adjustment knob of the manipulator, and the cells (immersed in medium in an open 60-mm petri dish) are placed on the microscope stage. A gentle stream of CO_2 maintains the pH optimum. Cells are brought into focus, and a distinct field is chosen for injection. The capillary is now lowered until it is nearly in focus, and an individual cell is approached by operating the manipulator lever for horizontal movements. The cell is injected by further lowering the capillary tip once it is directly above the cell. Both movements of the capillary are controlled by one hand (the right one), while the other hand exerts a gentle pressure on the syringe. A dent is seen on the cell surface when the capillary touches the cell. Microinjection itself is marked by a slight enlargement (swelling) of the cell. Moreover, the nucleus gains in contrast. Either the nucleus or the cytoplasm can be injected. After transfer of the sample, the capillary tip is brought just above the plane of the cell layer and moved to the next cell. About 150–200 cells can be injected within 10 min with some practice. After microinjection, cells are maintained as usual, checked at appropriate intervals for possible cytotoxic effects of the sample, and finally processed for evaluation.

The volume injected per cell can be estimated by measuring the distance covered by the meniscus within the pipette after microinjection of a certain number of cells and by determining the inner diameter of the capillary. By this method, we determined a mean injection volume of 1×10^{-11} to 2×10^{-11} ml per fibroblast culture cell, but up to 10^{-10} ml can be transferred. Another approach is to microinject a radiolabeled compound of high activity.

The injection volume per cell can be enlarged by a factor of 10^5 using as recipients multinucleated giant cells generated by fusion of tissue culture cells with appropriate agents (Graessmann et al., 1979) (Fig. 5).

To use these multinucleated cells, optimal fusion conditions have to be tested for each individual cell line. Variables are (a) the choice of the fusion agent, e.g., PEG 1000, PEG 6000, inactivated Sendai virus, (b) concentration, (c) time of action of the fusion agent, (d) density of cells before fusion, and (e) mode of washing of cells after fusion.

For a comparison with related techniques developed for the transfer of biomaterials into eukaryotic cells, the following list of general features of the microinjection system may be helpful:

1. Every tissue culture cell line tested so far has proved suitable for

Microinjection

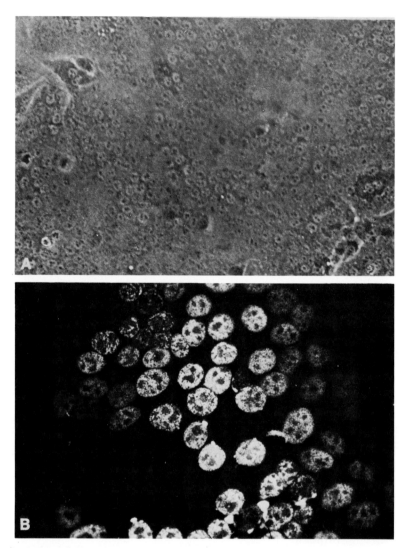

Figure 5. A. Partial view of a fused HeLa cell with several hundred nuclei. Arrows show unfused cells. Phase contrast, 200× magnification. B. T-antigen-positive multinucleated HeLa cell (partial view), fixed and stained 15 hr after SV40 DNA injection. The nuclei are located in different planes; therefore some are out of focus.

microinjection; suspension culture cells, e.g., lymphocytes, are prepared for microinjection by binding them to a substrate via suitable linkers.

2. Virtually no limitations exist regarding the material to be transferred. Intact virions, DNAs, RNAs, and proteins as well as small metabolites or substances unrelated to cellular metabolism can be introduced in purified form, without the involvement of helper macromolecules or chemical treatment. The number of molecules transferred is directly correlated with the concentration of the injection solution and the volume transferred per cell.

Table I. General Features of the Microinjection Technique

	Single adherent tissue culture cells	Suspension culture cells, fixed to dish	Fused cells; PEG 1000 as fusion agent
Recipient cells	Primary or secondary cells; all permanent cell lines	Lymphocytes	Permanent cell lines; lymphocytes
Material transferable	Cell organelles, viruses, DNA, RNA, proteins, etc.	Lymphocytes	Permanent cell lines; lymphocytes
Site of application	Nucleus/cytoplasm	Nucleus/cytoplasm	Cytoplasm
Injection volume per cell, ml	10^{-10}–10^{-11}	10^{-11}	10^{-6}–10^{-5}
Ways of analysis	Biologic/biochemical	Biologic	Biologic/biochemical
Efficiency	90–100%	90–100%	50–60%

Even intact cell organelles, e.g., cell nuclei (Graessmann, 1968) can be implanted into culture cells by microinjection.

3. The site of inoculation within the recipient cell (nucleus or cytoplasm) can be chosen by the investigator. With more emphasis on the technical aspects of microinjection even intramitochondrial injections seem possible (Diacumakos, 1980). This allows studies on intracellular transport and compartmentalization, compartment-dependent modification steps, etc.

4. A sample volume of about 2 μl is sufficient for microinjection; the rest of the material is retained for other experiments.

5. The number and localization of recipient cells are known.

6. Injected cells respond efficiently and remain as viable as their uninjected neighbors on the condition that the material transferred has no cytopathic effects (see Table I).

However, the microinjection technique is restricted to the use of cultured cells as recipients; it is hardly applicable to intact organisms. Yet it seems conceivable to isolate cells from individuals (e.g., bone marrow cells), to passage them *in vitro*, and to clone out lines from injected cells. The donor individual could then be the recipient for these cells.

References

Diacumakos, E. G., 1980, in: *Introduction of Macromolecules into Viable Mammalian Cells* (R. Baserga, C. Croce, and P. Pecora, eds.), New York, pp. 85–98.
Graessmann, A., 1968, Doctoral Dissertation, Freie Universitaet, Berlin.
Graessmann, A., 1970, *Exp. Cell. Res.* **60**:373–382.
Graessmann, A., and Graessmann, M., 1971, *Hoppe-Seyler's Z. Physiol. Chem.* **352**:527–532.
Graessmann, A., Graessmann, M., and Mueller, C., 1979, *Biophys. Res. Commun.* **88**:428–432.
Graessmann, A., Wolf, H., and Bornkamm, G. W., 1980, *Proc. Natl. Acad. Sci. USA* **77**:433–436.

Chapter 33

Techniques of Somatic Cell Hybridization by Fusion of Protoplasts

HORST BINDING and REINHARD NEHLS

1. Introduction

Somatic cell hybridization in higher plants has been developed to the point where it is a reliable technique for the formation of cell clones with recombinant genetic information of even far removed taxa (e.g., Kao, 1977; Binding and Nehls, 1978). Recombinant plants have only been obtained in related taxa, namely in the genus *Nicotiana* and other members of the Solanaceae family at high yields, in *Arabidopsis* + *Brassica* (Gleba and Hoffmann, 1979), *Daucus* + *Aegopodium* (Dudits et al., 1979), and within the genus *Daucus* (Dudits et al., 1977).

2. Plant Material

2.1. Plant Species

The established fusion techniques are appropriate for combination of protoplasts of any type to one another and even to mammalian cells. However, the regeneration capacities of isolated protoplasts appear to be very different in different taxonomic groups or even between genotypes of one species. In spite of nearly two decades of experience with protoplast regeneration (mosses; Binding, 1964), knowledge on appropriate procedures and on the processes involved is still rather limited. The species listed in Table I are characterized by high capacities of plant regeneration under the conditions currently available and hence may be recommended as model plants. Interestingly, the ability to regenerate only one parental protoplast type was sufficient for the initiation of cell divisions in fusion products in a number of combinations (Constabel et al., 1974).

HORST BINDING and REINHARD NEHLS • Botanisches Institut, Christian-Albrechts-Universität, Kiel, Federal Republic of Germany.

Table I. Species Recommeded as Model Plants for Somatic Hybridization Experiments

Family Species	Protoplast source	Reference for protoplast regeneration
Rutaceae		
Citrus sinensis	Embryogenic suspension	Vardi et al. (1975)
Apiaceae		
Daucus carota	Embryogenic suspension	Dudits et al. (1977)
	Shoots	Binding et al. (1981)
Solanaceae		
Datura innoxia	Shoots	Schieder (1975)
Nicotiana plumbaginifolia	Shoots	Gill et al. (1978)
Petunia hybrida	Shoots	Binding et al. (1981)
Solanum tuberosum 4x	Shoots	Shepard and Totten (1977)
2x	Shoots	Binding et al. (1978)
Asteraceae		
Sencio vulgaris	Shoots	Binding and Nehls (1980)
Poaceae		
Pennisetum americanum	Embryogenic suspension	V. Vasil and I. K. Vasil (1980)

2.2. Cultivation of Plant Material before Protoplast Isolation

2.2.1. Plants Grown on Soil

Experiments with protoplasts isolated from plants grown outdoors, in greenhouses, or in growth cabinets are decreasing in number. It has appeared from several investigations, especially with mesophyll protoplasts, that successful regeneration depends on the age of the plants, the age of leaves, watering, type of soil, light conditions, and even the season. Furthermore, microbial contaminations often could not be sufficiently removed before protoplast isolation. The use of soil-grown plants is therefore not recommended.

2.2.2. Culture of Plant Material in Vitro

In vitro culture of plant material used for protoplast isolation has several advantages, due especially to its axenic condition, the controlled environment, and the juvenile states established by the rhythm of subcultures (Binding, 1974c, 1975). A few commonly applied techniques of plant tissue culture will be considered here. Further information is available in several books [e.g., White (1963), Street (1973), Reinert and Bajaj (1977); for bibliography see Pierik (1979)].

2.2.2a. Procedures Providing Axenity. Organs of soil-grown plants are surface-sterilized by one of the techniques shown in Table II. Cut ends and

Table II. Techniques for Surface Sterilization of Plant Organs[a]

Agent	Concentration in distilled H_2O, %	Duration
$Ca(OCl)_2$	2	20 min
Domestos	5	30 min
Ethanol	70	10–30 sec
Chlorox	10	3–5 min
Na-dodeceyl sulfate	0.1	3–7 min
+ $HgCl_2$	0.1	

[a] Each treatment followed by thorough washing with sterile tap water.

lesions must be excised immediately after the incubation in sterilizing agents to get rid of potentially contaminated tissues. Both sterilized organs and material grown under axenic tissue culture conditions must be handled under sterile conditions. A laminar flow clean bench is best suited for this purpose. Forceps, razor blades, and other tools are sterilized by immersion into about 70% ethanol and heating at an alcohol or gas burner. Polystyrol culture vessels are commonly purchased in sterilized packings; glass vessels may be autoclaved (120°C, 15–20 min) or dry-heat-sterilized (180°C, 5 hr). Solutions and culture media are autoclaved, as long as they do not contain heat-sensitive substances. The latter (e.g., β-indolylacetic acid) are dissolved separately, passed through filters with pore sizes of 0.15 μm, and added to the autoclaved parts of the final solutions. Sucrose is autoclaved separately from inorganic salt solutions which may contain agar. Liquid culture media are either autoclaved or, more generally, filter-sterilized.

2.2.2b. Culture of Leaf Discs. Cell divisions could not be induced in axenic mesophyll protoplast preparations in a number of species. A method has been described to obtain rejuvenation of leaf cells prior to protoplast isolation [in *Brassica oleracea* (Gatenby and Cocking, 1977); in *Vicia narbonensis* (Donn, 1978)]. After surface sterilization of leaves and after removal of the lower epiderm, leaf discs are placed onto culture media MS, LS, or B5 (Tables IIIA–IIIC) containing a cytokinin and an auxin, the upper epiderm facing down. Protoplasts are then isolated after a few days of incubation.

2.2.2c. Culture of Plantlets and Shoots. *Culture Vessels.* Plantlets and shoots of most of the investigated species of the magnoliatae can easily be grown in petri dishes 5–10 cm in diameter. Polystyrol dishes must be sealed by strips of Parafilm or a comparable product to prevent evaporation. Larger sprouts are planted in taller vessels of various types, e.g., cylindric polystyrol boxes (available sterilized), jam glasses, wide-necked Erlenmeyer flasks, etc., preferably covered by halves of petri dishes.

Culture Media. The composition of culture media depends on the demands of the plant material. However, the large number of recipes, including numerous, sometimes minor, variations may be reduced to a few essential ones. Most commonly used are media B5 (Gamborg et al., 1968), MS (Murashige and Skoog, 1962), and LS (Linsmeyer and Skoog, 1965), the latter

Table IIIA. Culture Media[a]

Macroelements, mM						Trace elements, μM				
	MS/LS	NT	B5	V-47	8p		MS/LS/NT	B5	V-47	8p
$(NH_4)_2SO_4$	—	—	0.6	—	—	$MnSO_4$	100	43	30	45
NH_4NO_3	20	10	—	3.5	8	$ZnSO_4$	30	7	5	7
KNO_3	19	9.5	30	14.5	20	KI	5	4.5	1.5	4.5
$CaCl_2$	3.1	1.6	1	5	4	H_3BO_3	100	48	30	48
$MgSO_4$	1.4	4.4	1.8	2	1.2	Na_2MoO_4	1	1	0.5	1
KCl	—	—	—	—	4	$CuSO_4$	0.1	0.1	0.06	0.1
KH_2PO_4	1.3	5	—	0.5	1.3	$CoCl_2$	0.1	0.1	0.05	0.1
NaH_2PO_4	—	—	1.1	—	—					

Iron solutions (all media): *either* Fe-EDTA 0.1 mM *or* Sequestrene 28 mg/l

[a] MS: Murashige and Skoog (1962); LS: Linsmeyer and Skoog (1965); NT: Nagata and Takebe (1971); B5: Gamborg et al. (1968); V-47: Binding (1974a); 8p: Kao and Michayluk (1975).

differing from MS in having thiamin (0.4 mg/l) as the only vitamin (Table IIIB). Most of the investigated members of the magnoliatae do not need any phytohormone supplement; nevertheless, 2.5 μM 6-benzyladenin is favorable in several species, especially for shoot cultures. If the growth is not satisfactory, it is advisable to try the following ingredients, singly or in combination: cytokinin (6-benzyladenin, zeatin, or kinetin) 2.5 μM; α-naphthylacetic acid 10 μM; 2,4-dichlorophenoxyacetic acid 10 μM; coconut water 5%.

Environmental Conditions. Tissue culture rooms or growth cabinets are used for the establishment of appropriate physical conditions. Temperatures are adjusted to values between 18 and 30°C; those of 25°C are sufficiently

Table IIIB. Culture Media[a] —Organic Compounds

	Per liter	MS	NT	B5	V-47	8p	V-KM
Sucrose	g	30	10	20	17.1	0.25	0.25
Myo-inositol	mg	100	100	100	—	100	100
Glycine	mg	2	—	—	1.4	—	—
Thiamine HCl	mg	0.1	1	10	4	1	1
Nicotinic acid	mg	0.5	—	1	4	—	—
Pyridoxine HCl	mg	0.5	—	1	0.7	1	1
Biotin	mg	—	—	—	0.04	0.01	0.01
Folic acid	mg	—	—	—	0.5	0.4	0.4
Coconut water (ripe fruits, 60°C, 30 min)	% v/v	—	—	—	—	2	2
6 Benzyladenin	mg	Variable	1	Variable	0.5	—	0.5
Kinetin	mg	Variable	—	Variable	—	—	—
Zeatin	mg	Variable	—	Variable	—	0.5	—
α-Napthylacetic acid	mg	Variable	3	Variable	2	1	1
2,4-Dichlorophenoxyacetic acid	mg	—	—	0–2	—	0.2	0.1
β-Indolylacetic acid	mg	Variable	—	Variable	—	—	—

[a] MS: Murashige and Skoog (1962); NT: Nagata and Takege (1971); B5: Gamborg et al. (1968); V-47, Binding (1974a); 8p, Kao and Michayluk (1975); V-KM, Binding and Nehls (1977).

Table IIIC. Culture Media—Additional Organic Components of Medium 8p (and V-KM)[a]

Component	mg/liter	Component	mg/liter
D-Ca-Pantothenate	1	Na-Pyruvate	20
p-Aminobenzoic acid	0.02	Malic acid	40
Cholin chloride	1	Citric acid	40
Riboflavin	0.2	Fumaric acid	40
(not in V-KM)		(adjust stock solution to pH 5.5 by NH$_4$OH)	
Ascorbic acid	2		
Vitamin A	0.01		
Vitamin HB$_2$	0.01	Glucose	68.4 g/liter
Vitamin B$_{12}$	0.02	Casamino acid (vitamin-free)	250 mg/liter

Fructose, ribose, xylose, mannose, rhamnose, cellobiose, sorbitol, and mannitol: 250 mg each

[a] 8p: Kao and Michayluk (1975); V-KM: Binding and Nehls (1977).

tolerated in general. Low humidity prevents growth of mold in the room. It must be noted, however, that some investigators favor high humidity to reduce evaporation of the culture vessels. Beneficial light conditions for most of the species are established by white fluorescent tubes of intensities of about 3000 lux and day periods of 14 hr. Optimal conditions with respect to the yields of viable protoplasts must be determined for any types of plant sources individually. Optimal light intensities, for example, are about 3000 lux for tobacco (Binding, 1975) and 7000 lux for *Petunia hybrida* (Binding, 1974b).

Establishment of the Cultures. Cultures may be initiated from excised embryos, seeds, seedlings, shoot apices, or larger shoot tips and from any organ or tissue that is able to organize shoots under the culture conditions. Contamination by microorganisms is prevented by procedures given in Table II or by the excision of inner parts of plant organs under aseptic conditions. Subculture periods are 3–4 weeks. Roots should be removed at each transfer. Initiation of roots is prevented in most members of the magnoliatae by growth on auxin-free, but cytokinin-containing media. The genetic stability of the clones is best preserved by transferring upper parts of the shoots; increased proliferation is obtained by planting the decapitated lower parts.

2.2.2.d. Cell Suspension Cultures and Friable Callus. Cells of suspension cultures have been repeatedly used as protoplast sources since 1971 (Kao et al. 1971). For use in cell hybridization, small-scale cultures are sufficient. The description of culture techniques will therefore be confined to such cultures.

The establishment of cell suspension cultures begins with the induction of cell proliferation of explants of diverse origins on agar plates. Basically, the same culture media are used as for shoot cultures, except that cytokinins are kept at low levels or are omitted, whereas auxins are added. In most of the investigated cases, the presence of 5–10 μM 2,4-dichlorophenoxyacetic acid as the only growth hormone resulted in the formation of friable callus which itself is already useful for protoplast isolation. However, better results are commonly achieved with cell cultures initiated by the suspension of this type of callus. Clones which exhibit early segregation of the daughter cells can be obtained by passing the suspensions through cloth or steel sieves at each transfer. The cultures are grown in Erlenmeyer flasks or glass bottles and incubated on shakers of diverse types or rollers. Temperatures are adjusted to about 25°C. Presence and regimes of light depend on the types of suspensions. Generally, the cells are grown in the dark. Culture medium MS, LS, or B5 supplemented with 5–10 μM 2,4-dichlorophenoxyacetic acid is used in most laboratories.

Embryogenic suspension cultures are the only valid sources for protoplast regeneration in Citrus sinensis (Vardi et al., 1975) and in cereals (Brar et al., 1980, V. Vasil and I. K. Vasil, 1980, Lu et al., 1981). The establishment of embryogenic suspension cultures in cereals is rather sophisticated and will therefore be briefly described here following the procedures developed by V. Vasil and I. K. Vasil (1981a,b), starting from excised immature embryos. They are planted on LS medium containing 2.5 mg/liter 2,4-dichlorophenoxyacetic acid, 0.5 mg/liter thiamine, 3% sucrose, and 0.8% agar and grown in the dark at 27°C. Pale-yellow, small-celled callus is subcultured. Suspension cultures are obtained from this type of callus by incubation in liquid medium in 250-ml Erlenmeyer flasks at 150 rpm in the dark.

2.3. Differentiation Stages of Cells Used for Protoplast Isolation

In principle, all vital plant cells are appropriate as protoplast sources for cell hybridization experiments insofar as cell walls can be removed by the aid of enzyme preparations and as they regenerate under the applied conditions. Especially pure preparations can be obtained from the mesophyll of young but almost expanded leaves harvested preferably just before the light phase. They exhibit high regeneration capacities in a number of plant species, especially in the family of Solanaceae. More reliable plant regeneration, however, was achieved in most investigated species when juvenile cells were used. Good results have been gained in the class of magnoliatae by the use of cells of germlings (Eriksson et al., 1978) or shoot tips, including leaves below one-tenth of the fully developed sizes under the applied culture conditions (Binding et al., 1981). Satisfactory mitotic activities are found in preparations from cell suspension cultures especially when isolated at the log growth phase, but plant regeneration failed in many cases. Best suited for this purpose are protoplasts from embryogenic cultures, as has been demonstrated most strikingly in cereals (V. Vasil and I. K. Vasil, 1980, Lu et al., 1981).

3. Protoplast Isolation Techniques

Procedures for the isolation of protoplasts by enzymic methods are outlined in Table V. Mechanical isolation by cutting plasmolyzed cells has been used (Binding, 1966) and may be applied in a few extraordinary cases only.

3.1. Preparation of the Plant Material

3.1.1. Leaves and Shoot Tips

The leaves of terraneous plants are protected by a cuticula and cutinized cell walls of the epiderms. The easiest way to allow the penetration of the wall-digesting enzymes into the mesophyll is by stripping the lower epiderm of wilted leaves. Peeling is possible with leaves of soil-grown plants of a number of members of the magnoliatae. In some cases it can also be applied to leaves of shoots grown *in vitro*. In a number of magnoliatae and numerous liliatae, peeling is much more difficult or impossible. The leaves are hence cut or torn to strips of a few millimeters in breadth. Small leaves and shoot tip material are sliced by a razor blade to get pieces 1–3 mm square. The material is then either transferred directly to the enzyme solutions or preincubated in hypertonic solutions (0.4–0.7 M) of mannitol, sorbitol, or sucrose; 30–60 min of preincubation is especially indicated for leaves of *Nicotiana* species.

3.1.2. Friable Callus and Cell Cultures

No special preparation of friable callus and cell suspension cultures is necessary prior to the enzyme incubation. Sometimes it is useful to remove larger cell aggregates by pouring the suspensions in culture media or in the enzyme solution through miracloth or sieves with pore sizes of 40–100 μm, depending on the size of the cell.

3.2. Preparations of Enzyme Solutions

The isolation of viable protoplasts from all sources is nearly exclusively managed by enzymic digestion of the various cell wall components in a single-step procedure introduced by Power et al., (1970).

A range of commercial enzyme preparations is listed in Table IV, classified according to their main activities. Commercially available enzyme preparations containing mainly cellulase and pectinase activities (Table IV) are sufficient for release of protoplasts from most tissues and cell cultures, but there is still a large number of different types of tissues in which protoplast isolation is rather difficult, e.g., leaves of members of the order of caryophyllales, compact calluses, microspores, and others. Success in some cases has been obtained by using additional enzyme preparations of different types (Table IV). Representative enzyme combinations for protoplast isolation are given in Table V.

Free protoplasts persist only in iso- or hypertonic solutions. Mannitol, sorbitol, or sucrose in 0.4–0.7 M concentrations are frequently used as osmotic stabilizers. Addition of ions, especially of calcium, apparently results in better protoplast preparations.

The components of the enzyme solutions are dissolved (cf. Table V) in distilled water. Some of the commercial enzyme preparations contain so many

Table IV. Some Commercial Enzyme Preparations for the Isolation of Protoplasts

Pectinases	
Pectinase P-4625	Sigma Chemical Co., St. Louis, Missouri
Rohament P	Röhm, Darmstadt, FRG
PATE	Hoechst, Frankfurt, FRG
Macerozyme R-10	Kinki Yakult Co., Nishinomiya, Japan
Pectinol R-10	Rohm & Haas, Philadelphia, Pennsylvania
Macerase	CalBiochem, La Jolla, California
Cellulases	
Cellulase Onozuka R-10	Kinki Yakult Co., Nishinomiya, Japan
Cellulase C	Röhm, Darmstadt, FRG
Cellulysin	CalBiochem, La Jolla, California
Meicellase-P	Meiji Seiki Kaisha, Tokyo, Japan
Driselase pectinase activity)	Kyowa Hakko Kogyo Co., Tokyo, Japan
Hemicellulases	
Rhozyme HP 150	Röhm & Haas, Philadelphia, Pennsylvania
Hemicellulase	Sigma Chemical Co., St. Louis, Missouri
β-Glucuronidase	
Helicase	Industrie Biologique Francaise Gennevilliers, France

insoluble substances that it is advisable to remove them by centrifugation. The pH is adjusted to 5.4–6.5. The solutions are filter-sterilized.

3.3. Enzyme Incubation

The techniques of enzyme incubation depend on the protoplast sources, on the enzyme preparations and charges used, and on the custom of the individual researcher. Ranges of parameters usually applied are compiled in Table VI.

3.3.1. Incubation Mixtures

Peeled leaf discs are floated on the surface of the enzyme solution, upper epiderm upside in petri dishes. Cells and small tissues are suspended in the solutions, and cell suspension cultures may be simply mixed with equal volumes of enzyme solutions. The solutions are infiltrated into tissues by about 1 min exposure to weak vacuum. Not too much plant material must be suspended in a given volume. The formulation indicated in Table VI has been evaluated for mesophyll protoplasts of *Petunia*, but is also favorable in many other systems.

3.3.2. Incubation Procedures

Supply of oxygen to the incubation mixtures is an important factor for good protoplast yields. The suspensions are therefore either agitated by shaking or rolling, continuously or at least occasionally, or the incubation is

Table V. Incubation Mixtures for Protoplast Isolation from Various Sources

Leaves and shoot tips		Cell suspensions		Callus	
Macerozyme	0.5–1%	Macerozyme	0.5–0.8%	Macerozyme	0.5–1%
or Pectinase P-4625	0.5–2%	or Pectinase P-4625	0.5%	or PATE	1%
+		+		+ Rohament P	0.08%
Cellulase Onozuka R-10	1.5–3%	Cellulase Onozuka R-10	2–5%	Cellulase Onozuka R-10	0.5–5%
or Meicellase	1.5–3%	or Cellulysin	4%	or Driselase	0.2–0.5%
or Cellulysin	1–2%	+			
		Driselase	0.5–2%		
Dissolved in:					
5 mM Ca(NO$_3$)$_2$ } or { 5 mM Ca(NO$_3$)$_2$		B5 minerals } or { 5 mM CaCl$_2$		} or { 0.7 mM CaCl$_2$	
0.25 M mannitol				0.7 mM NaH$_2$PO$_4$		0.7 mM NaH$_2$PO$_4$
0.5 M mannitol	0.25 M sorbitol		0.55 M sorbitol	3- mM 2-(N-morpholino)-ethane sulfonic acid) (Mes)	3 mM Mes
				0.45 M glucose	0.7 M glucose

Adjusted to pH 5.4–6.5

Table VI. Preparation of Protoplast Suspensions

Peeled leaves	Dissected leaves and shoot tips	Cell suspensions
Placing onto enzyme solution	Suspension in enzyme solution (100 mg plant material in 10 ml of enzyme solution) and subjection to weak vacuum for 1 min	Mixing up with equal volumes of enzyme solution
Floating of leaf segments on enzyme solution in a petri dish	Incubation in shallow layers of enzyme solution (e.g., 12 ml in a 9-cm petri dish) or incubation in vessels agitated on culture rollers or shakers	

Dim light or darkness; 25–28°C, 4–12 hr or 25°C, 1–3 hr; followed by 10°C, up to 24 hr

carried out in shallow layers using petri dishes. The significance of light during incubation is still rather obscure. Dim light seems to be useful for green material. Cells free of chlorophyll are probably better incubated in the dark.

Optimal temperatures and incubation times must be found for any given system. They depend on species, types of tissue, enzyme products, charges, and concentrations and are, furthermore, interdependent. Usually, protoplasts of mesophyll and shoot tips are released within 4–12 hr at 25°C. For longer incubation times, it is advisable to transfer to lower temperatures (e.g., 10°C) after 1–3 hr at about 25°C.

3.4. Collection of Isolated Protoplasts

The steps of protoplast collection are summarized in Table VII. Protoplasts loosely clustered by remaining parts of tissues are suspended by

Table VII. Collection of Isolated Protoplasts

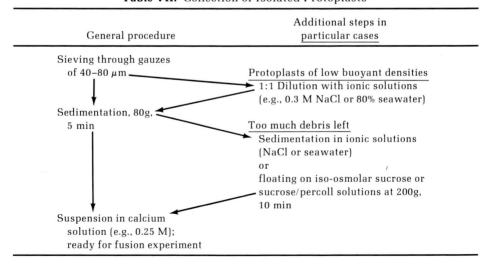

so-called teasing procedures, such as agitating leaf discs by a forceps, shaking the incubation vessels briefly but vigorously, or pipetting onto sieves if the remnants of tissues are small enough. The sieves may be made of synthetic fibers or steel fixed as bottoms of cylindric vessels. Steel gauze may also (after cutting them to discs of about 3 cm in diameter) be folded to form wide cones and held by a forceps eccentrically on top of a centrifuge tube, touching it at one side. They are kept sterile in 70% alcohol and washed in sterile distilled water before use. Protoplasts of chlorophyll-deficient cells sometimes have similar or even lower buoyant densities than the enzyme solutions. In those cases the incubation mixtures are diluted by iso-osmolar ionic solutions before centrifugation. Usually, the pellet can be used directly for the fusion experiment. If too much debris is left, the preparation must be purified by sedimentation or floating as indicated in Table VII.

3.5. Preparation of Protoplasts Lacking Particular Genetic Capacities

A few attempts have been made to incorporate subprotoplasts (miniprotoplasts) into experiments on somatic hybridization (Binding, 1976; Wallin et al., 1979). Subprotoplasts are formed spontaneously by fractionation of protoplasts (Binding and Kollmann, 1976; Binding, 1979). Enucleation of protoplasts has been obtained by density gradient centrifugation of a suspension in the presence of cytochalasin B (Wallin et al., 1978).

In a powerful method, Zelcer et al. (1978) inactivated the nuclei of protoplasts by X-rays. Freshly isolated protoplasts are suspended in a solution of salts of the NT medium in 0.55 M mannitol at densities of 10^5/ml and exposed to 5 kR in a Rich Seifert & Co. apparatus.

4. Protoplast Fusion

4.1. Ca^{++}–High pH–Polyethylene Glycol (PEG) Techniques

Fusion techniques based on the action of calcium ions and high pH (Keller and Melchers, 1973) and of PEG (Kao and Michayluk, 1974) have been used in several modifications and combinations. The processes involved are not yet fully understood. Nagata and Melchers (1978) found that in calcium solutions fusion occurred when the zeta potentials of the plasmalemmas approached 0 mV. Incubation in PEG solutions results mainly in adhesion of the protoplasts, facilitating the intrinsic process of fusion. Electron microscopic investigations indicated that the fusion process takes about 1 hr when polyethylene glycol is applied (Burgess and Fleming, 1974; Fowke et al., 1975, 1977). It is extremely important to induce fusion as soon as possible, because the reformation of cell walls will impede or even prevent the formation of fusion bodies within 10–60 min, depending on the plant species (Weber et al., 1976).

Table VIII. Ca^{++}-High pH Solution

Dissolve 0.25 M $Ca(NO_3)_2$ in (a) 0.1 M glycine; (b) 0.1 M NaOH
Mix up (a) and (b) in appropriate ratios to result in pH values of 9.5–10.5
Utilization must be within 24 hr

4.1.1. The Droplet Fusion Technique

This procedure is modified from Kao (1976). Pure protoplast suspensions of the presumptive fusion partners are united in a ratio of about 1:1 and collected in a small volume (less than 0.2 ml) of a 0.25 M $Ca(NO_3)_2$ solution at about pH 6, placed on the bottom of a petri dish or, preferably, on a glass coverslide, and allowed to settle for 5 min. Then, 150 μl of a 40% solution of PEG 1500 is gently added to the margin of the primary droplet, thus displacing the $Ca(NO_3)_2$ solution. After 10–15 min, another 150 μl of Ca^{++}-high-pH solution (Table VIII) is gently added in the same way. Surplus PEG may be removed by a capillary prior to this step. After another 30–45 min, the reaction mixture is slightly diluted by small droplets of culture medium amounting to about 0.3 ml. About 1 ml is then flushed over the sticking protoplasts. This washing solution is removed and the final amount of culture medium is added. At all steps, enough liquid must be left to cover the protoplasts completely. The method described above provides high percentages of fusion events and is appropriate especially with small populations of protoplasts to be fused. Difficulties may arise in obtaining separate clones, due to the firm clustering of the cells in one spot of the culture vessel.

4.1.2. The Phase Boundary Technique

Large populations of protoplasts can efficiently be fused in a discontinuous gradient of PEG and Ca^{++} solutions (Binding and Nehls, 1978). For this method, the parental protoplast types are mixed in a ratio of about 1:1, pelleted by centrifugation in 0.25 M $Ca(NO_3)_2$ solution at about pH 6, and suspended in 0.5 ml of the Ca^{++}-high-pH solution. The suspension is then gently placed onto 0.3 ml of a 40–45% solution of PEG 6000 covering just the bottom of a 12-ml test tube. The protoplasts settle at the phase boundary of the PEG and Ca^{++} solutions during the incubation. After 30–45 min, small droplets of culture medium (final amount: 1 ml) are allowed to trickle along the wall of the test tube and bring about the dilution of the fusion agents mostly by diffusion. After another 30–60 min, culture medium is added to a final volume of 12 ml. The compiled solutions are mixed up by pouring into another test tube. The protoplasts are sedimented by gentle centrifugation 1 hr later and plated at the appropriate densities in a 35-mm Petri dish.

4.1.3. Variations of the Ca^{++}-High pH–PEG Technique

To facilitate the approach of the Ca^{++} solution of high pH to the protoplasts sitting at the bottom of a petri dish, it might be advantageous to mix up the fusion agents in one solution instead of displacing the PEG. Haydu et al. (1977) obtained an increase of the fusion efficiencies by adding dimethyl sulfoxide, and Nagata (1978) by adding polyvinyl alcohol, to the PEG solution.

4.2. Other Fusion Techniques

A number of other approaches to protoplast fusion have been reported, but without giving rise to hybrid cell lines. They are mentioned here just for reference. Binding (1966) fused protoplasts in seawater to which an electric field was applied; Kameya (1973) took advantage of gelatin; Binding (1974b) used diluted seawater adjusted to high pH by ammonium hydroxide; Nagata et al. (1979) investigated the action of a phospholipid reversing the zeta potentials of the protoplast surface with respect to induction of specific fusion; and, most recently, Zimmermann and Scheurich (1981) devised a method to fuse protoplasts in alternating electric fields.

5. Techniques for the Regeneration of Fusion Products

5.1. Culture Techniques of Fusion Bodies

5.1.1. Nutrition

A number of different nutrient solutions have been developed for the culture of isolated protoplasts. For detailed information, the reader is referred to more detailed discussions (see for example I. K. Vasil and V. Vasil, 1980). Some media were effective in a greater variety of taxa, and a few of them are mentioned here (Table IIIA). They contain the inorganic salts of B5, NT, or V-47 media or others. Simple addition of vitamins, growth hormones, and nutritive sugar as indicated in the original formulas is sufficient in tobacco, *Petunia*, and some other species as well as in some cell suspension systems. The addition of several organic nutrients and growth factors promoted growth in these cases, and made it possible in many other species, especially if the protoplasts were derived from juvenile cells. The requirements are best met by the organic components of the 8p medium, e.g., in medium V-KM (Table IIIC).

High plating densities are necessary or, in some species, at least favorable for protoplast regeneration, indicating that not all nutritive demands are satisfied by the culture media even in the case of the 8p medium, but that the solutions are enriched by essential products of the cell metabolism. Appropriate ratios of cell mass/amount of medium are obtained by plating either the complete bulks of protoplasts of a fusion experiment, by growing fusion products individually in microdroplets, or by incorporating them into suspensions of feeder protoplasts.

5.1.2. Osmotic Stabilization

The isolated protoplasts are osmotically stabilized by sugars or sugar alcohols (glucose, mannitol, or sorbitol) at 0.4–0.7 M concentrations. In most species, low osmolarity is best. During culture, increase of the osmolarity of the culture medium by the formation of condensed water as well as changes of the osmolarity of the cells must be avoided. This is best managed by using growth chambers with small variations of temperature and continuous light or darkness.

5.1.3. Supply of Oxygen

Oxygen is provided by culturing the protoplasts in small droplets or thin layers in petri or Cuprak dishes. In general, continuous agitation of the cultures is damaging.

5.1.4. Light Conditions

Protoplasts of cell suspension cultures are commonly grown in the dark. Low light intensities not exceeding 3000 lux appear to be favorable for protoplasts of mesophyll and shoot tips. However, plating efficiencies were increased in a few cases by dark periods of 1 to 10 days prior to the exposure to dim light (e.g., Enzmann-Becker, 1973; Saxena et al., 1981).

5.1.5. Temperatures

Temperatures of about 25°C have been used in the majority of experiments on protoplast regeneration. After transfer to culture media, temporary exposure (1–24 hr) to lower temperatures of, e.g., 10°C is probably helpful for the stabilization of the protoplasts.

5.1.6. Plating Techniques

The choice of the plating technique depends on the aims of the experiment and on the markers used for recognition and selection of the fusion products.

5.1.6a. The Bulk Culture Technique. Usually, all protoplasts, if uniparental or fusion bodies, are cultured together in petri dishes of glass or polystyrol. Pretreatment of the vessels is not necessary. The suspensions are plated in amounts that enable just the spreading over the bottom of the dish by gently turning or shaking. This amount is about 0.7 ml for a dish 35 mm in diameter.

Optimal protoplast titers must be evaluated for special protoplast systems. Protoplasts of mesophyll are plated at densities of 5×10^3 to 1×10^4, those of juvenile cells of shoot tips at about 2×10^4 and more, and those of vigorously growing cell cultures at about 10^5 cells/ml.

5.1.6b. Microdroplet Cloning Technique. Culture of isolated fusion bodies in small droplets has been developed for somatic hybridization experiments by Kaoiter and has been improved by Gleba and Hoffmann (1978). All protoplasts of a fusion experiment are cultured together in suspension as long as the fusion products are recognized by, for instance, the content of different types of plastids. Usually, those characteristics disappear within 1–3 days of culture. The fusion products are picked from the suspension by micropipettes or capillaries as long as they are recognizable, but, if possible, after the formation of cell walls. They are then placed in about 1-μl droplets into Cuprak dishes. The procedure can be done by hand (Gleba and Hoffmann, 1978; Nehls, 1981), but may be also managed by the aid of a micromanipulator.

5.1.6c. The Nurse Culture Cloning Technique. The nurse culture technique used by Menczel et al. (1978) combines the advantages of both methods described earlier. Fusion products are transferred from a bulk culture a few days after plating into a suspension of protoplasts which carry a marker enabling the recognition of the hybrid clones when cell clusters are formed. The marker used was chlorophyll deficiency.

5.2. Subculture of Regenerated Cell Clusters

Protoplast regenerants may be left in the original medium for 14 days or longer until they can survive the transfer onto low-osmotic agar media for organogenesis or into low-osmotic nutrient solutions for propagation in cell

Table IX. Steps in Subculturing Protoplast Regenerants

Protoplasts from shoot tips [modified from Binding and Nehls (1980)]	Protoplasts from embryogenic suspension cultures of cereals (V. Vasil and I. K. Vasil, 1980)
Protoplast culture in liquid V-KM	Protoplast culture in liquid 8p*
Dilution every 1–3 days by liquid or soft agar medium V-KM	
Layering onto solid agar medium V-KM from the tenth day on	
Transfer of calluses (size 1–2 mm) onto low-osmotic solid agar; e.g., B5 + 5% coconut water + 2.5 μM 6-benzyladenin for shoot formation	Subculture on corresponding low-osmotic medium after 14 days
	Transfer of calluses (size 1–2 mm) to corresponding solid medium without growth substances for the formation of embryoids and plantlets
Transfer of excised shoots onto hormone-free or β-indolylacetic acid-containing media for root formation	

suspension culture. However, the high mitotic activities, embryogenic capacities, or potencies to form adventitious shoots are more reliably preserved by a series of dilutions of the culture media. As examples, protocols are outlined in Table IX for protoplasts of shoot tips (Binding and Nehls, 1980) and of embryogenic suspension cultures (V. Vasil and I. K. Vasil, 1980, Lu et al., 1981). Embedding of the cells or cell clusters in soft agar media (Takebe et al., 1971) is suitable for the establishment of homogeneous suspensions giving best controlled conditions. The moment for the addition of agar media depends on the plant species and on experimental conditions. Usually, it is not tolerated before the third day of culture. Preference is given to low agar concentrations of 0.25–0.3%. The low agar concentrations give higher survival rates and facilitate further steps of dilutions.

5.3. Formation of Shoots and Plants

There are many observations on the formation of somatic embryos and adventitious shoots in the literature (Street, 1979). It is impossible to summarize here the information obtained or to present common rules. Plants of a number of species have been grown from cells originating from nearly any type of explant. It has been discussed repeatedly if it is true that any plant cell in all species is totipotent (Potrykus et al., 1976). It has emerged from several investigations on protoplast regeneration that plant formation is most reliably obtained when juvenile cells of the region of the shoot meristems (Binding et al., 1981) and thoroughly propagated embryogenic suspension cultures (V. Vasil and I. K. Vasil, 1980, Lu et al., 1981) are used as protoplast sources, when the regenerants are cultured under optimal growth conditions and are transferred to media stimulating (or not suppressing?) organogenesis as soon as possible. Most favorable for embryogenesis is reduction or omission of 2,4-dichlorophenoxyacetic acid in low osmotic media; efficient formation of adventitious shoots has been obtained on agar medium B5 or MS with 2.5 μM 6-benzyladenin. Addition of coconut water was frequently favorable, and in some cases essential.

Root formation is induced at shoot cuttings in many species on medium B5 or MS lacking phytohormones or supplemented with auxin, preferably β-indolylacetic acid. The plantlets are transferred to soil as soon as the roots are 1–3 mm in length. The soil must be sopping wet when planting. Normally, covering by a transparent shelter is necessary just for 1 or a few days. Best success is obtained when using small shoots (1–2 cm in length) for plant regeneration.

6. Selection and Analysis of Fusion Products

Several characteristics of the parental protoplasts and fusion products have been used in somatic hybridization experiments. They have been

Table X. Characteristics and Techniques for the Selection and Analysis of Fusion Products

Characteristics and techniques	Enrichment by centrifugation	Recognition as fusion bodies	Recognition as regenerants	Self-acting selection	Analysis of fusion products	Progeny analysis	Selected references[a]
Heritable markers							
Light sensitivity				+		+	22
Chlorophyll deficiency			+			+	28, 31
Drug resistance				+		+	6, 13, 21
Auxotrophy				+		+	12
Medium requirement				+			27
Hormone autotrophy				+		+	32
Male sterility			+			+	1, 17
Organogenic potency			+				21
Morphology	(+)		+			+	4, 19
Heterosis of hybrids			+				30
Staining of nuclei		+	(+)		(+)	(+)	4, 7
Chromosome morphology			+		+	+	4, 10, 18, 19
Isozyme patterns			+		+	+	11, 21, 30, 32, 33
Fraction I protein			+		+	+	2, 5, 16, 20, 23
DNA/RNA hybridization			+		+	+	8
Restrictase patterns			+		+	+	3, 8
Secondary metabolites			+		+	+	25
Induced markers							
Buoyant densities	+						15
Fluorescence		+					9
Biochemical inhibition				+			24
X-ray inactivation				+			1, 13, 34
Markers of cell differentiation							
Plastid differentiation		+					4, 10
Pigmentation of vacuoles		+					26, 29
Cytomorphology		+					4, 10, 29
Buoyant density	+						14

[a] 1. Aviv and Galun (1980); 2. Aviv et al. (1980); 3. Belliard et al. (1979); 4. Binding and Nehls (1978); 5. Chen et al. (1977); 6. Cocking et al. (1974); 7. Constabel et al. (1975); 8. Dudits et al. (1979); 9. Galbraith and Mauch (1980); 10. Gleba and Hoffmann (1978); 11. Gleba and Hoffmann (1980); 12. Glimelius et al. (1978); 13. Gressel et al. (1981); 14. Harms and Potrykus (1978); 15. Harms and Potrykus (1979); 16. Iwai et al. (1980); 17. Izhar and Tabib (1980); 18. Kao (1977); 19. Krumbiegel and Schieder (1979); 20. Kung et al. (1975); 21. Maliga et al. (1978); 22. Melchers and Labib (1974); 23. Melchers et al. (1978); 24. Nehls (1981); 25. Ninnemann and Jüttner (1981); 26. Potrykus (1971); 27. Power et al. (1977); 28. Power et al. (1979); 29. Reinert and Gosch (1976); 30. Schieder (1978a); 31. Schieder (1978b); 32. Smith et al. (1976); 33. Wetter (1977); 34. Zelcer et al. (1978).

compiled and discussed in a number of papers (e.g., Cocking, 1978; Binding and Nehls, 1979; Schieder and Vasil, 1980). Production of mutant cell lines appropriate for this purpose has been summarized by Maglia (1980). The choice depends on what is available in the investigated species, and on the aims of the experiment. Therefore, just the common features are mentioned here. Markers are listed in Table X.

Regeneration of fusion bodies usually gives rise to a great variety of subclones (Binding and Nehls, 1981). They may contain either uniparental or hybrid nuclei, the latter carrying either complete sets of the parental chromosomes, or increased or reduced numbers of chromosomes (e.g., Kao, 1977; Binding and Nehls, 1978; Hoffmann and Adachi, personal communication; Krumbiegel and Schieder, 1979). Cell organelles segregate soon or later (Smith et al., 1976). Recombinant cell organelles have been found in the case of mitochondria (Belliard et al., 1979). Early one-step selection of fusion products and subcloning of the whole cell progeny are indicated to obtain the complete spectrum of diverse genetic types. Prolonged application of selective pressure or late selection by particular pairs of markers results in the preservation only of subclones containing the responsible genetic traits.

Analyses of the regenerated cell clones and plants are carried out in order to verify the hybrid or cybrid nature so that one is not misled by modifications or mutations, and to get information on the genetic constitution as well as on the expression of the genetic constituents.

The nuclear situation is checked by analyses of karyotypes and isozyme patterns as well as by DNA/mRNA hybridization (Dudits et al., 1979). The plastids are identified by DNA restrictase patterns, and by fraction I protein analyses giving, in addition, information on nuclear genes (Poulsen et al., 1980). DNA restrictase patterns are also useful in mitochondria.

References

Aviv, D., and Galun, E., 1980, Restoration of fertility in cytoplasmic male sterile (CMS) *Nicotiana sylvestris* by fusion with X-irradiated *N. tabacum* protoplasts, *Theor. Appl. Genet.* **58**:121–127.

Aviv, D., Fluhr, R., Edelman, M., and Galun, E., 1980, Progeny analysis of the interspecific somatic hybrids: *Nicotiana tabacum* (CMS) + *Nicotiana sylvestris* with respect to nuclear and chloroplast markers, *Theor. Appl. Genet.* **56**:145–150.

Belliard, G., Vedel, F., and Pelletier, G., 1979, Mitochondrial recombination in cytoplasmic hybrids of *Nicotiana tabacum* by protoplast fusion, *Nature* **281**:401–403.

Binding, H., 1964, Regeneration und Verschmelzung nackter Laubmoosprotoplasten, *Z. Naturforsch.* **19b**:755.

Binding, H., 1966, Regeneration und Verschmelzung nackter Laubmoosprotoplasten, *Z. Pflanzenphysiol.* **55**:305–321.

Binding, H., 1974a, Cell cluster formation by leaf protoplasts from axenic cultures of haploid *Petunia hybrida* L., *Plant Sci. Lett.* **2**:185–188.

Binding, H., 1974b, Fusionsversuche mit isolierten Protoplasten von *Petunia hybrida* L., *Z. Pflanzenphysiol.* **72**:422–426.

Binding, H., 1974c, Regeneration von haploiden und diploiden Pflanzen aus Protoplasten von *Petunia hybrida* L., *Z. Pflanzenphysiol.* **74**:327–356. In (English transl., F. de Bruijn, 1980, *Plant Molec. Biol. Newsletter* **1**:77–95.]

Binding, H., 1975, Reproducibly high plating efficiencies of isolated mesophyll protoplasts from shoot cultures of tobacco, *Physiol. Plant.* **35**:225–227.

Binding, H., 1976, Somatic hybridization experiments in solanaceous species, *Mol. Gen. Genet.* **144**:171–175.

Binding, H., 1979, Subprotoplasts and organelle transplantation, in: *Plant Cell and Tissue Culture* (W. R. Sharp, P. O. Larsen, F. F. Paddock, and V. Raghavan, eds.), Ohio State University Press, Columbus, Ohio, pp. 789–805.

Binding, H., and Kollmann, R., 1976, The use of subprotoplasts for organelle transplantation, in: *Cell Genetics in Higher Plants* (D. Dudits, G. L. Farkas, and P. Maliga, eds.), Akadémiai Kiadó, Budapest, pp. 191–206.

Binding, H., and Nehls, R., 1977, Regeneration of isolated protoplasts to plants in *Solanum dulcamara* L., *Z. Pflanzenphysiol.* **85**:279–280.

Binding, H., and Nehls, R., 1978, Somatic cell hybridization of *Vicia faba* + *Petunia hybrida*, *Mol. Gen. Genet.* **164**:137–143.

Binding, H., and Nehls, R., 1979, Recombination: Asexual recombination in higher plants, *Progr. Bot.* **41**:173–184.

Binding, H., and Nehls, R., 1980, Protoplast regeneration to plants in *Senecio vulgaris* L., *Z. Pflanzenphysiol.* **99**:183–185.

Binding, H., and Nehls, R., 1981, Recombination: Recombination in higher plants, *Progr. Bot.* **43**:132–138.

Binding, H., Nehls, R., Schieder, O., Sopory, S. K., and Wenzel, G., 1978, Regeneration of mesophyll protoplasts from dihaploid clones of *Solanum tuberosum*, *Physiol. Plant.* **43**:52–54.

Binding, H., Nehls, R., Kock, R., Finger, J., and Mordhorst, G., 1981, Comparative studies on protoplast regeneration in herbaceous species of the dicotyledoneae class, *Z. Pflanzenphysiol.* **101**:119–130.

Brar, D. S., Rambold, S., Constabel, F., and Gamborg, O. L., 1980, Isolation, fusion and culture of Sorghum and corn protoplasts, *Z. Pflanzenphysiol.* **96**:269–275.

Burgess, J., and Fleming, E. N., 1974, Ultrastructural studies of the aggregation and fusion of plant protoplasts, *Planta* **118**:183–193.

Chen, K., Wildman, S. G., and Smith, H., 1977, Chloroplast DNA distribution in parasexual hybrids as shown by polypeptide composition of fraction I protein, *Proc. Natl. Acad. Sci. USA* **74**:5109–5112.

Cocking, E. C., 1978, Selection and somatic hybridization, in: *Frontiers of Plant Tissue Culture* (T. A. Thorpe, ed.), Univ. of Calgary, pp. 151–158.

Cocking, E. C., Power, J. B., Evans, P. K., Safwat, F., and Frearson, E. M., 1974, Naturally occurring differential drug sensitivities of cultured plant protoplasts, *Plant Sci. Lett.* **3**:341–350.

Constabel, F., Dudits, D., Gamborg, O. L., and Kao, K. N., 1975, Nuclear fusion in intergeneric heterokaryons. A note, *Can. J. Bot.* **53**:2092–2095.

Donn, G., 1978, Cell division and callus regeneration from leaf protoplasts of *Vicia narbonensis*, *Z. Pflanzenphysiol.* **86**:65–75.

Dudits, D., Hadlaczky, G., Levi, E., Fejer, O., Haydu, Z., and Lazar, G., 1977, Somatic hybridization of *Daucus carota* and *D. capillifolius* by protoplast fusion, *Theor. Appl. Genet.* **51**:127–132.

Dudits, D., Hadlaczky, G., Bajszar, G., Lazar, G., and Horvath, G., 1979, Plant regeneration from intergeneric cell hybrids, *Plant Sci. Lett.* **15**:101–112.

Enzmann-Becker, G., 1973, Plating efficiencies of protoplasts of tobacco in different light conditions, *Z. Naturforsch.* **28c**:470–471.

Eriksson, T., Glimelius, K., and Wallin, A., 1978, Protoplast isolation, cultivation and development, in: *Frontiers of Plant Tissue Culture* (T. A. Thorpe, ed.), Univ. of Calgary, pp. 131–139.

Fowke, L. C., Rennie, P. J., Kirkpatrick, J. W., and Constabel, F., 1975, Ultrastructural characteristics of intergeneric protoplast fusion, *Can. J. Bot.* **53**:272–278.

Fowke, L. C., Constabel, F., and Gamborg, O. L., 1977, Fine structure of fusion products from soybean cell culture and pea leaf protoplasts, *Planta* **135**:257–266.

Galbraith, D. W., and Mauch, T. J., 1980, Identification of fusion of plant protoplasts: II.

Conditions for the reproducible fluorescence labelling of protoplasts derived from mesophyll tissue, Z. Pflanzenphysiol. **98:**129–140.

Gamborg, O. L., Miller, R. A., and Ojima, K., 1968, Nutrient requirement of suspension cultures of soybean root cells, Exp. Cell Res. **50:**151–158.

Gatenby, A. A., and Cocking, E. C., 1977, Callus formation from protoplasts of narrow stem kale, Plant Sci. Lett. **8:**275–280.

Gill, R. Rashid, A., and Maheshwari, S. C., 1978, Regeneration of plants from mesophyll protoplasts of Nicotiana plumbaginifolia Viv., Protoplasma **96:**375–379.

Gleba, Y. Y., and Hoffmann, F., 1978, Hybrid cell lines of Arabidopsis thaliana and Brassica campestris: No evidence for specific chromosome elimination, Mol. Gen. Genet. **165:** 257–264.

Gleba, Y. Y., and Hoffmann, F., 1980, "Arabidobrassica": A novel plant obtained by protoplast fusion, Planta **149:**112–117.

Glimelius, K., Eriksson, T., Grafe, R., and Müller, A. J., 1978, Somatic hybridization of nitrate reductase-deficient mutants of Nicotiana tabacum by protoplast fusion, Physiol. Plant. **44:**273–277.

Gressel, J., Ezra, G., and Jain, S. M., 1981, Genetic and chemical manipulation of crops to confer tolerance to chemicals, in: Chemical Manipulation of Crop Growth and Development (J. S. McLaren, ed.), Butterworth, London, pp. 1–13.

Harms, C. T., and Potrykus, I., 1978, Enrichment for heterokaryocytes by the use of iso-osmotic density gradients after plant protoplast fusion, Theor. Appl. Genet. **53:**49–55.

Harms, C. T., and Potrykus, I., 1979, Induced buoyant density shift of protoplasts and somatic hybridization studies with Citrus and tobacco, in: Abstr. 5th Internat. Protoplast Symp., Szeged, p. 64.

Haydu, Z., Lázár, G., and Dudits, D., 1977, Increased frequency of polyethylene glycol induced protoplast fusion by dimethylsulfoxide, Plant. Sci. Lett. **10:**357–360.

Hoffmann, F., and Adachi, T., 1981. Arabidobrassica: Chromosomal recombination and morphogenesis in asymmetric intergeneric hybrid cells, Planta **153:**586–593.

Iwai, S., Nagao, T., Nakata, K., Kawashima, N., and Matsuyama, S., 1980, Expression of nuclear and chloroplastic genes coding for fraction I protein in somatic hybrids of Nicotiana tabacum + rustica, Planta **147:**414–417.

Izhar, S., and Tabib, Y., 1980, Somatic hybridization in Petunia. Part 2: Heteroplasmic state in somatic hybrids followed by cytoplasmic segregation into male sterile and male fertile lines, Theor. Appl. Genet. **57:**241–245.

Kameya, T., 1973, The effects of gelatin on aggregation of protoplasts from higher plants, Planta **115:**77–82.

Kao, K. N., 1976, A method for fusion of plant protoplasts with polyethylene glycol, in: Cell Genetics in Higher Plants (D. Dudits, G. L. Farkas, and P. Maliga), Akadémiai Kiadó, Budapest, pp. 233–237.

Kao, K. N., 1977, Chromosomal behaviour in somatic hybrids of soybean–Nicotiana glauca, Mol. Gen. Genet. **150:**225–230.

Kao, K. N., and Michayluk, M. R., 1974, A method for high-frequency intergeneric fusion of plant protoplasts, Planta **115:**355–367.

Kao, K. N., and Michayluk, M. R., 1975, Nutritional requirements for growth of Vicia hajastana cells and protoplasts at a very low population density in liquid media, Planta **126:**105–110.

Kao, K. N., Gamborg, O. L., Michayluk, M. R., and Keller, W. A., 1971, Cell divisions in cells regenerated from protoplasts of soybean and Haplopappus gracilis, Nature New Biol. **232:**124.

Kao, K. N., Constabel, F., Michayluk, M. R., and Gamborg, O. L., 1974, Plant protoplast fusion and growth of intergeneric hybrid cells, Planta **120:**215–227.

Keller, W. A., and Melchers, G., 1973, The effect of high pH and calcium on tobacco leaf protoplast fusion, Z. Naturforsch. **28c:**737–741.

Krumbiegel, G., and Schieder, O., 1979, Selection of somatic hybrids after fusion of protoplasts from Datura innoxia Mill. and Atropa belladonna L., Planta **145:**371–375.

Kung, S. D., Gray, J. C., Wildman, S. G., and Carlson, P. S., 1975, Polypeptide composition of fraction I protein from parasexual hybrid plants in the genus Nicotiana, Science **187:**353–355.

Linsmeyer, E. M., and Skoog, F., 1965, Organic growth factor requirements of tobacco tissue cultures, *Physiol. Plant.* **18**:100–127.

Lu, C. -Y., Vasil, V., and Vasil, I. K., 1981, Isolation and culture of protoplasts of *Panicum maximum* Jacq. (guinea grass): Somatic embryogenesis and plantlet formation Z. *Pflanzenphysiol.* **104**:311–318.

Maliga, P., 1980, Isolation, characterization, and utilization of mutant cell lines in higher plants. in: *Internat. Rev. Cytol., Suppl. 11A* (I. K. Vasil, ed.), Academic Press, New York, pp. 225–250.

Maliga, P., Kiss, Z. R., Nagy, A. H., and Lázár, G., 1978. Genetic instability in somatic hybrids of *Nicotiana tabacum* and *Nicotiana knightiana*, *Mol. Gen. Genet.* **163**:145–152.

Melchers, G., and Labib, G., 1974, Somatic hybridization of plants by fusion of protoplasts. I. Selection of light resistant hybrids of "haploid" light sensitive varieties of tobacco, *Mol. Gen. Genet.* **135**:277–294.

Melchers, G., Sacristán, M. D., and Holder, A. A., 1978, Somatic hybrid plants of potato and tomato regenerated from fused protoplasts, *Carlsberg Res. Commun.* **43**:203–218.

Menczel, L., Lázar, G., and Maliga, P., 1978, Isolation of somatic hybrids by cloning *Nicotiana* heterokaryons in nurse cultures, *Planta* **143**:29–32.

Murashige, T., and Skoog, F., 1962, A revised medium for rapid growth and bioassays with tobacco tissue cultures, *Physiol. Plant.* **15**:473–497.

Nagata, T., 1978, A novel cell-fusion method of protoplasts by polyvinylalcohol, *Naturwiss.* **65**:263.

Nagata, T., and Melchers, G., 1978, Surface charge of protoplasts and their significance in cell–cell interaction, *Planta* **142**:235–238.

Nagata, T., and Takebe, I., 1971, Plating of isolated tobacco mesophyll protoplasts on agar medium, *Planta* **99**:12–20.

Nagata, T., Eibl, H., and Melchers, G., 1979, Fusion of plant protoplasts induced by a positively charged synthetic phospholipid, Z. *Naturforsch.* **34c**:460–462.

Nehls, R., 1978, The use of metabolic inhibitors for the selection of fusion products of higher plant protoplasts, *Mol. Gen. Genet.* **166**:117–118.

Nehls, R., 1981, Versuche zur Regeneration und Hybridisierung von Protoplasten höherer Pflanzen und zur Selektion von Fusionsprodukten, Diss., Univ. Kiel, pp. 1–91.

Ninnemann, H., and Jüttner, F., 1981, Volatile substances from tissue culture of potato and tomato and their somatic fusion products—Comparison of gas chromatic patterns for identification of hybrids, Z. *Pflanzenphysiol.* **103**:95–107.

Pierik, R. L. M., 1979, *In vitro Culture of Higher Plants—Bibliography*, Wageningen, pp. 1–149.

Potrykus, I., 1971, Intra and interspecific fusion of protoplasts from petals of *Torenia baillonii* and *Torenia fournieri*, *Nature New Biol.* **231**:57–80.

Potrykus, I., Harms, C. T., and Lörz, H., 1976, Problems in culturing cereal protoplasts, in: *Cell Genetics in Higher Plants* (D. Dudits, G. L. Farkas, and P. Maliga), Akadémiai Kiadó, Budapest, pp. 129–140.

Poulsen, C., Porath, D., Sacristán, M. D., and Melchers, G., 1980, Peptide mapping of the ribulose bisphosphate carboxylase small subunit from the somatic hybrid of tomato and potato, *Carlsberg Res. Commun.* **45**:249–267.

Power, J. B., Cummings, S. E., and Cocking, E. C., 1970, Fusion of isolated plant protoplasts, *Nature* **225**:1016–1018.

Power, J. B., Berry, S. F., Frearson, E. M., and Cocking, E. C., 1977, Selection procedures for the production of inter-species somatic hybrids of *Petunia hybrida* and *Petunia parodii*. I. Nutrient media and drug sensitivity complementation selection, *Plant Sci. Lett.* **10**:1–6

Power, J. B., Berry, S. F., Chapman, J. V., Sink, K. C., and Cocking, E. C., 1979, Somatic hybrids between unilateral cross-incompatible *Petunia* species, *Theor. Appl. Genet.* **5**:97–99.

Reinert, J., and Bajaj, Y. P. S., 1977, eds., *Plant Cell, Tissue and Organ Culture*, Springer Verlag, Berlin, pp. 1–803.

Reinert, J., and Gosch, G., 1976, Continuous division of heterokaryons from *Daucus carota* and *Petunia hybrida* protoplasts, *Naturwiss.* **63**:534.

Saxena, P. K., Gill, R., Rashid, A., and Maheshwari, S. C., 1981, Plantlet formation from isolated protoplasts of *Solanum melongena* L., *Protoplasma* **106**:355–359.

Schieder, O., 1975, Regeneration of haploid and diploid *Datura innoxia* Mill. mesophyll protoplasts to plants, *Z. Pflanzenphysiol.* **76**:462–466.

Schieder, O., 1978a, Somatic hybrids of *Datura innoxia* Mill. + *Datura discolor* Bernh. and of *Datura innoxia* Mill. + *Datura stramonium* L. var. tatula L. I. Selection and characterization, *Mol. Gen. Genet.* **162**:113–119.

Schieder, O., 1978b, Genetic evidence for the hybrid nature of somatic hybrids from *Datura innoxia* Mill., *Planta* **141**:333–334.

Schieder, O., Vasil, I. K., 1980, Protoplast fusion and somatic hybridization, in: *International Review in Cytology*, Suppl. 11B (I. K. Vasil, ed.), pp. 21–46, Academic Press, New York.

Shepard, J. P., and Totten, R. E., 1977, Mesophyll cell protoplasts of potato; isolation, proliferation, and plant regeneration, *Plant Physiol.* **60**:313–316.

Smith, H. H., Kao, K. N., and Combatti, N. C., 1976, Interspecific hybridization by protoplast fusion in *Nicotiana*, *Heredity* **67**:123–128.

Street, H. E., 1973, ed., *Plant Tissue and Cell Culture*, Blackwell, Oxford, pp. 1–503.

Street, H. E., 1979, Embryogenesis and chemically induced organogenesis, in: *Plant Cell and Tissue Culture* (W. R. Sharp, P. O. Larsen, E. F. Paddock, and V. Raghavan, eds.), Ohio State Univ. Press, Columbus, Ohio, pp. 123–153.

Takebe, I., Labib, G., and Melchers, G., 1971, Regeneration of whole plants from isolated mesophyll protoplasts of tobacco, *Naturwiss.* **58**:318–330.

Vardi, A., Spiegel-Roy, P., and Galun, E., 1975, *Citrus* cell culture: Isolation of protoplasts, plating densities, effect of mutagens and regeneration of embryos, *Plant Sci. Lett.* **4**:231–236.

Vasil, I. K., and Vasil, V., 1980, Isolation and culture of protoplasts, in: *Internat. Rev. Cytol.*, Supl. 11B (I. K. Vasil, ed.), Academic Press, New York, p. 1–19.

Vasil, V., and Vasil, I. K., 1980, Isolation and culture of cereal protoplasts. Part 2: Embryogenesis and plantlet formation from protoplasts of *Pennisetum americanum*, *Theor. Appl. Genet.* **56**:97–100.

Vasil, V., and Vasil, I. K., 1981a, Somatic embryogenesis and plant regeneration from suspension cultures of pearl millet (*Pennisetum americanum*), *Ann. Bot.* **47**:669–678.

Vasil, V., and Vasil, I. K., 1981b, Somatic embryogenesis and plant regeneration from tissue cultures of *Pennisetum americanum* x *P. purpureum* hybrid, *Am. J. Bot.* **68**:864–872.

Wallin, A., Glimelius, K., and Eriksson, T., 1978, Enucleation of plant protoplasts by cytochalasin B, *Z. Pflanzenphysiol.* **87**:333–340.

Wallin, A., Glimelius, K., and Eriksson, T., 1979, Formation of hybrid cells by transfer of nuclei via fusion of miniprotoplasts from cell lines of nitrate reductase deficient tobacco, *Z. Pflanzenphysiol.* **91**:89–94.

Weber, G., Constabel, F., Williamson, F., Fowke, L., and Gamborg, O. L., 1976, Effect of preincubation of protoplasts on PEG-induced fusion of plant cells, *Z. Pflanzenphysiol.* **79**:459–464.

Wetter, L. R., 1977, Isoenzyme patterns in soybean–*Nicotiana* somatic hybrid cell lines, *Mol. Gen. Genet.* **150**:231–235.

White, P. R., 1963, *The Cultivation of Animal and Plant cells*, 2nd ed., Ronald Press, New York, pp. 1–228.

Zelcer, A., Aviv, D., and Galun, E., 1978, Interspecific transfer of cytoplasmic male sterility by fusion between protoplasts of normal *Nicotiana sylvestris* and X-ray irradiated protoplasts of male-sterile *N. tabacum*, *Z. Pflanzenphysiol.* **90**:397–407.

Zimmermann, U., and Scheurich, P., 1981, High frequency fusion of plant protoplasts by electric fields, *Planta* **151**:26–32.

Chapter 34
Techniques for Chromosome Analysis

VAITHILINGAM G. DEV and RAMANA TANTRAVAHI

1. Introduction

Chromosome identification methods are widely used in somatic cell genetic studies, e.g., for ascertaining the normal karyotype and changes associated with transformation and neoplasia, identification of parental contribution of chromosomes in somatic cell hybrids and for gene mapping, and for verifying the cross-contamination of cell lines.

Before the introduction of quinacrine banding by Casperson et al. (1971), karyotyping was based mainly on the morphologic variation in the size of the chromosomes and the position of the centromere. Conventional staining of metaphase preparations with dyes such as orcein or Giemsa produced an even staining of the chromosomes which, at best, permitted the grouping of chromosomes, as in the case of humans. In the case of mouse, only the largest and smallest of the chromosomes could be distinguished. The limitation of such staining is illustrated in Fig. 1. The discovery of quinacrine and several later banding methods now enable us to unambiguously identify the chromosome complements in somatic cells and in somatic hybrid cells. By far the most commonly used method is Giemsa banding. However, under several circumstances other banding techniques may be not only useful but necessary for chromosome identification. Sometimes combinations of staining methods, e.g., Q-banding followed by C-banding, will be necessary in studies of interspecific somatic cell hybrids. A technique like Giemsa-11 is useful in identifying interspecific chromosome translocations. This chapter will describe in detail harvesting cell cultures for chromosome preparation and the various methods for chromosome identification. The mechanism of differential staining produced by these methods will also be discussed briefly.

VAITHILINGAM G. DEV • Department of Medical Genetics, University of South Alabama, College of Medicine, Mobile, Alabama 36617.   RAMANA TANTRAVAHI • Cytogenetics Laboratory, Sidney Farber Cancer Institute and Department of Medicine, Harvard Medical School, Boston, Massachusetts 02115.

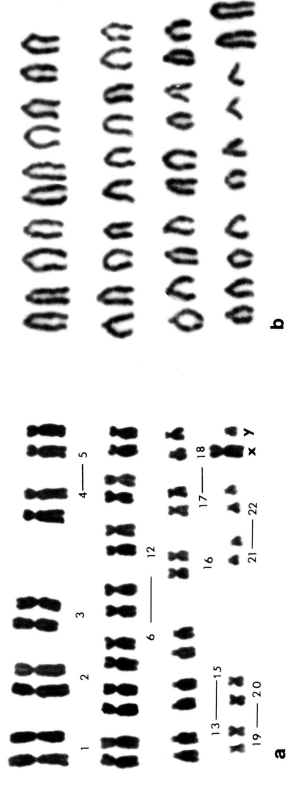

Figure 1. Unbanded karyotypes of (a) human and (b) mouse.

2. Harvesting and Chromosome Preparation

Since cell and tissue culture methods have been extensively covered in various textbooks (Paul, 1975) and in reviews, we are not suggesting here any special method of culturing for chromosome preparation. Metaphase is a stage in mitosis when chromosomes appear as visible structures. Harvesting cells which yield good mitotic chromosomes is essential for banding. In principle, cells in log phase are treated with one of the spindle inhibitors, such as Colcemid, to accumulate a sufficiently large number of metaphase cells. The cells are then swelled with a hypotonic treatment, fixed with acetic acid–methanol fixative, and spread on glass slides. The procedure is described below.

2.1. Reagents

1. Colcemid stock solution 10 μg/ml (can be purchased from GIBCO).
2. Trypsin, 0.25% in Hanks BSS.
3. 0.075 M KCl.
4. Clean slides kept in distilled water at 4°C.
5. Fixative, 3:1 methanol:acetic acid.

2.2. Procedure

1. Select semiconfluent cultures that are in log phase. Observe for mitotic cells that are rounded and loosely attached. Add Colcemid (0.1 ml of Colcemid to every 10 ml of media). The length of time for exposure to Colcemid varies with different kinds of cells. For fast-growing cells such as mouse L cells, human HeLa cells, Chinese hamster CHO cells, or somatic cell hybrids, about 30 min exposure is sufficient. The main function of Colcemid in such short exposure is not to accumulate metaphases, but rather to destroy the spindle fibers so as to get good chromosome spreading. For primary cultures from adult tissue or amniotic fluid cells, an exposure time of 2.5–4 hr may be necessary.
2. Pour the media into a 15-ml centrifuge tube. Add trypsin to the culture flask just to cover the cells and place the culture in the incubator until the cells detach from the monolayer (2–3 min is usually sufficient).
3. Add 5 ml of media to neutralize the trypsin and collect the detached cells along with the media and add to the media that was previously collected in the centrifuge tube.
4. Centrifuge at 1000 rpm for 10 min.
5. Discard the supernatant. Break the cell button (pellet) by tapping the bottom of the tube. Add, drop by drop, about 10 ml of hypotonic solution and mix by pipetting gently. Leave at room temperature for 10 min.
6. Centrifuge and discard the supernatant. Break the pellet and add about

10 ml of freshly prepared fixative. Leave at room temperature for 1 hr or overnight in the refrigerator.

7. Change the fixative twice, each time centrifuging and discarding the supernatant. Finally, suspend the cells in 0.5 ml of fresh fixative.

8. Remove a slide from water and shake off the excess water. Drop three drops of cell suspension onto the cold, wet slide, proceeding from left to right, holding the slide at an angle. Place the slide at 30° angle and let air dry. *Do not flame dry.* Store the slides in the refrigerator.

2.3. Technical Notes

1. Cleaning slides: Clean the slides with a laboratory detergent. A clean slide is a very important factor in good metaphase spreading and chromosome banding. Special attention is needed in this regard. Scrub the slides individually with fingers and rinse off the soap thoroughly. Rinse once in distilled water and store the slides in distilled water at 4°C.

2. If the chromosomes appear contracted and the sister chromatids stay apart, decrease the time of exposure or concentration of Colcemid. Insufficient metaphases mean either the time in Colcemid was not sufficient or the culture was not in log phase. Change the media in culture the day before harvesting. Suspension cultures need to be split 24 hr before harvesting.

3. If the spreading of the chromosomes is poor, increase the time in hypotonic solution and incubate at 37°C.

3. Quinacrine Banding (Q-Bands)

Metaphase chromosome preparations stained either with quinacrine mustard or quinacrine dihydrochloride and examined under ultraviolet illumination show characteristic bright and dull bands along the length of chromosomes. These patterns are consistent and reproducible. The observation that brilliant quinacrine fluorescence was produced in almost exclusively AT-rich DNA in *Drosophilia* (Ellison and Barr, 1972) was followed by studies of fluorescence from mixtures of AT-rich and GC-rich DNA. While AT-rich DNA produced bright fluorescence, the GC-rich DNA quenched the fluorescence (Weisblum and deHaseth, 1972). Thus, the bright and dull bands produced on chromosomes may represent the underlying differences in the base composition. The role of chromosomal proteins and the condensation of chromatin and other factors are also implicated in the banding produced by quinacrine (Comings, 1977, 1978). Figure 2 illustrates a human karyotype prepared by this method. The procedure is simple and a quick way of producing chromosome banding. The procedure does not involve any pretreatment of chromosome preparation. The advantages of Q-banding are (1) the same slides can be used for a second or even a third staining, (2) information on the quinacrine bright polymorphisms of human chromosomes

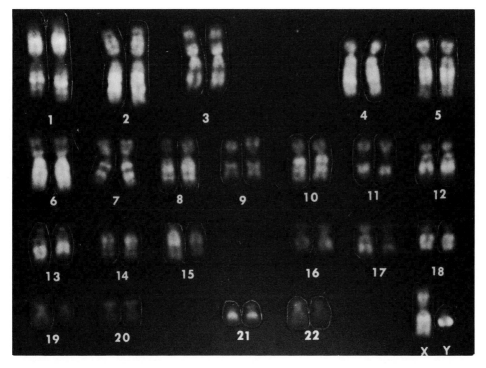

Figure 2. Quinacrine-banded karyotype of a human male. Note the brightly fluorescing Y chromosome and the Q-bright polymorphism on chromosomes 13, 21 and 22. [Courtesy Uma Tantravahi.]

is extremely useful in certain types of analyses, and (3) the distal part of human Y is brilliant and is readily recognizable. The disadvantages are (1) the staining is not permanent, (2) observations have to be recorded on a photographic film and analysis done with prints, and (3) under UV radiation there are problems, at times, of fading, resulting in uniformly dull chromosomes. Our Q-banding method, a modification of that developed by Casperson et al (1971), is described below.

3.1. Reagents

1. Quinacrine mustard or quinacrine dihydrochloride solution: To 50 ml of glass-distilled water in coplin jar add 2.5 mg of quinacrine mustard or 1 g of quinacrine dihydrochloride, mix thoroughly, wrap the jar with aluminum foil, and store in the refrigerator. Quinacrine mustard solution is stable up to 6 months. Quinacrine dihydrochloride solution is good up to two weeks at 4°C.

2. Tris-maleate buffer: Dissolve 24.2 g TRIZMA BASE (Sigma Chemical Co.) and 23.3 g maleic acid (Eastman Chemicals) in 1 liter of distilled water. For a working solution adjust 25 ml of this stock solution to pH 5.6 with

addition of 0.2 N NaOH and, using distilled water, dilute the resultant mixture to 100 ml. This solution may be stored at 4°C up to 2 weeks.

3.2. Procedure

1. Dehydrate the slide for 5 min in 3:1 methanol:acetic acid fixative and air dry.
2. Place the slide in the staining solution for 8–10 min and then rinse for 2 min in running tap water.
3. Rinse the slide in Tris-maleate buffer briefly and layer a drop of buffer on the slide. Place a coverslip of thickness 0 or 1 (22 × 55 mm) over the preparation.
4. Squeeze off excess buffer with a paper towel and seal the edges with clear nail polish or rubber cement.
5. Examine and photograph metaphases that are well spread and well banded, using a 63× or 100× oil immersion objective, using a filter combination of BG 38, 490 nm excitation filter, and 510 nm barrier filter. After photography, remove the coverslip, taking care not to get the oil on the slide, and rinse the slide briefly in distilled water. Pass the slide through a successive series of 50%, 70%, 95% ethanol, refix in 3:1 methanol:acetic acid fixative, dry, and store in the refrigerator until needed for further staining.

3.3. Technical Notes

1. Slides that are stored in the refrigerator have yielded good banding even after 3 years.
2. Photography is best done with Kodak Panatomic X film with an exposure time of 15–25 sec. Recommended developing solution is microdol × 1:3 dilution. Develop according to manufacturer's direction at 24°C for 11 min. For a detailed description and helpful comments on photography of Q-banding refer to Breg (1972).

4. Giemsa Banding (G-Bands)

Several methods are available for Giemsa banding: All of these methods involve a treatment of chromosomes with either a proteolytic enzyme such as trypsin (Seabright, 1972), mild denaturation of chromosomes with NaOH (Schnedl, 1971), or heating the slides in a saline citrate solution at 65°C (Sumner et al., 1971). Modification of chromosomal proteins or selective denaturation and renaturation of chromosomal DNA or both have been postulated as the mechanism of G-banding (Schnedl, 1974, Comings, 1978). The precise mechanism is still unclear. G-bands correspond to Q-bands with a few exceptions. In the human karyotype, quinacrine produces brilliant fluorescence in the distal part of the Y chromosome, while G-banding stains the heterochromatic regions of chromosomes 1 and 16 darkly. Quinacrine does not produce bright fluorescence in these regions. The polymorphic Q-bright regions do not stain darkly with Giemsa. In the mouse the centromeric regions stain darkly with Giemsa but show dull fluorescence with quinacrine. The advantages of the G-banding method are (1) the staining is permanent, (2) no fluorescent microscopy is necessary, (3) the result is easily photographed. A special usefulness of G-banding is in the analysis of mouse–human and mouse–Chinese hamster somatic cell hybrids. The darkly staining centromeric heterochromatin of mouse chromosomes enables one to easily discriminate mouse chromosomes from human and hamster chromosomes. Although the different G-banding methods work well in different laboratories, the following method was found to produce consistently good results in somatic cell hybrids. A G-banded mouse karyotype is presented in Fig. 3.

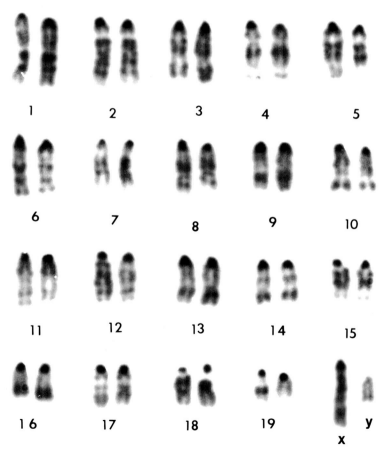

Figure 3. Giemsa-banding karyotype of mouse. Note the darkly stained centromeric regions.

4.1. Reagents

1. Phosphate-buffered saline (PBS): Dissolve 8 g of NaCl, 0.2 g of KH_2PO_4, and 1.16 g of Na_2HPO_4. Store this solution at 4°C.
2. Stock trypsin solution: 1.25 g of Difco 1:250 trypsin in 200 ml of distilled water that has been acidified to pH 2.5 with concentrated HCl. Dispense into 1-ml quantities and store in freezer.
3. Phosphate–citrate buffer, pH 6.8:9.1 ml of 0.1 M citric acid plus 40.9 ml of 0.2 M Na_2HPO_4 and dilute to 100 ml with distilled water. Buffer tablets obtained from Bio Medical Specialties (Santa Monica, California) can be substituted for the phosphate–citrate buffer.
4. Gurr's improved Giemsa R66 obtained from Bio Medical Specialties.

4.2. Procedure

1. Prepare Giemsa staining solution by diluting 4 ml Giemsa stain into 100 ml in phosphate–citrate buffer.
2. Prepare working trypsin solution by adding 1 ml of trypsin stock to 49 ml of PBS. Adjust to pH 7 with 0.2 M in Na_2HPO_4.
3. Immerse the slides in trypsin working solution for 5–25 sec at room temperature.
4. Rinse in PBS for 30 sec and for another minute in PBS kept in a second staining jar.
5. Stain in the working Giemsa stain solution for 4–5 min.
6. Rinse in distilled water for 5–10 sec and air dry.
7. If a long time of storage of the slide is needed, clear the slide in xylene and mount a coverslip with mounting medium (Permount or Deepex).

4.3. Technical Notes

1. The working solution of trypsin should be prepared last, and all slides should be pretreated within 10 min after the pH is adjusted.
2. Speed and accuracy are essential, especially in regard to the trypsin and staining solutions.
3. Time of trypsin treatment may vary with each harvest and age of slides. A test slide needs to be processed.
4. For consistent results the slides should be aged for 4–7 days in a 40°C dry incubator. Slides that are stored in a refrigerator can also be banded by incubating overnight at 60°C. Satisfactory banding can be obtained on fresh slides by heating them for 2 hr at 90–95°C.
5. Excess treatment in trypsin results in fuzzy chromosomes that appear "chewed up" and under treatment produces even, dark staining without any appreciable differential staining.

6. Quality of metaphases determines banding quality. Contracted, poorly fixed chromosomes result in poor banding.

7. Kodak technical Pan film 2415 and developing in Kodak HC110 developer, dilution F according to the manufacturer's instruction are found to be ideal for photography of G-bands. Print on a contrast grade 2 paper.

5. Reverse Banding with Chromomycin A_3/Methyl Green (R-Bands)

Banding patterns that are reverse to Q- and G-bands can be produced when chromosome preparations are stained with acridine orange (Bobrow et al., 1972) or Giemsa (Dutrillaux and Lejeune, 1971; Sehested 1974) after the slides are incubated in hot buffers. At high temperatures the AT-rich DNA sequences denature, producing single-stranded DNA. When acridine binds to single-stranded DNA it produces reddish fluorescence; it produces yellow green fluorescence in native DNA. The mechanism of Giemsa R-banding is obscure, because the mode of binding of Giemsa to single-stranded versus double-stranded DNA remains unknown (Comings, 1977, 1978).

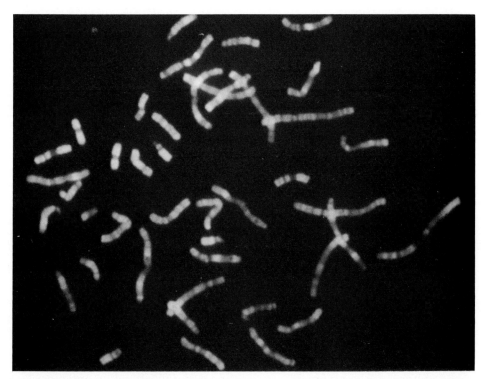

Figure 4. A human metaphase cell stained with chromomycin A_3/methyl green, showing a reverse banding pattern.

The R-Bands can also be produced by using a combination of dyes (Sahar and Latt, 1978). R-Bands produced by acridine orange or by combination of dyes require fluorescence microscopy, while R-bands produced by Giemsa can be examined and photographed under bright field optics. R-Bands are extremely useful to study chromosomes and chromosome regions that are dull with quinacrine and pale with Giemsa, because the same regions show bright fluorescence with acridine orange or chromomycin A_3/methyl green, and dark staining with Giemsa. Because of rapid fading, photography of fluorescent R-bands requires relatively fast films. Giemsa R-bands, like G-bands, are permanent and the same slides cannot be used for a second procedure.

The method using chromomycin A_3/methyl green where enhanced R-banding is produced by energy transfer (Latt et al., 1980) is highly recommended. This method does not involve incubation at high temperature and can easily be combined with other methods and is described below. A human metaphase stained with this method is presented in Fig. 4.

5.1. Reagents

1. 0.14 M Phosphate buffer with 500 μM $MgCl_2$. Dissolve 9.66 g of $NaH_2PO_4 \cdot H_2O$ into distilled water and make up to 500 ml. This can be stored up to 4 weeks in a refrigerator. Dissolve 9.95 g of Na_2HPO_4 in distilled water and make up to 500 ml. This can be stored up to 4 weeks in a refrigerator. Titrate NaH_2PO_4 with Na_2HPO_4 to 6.8 pH, and add 10 mg $MgCl_2$ for every 100 ml buffer. The buffer is good for 2 weeks in refrigerator.

2. 500 μM Chromomycin A_3. Dissolve 10 mg of chromomycin A_3 (Cal-Biochem) in 16 ml of phosphate buffer. Work in subdued light and stir with a magnetic stirrer for 10–15 min. Dispense 0.2-ml aliquots into freezing vials and freeze.

3. 0.15 M NaCl/0.005 M HEPES buffer pH 7.0. HEPES powder is purchased from GIBCO and stored at room temperature. To 1.75 g of NaCl and 238 mg of HEPES add distilled water and make up to 200 ml. Titrate to pH 7.0 with 1 N NaOH.

4. 100 μM Methyl Green in NaCl/HEPES buffer. Dissolve 2.6 mg of methyl green (Eastman Co.; stored at room temperature) in 50 ml of NaCl/HEPES buffer *just before use*. Methyl Green is unstable in this solution.

5.2. Procedure

1. Place the slide in a humid chamber (a petri dish with moist toweling and two glass rods) and layer with 0.14 M phosphate buffer pH 6.8 with 500 μM $MgCl_2$ for 10 min.

2. Shake off the buffer, return the slide to humid chamber, and place 2–3 drops of 500 μM chromomycin A_3 made in phosphate buffer. Cover with a coverslip and stain for 10 min.

3. Shake off the coverslip and layer the slide with 0.15 M NaCl/0.005 M HEPES buffer pH 7.0. Discard the buffer and repeat the process.

4. Place the slide in a humid chamber and flood the slide with 100 μM freshly prepared methyl green made in NaCl/HEPES buffer pH 7.0 and stain for 10 min at room temperature. (Slides can also be stained in Methyl Green kept in a coplin jar.)

5. Rinse the slide twice in NaCl/HEPES buffer; dry the slide.

6. Mount the slide in glycerol using 0 or 1 thickness 22 $\times$ 50 mm coverslip.

7. Squeeze out excess glycerol in several folds of paper towels; remove excess glycerol from the slide with a warm, moist toweling.

8. Seal the edges with nail polish or rubber cement.

9. Observe and photograph as described in the Q-banding method. Kodak Plus X Pan film is used with an exposure time of 3–5 sec.

10. Remove the coverslip and wash the slide in two changes of PBS and distilled water and dry.

11. Pass the slide through 50%, 70%, 95%, and absolute ethanol (5 min each).

12. Refix the slide in 3:1 (v/v) methanol:acetic acid for 10 min. The slide can be used for a second staining procedure such as Q-, C-, or NOR-banding.

6. Constitutive Heterochromatin Banding (C-Bands)

Constitutive heterochromatin is present in varying amounts in virtually all the mammalian genomes. This heterochromatin consists of satellite DNA of highly repetitive short DNA sequences. The satellite DNA in the heterochromatin of mouse chromosomes was first demonstrated by Pardue and Gall (1970) by *in situ* hybridization. By following the pretreatments required for *in situ* hybridization and then staining with Giemsa, Arrighi and Hsu (1970) localized the constitutive heterochromatin regions in the human chromosomes. The regions that stain darkly by this method are called C-bands. In the human the major sites of C-bands are on the secondary constrictions of chromosomes 1, 9, and 16 and on the distal part of the long arm of the Y chromosome. In addition to these sites, C-bands are also seen on the centromeric regions of all the chromosomes, although to a lesser degree. Most of the chromosomes in the mouse karyotype except the Y contain the C-band material at the centromeric region. The prominent C-bands of the mouse chromosomes facilitate their recognition in the interspecific somatic cell hybrids. The heritable polymorphism in the size of the C-bands enables one to identify the parental origin of several pairs of homologs in inbred strains of mice (Dev et al., 1973). Although C-banding can be used to distinguish chromosomes of one species from the other in somatic cell hybrids, a general banding method such as Q- or R-banding is necessary to identify the particular chromosome within the species.

Alkali denaturation using NaOH was the method of Arrighi and Hsu

(1971), while Sumner (1972) used barium hydroxide and Dev et al. (1973) used formamide as denaturing agents. The mechanism involved in C-band production is thought to be by denaturation of DNA followed by reassociation of highly repetitive DNA sequences in hot SSC. Upon staining with Giemsa, the regions containing reassociated sequences stain darkly. However, several authors demonstrated considerable loss of unique-sequence DNA by these procedures. The loss of DNA may, at least in part, be responsible for the formation of C-bands (Comings 1978).

Constitutive heterochromatin in plant as well as mammalian genomes is rich in 5-methylcytosine. When chromosomes are denatured by ultraviolet irradiation and allowed to react with anti-5-methylcytosine antibodies, certain regions show considerable binding (Miller et al., 1974; Schreck et al., 1974).

Centromeric heterochromatin can also be demonstrated by sequential staining of metaphase chromosomes with AT-specific DNA ligand distamycin followed by staining with DAPI (Diamidinophenylindol), an AT-specific fluorochrome (Schweizer et al., 1978). Similar combinations of dyes are Hoechst 33258 and Netropsin (Sahar and Latt, 1980). The mechanism involved in the above two methods is one of dye competition for binding sites.

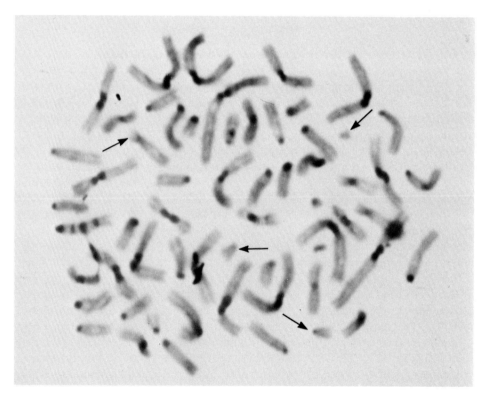

Figure 5. A C-banded metaphase from a mouse–human somatic hybrid cell. The arrows indicate some of the human chromosomes. Note that the centromeric regions of the mouse chromosomes are more darkly stained than are those of the human chromosomes.

A C-banding method using barium hydroxide is described here. We favor this method over the others because of the simplicity and reproducibility. Like any other banding method, we have found the following modification necessary to obtain good results in our laboratories. A C-banded metaphase cell is presented in Fig. 5.

6.1. Reagents

1. Barium hydroxide: A saturated solution in distilled water is prepared just before use.
2. Ethanol, 95% and 70%.
3. 0.85% saline.
4. Giemsa stain: 4% made in phosphate–citrate buffer, pH 6.8. Buffer tablets can obtained from Bio-Medical Specialities, who supply Gurr R66 Giemsa stain.
5. 2× SSC: 0.3 M sodium chloride and 0.03 sodium citrate. A stock solution 20× SSC is prepared by dissolving 35 g of NaCl and 17.79 g of trisodium citrate in 200 ml of distilled water. Make a 1:10 dilution with distilled water just before use.

6.2. Procedure

1. After the slide is photographed for Q-bands or R-bands, remove the coverslip. Slides that were stored for even up to 3 years yielded good results.
2. Dip the slide 4–6 times in a coplin jar containing 95% ethanol.
3. Dip the slide 4–6 times in saline. This is prepared fresh.
4. Prepare a saturated solution of barium hydroxide with distilled water in a beaker or coplin jar. Allow the salt to sediment and transfer the supernate into a second coplin jar.
5. Place the slide in barium hydroxide for 2–3 min.
6. Dip the slide 4–6 times in 70% ethanol.
7. Repeat step 6 with 70% ethanol in a second coplin jar.
8. Dip the slide 4–6 times in saline in a second coplin jar.
9. Incubate the slide for 4 hr in 2× SSC in a coplin jar set in a water bath at 60°C.
10. Stain the slide for 2–3 min in 4% Giemsa.
11. Rinse in a beaker of distilled water and dry.
12. Wash in xylene for 5 min and mount in Permount if long-term storage is necessary.
13. Slides can be observed and photographed under bright-field optics.
14. Photograph the same metaphases previously photographed for Q-bands or R-bands on Kodak Technical Pan film 2415. Develop in D19 at 20°C for 4 min. Print on grade 2 paper.

6.3. Technical Notes

The time of treatment in barium hydroxide is the most critical step in the whole procedure. Leaving the slide too long in barium hydroxide produces uniformly pale-looking chromosomes. Not treating the slide for enough length of time produces uniformly dark (looking like Giemsa-stained) chromosomes.

Prior UV exposure of metaphases for observation and photography for Q-bands or R-bands affects the time required (less time) in barium hydroxide. The 2–3 minutes in Ba(OH)$_2$ recommended is after a cell is exposed for 25 sec for observation and photography of Q-bands. Human and mouse slides that were not Q- or R-banded may require 7–10 min of barium hydroxide treatment.

7.1. Staining for Active Nucleolus Organizers

Nucleolus organizer regions appear as unstained regions in metaphase chromosomes and are called secondary constrictions. These regions contain multiple copies of genes that code for 18S plus 28S ribosomal RNA (rRNA) (Henderson et al., 1972; Elsevier and Ruddle, 1975). The number of chromosomes that carry these genes vary from one pair to several pairs, depending on the species. A silver staining method developed by Howell et al. (1975) selectively stained the NOR which are located on human acrocentric chromosomes. NOR of several species were demonstrated by this method (Goodpasture and Bloom, 1975; Tantravahi et al., 1976; Dev et al., 1977).

In mouse–human somatic cell hybrids that synthesize only mouse rRNA (Eliceiri and Green, 1969), the mouse NORs were stained and human NORs were not (Miller et al., 1976). In the somatic cell hybrids that synthesize only human rRNA, only the human NORs were stained (Miller et al., 1976). Both parental NORs were stained when both types of rRNAs were synthesized (Miller et al., 1978; Weide et al., 1979). Thus, silver staining indicates the active NORs and can be used to study the rRNA gene activity on a cell-by-cell basis.

The mechanism of this selective binding of silver to the NOR is not completely understood. However, there is evidence that the proteins, and not the DNA, in these regions are responsible for the staining. Examples of rRNA gene activity as shown by staining on somatic hybrid cells are shown in Fig. 6.

The most commonly used method is the one published by Goodpasture and Bloom (1975). A method recently reported by Howell and Black (1980) is found to be simpler and superior to the other methods and is described here.

7.1. Reagents

1. 50% silver nitrate: Dissolve 5 g silver nitrate in 10 ml of distilled water and filter through a Millipore filter (unfiltered solution also has been found to be satisfactory). Store in a glass container wrapped with aluminum foil. Stable up to 1 year.

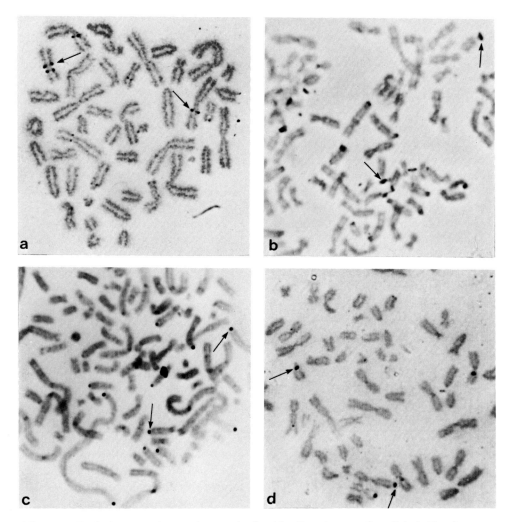

Figure 6. Nucleolus organizer regions stained with silver in somatic cell hybrids. Arrows indicate some of the silver-stained NORs. (a) Mouse–human somatic cell hybrid where only the mouse NORs stained. (b) Mouse–human somatic cell hybrid where only the human NORs stained. (c) Hamster–human cell hybrid where only hamster NORs stained. (d) Hamster–mouse hybrid where both parental NORs stained.

2. Gelatin developer: This solution is prepared by dissolving 2 g. of powdered gelatin in 99 ml of distilled water and 1 ml of formic acid.

7.2. Procedure

1. Add four drops of silver nitrate and one drop of gelatin on the slide and place a coverslip over it.
2. Place the slide on a hot plate or incubator which is maintained at 56–60° until the slide turns golden brown (usually in 3 min).

3. Remove the coverslip and rinse in distilled water as quickly as possible and air dry.

4. Photography is done as in the C-banding method.

7.3. Technical Notes

1. Mounting a coverslip is not necessary. Sometimes mounting a coverslip displaces the silver grains.
2. If the rest of the chromosomes except the NOR are pale, counterstain the slide with 1% Giemsa for 1 min.
3. Silver nitrate leaves an indelible mark on skin, clothes, and on work areas. Use extreme care not to get the solution on you, countertop, and floor.
4. The silver precipitate from the slides can be removed for a second staining by dipping the slide in 1% solution of potassium ferricyanide prepared in hypo solution, which is used for fixing photographic paper.
5. This method should be combined with a general banding method for identification of chromosomes.

8. Giemsa-11 Staining

Staining chromosome preparations with Giemsa stain at high alkaline pH produces magenta color of mouse chromosomes and a blue staining of human chromosomes (Bobrow and Cross, 1974). The mechanism involved in the production of this differential staining remains unknown. This method is very useful for rapid screening of a large number of mouse–human somatic cell hybrids to detect translocation of interspecific chromosomal material too small to be detected by other banding methods. A modified alkaline-Giemsa method reported by Friend et al. (1976) is presented here. A somatic hybrid cell stained by this method is shown in Fig. 7.

8.1. Reagents

1. 0.05 M Na_2HPO_4 adjusted to pH 11–11.3 by the addition of NaOH.
2. Giemsa-11 stain: Prepare fresh by adding approximately 1 ml of Giemsa stain to 50 ml of the above buffer, prewarmed to 37°C.

8.2. Procedure

1. Soak the slides for 1–2 hr in six changes of distilled water.
2. Stain for 3–5 min in freshly prepared alkaline Giemsa solution at 37°C.
3. Rinse briefly in distilled water and air dry.

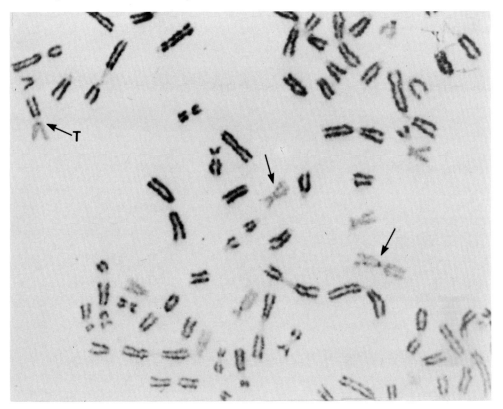

Figure 7. Giemsa-11 staining of a somatic hybrid cell. The arrows indicate some of the lightly stained human chromosomes and a translocation (indicated by T) between a human and a mouse chromosome. [Courtesy R. S. Kucherlapati and by permission of Academic Press (Friend, et al., 1976).]

8.3. Technical Notes

1. Mix the staining solution gently to prevent formation of surface film on the slide.
2. Use freshly prepared stain solution for every batch of three slides.
3. Slides that were previously stained with quinacrine should be soaked in 3:1 methanol: acetic acid fixative for 1–2 hr with 2–3 changes of fixative.
4. Photography is done as in the C-banding method.

Acknowledgments

The authors thank Debbie Ryan for excellent technical assistance in preparing the illustrations, and Betty Thompson and Beverly Bethell for their assistance in manuscript preparation. Special thanks go to George Widney for critical reading and for improvements in the style of the manuscript.

References

Arrighi, F. E., and Hsu, T. C., 1971, Localization of heterochromatin in human chromosomes, *Cytogenetics* **10**:81–86.

Bobrow, M., and Cross, J., 1974, Differential staining of human and mouse chromosomes in interspecific cell hybrids, *Nature* **251**:77–79.

Bobrow, M., Collacott, H. E., and Madan, K., 1972, Chromosome banding with acridine orange, *Lancet* **2**:1311.

Breg, W. R., 1972, Quinacrine fluorescence for identifying metaphase chromosomes, with special reference to photomicrography, *Stain Technology* **47**:87–93.

Casperson, T., Lomakka, G., and Zeck, L., 1971, The 24 fluorescence patterns of the human metaphase chromosomes—distinguishing characters and variability, *Heriditas* **67**:89–102.

Comings, D. E., 1977, Chromosome banding and chromosomal proteins, in: *Molecular Human Cytogenetics, ICN–UCLA Symposium on Molecular and Cellular Biology*, Volume VII (R. S. Sparkes, D. E. Comings, and C. F. Fox, eds.), Academic Press, New York, pp. 65–74.

Comings, D. E., 1978, Mechanism of chromosome banding and implications for chromosome structure, *Ann. Rev. Genet.* **12**:25–46.

Comings, D. E., Avelino, E., Okada, T. A., and Wyndat, H. E., 1973, The mechanism of banding of chromosomes, *Exp. Cell Res.* **77**:469–493.

Dev, V. G., Miller, D. A., Allderdice, P. W., and Miller, O. J., 1972, Method for locating centromeres of mouse meiotic chromosomes and its application to T163H and T70H translocations, *Exp. Cell Res.* **73**:259–262.

Dev, V. G., Miller, D. A., and Miller, O. J., 1973, Chromosome markers in Mus musculus: Strain difference in c-banding, *Genetics* **75**:663–670.

Dev, V. G., Tantravahi, R., Miller, D. A., and Miller, O. J., 1977, Nucleolus organizers in mus musculus subspecies and in the Rag cell line, *Genetics* **86**:389–398.

Dutrillaux, B., and Lejeune, J., 1971, Sur un novelle technique d'analyse du caryotype humain, *C. R. Acad. Sci. (Paris)* **27**:2638–2640.

Eliceiri, G. L., and Green, H., 1969, Ribosomal RNA synthesis in human–mouse hybrid cells, *J. Mol. Biol.* **4**:253–260.

Ellison, J. R., and Barr, H. J., 1972, Quinacrine fluorescence of specific regions: Late replication and high A:T content in Samoaia leonenzis, *Chromosoma (Berl.)* **36**:375–390.

Elsevier, S. M., and Ruddle, F. H., 1975, Location of genes coding for 18S and 28S ribosomal RNA within the genome S1 Mus musculus, *Chromosoma (Berl.)* **52**:219–228.

Friend, K. K., Dorman, B. P., Kucherlapati, R. S., and Ruddle, F. H., 1976, Detection of interspecific translocations in mouse–human hybrids by alkaline-Giemsa staining, *Exp. Cell Res.* **99**:31–36.

Goodpasture, C., and Bloom, S. E., 1975, Visualization of nucleolar organizer regions in mammalian chromosomes using silver staining, *Chromosoma (Berl.)* **53**:37–50.

Henderson, A. S., Warburton, D., and Atwood, K. C., 1972, Location of rDNA in the human chromosome complement, *Proc. Natl. Acad. Sci. USA* **69**:3394–3398.

Howell, W. M., and Black, D. A., 1980, Controlled silver staining of nucleolus organizer regions with a protective colloidal developer: in a one step method, *Experientia* **36**:1014–1015.

Howell, W. M., Denton, T. E., and Diamond, J. R., 1975, Differential staining of the satellite regions of human acrocentric chromosomes, *Experientia* **31**:260–262.

Jones, K. W., and Corneo, G., 1971, Locations of satellite and homogeneous DNA sequences on human chromosomes, *Nature New Biol.* **233**:268–271.

Latt, S. A., Jurgens, L. A., Matthews, D. J., Gustashan, K. M., and Sahar, E., 1980, Energy transfer-enhanced chromosome banding. An overview, *Cancer Genet. Cytogenet.* **1**:187–196.

Miller, D. A., Dev, V. G., Tantravahi, R., and Miller, O. J., 1976, Suppression of human nucleolus organizer activity in mouse–human somatic hybrid cells, *Exp. Cell Res.* **101**:235–243.

Miller, O. J., Schnedl, W., Allen, J., and Erlanger, B. F., 1974, 5-Methylcytosine localized in mammalian heterochromatin, *Nature* **251**:636–637.

Miller, O. J., Miller, D. A., Dev, V. G., Tantravahi, R., and Croce, C. M., 1976, Expression of human and suppression of mouse nucleolus organizer activity in mouse–human somatic cell hybrids, *Proc. Natl. Acad. Sci. USA* **73**:4531–4535.

Miller, O. J., Dev, V. G., Miller, D. A., Tantravahi, R., and Eliceiri, C. L., 1978, Transcription and processing of both mouse and hamster ribosomal RNA genes, *Exp. Cell Res.* **115**:457–460.

Pardue, M. L., and Gall, J. G., 1970, Chromosomal localization of mouse satellite DNA, *Science* **168**:1356–1358.

Paul, J., 1975, *Cell and Tissue Culture*, 5th ed., Churchill Livingstone, Edinburgh.

Sahar, E., and Latt, S. A., 1978, Enhancement of banding patterns in human metaphase chromosomes by energy transfer, *Proc. Natl. Acad. Sci. USA* **75**:5650–5654.

Sahar, E., and Latt, S. A., 1980, Energy transfer and binding competition between dyes used to enhance staining differentiation in metaphase chromosomes, *Chromosoma (Berl.)* **79**:1–28.

Schnedl, W., 1971, Analysis of the human karyotype using a reassociation technique, *Chromosoma (Berl.)* **34**:448–454.

Schnedl, W., 1974, Banding pattern in human chromosomes, in: *Methods in Human Cytogenetics* (H. G. Schwarzacher, U. Wolf, and E. Passarge, eds.), Springer, Berlin, pp. 95–117.

Schreck, R. R., Warburton, D., Miller, O. J., Beiser, S. M., and Erlanger, B. F., 1973, Chromosome structure as revealed by a combined chemical and immunochemical procedure, *Proc. Natl. Acad. Sci. USA* **70**:804–807.

Schreck, R. R., Erlanger, B. F., and Miller, O. J., 1974, The use of antinucleoside antibodies to probe the organization of chromosomes denatured by ultraviolet irradiation, *Exp. Cell Res.* **83**:31–39.

Schweizer, D., Ambros, P., and Andrle, M., 1978, Modifications of DAPI banding on human chromosomes by prestaining with DNA-binding oligopeptide antibiotic, Distamycin A, *Exp. Cell Res.* **111**:327–332.

Seabright, M., 1971, A rapid banding technique for human chromosomes, *Lancet* **2**:971–972.

Sehested, J., 1974, A simple method for R-banding of human chromosomes, showing a pH-dependent connection between R and G bands, *Humangenetik* **21**:55–58.

Sumner, A. T., 1972, A simple technique for demonstrating centromeric heterochromatin, *Exp. Cell Res.* **75**:304–306.

Sumner, A. T., Evans, H. J., and Buckland, R. A., 1971, New technique for distinguishing between human chromosomes, *Nature New Biol.* **232**:31–32.

Tantravahi, R., Miller, D. A., Dev, V. G., and Miller, O. J., 1976, Detection of nucleolus organizer regions in chromosomes of chimpanzee, gorilla, orangutan and gibbon, *Chromosoma (Berl.)* **56**:15–27.

Weide, L. G., Dev, V. G., and Rupert, C. S., 1979, Activity of both mouse and chinese hamster ribosomal RNA genes in somatic cell hybrids, *Exp. Cell Res.* **123**:424–429.

Weisblum, B., and deHaseth, P., 1972, Quinacrine, a chromosome stain specific for deoxyadenylate-deoxythymidylate rich regions in DNA, *Proc. Natl. Acad. Sci. USA* **69**:629–632.

Chapter 35

Genetic Analysis of Hybrid Cells Using Isozyme Markers as Monitors of Chromosome Segregation

STEPHEN J. O'BRIEN, JANICE M. SIMONSON, and MARY EICHELBERGER

The construction of somatic cell hybrids between different mammalian species has provided an important technique for the development of genetic maps of a variety of biologic species. The human genetic map has been derived almost exclusively from genetic analysis of rodent × human cells (Ruddle and Creagan, 1975; McKusick and Ruddle, 1977). The murine genetic map originally progressed largely through sexual genetic analysis between inbred strains homozygous for different alleles at various loci (Miller and Miller, 1972; Davisson and Roderick, 1980). However, the addition to the map of numerous biochemical loci which did not vary between inbred strains has more recently been provided by analysis of mouse × hamster cell hybrids which preferentially segregate murine chromosomes (Lalley et al., 1978a,b; Francke and Taggart, 1980; Womack, 1980). In addition, fairly extensive biochemical genetic maps of chimpanzee, gorilla, orangutan, and the domestic cat have been prepared using similar technologies (Pearson et al., 1979, 1981; O'Brien and Nash, 1982).

The majority of genes which have been mapped in these species are structural genes for soluble enzymes which can be histochemically visualized following gel electrophoresis. During evolutionary divergence of biologic species, nucleotide substitutions accumulate in genomic DNA, become polymorphic, and are subsequently fixed. When such nucleotide substitutions occur in regions of enzyme structural genes in which amino acid substitutions are not deleterious to enzyme activity, charge changes can be

---

STEPHEN J. O'BRIEN, JANICE M. SIMONSON, and MARY EICHELBERGER • Section of Genetics, Laboratory of Viral Carcinogenesis, National Cancer Institute, Frederick, Maryland 21701.

incorporated into the enzyme's primary structure. Thus, comparison of homologous enzymes of related species often shows differences upon aqueous starch or acrylamide gel electrophoresis. Such isozyme differences are common between many biologic species for virtually hundreds of isoenzyme (isozyme) systems. It has therefore been possible to monitor chromosome segregation in a collection of hybrid clones by isozyme staining of virtually any enzyme whose mobility varies between the two parental species.

Interspecific somatic cell hybrids provide an important technology for gene mapping because it is possible to generate hybrids which preferentially segregate the chromosomes of one parental species in different combinations among the hybrid clones. Concordant expression and loss of any two isozymes form the basis for identification of a syntenic group in the segregant parent. The syntenic groups, which are analogous to linkage groups derived from sexual genetic crosses (Ruddle et al., 1971; McKusick and Ruddle, 1977), presumably represent groups of loci which reside on individual chromosomes. The empirical definition of a syntenic group of multiple loci depends upon two important observations: (1) the concordant appearance of the markers in a hybrid panel, and (2) substantial discordancy with all the other markers followed in the same cross. Assignment of specific syntenic groups to individual chromosomes can then be achieved by demonstration of concordant segregation of the syntenic group with a particular G-banded chromosome.

The majority of isozyme systems employed in our laboratory are resolved on vertical starch gels using the Buchler Instruments (Fort Lee, New Jersey) apparatus at 4°C. It is possible to employ other media (acrylamide, cellulose acetate filters, agarose gels, and isoelectric focusing) for many of these enzymes; however, we find starch to be the most efficient and flexible for a variety of enzymes. A more thorough discussion of the advantages and disadvantages of different gel media has been published elsewhere (Brewer, 1970; O'Brien and MacIntyre, 1976). The method for preparing starch includes cooking and subsequent aspiration of 12–15% hydrolyzed starch (Electrostarch Co., Madison, Wisconsin) in an appropriate gel buffer with various modifications specified by the manufacturer for each lot of starch. We have described the specific step-by-step procedure for starch preparation used in our laboratory elsewhere (O'Brien and Barile, 1981).

In order to analyze a series of hybrid clones, each line is expanded to 5–10 confluent T-75 tissue culture flasks, trypsinized, and centrifuged. The cell pellet is resuspended in 3–5 volumes of hypotonic buffer (0.01 M Tris, 0.001 M Na_2 EDTA, pH 7.1) and frozen-thawed 3× either in a −70°C freezer or in a dry ice/ethanol bath. The crude extract is centrifuged at 27,000 for 20 min at 4°C. The supernatant can then be stored at −70°C for months. For electrophoresis, 15–35 μl of extract is applied to the gel and separated overnight at 4°C as described (O'Brien and Barile, 1981). After electrophoresis, the gel is sliced horizontally into three slabs, each of which is stained for a different isozyme. The selected isozymes are visualized by application of a histochemical stain to the gel slab. Buffer systems and biochemical recipes for over 100 isozymes resolved in mammalian cells have been published (Nichols and Ruddle, 1973; Harris and Hopkinson, 1976; Siciliano and Shaw, 1976) and will not be recapitulated here. Virtually all the stains involve a soluble colorless enzyme substrate and a soluble dye which produces a colored precipitate upon a chemical reaction with the enzyme's product.

The isozymes and the derived genetic maps in various species provide a useful basis for determination of the segregant chromosome constitution of a series of hybrid clones. In our laboratory, we routinely run five gels daily and usually develop 3–4 isozyme systems per gel. Thus, it becomes possible to type 70 hybrid extracts for the complete human chromosome complement in less than 2 man-weeks. To provide the same data by G-banded karyologic analysis may take up to 6 months. Thus, it is especially important to develop the isozyme screens in a mammalian gene mapping laboratory. Finally, the isozyme screening provides an important preliminary procedure for selecting hybrids with useful combinations of segregant chromosomes. A typical cell fusion generating 100 clones often has 50% that are useless, either because they have segregated all the segregant parental chromosomes or conversely have retained them all.

The scoring of isozyme patterns in hybrid cells is theoretically quite straightforward. Actually, however, there are several types of complications which can occur that will interfere with and often preclude unambiguous scoring. We shall attempt to address each of these situations in this chapter, with specific emphasis on our experience with hybrids segregating human chromosomes and with hybrids segregating feline chromosomes. The same principles apply to hamster × mouse hybrids that were used to map murine genes; however, we refer the reader to Lalley et al. (1978a,b) for specific examples in this species. The human observations are derived from work in our laboratory (Lemons et al., 1977, 1978; O'Brien et al., 1980; O'Brien and Nash, 1982) as well as scores of other laboratories which have employed rodent × human cell hybrids. The feline results are derived from our experience in construction of the biochemical genetic map of the domestic cat (O'Brien et al., 1980; O'Brien and Nash, 1982).

We present a typical standard result of a stain for adenosine deaminase in certain rodent × cat hybrids in Fig. 1. These hybrids preferentially segregate feline chromosomes and retain the rodent genome. The rodent parents were RAG, a BALB/c mouse line (Klebe et al., 1970), and E36, a Chinese hamster line (Gillin et al., 1972). Two feline parental fibroblast lines are shown, FC2 and CRFK. Mouse × cat hybrids (MXC) express the mouse ADA plus the feline form. ADA is polymorphic for two alleles, a and b, in the cat (O'Brien, 1980), and the feline parent of both MXC and HXC (hamster × cat) was heterozygous for a/b. Thus, the hybrids that retain the feline chromosome that encodes ADA can express either one or both parental alleles (compare MXC18C23D to MXC18C23I), depending on which feline chromosomes are present. With certain polymorphic isozyme systems, one allele will comigrate with the rodent form of the homologous enzyme, which makes scoring impossible. An example of this in the cat system is the esterase-D (ESD)

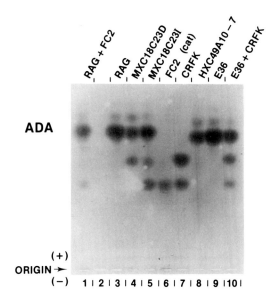

Figure 1. Starch gel electropherogram which has been stained for adenosine deaminase (ADA). The parental cells include murine RAG (Klebe et al., 1970), E36 Chinese hamster (Gillen et al., 1972), CRFK and FC2 cat (O'Brien and Nash, 1981). Artificial mixtures of parental extracts are loaded in the extreme lanes. Mouse × cat (MXH) and hamster × cat (HXC) hybrids are as indicated. These results form the basis of the ADA diagram in Fig. 3.

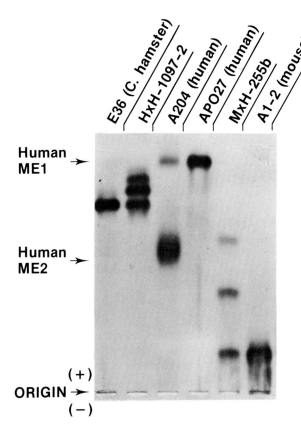

Figure 2. Starch gel electropherogram which has been stained for malic enzyme (ME1 and ME2). The parent cells include murine A1–2 (Lemons et al., 1977), human A204 and AP027, and Chinese hamster E36. Somatic cell hybrids between human and either rodent exhibit five bands characteristic of a tetramer.

system (see below). The HXC hybrid is negative for the feline isozyme of ADA.

Adenosine deaminase is a monomeric enzyme. In such cases, positive hybrids exhibit both parental bands and no additional isozyme bands. However, it should be noted that 72% of the isozyme systems are multimeric subunit enzymes (Hopkinson et al., 1976). Thus heterospecific hybrid cells often produce heteropolymeric isozymes (with subunits from each parent species). A dimer would then produce three bands (e.g., phosphogluconate dehydrogenase, PGD), a trimer four bands (e.g., purine nucleoside phosphorylase, NP), a tetramer five bands (e.g., malic enzyme, soluble, ME1). We present in Fig. 2 a gel containing mouse × human and Chinese hamster × human hybrids developed for malic enzyme (ME1). ME1 is a tetramer, so hybrid cells exhibit up to five ME1 bands. The lightness or apparent absence of the human parental bands is a reflection of the dosage of human chromosome 6 (which encodes human ME1) in these hybrids. In general, the dosage of a segregant chromosome seldom exceeds 0.8 chromosome equivalent per cell, less than half the dosage of the rodent counterpart. Finally, there is a second human malic enzyme (ME2), a mitochondrial form which occurs in some extracts (but not all) and which is encoded by a gene on a separate chromosome from that which specifies ME1.

Table I is a list of 41 isozyme systems which we find useful in genetic analysis in our laboratory. These 41 systems are selected because they are readily resolved in hybrid cells, and because they include representative genes mapped to every human and feline chromosome. The chromosomal location of each gene is indicated for both man and cat. Also, the buffer systems used for each isozyme are indicated.

Figures 3 and 4 present composite electropherograms of most of the enzyme systems listed in Table I. In cases of enzyme polymorphism, the relative mobility of the most common allele is presented. The allele frequencies of human and feline polymorphic gene enzyme systems are presented in Table I (Harris and Hopkinson, 1972; O'Brien et al., 1980; O'Brien 1980). In some cases (like *ADA* in cat crosses), polymorphism presents no interference with reliable scoring. However, in others, like *GLO* in humans, one allozyme (allelic isozyme) comigrates with the rodent form. In these systems, the allozyme genotype of the segregant parent must be specifically arranged for accurate scoring. These systems are so indicated in Table I.

The feasibility of scoring each of the 41 systems in four classes of crosses (mouse × human, hamster × human, mouse × cat, hamster × cat) is also presented in Table I. In general, these columns can be derived by examination of Figs. 3 and 4. A negative score indicates comigration of the homologous enzymes of the two species (e.g., PP in mouse × human). The usefulness of these feasibility scores in Table I comes in selection of the isozyme systems to run in a genotype characterization of a group of hybrid colonies. We have intentionally been conservative in these columns by giving a negative score to systems that are equivocal or barely possible (e.g., ME1 in hamster × cat). Thus, certain of the systems, which we do not routinely score, *can* be resolved by suitable empirical modification (e.g., isoelectric focusing).

Table I. Isozyme Markers of Use in Following Chromosome Segregation[a]

Gene symbol human/cat (mouse)	Enzyme	Buffer system[b]	Hetero-polymer formation[c]	Human Analysis					Cat Analysis			
				Chromosome	Alleles	Feasibility with most common allele			Chromosome	Alleles	Feasibility with most common allele	
						Ha × H	M × H				Ha × C	M × C
PGM1 (Pgm-2)	Phosphogluco-mutase-1	TC	−	1	1 (0.75), 2 (0.25)	+	+		C1	a	+	+
PGD (Pgd)	6-phospho-gluconate dehydrogenase	TEB	+	1	A	+	−		C1	a (0.38), b (0.62)	+	+
PEPC (Dip-2)	Peptidase-C	TEB	−	1	1	+	+		—	N.D.	−	−
IDH1 (Idh-1)	Isocitrate dehydrogenase (soluble)	TC	+	2	1	+	+		C1	a	+	+
MDH1 (Mor-2)	Malate dehydro-genase (soluble)	TEB	+	2	1	+	+		A3	a	+	+
ACP1 (Acp-1)	Acid phosphatase	Pi	−	2	A (0.40), B (0.60)	+	+		A3	a	+	+
ACY1	Amino acylase-1	TEB	+	3	N.D.	+	+		—	N.D.	+	+
PEPS (Pep-1)	Peptidase-S	TEB	+	4	1	+	+		B1	N.D.	−	+
PGM2 (Pgm-1)	Phosphogluco-mutase-2	TC	−	4	1	+	+		—	N.D.	N.D.	N.D.
HEXB	Hexosaminidase-B	TEB	−	5	1	+	−		—	—	−	−
GLO (Glo-1)	Glyoxylase	TH	+	6	1 (0.60), 2 (0.40)	+	+		B2	a	+	−
PGM3	Phosphogluco-mutase-3	TC	−	6	1 (0.75), 2 (0.25)	+	−		B2	a (0.74), b (0.36)	+	+
ME1 (Mod-1)	Malic enzyme	TEB	+	6	1	+	+		B2	a (0.94), b (0.06)	−	+
GUSB (Gus-1)	β-Glucuronidase	TEB	+	7	1	+	+		—	a	−	−
GSR (Gr-1)	Glutathione reductase	TEB	+	8	1 (0.87), 2 (0.13)	+	−		C2	a	+	−

Genetic Analysis of Hybrid Cells

Symbol	Name	Col1	Col2	Col3	Col4	Col5	Col6	Col7	Col8	Col9	Col10	Col11
AK1	Adenylate kinase	TC	–	9	1	+	+	+	U5	a	+	+
ACO1	Aconitase-1	TC	–	9	1 (0.85), 2 (0.15)	–	–	–	–	–	N.D.	N.D.
PP (Pyp)	Inorganic pyrophosphatase	TEM	+	10	1	+	+	+	D4	a	+	+
HK (Hk-1)	Hexokinase-1	TEB	–	10	1	–	–	–	D2	a	–	+
GOT1 (Got-1)	Glutamate-oxaloacetate transaminase	TEB	+	10	1	+	+	+	–	a	N.D.	+
LDHA (Ldh-1)	Lactate dehydrogenase-A	TEB	+	11	1	+	+	+	A2	a	+	+
ACP2 (Acp-2)	Acid phosphatase-2	Pi	+	11	1	+	+	+	A2	N.D.	+	–
TPI (Tpi)	Triose phosphate isomerase	TEM	+	12	1	+	+	–	B4	a	+	+
LDHB (Ldh-2)	Lactate dehydrogenase-B	TEB	+	12	1	+	+	+	B4	a	+	+
PEPB (Pep-2)	Peptidase-B	TEB	–	12	1 (0.90), 2 (0.10)	–	++	++	B4	a	+	++
ESD (Es-10)	Esterase-D	TC	+	13		–	++	++	A1	a (0.78), b (0.22)	–	+
NP (Np-1)	Purine nucleoside phosplorylase	TEB	+	14	1	+	+	+	B3	a	+	+
MPI (Mpi-1)	Mannose phosphate isomerase	TEB, TC	–	15	1	+	+	+	B3	a	+	+
PKM2 (Pk-3)	Pyruvate kinase (mitochondrial)	TEB	+	15	1	N.D.	+	+	B3	N.D.	–	+
HEXA	Hexoseaminidase A	TEB	–	15	1	+	+	+	B3	a	+	–
APRT (Aprt)	Adenine phosphoribosyl transferase	TG	+	16	1	–	–	+	–	–	+	+
D1A4	Diaphorase-4	TEB	+	16	N.D.	++	++	++	–	N.D.	–	++
GALK (Glk)	Galactokinase	TEB	+	17	1	++	++	++	–	N.D.	–	++
PEPA (Pep-1)	Peptidase-A	TEB	+	18	1 (0.75), 2 (0.20), 8 (0.05)	–	+	+	U4	a (0.61), b (0.39)	+	–

(Cont.)

Table I. Isozyme Markers of Use in Following Chromosome Segregation[a] (Continued)

Gene symbol human/cat (mouse)	Enzyme	Buffer system[b]	Hetero-polymer formation[c]	Human analysis			Cat analysis			
				Chromosome	Alleles	Feasibility with most common allele		Chromosome	Alleles	Feasibility with most common allele	
						Ha × H	M × H			Ha × C	M × C
GPI (Gpi-1)	Glucose phosphate isomerase	TEB	+	19	1	+	+	U2	a (0.95), b (0.05)	+	−
PEPD (Pep-4)	Peptidase-D	TEB	+	19	1	+	−	—	a	−	−
ADA	Adenosine deaminase	TC	−	20	1 (0.95), 2 (0.05)	+	+	U1	a (0.55), b (0.39), c (0.06)	+	+
SOD1	Superoxide dismutase-1	TEB	+	21	1	+	+	C2	a	−	+
ACO2	Aconitase-2	TC	−	22	1	+	−	—	N.D.	N.D.	N.D.
HPRT (Hprt)	Hypoxanthine guanine phosphoribosyl transferase	TG	+	X	1	+	−	X	N.D.	+	−
G6PD (G6pd)	Glucose-6-phosphate dehydrogenase	TEB	+	X	A (0.35), B (0.65)	+	+	X	a	+	+

[a] Gene symbols are for homologous enzymes recommended by the nomenclature committee of the Vth International Workshop on Human Gene Mapping (Shows et al., 1979).

[b] Buffer systems include: (a) TC pH 7.1: 0.14 M Tris, 0.043 M citric acid, pH 7.1. The conductivity of TC-1× was adjusted with distilled H$_2$O or TC-5× (a 5× concentrate of TC-1×) to 3.3 mΩ. TC-1× is used full strength in the cathode chamber and diluted to TC-0.8× in the anode buffer. The gel buffer is TC-0.07×. (b) TEB pH 8.6: TEB-1× (0.18 M Tris, 0.004 M EDTA, 0.1 M boric acid, pH 8.6, conductivity 1.1 mΩ) is used in the cathode and 0.8× in the anode. The gel buffer is TEB-0.1×. For NADP-dehydrogenases, 1 ml of NADP (20 mg/ml) is added to the cathode buffer for both TC and TEB systems. (c) TH pH 8.0: 0.2 M Tris-histidine is titrated with histidine-HCl to pH 8.0 TH-1× is the cathode buffer, and TH-0.8× is used in the anode. The gel buffer is TH-0.05×. (d) TEM: electrode, 0.1 M Tris, 0.01 M EDTA, 0.1 M maleic acid, 0.01 M MgCl$_2$, pH 7.4; gel 0.1× of electrode buffer; Pi: electrode 0.1 M Na phosphate pH 6.5; gel 0.1× of electrode buffer. (e) TG: electrode 0.005 Tris, 0.039 M glycine pH 8.9; gel 0.37 M Tris HCl pH 8.9.

[c] The formation of heteropolymers as an indication of subunit enzymes (dimers, trimers, tetramers, etc.) is indicated by + in column 4 (Hopkinson et al., 1976). Chromosome assignment for each isozyme mapped is indicated. The presence of genetic polymorphism in natural populations is indicated by the listing of two or more alleles and their frequencies under the column heading "alleles." Only those polymorphisms for which the second most common allele has a frequency greater than 0.05 are indicated as being polymorphic. In two cases, G6PD and ACO1 in man, the allelic frequencies in Negro populations is presented since the Caucasian populations are not polymorphic at these loci.

[d] The feasibility of resolution of human and cat isozymes in hybrids made with individuals homozygous for the most common allele at each locus is indicated. A + indicates that this system is easily scored in hybrids and a − indicates that it is not for a number of possible reasons outlined in the text.

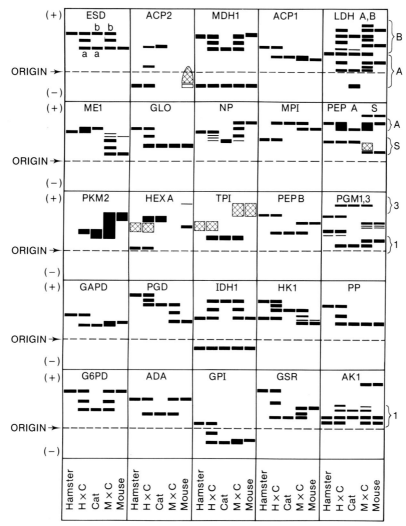

Figure 3. Composite electropherograms of isozyme systems used to resolve extracts of parental and hybrid cells segregating feline chromosomes (MXC, mouse × cat; and HXC, Chinese hamster × cat). The parental cells include: cat, CRFK and FC1; mouse, RAG; and Chinese hamster, E36. For systems that are polymorphic in the cat, the most common allozyme is represented. Relative positions of alternative alleles are presented in O'Brien (1980).

The isozyme systems presented here provide a thorough and expedient group of genetic markers for use in analysis of cell hybrids. They represent the combined efforts of virtually hundreds of laboratories over the past 20 years and we gratefully acknowledge their contributions (Harris and Hopkinson, 1976; Shows et al., 1979).

Finally, the accumulation of genetic markers for analysis of cell hybrids appears to be in a transitional phase. Two new classes of chromosome markers are now being applied to mapping genes in hybrid mammalian cells.

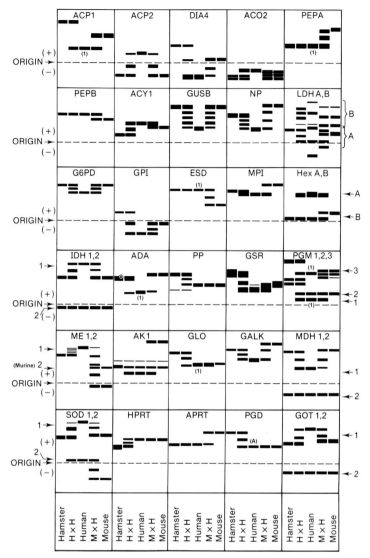

Figure 4. Composite electropherograms of isozyme systems used to resolve extracts of parental and hybrid cells segregating human chromosomes (HXH hamster × human; MXH mouse × human). For systems that are polymorphic in man, the allele designation is presented. The relative position of alternative human alleles of these systems is presented in O'Brien et al. (1980).

The first is the use of monoclonal antibodies directed against ubiquitous cell surface antigens the games for which are located on specific human and feline chromosomes (Barnstable et al., 1978). The second important reagent is the development of molecularly cloned restriction enzyme fragments which encode unique DNA sequences located on different human chromosomes. These molecular clones can be specific for a known gene (e.g., β-globin) or simply a random fragment of unknown function selected from a recombinant

DNA library of cellular DNA (Skolnick and Francke, 1981). Once assigned to specific human or feline chromosomes, both the monoclonal antibody and the arbitrary restriction fragments should become as useful (if not more so) than the isozyme systems described here.

References

Barnstable, C. J., Bodmer, W. F., Brown, G., Galfre, G., Milstein, C., Williams, A. F., and Ziegler, A., 1978, Production of monoclonal antibodies to group A erythrocytes, HLA and other cell surface antigens—New tools for genetic analysis, Cell **14:**9–20.
Brewer, G. J., 1970, An introduction to isozyme techniques, Academic Press, New York.
Davisson, M. T., and Roderick, T. H., 1980, Linkage map of the mouse (Mus musculus), Genetic Maps **1:**225–233.
Francke, U., and Taggart, R. T., 1980, Comparative gene mapping: Order of loci on the X chromosome is different in mice and humans, Proc. Natl. Acad. Sci. USA **77:**3595–3599.
Gillin, F. D., Roufa, D. J., Beaudet, A. L., and Caskey, C. T., 1972, 8-Azaguanine resistance in mammalian cells, I. Hypoxanthine-guanine phosphoribosyl transferase, Genetics **72:**239 252.
Harris, H., and Hopkinson, D. A., 1976, Handbook of Enzyme Electrophoresis in Human Genetics, North-Holland, Amsterdam.
Hopkinson, D. A., Edwards, Y. H., and Harris, H., 1976, The distribution of subunit numbers and subunit sizes of enzymes: A study of the products of 100 human gene loci, Ann. Hum. Genet. Lond. **39:**383–410.
Klebe, R. J., Chen, T., and Ruddle, F. H., 1970, Controlled production of proliferating somatic cell hybrids, J. Cell Biol. **45:**74–82.
Lalley, P. A., Francke, U., and Minna, J. D., 1978a, Homologous genes for enolase, phosphogluconate dehydrogenase, phosphoglucomutase, and adenylate kinase are syntenic on mouse chromosome 4 and human chromosome 1p, Proc. Natl. Acad. Sci. USA **75:**2382–2386.
Lalley, P. A., Minna, J. D., and Francke, U., 1978b, Conservation of autosomal gene synteny groups in mouse and man, Nature **274:**160–163.
Lemons, R. S., O'Brien, S. J., and Sherr, C. J., 1977, A new genetic locus, BEVI, on human chromosome six which controls the replication of baboon type C virus in human cells, Cell **12:**251–262.
Lemons, R. S., Nash, W. G., O'Brien, S. J., Benveniste, R. E., and Sherr, C. J., 1978, A gene (Bevi) on human chromosome 6 is an integration site for baboon type C DNA provirus in human cells, Cell **14:**995–1005.
McKusick, V. A., and Ruddle, F. H., 1977, The status of the gene map of human chromosomes, Science **196:**390–405.
Miller, D. A., and Miller, O. J., 1972, Chromosome mapping in the mouse, Science **178:**949–955.
Nichols, E. A., and Ruddle, F. H., 1973, A review of enzyme polymorphism, linkage and electrophoretic conditions of mouse and somatic cell hybrids in starch gels, J. Histochem. Cytochem. **21:**1066–1081.
Nichols, E. A., and Ruddle, F. H., 1979, A review of enzyme polymorphism, linkage and electrophoretic conditions for mouse and somatic cell hybrids in starch gels, J. Histochem. Cytochem. **21:**1066–1081.
O'Brien, S. J., 1980, The extent and character of biochemical genetic variation in the domestic cat, J. Hered. **71:**2–8.
O'Brien, S. J., and Barile, M. F., 1982, Isozyme resolution in mycoplasmas, in: Methods in Mycoplasmology (S. Razin and J. G. Tully, eds.), in press.
O'Brien, S. J., and MacIntyre, R. J., 1977, Genetics and biochemistry of enzymes and specific proteins of Drosophila, in: Genetics and Biology of Drosophila, Volume IIa (T. R. F. Wright and M. Ashburner, eds.), Academic Press, New York, pp. 395–551.

O'Brien, S. J., and Nash, W. G., 1982, Genetic mapping in mammals: chromosome map of the domestic cat, *Science* **216:**257–265.

O'Brien, S. J., Nash, W. G., Simonson, J. M., and Berman, E. J., 1980, Establishment of a biochemical genetic map of the domestic cat (*Felis catus*), in: *Feline Leukemia Virus* (W. D. Hardy, M. Essex, and A. J. MacClelland, eds.), Elsevier, North-Holland, New York.

O'Brien, S. J., Shannon, J. E., and Gail, M. F., 1980, A molecular approach to the identification and individualization of human and animal cell cultures: Isozyme and allozyme genetic signatures, *In Vitro* **16:**119–135.

Pearson, P. L., Roderick, T. H., Davisson, M. T., Garver, J. J., Warburton, D., Lalley, P. A., and O'Brien, S. J., 1979, Comparative mapping. Report of the International Committee, *Cytogenet. Cell. Genet.* **25:**82–95.

Pearson, P. L., Roderick, T. H., Davisson, M. T., Lalley, P. A., and O'Brien, S. J., 1981, Comparative mapping: Report of the International Committee, *Cytogenet. Cell Genet.*, in press.

Ruddle, F. H., Chapman, V. M., Ricciuti, F., Murnane, M., Klebe, R., and Meera-Khan, P., 1971, Linkage relationships of seventeen human gene loci as determined by man–mouse somatic cell hybrids, *Nature New Biol.* **232:**69–73.

Ruddle, F. H., and Creagan, R. P., 1975, Parasexual approaches to the genetics of man, *Ann. Rev. Genet.* **9:**407–486.

Shows, T. B., 1977, Genetic and structural dissection of human enzymes and enzyme defects using somatic cell hybrids, in: *Isozymes, Current Topics in Biological and Medical Research*, Volume 2 (M. C. Ratazzi, J. G. Scandalios, and G. S. Whitt, eds.), Alan R. Liss, Inc., New York.

Shows, T. B., et al., 1979, International systems for human gene nomenclature, *Cytongenet. Cell Genet.* **25:**96–116.

Siciliano, M. J., and Shaw, C. R., 1976, Separation and localization of enzymes in gels, in: *Chromatographic and Electrophoretic Techniques*, Volume 2, 4th ed. (I. Smith, ed.), W. Heinemann Medical Books Publ. Inc., Chicago, Illinois. pp. 185–209.

Skolnick, M. H., and Francke, U., 1981, Report of the committee on human gene mapping by recombinant DNA techniques, *Cytogenet. Cell Genet.* (in press).

Womack, J. E., 1980, Biochemical loci of the mouse, *Mus musculus, Genetic Maps* **1:**218–224 (in press).

Chapter 36

Future Perspectives in Somatic Cell Genetics

DORIS L. SLATE and FRANK H. RUDDLE

Somatic cell genetic techniques have proven valuable in determining the arrangement of genes on chromosomes and in studying the regulation of eukaryotic gene expression. Studies with somatic cells taken from patients with inherited diseases have revealed the biochemical bases for many human disorders, and have provided potential strategies for diagnostic and therapeutic approaches. The generation of mutants or variants of somatic cells in culture can facilitate the analysis of complex physiologic systems via complementation analysis.

Recent developments in cell and molecular biology now permit the study of gene mapping and regulation at many levels, from the whole chromosome to the nucleotide level. In this chapter, we will attempt to summarize the current status of somatic cell genetics and speculate about possible future developments in the field.

Five major systems have been developed for the transfer of genetic material from one donor cell or cell line to recipient somatic cells. These systems result in the transfer of varying amounts of genetic material, and can be used for gene mapping, studies of gene regulation, and complementation analysis, for example. The major techniques can be classified as whole-cell fusion, cell fragment (karyoplast or cytoplast) fusion, microcell-mediated gene transfer, chromosome-mediated gene transfer, and DNA-mediated gene transfer (Fig. 1).

Whole-cell fusion results in the formation of heterokaryons or homokaryons containing the complete genomes of the two (or more) parental cells. Intraspecific cell hybrids may retain their new complement of chromosomes intact, while interspecific hybrids tend to segregate the chromosomes of one input species, thus permitting the mapping of genes to specific chromosomes by concordance testing (Ruddle and Creagan, 1975). Fusion of cells with distinct differentiated phenotypes has been used to study gene regulation and cell differentiation (Ringertz and Savage, 1976). The "parasexual" nature of

---

DORIS L. SLATE • Department of Biology, Yale University, New Haven, Connecticut 06511. Present address: Cancer Metastasis Research Group, Pfizer Central Research, Groton, Connecticut 06340.   FRANK H. RUDDLE • Department of Biology, Yale University, New Haven, Connecticut 06511.

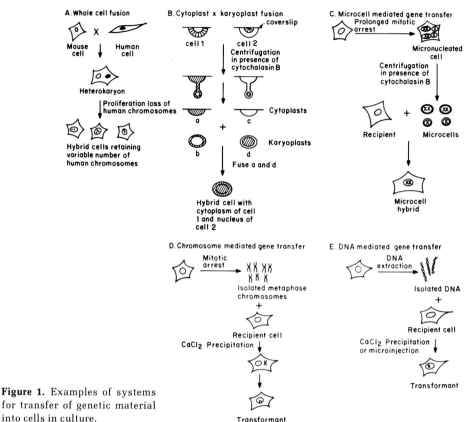

Figure 1. Examples of systems for transfer of genetic material into cells in culture.

whole-cell fusion allows gene mapping and the establishment of dominance/recessiveness relationships for differentiated traits in cell hybrids.

Fusions involving cell fragments termed karyoplasts (or minicells) or cytoplasts (or anucleate cells) to each other or to other intact cells result in the transfer of nuclear or extranuclear genetic material, respectively, to a recipient. Karyoplast–cytoplast fusions have permitted the mapping of genes which control mitochondrial functions (see Chapter 12 by Wallace, Chapter 9 by Eisenstadt and Kuhns, and Chapter 11 by Doersen and Stanbridge, for example), and have also been used to study the regulation of nuclear gene expression in the presence of different cytoplasmic environments.

Microcell (or microkaryoplast)-mediated gene transfer (see Chapter 23 by Fournier and Chapter 22 by Crenshaw and Murrell) permits the transfer of small numbers of chromosomes to recipient cells. This is accomplished by fusing "microcells" containing chromosomes (surrounded by nuclear membrane), a small rim of cytoplasm, and a plasma membrane to recipient cells. Microcells can be generated by treating cells with mitotic inhibitor for several days, or with other treatments, and can be size-fractionated to control the amount of genetic material transferred. Microcell-mediated gene transfer can be used for gene mapping or the analysis of gene regulation. (See Chapters 22–24, this volume.)

Chromosome-mediated gene transfer (see Chapter 26 by McBride) is accomplished by precipitating isolated donor metaphase chromosomes onto recipient cells, which then take them up by phagocytosis. This type of gene transfer usually results in retention of chromosome fragments. Both stable and unstable "transformants" are observed following chromosome-mediated gene transfer (Klobutcher and Ruddle, 1979), with maintenance or acquisition of centromere function by the transferred genetic material required for stable retention.

DNA-mediated gene transfer is similar to chromosome-mediated gene transfer, but isolated DNA is used and taken up by recipient cells (see Chapter 29 by Bollon and Silverstein). Integration of the transferred DNA into the recipient genome results in stable retention of the introduced material. Whole-cell DNA or cloned fragments (alone or in plasmid vehicles; see Chapter 31 by Hsiung and Kucherlapati and Chapter 30 by Linnenbach and Croce) can be used as donor material. A recent exciting refinement in this area involves the use of microinjection techniques (see Chapter 32 by Graessmann and Graessmann) to introduce specific DNAs into the nucleus or cytoplasm of recipient cells. This technique is already proving to be extremely valuable in studying the regulation of gene expression. The introduction of cloned eukaryotic genes, normally subject to hormonal induction, into cells via DNA-mediated transfer has been reported; following transfer, the genes are only expressed if the appropriate hormone is present (Hynes et al., 1981; Kurtz, 1981).

The rapid growth of recombinant DNA technology has not only provided purified, defined sequences for gene transfer, but has also permitted rapid gene mapping. Genes that were mapped by classical somatic cell genetic techniques involving cell fusion were mainly those that coded for constitutive "housekeeping" functions (mostly enzymes) or certain proteins inducible in cell hybrids. With the availability of cloned genes, it is now possible to map genes by direct *in situ* hybridization to metaphase chromosomes (Harper et al., 1981) or by restriction endonuclease analysis (see Chapter 31 by Hsiung and Kucherlapati). For restriction analysis, a panel of interspecific cell hybrids is first characterized for chromosome and isozyme retention; DNA from each clone is then isolated and digested with one or more restriction endonuclease. The digested DNA is then electrophoresed through an agarose gel and transferred to nitrocellulose filters. The cloned DNA sequence to be mapped is radiolabeled and hybridized to the DNA fragments on the filters. If a species difference in hybridization patterns can be found, then the analysis of the hybridization patterns of the hybrid cell DNAs should allow the chromosomal location of the cloned DNA sequence to be determined. This type of mapping has now been accomplished for a wide variety of genes (D'Eustachio et al., 1980; Owerbach et al., 1981).

The use of cell fusion techniques to generate "hybridomas" (by fusing myeloma cells to antibody producing B cells), producing large amounts of monoclonal antibodies, will open up new areas of somatic cell genetics. For example, generating antibodies against a cell surface antigen coded by a specific chromosome, coupled with the use of cell sorter technology, will permit the isolation of populations of hybrid cells retaining or lacking that chromosome. Variants expressing elevated or depressed levels of the surface antigen could also be isolated. The widespread use of monoclonal antibody technology will have profound implications for basic research, diagnosis, and possibly for clinical therapy of disease.

The development of cell sorter and flow cytometry techniques will have a major impact on the field of somatic cell genetics. Hybrid cells can be identified and isolated (see Chapter 7 by Jongkind and Verkerk), individual cells can be easily cloned, and expression of cell surface antigens or other proteins can be easily scored with the current generation of sorters and the appropriate stains and/or antibodies. One very exciting application for this technology is the isolation of pure populations of individual chromosomes from a given species (see Chapter 25 by Wray and Stubblefield); this can be done on the basis of relative DNA content. By isolating chromosome-specific DNA and cloning it, probes for each individual chromosome can be obtained and used for gene mapping or gene regulation studies in gene transfer experiments.

The use of liposomes or red cell ghosts for the introduction of macromolecules (including DNA) into recipient cells is also advancing rapidly (see Chapter 28 by Straubinger and Papahadjopoulos and Chapter 27 by Rechsteiner). These vehicles permit the transfer of large amounts of defined materials into cells, and will also increase our knowledge of the process of membrane fusion.

The rapidly expanding field of microinjection of macromolecules into individual cells, described in the DNA-mediated gene transfer section above, will broaden our knowledge of cellular regulation at many levels: transcription of introduced DNA, translation and stability of introduced mRNA, and function of introduced proteins. The construction of complex vectors (see Chapter 30 by Linnenbach and Croce) for DNA transfer, complete with replicon, centromere function, selectable markers, etc., will permit introduction and stable retention of defined pieces of genetic material for analysis of gene function.

Studies of cellular differentiation can be made using mouse teratocarcinoma cells *in vitro* as a model system. These cells can differentiate *in vitro* and can also be fused with blastocysts to study gene expression in chimeric mice. Gene transfer experiments can be performed with the teratocarcinoma cells before introduction into blastocysts, and the fate of transferred material followed through development. Microinjection of DNA into fertilized mouse eggs followed by implantation of the eggs into foster mothers can also be used to study gene expression and germ line transmission of the transferred DNA (Gordon et al., 1980).

Somatic cell genetic techniques which have been developed for animal cells will also be expanded for the genetic and developmental analysis of plants (see Chapter 33 by Binding and Nehls). Advances in the areas of plant genetics and biochemistry are being made rapidly; the ability to monitor inheritance of genetic material introduced into protoplasts after regeneration of complete plants makes this area a fascinating one. Plant cloning vectors, such as the Ti plasmid from *Agrobacter tumifaciens*, are being developed for gene transfer studies in several species.

Now that many structural genes for proteins, cell surface receptors, hormones, etc., have been mapped, the ability to isolate or create mutants for these genes in cultured cells will permit analysis of their roles in cell function. For example, given a monoclonal antibody against a specific cell surface receptor, cells lacking the receptor could be isolated and studied for altered function.

Somatic cells can be used as models for processes which occur *in vivo*; for example, the development of resistance to chemotherapeutic agents in tissue culture cells via gene amplification (Schimke et al., 1981) has stimulated the search for a similar change in human tumor cells in the clinic. Cells from patients with metabolic disorders can be studied in culture to reveal their biochemical abnormalities and to develop possible therapies.

In summary, somatic cell genetics will provide fertile ground for studying gene organization, expression, and regulation over the coming decades. The ability to manipulate and introduce genetic material into cells in culture and study expression both at the *in vitro* cellular level and the *in vivo* organismal level will make the classical techniques of somatic cell genetics, and their more recent molecular applications, very powerful tools in biology.

References

D'Eustachio, P., Pravtcheva, D., Marcu, K., and Ruddle, F. H., 1980, Chromosomal location of the structural gene cluster encoding murine immunoglobulin heavy chains, *J. Exp. Med.* **151**:1545–1550.

Gordon, J. W., Scangos, G. A., Plotkin, D. J., Barbosa, J. A., and Ruddle, F. H., 1980, Genetic transformation of mouse embryos by microinjection of purified DNA, *Proc. Natl. Acad. Sci. USA* **77**:7380–7384.

Harper, M. E., Ullrich, A., and Saunders, G. E., 1981, Localization of the human insulin gene to the distal end of the short arm of chromosome 11, *Proc. Natl. Acad. Sci. USA* **78**:4458–4460.

Hynes, N. E., Kennedy, N., Rahmsdorf, U., and Groner, B., 1981, Hormone-responsive expression of an endogenous proviral gene of mouse mammary tumor virus after molecular cloning and gene transfer into cultured cells, *Proc. Natl. Acad. Sci. USA* **78**:2038–2042.

Klobutcher, L. A., and Ruddle, F. H., 1979, Phenotype stabilization and integration of transferred material in chromosome mediated gene transfer, *Nature* **280**:657–660.

Kurtz, D. T., 1981, Hormonal inducibility of rat $\alpha_{2\mu}$ globulin genes in transfected mouse cells, *Nature* **291**:629–631.

Owerbach, D., Rutter, W. J., Shows, T. B., Gray, P., Goeddel, D. V., and Lawn, R. M., 1981, Leucocyte and fibroblast interferon genes are located on human chromosome 9, *Proc. Natl. Acad. Sci. USA* **78**:3123–3127.

Ringertz, N. R., and Savage, R. E., 1976, in: *Cell Hybrids*, Academic Press, New York.

Ruddle, F. H., and Creagan, R. P., 1975, Parasexual approaches to the genetics of man, *Ann. Rev. Genet.* **9**:407–486.

Schimke, R. T., Brown, P. C., Kaufman, R. J., McGrogan, M., and Slate, D. L., 1981, Chromosomal and extrachromosomal localization of amplified dihydrofolate reductase genes in cultured mammalian cells, *Cold Spring Harbor Symp. Quant. Biol.* **45**:785–797.

Index

Acid phosphatase, 518, 519
Aconitase, 519
Adenine phosphoribosyl transferase (APRT), 3, 313, 519
  screening for deficiency in, 15
Adenosine deaminase, 515, 517, 520
Adenosine 5'-monophosphate (AMP), 13
Adenosine 5'-triphosphate (ATP), in microcytosphere production, 238
Adenylate kinase, 519
Amino acylase, 518
Antibiotic resistance (see also Drug resistant cell mutants)
  hereditary markers of, 124–125
  mitochondria in transfer of, 203–209
Antibodies, in hybrid clone isolation, 18, 522
Antimycin resistance, 124, 127
Apoptotic cell destruction, 129
ATPase activity
  mitochondrial encoding of, 123–126
  in oligomycin (rutamycin) resistant mutants, 127–129, 134–135
Azaguanine resistant mutants, 126

Bacterial ribosomes, antibiotic sensitivity of, 159
Banding techniques, chromosomal (see also Chromosome identification), 352–358, 493–509
Bead labeling, polystyrene (see also Fluorescent labeling) 82–85, 192–193, 263
Blotting analysis (Southern), 436–440, 455–456
Bromodeoxyuridine (BrdU)
  in cell mutant resistance, 12
  in mitochondrial mutagenesis, 125, 160
  in nuclear fluorescence selection, 51–52

Calcium ion (Ca^{++})
  cell viability and exclusion of, 25–27
  in cytotoxicity, 25–27

Calcium ion (Ca^{++}) (cont.)
  exclusion of, in PEG cell fusion methods, 23–28
  in plant protoplast fusion, 481–483
Calcium phosphate coprecipitation
  in DNA transfer preparation, 416, 422, 430–431
  in metaphase chromosome uptake, 380–381
CB, see Cytochalasin B
C-bands, see Constitutive heterochromatin
Cell(s)
  apoptosis of, 329
  CAP resistant, see Chloramphenicol resistance
  drug resistant, see Drug resistant cell mutants/variants
  enucleation of, see Cell enucleation
  erythrocyte, see Red-blood cells
  fluorescent labeling of, 81–98, 101–109, 262–265
  fusion of, cell-to-cell, see Cell hybrids/hybridization
  fusion of, with cytoplast/karyoplast fragments, see Cell fusion
  liposome interactions with, 405–410
  membrane of, see Cell membrane
  microinjection of (see also Microinjection), 463–470
  minisegregant (see also Minisegregant cells), 329–346
  mitotic, see Mitotic cells
  oligomycin resistant, see Oligomycin (rutamycin) resistant mutants/variants
  reconstituted (see also Cell fusion), 167, 191, 261–262, 269
  red-blood, see Red-blood cells; Red-blood cell ghosts
  rutamycin resistant (see also Oligomycin (rutamycin) resistant mutants/variants), 123–125
  spontaneous disintegration of, 329
  tumorigenic, see Tumor cells
  viable, characteristics of, 25
  yeast, see Yeast cells

Chromosome(s) (cont.)
  transfer of genes via, see Gene transfer
Chromosome identification (see also
    Karyotyping)
  acridine orange banding in, 501
  banding techniques in, 352–358, 493–509
  C-bands in, 503–506
  cell culture preparation in, 350–352,
    495–496
  constitutive heterochromatin in, 503–506
  chromomycin A_3/methyl green in, 501–503
  development of methods in, 352–358, 493
  Giemsa staining in, 498–501
  Giemsa-11 staining in, 508–509
  G/Q banding in, 236–258
  of metaphase chromosomes, 352–358
  nucleolus organizer regions (NOR) in, 506
  Q-banding, 356–358, 496–498
  reverse (R) banding in, 501–503
  utilization of, 493
Constitutive heterochromatin (C-bands)
  in chromosome identification, 503–506
  location of, 503–505
  staining procedure for, 505–506
Co-transformation, of unlinked genes,
    424–425
Cybrid(s)
  characterized, 90, 191, 281
  in chloramphenicol (CAP) resistant mutant
    formation, 167–170
  fluorescent isolation of, 90
  formation of, see Cybrid formation
  isolation of fibroblast, 90–93
  mitochondrial mass growth in, 133–134
  mitochondrial respiration in, 131–133
  in oligomycin (rutamycin) resistant
    mutagenesis, 126, 128, 133–134
Cybrid formation
  characterized, 90, 118, 189
  diagrammed, 167
  cell enucleation in, 190, 194
  cell-to-cytoplast fusion in, 194–195
  cytoplasmic incompatibility in, 198
  cytoplast-to-cell ratio in, 190, 192, 197–198
  fusion efficiency in, 195–198
  genetic markers in selection for, 191–192
  initial fusion in, 191, 196–197
  poisoning in selection for, 192
  polystyrene bead labeling in, 193
  selection methods in, 191–192
  tumorigenicity suppression and, 198
Cybridoids, 191
Cytochalasin B, 67–68, 76–77, 92, 125, 167–
    168, 190, 194, 247, 256, 282, 292, 299,
    303–304, 313–319
Cytoplasm, see Cytoplast(s)
Cytoplasmic inheritance

Cytoplasmic inheritance (cont.)
  antibiotic resistance as, 123–128, 131–134
  characterized, 189
  chloramphenicol (CAP) resistance as, 166–
    171, 176–184, 189
  cytoplasmic fusion and transfer in, 189–
    190, 198–199
  erythromycin (ERY) transfer as, 149–153
  mitochondrial involvement in, 211–212
  in mitochondrial mutagenesis, 125, 126
  nuclear genetic control vs., 189, 255
  rhodamine (R6G) modification of, 211–217
Cytoplasmic protein synthesis, 190
  inhibition of, 144–146
Cytoplasmic vesicles, (see also
    Microcytospheres), 237
Cytoplast(s)/Cytoplasm
  biological characteristics of, 190–191, 258
  cell stage heterogeneity of, 191
  composition of, 190
  in cybrid fusion (see also Cytoplast fusion),
    90–93, 194–196
  decontamination/purification of, 75
  donor identification for, 262–263
  fluorescent labeling of, 92–93, 262–263
  fusion/hybridization with, see Cytoplast
    fusion
  membrane bound organelles and, 237–242
    (see also Microcytospheres)
  markers for, in cybrid selection, 191–192
  nuclear component ratio with, 191
  nuclear gene regulation and, 189, 255
  preparation of, 75, 77–78, 168, 190–191,
    245, 256–258
  protein synthesis by, 90, 258
  recovery of, 70–75, 77, 256–258
  viral replication by, 258
Cytoplast fusion
  in cell reconstitution, 191
  in cybrid formation, 90–93, 194–196
  efficiency of initial, 196–197
  gene expression consequences in, 189–190,
    198–199
  with karyoplasts, 167, 191, 261–262
  utility of, 189–190
  viability of, 197–198

Dicyclohexylcarbodiimide (DCCD) binding
  C-14 labeling and assay of, 116–118
  in mitochondrial ATPase assay, 115–117
  in oligomycin (rutamycin) resistance 111,
    112, 115, 118
Density gradient techniques, in enucleation,
    68, 76–78, 194
Dimethyl sulfoxide (DMSO)
  chemical and pharmacological properties
    of, 36–37

Index

Dimethylsulfoxide (DMSO) (cont.)
  cutaneous membrane penetration by, 36
  in monolayer-PEG fusion protocol, 40–41
  in polyethyleneglycol (PEG) induced cell
      fusion, 35–46
Diphtheria toxin, 16
DNA (Deoxyribonucleic acid)
  agarose gel electrophoresis of, 454–455
  autoradiography of, 458–460
  blotting analysis of, 436–440, 455–456
  carrier, preparation of, 420–421
  cellular, isolation of, 435, 450–451
  constitutive heterochromatin and, 503–504
  in gene mapping, see Gene mapping
  in gene transfer, see Gene transfer
  human mitochondrial, 124
  isolation, of cellular, 435, 450–451
  lipid vesicle encapsulation and delivery of
      (see also Liposomes), 399–400, 403–404
  in mammalian cell transformation, 420–425
  mitochondrial, see Mitochondrial DNA
  nick translation of, 435–436, 450, 457–458
  recombinant, 382, 430
  replication of, 418–420
  restriction endonuclease analysis of, 450,
      453–454, 527
  in yeast cell transformation, 123–124,
      416–420
Drug resistant cell mutants/variants
  antibiotic markers of (see also specific
      antibiotics by name), 124–125
  cell line isolation of, 5–7
  cloning of, 6
  HAT medium in selection of, 12–17
  purine/pyrimidine analogs in selection of,
      1–7
Efrapeptin (EF) resistant mutants, 203–209
Endocytosis, of liposomes, 400
Enucleation, see Cell enucleation
Enzymes (see also Isozyme markers)
  characteristics of, 513–514
  in chromosome segregation, 518–520
  flow sorted cell assay of, 95–98
  genetic markers for, 513–514
Erythrocytes, see Red blood cells
  ghosts of, see Red blood cell ghosts
Erythromycin (ERY)
  in antibiotic resistance inheritance (see also
      Erythromycin resistance), 124
  cell growth inhibition by, 153
  mitochondrial protein synthesis inhibition
      by, 139–154
  mutant cell lines resistant to, see
      Erythromycin (ERY) resistant cell lines
Erythromycin (ERY) resistance, 124, 149–153
  cell hybrid transfer of, 151–153
  cytoplasmic transfer of, 150–151

Erythromycin (ERY) resistant cell lines
  characteristics of, 143–144
  ethyl methane sulfonate in production of,
      142–143
  mycoplasma contamination in, 149
  pH factor in proliferation of, 140–141
  production and selection of, 140–143

Flow cytometry (see also Flow sorting
      methods)
  chromosome sorting by, 338–339, 359–364
  karyotype sorting by, 360–364
Flow sorting methods (see also Flow
      cytometry), 81–98
  in cell fusion product assay, 93–98
  for cybrids, 90, 93
  for cytoplasts, 92
  in enzyme activity assay, 95–98
  fluorescent bead labeling in, 82–85, 90–98
  heterokaryon isolation by, 86–87
  in nonselective hybrid cell isolation, 87–90
  for tetraploid cells, 88–90
Fluorescence-activated cell sorter (FACS),
      270, 272, 276, 282–286
Fluorescent labeling
  benzimidazole dyes in, 84–86
  in cell enucleation, 193–196
  in cell fusion, 86–87, 101–109, 262–263
  in cybrid flow sorting, 81–93
  of cytoplasts, 92–93, 262–263
  of fibroblasts heterokaryons, 86–87
  in hybrid cell identification, 262–265
  in membrane fusion monitoring, 101–109
  polystyrene beads in, 82–84, 192–196, 263
  probe dye synthesis (F18, R18) in, 103–104
  resonance energy transfer in, 101, 105
  rhodamine-fluorochrome, 263–265
  stearylamine in, 84, 87
Fusion cybrid, 191
Fusion index (FI), 25–28, 31
Fusogenic agents, 41–44 (see also specific
      types by name, e.g., Dimethyl sulfoxide
      (DMSO); Polyethylene glycol (PEG))

Galactose, in CAP toxicity, 166
GAMA medium, in cell hybrid selection, 15
G-bands, see Giemsa banding
Gene cloning, 18; via yeast cells, 415–420
Gene mapping
  antibiotic resistances in, 124–125
  banding techniques in, 352–358, 493–509
  blot hybridization in, 450
  cell fusion and, 118, 291
  chromosome mediated gene transfer in,
      382–383
  DNA sequences in, 382, 449–450
  enzyme electrophoresis in, 513–515

Gene mapping (cont.)
  erythromycin (ERY) resistance in, 154
  HAT medium selection in, 17
  hybrid mammalian cells in, 118, 291, 513–523
  isozyme markers in, 513–523
  liquid hybridization in, 450
  in mammalian cells, 118, 512–513
  metaphase chromosome analysis in, 365–371
  microcell hybrid clones in, 309, 324–325
  mouse-human cell hybrids in, 291, 513, 515–523
  mitochondrial genome in, 111, 123–124
  monoclonal antibodies in, 522–523
  oligomycin (rutamycin) resistance inheritance in, 111, 112
  restriction enzyme fragments in, 522–523
  species specific gene transfer in, 449–450
Gene transfer
  in cell fusion, 255, 261, 291, 399
  chromosome mediated, 375–383
  cloning and, 18
  DNA mediated, 375, 382, 399, 405–410, 415–424
  by embryo microinjection and reimplantation, 425–426
  foreign sequence detection in, 430–435, 449–460
  HAT medium in, 18, 381
  liposome mediated, 385–397, 399–410
  metaphase chromosomes in, 375–383
  microcell mediated, 309–310, 315, 525, 526
  mutant cell line detection of, 423–424
  prokaryotic genes in, 424
  red blood cell ghosts in, 385–397
  transformant isolation/selection in, 381–382
  of unlinked genes, 424–425
  yeast genes in, 415–416
Genome, reconstruction of recombinant plasmid (see also mitochondrial genome), 429–430
Giemsa banding (G-bands), 493
  mechanism of, 501
  procedure in, 498–501
Giemsa-11 banding/staining, 508–509
Glass slides, in microinjection, 465–466
Glass microcapillaries, in microinjection, 464–465
G/Q chromosome banding, 352–358
Growth substrate modification, in cell enucleation, 75

HAM medium (hypoxanthine-aminopterin-methyldeoxycytidine), 15

HAT medium (hypoxanthine-aminopterin-thymidine)
  alternate modifications of, 15–16
  in cell line decontamination, 4
  in drug-resistant cell mutant selection, 2–3, 11–17, 212
  in gene mapping, 1–18
  in gene transfer, 18, 381
  in hybridoma isolation, 18
  modifications of, 14–17
  operational principle in screening by, 12–13
  preparation of, 4
  in screening of hybrid cell lines, 12, 17
Heterokaryons
  as binucleate cells, 51–52
  biochemical agents in selection of, 47–64
  cell concentration in recovery of, 58–61
  cell treatment in selection of, 52–57
  characterized, 47, 525–526
  cytotoxicity in recovery of, 58–60
  fluorescent labeling of, 86–87
  hybrid clones vs., 47
  isolation of fibroblast, 85–87
  microcell, 310
  nuclear fusion in selection of, 51–52
  recovery of, conditions for, 52–61
HGPRT activity, 13–18
Homokaryons, 525
HSV-1 (Herpes simplex virus, type 1)
  in cell fusion, 102, 104
  in fluorescent screening, 102, 104, 105
HOT medium, 15–16 (see also ouabain)
HPRT (hypoxanthine phosphoribosyl transferase)
  in CAP cell nucleus marking, 168–169
  in cybrid fraction sorting, 90
  in somatic cell variation, 1–2
Hybridomas, 18

Interferon production
  after enucleation, 190
  in yeast cloning, 416
Interspecific cell hybrids
  characterized, 449, 525
  elimination of incompatibility in, 215
  in gene mapping, 514
Intraspecific cell hybrids
  biochemical isolation of, 63
  characterized, 449, 525
Iodoacetamide (IAM), 47–64
  in heterokaryon selection, 48–49, 53–57
  preparation of, 52
Iodoacetate
  in cell surface membrane shedding, 221
  in cybrid selection, 191
Isozyme markers
  characteristics of, 513–514

Index

Isozyme markers (cont.)
  in chromosome segregation, 517–523
  electrophoresis of, 513–515
  in gene mapping, 513–515
  in hybrid cell line genetic analysis, 513–523
  preparation of, 514–515
Karyobrid, characterized, 191, 281
Karyoplasts (see also Cell nuclei)
  cell enucleation recovery of, 70, 75–76, 168, 190–191, 256–260, 269–271, 273–277
  in cell fusion, 261–262, 273, 277–279, 281–288, 526
  cell sorter separation of, 270, 272, 276–277, 280
  characteristics of, 256, 260
  cytoplast fusion with, 167, 191, 261–262, 273, 526
  flow cytometry sorting of, 360–364
  fluorescent labeling of, 262, 263
  in genetic determination by fusion, 118
  in hybrid formation, 167
  preparation of, 245, 258–259, 269–271; by bullet method, 258–259; by cytochalasin B treatment, 245; cell enucleation in, see Cell enucleation
  protein synthesis assay of, 272
  RNA synthesis assay of, 272
  separation/purification of, 75–78, 168, 190–191, 259–260, 269–280, 282–288
  size of, 276–277
  small, 269–280, 282–285
  tantalum in separation of, 270–272, 274–276
  temperatures in viability of, 280
  viability determination for, 272, 274, 279
  whole cell fusion with, 281–288
Karyotyping (see also Chromosome identification; Gene mapping)
  banding results in, 499
  development of, 499
  human chromosome analysis in, 360–364

Lactic acid production, 130–131
Large unilamellar vesicles (LUV), 402
  lipid composition and delivery efficiency of, 406–407
  macromolecular deliveries via, 405
Latex spheres, see Polystyrene bead labeling
Lectins, in cell enucleation (see also Phytohemagglutinin), 75
Lipids, in liposome preparation, 401
Liposome(s)
  cell incubation with, 408–409
  cell-induced leakage of, 407
  cellular interaction with, 405–406
  cell uptake mechanisms of, 400
  delivery efficiency of, 405–408

  DNA encapsulation and delivery by, 399–410
  gene transfer via, 399–410
  large unilamellar vesicles (LUV) as, 402, 405–407
  lipids in formation of, 401
  as macromolecule carriers, 386, 400
  microinjection of, 386
  multilamellar vesicles (MLV) as, 402
  plant cell modifications via, 409–410
  preparation of, 401–402
  purification/separation of, 404–405
  small unilamellar vesicles (SUV), 402
  targeting of cell surface components with, 410
  vesicle composition and intracellular delivery by, 406–407
Liposome mediated gene transfer, 399–410

Macromolecules (see also specific types by name, e.g., DNA, Proteins, RNA)
  loading of, in red blood cell ghosts, 389–390
  microinjection transfer of, 385–391
  preparation of, for red blood cell ghost loading, 389
  stability of, 391–392
Mammalian cells
  chloramphenicol resistance in, 160–161
  DNA transformation in, 420–425
  gene mapping with, 118, 512–513
  interspecific fusion between, 178–181
Membrane, plasma, see Cell membrane
Membrane fusion, fluorescent probe monitoring of, 101–109
Metaphase chromosomes
  C-band (constitutive heterochromatin) staining, 504–505
  in cell fusion, 291–292, 379–382
  DNA and RNA values in, 350
  in gene mapping, 365–371
  in gene transfer, 375–383
  identification of, 352–358
  isolation/preparation of, 293–297, 349–352, 376–379
  in microkaryoplasts, 293
  micronucleation of, 293–297
  nucleolus organizer regions (NOR) in, 506
  pH in isolation of, 349–350, 376, 379
  sedimentation sizing of, 358–359, 378
Microcapillaries, preparation of glass, 464–465
Microcells (see also Microkaryoplasts)
  characterized, 309–310
  chromosome transfer mediation by, 309–325; diagrammed, 310
  decontamination of, 320–323
  fusion of, with recipient cells, 313, 323–324

Cell enucleation
  cell component separation in, 75–76, 245, 256, 269
  of chloramphenicol (CAP) resistant cells, 168
  collagen in, 75
  contamination in, 75, 269
  in cybrid formation, 190, 194
  cytochalasin B (CB) in, 67–68, 168, 193–194, 237, 241, 245, 256, 269, 299–300, 363–364
  cytoplast recovery in, 70, 77, 256–258
  density gradient techniques in, 68, 76–78, 194
  efficiency of, 196–197, 257, 273, 299–301
  growth substrate modification in, 75
  isopycnic fluid layer in, 68, 76
  in karyoplast recovery/preparation, 70, 75–76, 168, 190–191, 256–260, 269–271, 273–277
  of mitotic cells (mitoplasts), 245–246
  monolayer techniques in, 67–78, 273, 316–319, 323–324, 393 (see also Monolayer cell enucleation techniques)
  of plant protoplasts, 481
Cell fusion, somatic (see also Cell hybrid(s)/hybridization)
  attachment vs. suspension in, 23
  calcium ion affects in, 24–27, 35, 481–483
  cell culture conditions in, 39–40
  cell density in, 38–39
  cell-to-cell, see Cell hybrid(s)/hybridization
  characterized, 90, 167, 261, 291, 399, 526
  chemical induction of, summarized, 23, 35
  chloramphenicol (CAP) resistance in, 178–181
  cybrids in, see Cybrid(s); Cybrid formation
  cytoplasts in, see Cytoplast-cell fusion
  cytotoxicity in (see also Polyethyleneglycol (PEG)-induced cell fusion), 23–27, 32–33, 38, 41–42, 57–58
  diagram of, 167, 526
  dimethyl sulfoxide (DMSO) in, 35–46
  donor source identification in, 262–263
  erythrocyte ghosts in, 392–394
  drug enhancement of, 11–12
  efficiency of, 39, 261–262, 265
  flow sorting assays in, 93–98
  fluorescent labeling in, 86–87
  fluorescent probe monitoring in, 101–109, 262–263
  fusogen choices in (see also specific fusogens by name, e.g., Dimethyl sulfoxide (DMSO); Polyethyleneglycol (PEG); Sendai virus), 11, 23, 35, 392
  genetic mapping and, 118, 291
  heterokaryon production by, 47–58

Cell Fusion (cont.)
  karyoplasts in, 261–262, 273, 277–279, 281–288, 526
  of mammalian species, 178–181
  metaphase chromosome uptake in, 291, 379–381
  microcells in, 313, 323–324
  microkaryoplasts in, 291–304
  minisegregant cells in, 344–345
  mitochondria in, 203–205
  mitochondrial incompatability in, 181
  mitoplast-interplast, 245–253
  murine-human, 179, 215, 262–264, 291
  nuclear directed genetic modification in, 255, 265
  nuclear hybrids in, 281–288
  oligomycin resistance in, 118–119
  optimum conditions in, 37–38, 57–58
  pH value effects in, 38
  phytohemagglutinin (PHA) in, 23, 28–29, 323
  plant protoplasts in, 471–488
  quantification of, 25–26
  red blood cell ghosts in, 392–394
  Sendai virus in, 11, 23, 28, 57, 168, 194, 262, 269, 273, 277–278, 323–324, 392
  suspension method in, 28–31, 40–41, 305, 392
  techniques for, listed, 399
  transformation (see also Cell transformation), 415–424
  viability assessment of, 25–26
  of whole cells, see Cell hybrid(s)/hybridization
Cell hybrid(s)/hybridization, somatic (see also Cell fusion)
  adenine-alanosine (AA) medium in selection of, 15
  antibodies in isolation of, 18, 522
  auxotrophic mutant selection in, 12
  biochemical agents in selection of, 48–49, 62–64
  calcium ion effects on, 23–27, 481–483
  characterization of, 90, 167, 291, 526
  combined mutant selection technique for, 12
  DNA sequence detection in, 449–460
  drug resistant markers in selection of, 1–7, 12–17, 124–125, 262
  drug resistant types, see Drug-resistant cell mutants/variants
  flow sorting of, 81–90
  formation of, diagrammed, 167, 526
  fluorescent labeling, 81–87, 101–109, 262–265
  fusing agents in (see also specific fusogens mentioned by name, e.g.,

Cell hybrid(s) (cont.)
  fusing agents (cont.)
    Polyetheleneglycol [PEG]), 11, 23, 35, 392
  gene expression modification in, 255–256
  in gene mapping, 118, 291, 309, 324–325, 513–523
  half selection, 12
  HAM medium in, 15
  HAT medium in, 2–3, 12–14, 212
  heterokaryon selection in, 47–65
  incompatibility in interspecific, 215
  index of, 25, 30
  interspecific, 213, 449, 514, 525
  intraspecific, 449, 525
  isolation/selection of, 12, 47–63, 82–84, 87–90, 262, 281–288
  isozyme marker analysis of, 513–523
  oligomycin (rutamycin) resistance in, 118–119
  parent cell characteristics in, 262–263
  plant protoplast in, see Plant protoplast fusion
  rhodamine 6G influences in, 211–217, 288
  nuclear-directed modifications of, 265
  nucleolus organizer regions in, 506–507
  oligomycin (rutamycin) resistance in, 118–119
  parent cell characteristics in, 262–263
  plant protoplasts in, see Plant protoplast fusion
  purine/pyrimidine analog resistance in selection of, 1–7
  research utility of, 269, 344–346
  suspension procedure in, 28–30, 41, 392
  temperature in, 12, 24
Cell membrane(s)
  antigen determinant modulation of, 232–233
  biological model of, 101
  cytoskeletal element modification and, 222, 226, 232–233
  fluorescent monitoring of fusing by, 101–109
  shedding of, see Cell membrane vesicle shedding
  vesiculation of, 221–222, 242
Cell membrane vesicle shedding, 221–233, 237–242
  deuterium oxide effects on, 225, 232
  drugs in the induction of, 221–222, 237
  inhibition of, 222, 227–232
  nucleotide effects on, 224, 232
  temperature in, 222, 226–227, 233
Cell-mitochondria fusion, 204–205
Cell nuclei (see also Karyoplasts)
  fluorescent labeling of, 262

Cell nuclei (cont.)
  in genetic modification, 255–256
  hybridization of, 255
  index of, 26
  pyknosis of, 27
Cell protein determination, 93, 95–98
Cell transformation (see also Cell fusion; Gene transfer)
  characteristics of, 415–416, 425
  DNA mediated, 415–425, 430–431
  of mammalian cells, 420–426
  potential applicability of, 425–426
  transformant identification in, 425
  viral antigens as vectors in, 429–445
  in yeast, 418–420
Chloramphenicol (CAP), structure of, 160
Chloramphenicol (CAP), resistance
  antibiotic transference in, 104, 208–209
  in bacterial ribosomes, 159
  cell line characteristics in, 165
  cell-mitochondrial transfer in, 203–209
  cell nucleus selection for, 168
  coding site determination in, 168
  cybrid formation and, 166–171
  cytoplasmic expression of, 168–171, 176–181
  as hereditary, 124, 126, 128, 168–171
  in human cell lines, 169
  in mammalian cells, 160–166
  mitochondrial DNA control and, 171–178, 189
  mitochondrial protein synthesis deficiency and, 166
  mitotic segregation of, 170
  nucleotide fragment location of, 173–176
  in respiratory deficiency of mutants, 131–132
  selection of mutants for, 165–166
Cholesterol, in cell induced leakage reduction, 407
Chromosome(s)
  band staining of, see Chromosome identification
  C-band locations on, 503
  cytoplasmic inheritance vs., 189
  flow cytometry sorting, 338–339, 359–364
  human F-group, isolation of, 363–364
  human karyotype, 360–364
  isozyme marker segregation of, 517–523
  metaphase, see Metaphase chromosomes
  microcell mediated transfer of, 309–310, 315
  microkaryoplast isolation with, 302–303
  nuclear vs. cytoplasmic, 189
  pH in isolation of, 349–351, 376–379
  premature condensation of, 246, 251–252
  syntenic group segregation, 514

Microcells (cont.)
  gene transfer via, 309–310, 315, 525;
    diagrammed, 526
  gravity sedimentation of, 322
  membrane filtration of, 321
  mononucleation of, 309–320
  preparation of, 310–323
Microcytospheres, preparation of, 237–242
Microkaryoplasts (see also Microcells)
  in cell fusion, 291–304
  characterized, 292
  gene transfer via, 291–292
  isolation of, 299–301, 302–303, 304
  metaphase chromosomes in, 293
  production of: by abnormal cytokinesis, 299; with colchicine/Colcemid, 292–299; with cytochalasin B, 299–301, 303–304
  purification of, 304
  suspension fusion of, 305
  whole cell fusion with, 301, 304–305
Microinjection technique
  characterized, 463
  macromolecule transfer by, 385–391, 399, 463–470
  procedure in, 463–470
  in red blood cell ghosts, 385–397
Micronucleation
  cell determination in, 303
  cell fragmentation in, 293–297
  induction of: by colchicine/Colcemid treatment, 292–299, 311–315; with cytochalasin B, 299–301, 303–304, 315
  metaphase chromosomes in, 293–297
  in microcell preparation, 309–320
  mitotic arrest in, 311–313
  mitotic cell formation in, 293–299, 312–314
  monolayer techniques in, 316–320
  replating technique in, 313–315
  suspension method in, 319–320
Micronuclei
  in chromosome transfer, 309–310
  formation of, 292, 309, 312–313
Microplast-mediated gene transfer, 291–292
Minicells, see Karyoplasts
Minisegregant cells
  chromosome content count for, 338–339
  DNA content of, 336–338
  extrusion of, 330–333; modifications for, 339–344
  human chromosomes in fusions with 345–346
  in murine-human cell hybridization, 344–346
  in murine neuroblastoma fusions, 344–346
  prematurely condensed chromosomes

Minisegregant cells (cont.)
  prematurely condensed chromosomes (cont.)
    (PCCs) in, 339
  separation by size, 333–336
  whole cell fusions with, 344–346
Mitochondria
  in antibiotic resistance transferance, 203–209
  ATPase activity of, 111, 112, 115–116, 123–126
  cell fusion/transformation with intact, 204–205
  cell line sources for, 203–204
  genome of, see Mitochondrial genome
  hereditary influences of (see also Cytoplasmic inheritance) 211–212
  inactivation of, in cell fusion, 211–217, 281–288
  mutagenesis of, see Mitochondrial mutagenesis
  in oligomycin resistance inheritance, 111–119, 123–135
  preparation/isolation of, 114–115, 203–204
  protein synthesis by, see Mitochondrial protein synthesis
  pyruvate toxicity in, 166
  rhodamine 6G toxicity in, 214–215, 288
  rhodamine 123 labeling of, 263–264, 282–285
  ribosomes of, see Mitochondrial ribosomes
Mitochondrial DNA
  antibiotic resistance transfer via, 203–209
  cell fusion with, 205
  cell line polymorphisms in, 171
  in chloramphenicol (CAP) resistance control, 171–178, 189
  depletion of, 160
  hybrid chromosome responses to, 215–217
  interspecies incompatibility of, 181
  interspecific exchanges of CAP resistant, 178–181
  mutagenesis of, 160–165
  restriction enzyme analysis of, 286–287
  rhodamine 6G effects on, 211–217
Mitochondrial genomes
  cytoplasmic inheritance role of, 211–212
  genetic information in, of mammalian vs. yeast cells, 123–124
  oligomycin resistance location in, 127
Mitochondrial mutagenesis
  antimycin resistant, 125
  chloramphenicol (CAP) in, 160, 165
  cybrid mass growth and, 133–134
  cycloheximide in, 160
  cytochalasin B (CB) in, 125
  in erythromycin (ERY) resistant cell line selection, 140–144

Mitochondrial mutagenesis (cont.)
  ethidium bromide in, 125, 141–143, 160, 165, 212
  induction of, in mammalian cells, 125
  lactic acid production in, 130–132
  manganese in, 125
  metabolic defects in antibiotic resistant, 130–131
  in oligomycin (rutamycin) resistant mutants, 125–135
  respiratory impairment in antibiotic resistant, 132–134
Mitochondrial protein synthesis
  in antibiotic resistant mutants, 124, 132
  cell free, 145–146
  chloramphenicol (CAP) in inhibition of, 144–145, 153, 160–166
  cycloheximide in inhibition of, 144–145
  after enucleation, 190
  erythromycin (ERY) in inhibition of, 139–154
  by isolated mitochondria, 147–148
  mycoplasma contamination and, 149
Mitochondrial ribosomes, antibiotic sensitivities of, 159–166
Mitotic cells/Mitoplasts
  characteristics of, 246, 248–251
  enucleation of, 245–246
  extrusion division induction in, 330–332
  Feulgen staining of, 331–332, 336, 338, 344
  fusion of, with interphase cells, 252
  minisegregant cell derivation from 330–336
  premature chromosome condensation (PCC) induction in, 246, 251–252
  preparation of, 247–248, 293–297, 330–332, 350–352
  RNA synthesis in, 251
Monolayer cell enucleation techniques, 67–78, 273, 316–319
  acrylic flask inserts in, 71–72
  cell fusion applications of, 273, 323–324, 393
  centrifuge tube method in, 73
  centrifugal force in, 73–74
  concanavalin-A coating in, 318–319
  contamination in, 74, 320
  coverslip (plastic disc) method, 68–70, 190, 316–317
  culture dish method, 73
  cytochalasin B (CB) in, 67–78, 269–271
  flask techniques in, 70–73, 318
  glass slide method in, 73
  growth substratum in, 68, 75
  plastic "bullet" method in, 317–318
  suspension method vs., 319–320 (see also Suspension fusion method)
Multilamellar vesicles (MLV), 402

Mycoplasma contamination, 4–5, 149
  tests for, 204, 270, 281

NADH (nicotinamide adenine dinucleotide, reduced), 132
NADH cytochrome c reductase deficiency, 132, 134
NADH oxidase, CAP inhibition of, 153, 165
Nick translation, of DNA sequences, 435–436, 450, 457–458
Nuclear fluorescence, 51–52
Nuclear genetic control, cytoplasmic inheritance vs., 189, 255
Nuclear hybrids
  isolation/selection of, 281–288; diagrammed, 283, 285
  whole cell hybrids vs., 288
Nuclear markers, in hybrid selection, 191–192
Nuclear transplantation, 255–265
Nucleic acid, 399, 402–404
Nucleolus organizer regions (NOR)
  characteristics of, 506
  metaphase chromosomes and, 506
  staining for, 506–508
Nucleotide activity, in cell membrane shedding, 224, 232
Nucleus, see Cell nucleus; Karyoplasts
Nystatin methyl ester (NME), 16

Oligomycin (rutamycin)-resistant cell mutants/variants (see also Drug-resistant cell mutants/variants)
  ATPase activity in, 117–118, 126–129
  cell inheritance in, 111–119
  cybrids and, 126, 128, 133–134
  cytoplasmic inheritance in, 123–128, 131–134
  hereditary transfer in, 118–119
  in human cell lines, 128
  isolation of, 125–126
  lactic acid production by, 130–131
  Liebovitz medium in growth of, 131
  in mammalian cell lines, 127–130
  respiration defect in growth of, 131–134
  selection and growth of, 112–113
  stability of, 113–114
Ouabain
  in erythromycin resistant cell hybridization, 151
  HAT medium in combination with, 15–16
  in hybrid cell resistance selection, 7, 15–16

Phenylglyoxal, as heterokaryon selection agent, 49
Phospholipid vesicles, 399 [see also Liposome(s)]
  infectiousness of, 406–407

Phytohemagglutinin (PHA)
  concentration effects of, 31
  in polyethylene glycol (PEG)-induced fusion enhancement, 23, 28–29, 323
  in polykaryon formation enhancement, 31–32
Plant protoplast fusion/hybridization
  calcium ion-high pH-PEG methods in, 481–483
  cell enucleation for, 481
  collection of protoplasts for, 480–481
  cultivation of plant material for, 474–476; in soil, 471; in suspension cultures, 475–476; in vitro 472–476
  embryogenic suspension cultivation in, 476
  enzymic digestion of materials for, 476–480
  isolation of protoplasts for, 476–481; cell differentiation stages in, 476
  purification of protoplasts for, 480–481
  regeneration of protoplast culture in, 483–485
    selection and analysis of resultant characteristics in, 486–488
Plasmid rescue procedure, 18
Polio RNA, 405–406
Polyene antibiotics
  in drug resistant hybrid cell selection, 16
  as HAT medium supplements, 16
Polyethylene glycol (PEG)
  as cell fusing agent/fusogen (see also Polyethylene glycol (PEG) induced cell fusion), 11, 23–33, 168, 194, 262, 304–305
  concentration in fusion with, 23
  in plant protoplast fusion, 481–483
  toxicity decreasing techniques for, 23–33
Polyethylene glycol (PEG)-induced cell fusion
  calcium ion (Ca^{++}) exclusion effects in, 23–30
  concentration optimum in, 37
  cytotoxicity in, 23–27, 32–33, 38, 41–42, 57–58
  dimethyl sulfoxide (DMSO) enhancement of, 35–46
  exposure duration in, 38
  fusion efficiency improvements in, 23–28, 37–40
  HEPES (hydroxyethylpiperazine) buffer in, 39
  in heterokaryon rescue, 57–58
  monlayer protocol in, 40–41, 273, 323
  morphological changes due to, 41–44
  of plant protoplasts, 481–483
  temperature in, 38
  suspension protocol in, 41
  toxicity of, 23–27, 32–33, 38, 41–43, 262

Polykaryon formation
  phytohemagglutinin (PHA) enhancement of, 31–32
  quantification of, 25–26
  suspension fusion and, 30–31
  time interval in PEG induced, 27–28
Polylysine, 75
Polystyrene bead labeling (see also Fluorescent labeling)
  in cybrid formation, 192–195
  in cybrid fusion efficiency, 195–196
  in hybrid cell identification, 82–84, 192–194, 262–265
  of karyoplasts, 262–263
Prematurely condensed chromosomes (PCCs)
  induction of, 246; with mitoplasts, 251–252
  in minisegregant cell chromosome assay, 339
Prokaryotic genes, 424
Protamine sulfate, 75
Proteins
  as cell enucleation growth substrate, 75
  mitochondrial encoding of, 123–124
  synthesis of, 190, 258, 272
  uptake of, in red blood cell hemolysis, 387
Proteolipid, 127
Purine(s)
  biosynthesis and interconversion of, 12–13
  in mitochondrial mutagenesis, 125, 160
Purine/pyrimidine nucleoside analogs
  development of, 2–3
  in drug-resistant cell mutant selection, 1–7
  preparation of, 4
  table of, 2
Pyknotic nuclei, in cell viability assessment, 26
Pyrimiidines (see also Purine/Pyrimidine nucleoside analogs), 12, 13
Pyruvate, in mitochondrial toxicity reduction, 166, 212
Pyruvate decarboxylase deficiency, 128, 132, 134

Quinacrine (Q) banding
  application of, 356–357, 496–497
  in chromosome identification, 356–358, 493, 496–498
  Giemsa banding vs., 498–499
Recombinant DNA
  in gene mapping, 382
  isolation and characterization of, 430
Red blood cell(s)/Erythrocytes
  characterized, 386
  ghosts in, in macromolecule transfer, see Red blood cell ghosts
  hemolysis of, 386–387, 389

Red blood cell(s) (cont.)
    macromolecule transfer via, see Red-blood cell mediated injection
Red blood cell ghosts
    fusion of, with cultured mammalian cells, 392–394
    identification of post-fusion, 394–397
    macromolecule loading into, 389–390; dialysis method, 389; preswell method, 389–390; rapid lysis method, 389
    macromolecule stability within, 391–392
    microinjection of, 385–397
    properties of loaded, 391
    removal of excess, 393–394
    unloading of, 391
Red blood cell mediated injection
    cell hemolysis in, 386–387
    DNA molecule size in, 387–388
    microneedle injection vs., 386
    preparation of cells for, 388–389
    results with, 394–397
Resonance energy transfer, 101, 102, 105
Respiration, cellular, mitochondrial protein synthesis inhibition and, 153
Restriction endonuclease analysis, 450, 453–454, 527
Restriction enzyme fragments, gene specific, 522–523
Reverse (R) banding, in chromosome identification, 501–503
Rhodamine-fluorochrome labeling, 263–265, 282–285
RNA (ribonucleic acid)
    chloramphenicol (CAP) resistance in mitochondrial, 173–176
    encapsulation and delivery of polio, 405
    karyoplast synthesis of, 272
    in liposome-mediated gene transfer, 400, 405
    mitoplast synthesis of, 251
    stability of, in loaded red-blood cell ghost, 391–392
    uptake of, in red blood cell hemolysis, 387
Rutamycin, see Oligomycin (rutamycin) resistant cell mutants

Sendai virus, inactivated
    as cell fusion agent, 11, 23, 28, 57, 168, 194, 262, 269, 273, 277–278, 292, 301, 323–324, 392
    preparation of, 260, 392–393
Simian virus 40 (SV40) tumor antigen, in viral DNA gene transfer, 429–445
Small unilamellar vesicles (SUV), 402
Somatic cell(s), see Cells

Stearylamine, in fluorescent labeling, 84, 87
Subprotoplasts, 481
Succinate
    in mitochondrial mutagenic respiration, 132–133
    in mitochondrial inhibitor toxicity, 166
Sucrose velocity gradients, in chromosome fractionation, 358–359
Suspension fusion method
    attachment vs., 23
    in cell enucleation, 76–78
    with microkaryoplasts, 305
    monolayer procedures vs., 319–320
    PHA/PEG method in, 28–29
    plating efficiency in, 30–31
    polykaryon characterization in, 30
    procedure in, 28–30, 41, 392
    viability index for, 30
Syntenic groups, characterized, 514

Tantalum
    in cytoplast preparation, 190–191
    in karyoplast preparation, 270–276, 280
Tetraploid cells, flow sorting of, 88–90
Thymidine kinase (TK) activity
    in chloramphenicol (CAP) cell nucleus marking, 168–169
    in gene mapping, 17
    in gene transfer, 18
    herpes simplex virus-1 and, 17, 18
    in mitochondrial mutagenesis, 125
    screening for mutants deficient in, 14
Tumorigenic cells
    antigen modification in, 232–233
    cell membrane shedding by, 222
    suppression of, in cybrid formation, 198

Viability index (VI)
    calcium ion (Ca^{++}) effects on, 26–27
    determination of, in cell fusion, 25–26
    PHA concentration effects on, 31
    in suspension cell fusion, 30
    time factor and, 27–28

Yeast (Saccharomyces cerevisiae)
    chloramphenicol (CAP) resistance coding in, 173–174
    gene cloning system for, 415–416
    as mitochondrial biogenics model, 111–112, 123–124
    oligomycin resistance in, 111–112, 127
    spheroplast preparation in, 416–417
    spheroplast transformation by DNA, 416–418